WITHDRAWN
UTSA LIBRARIES

VISUAL
CRYPTOGRAPHY
AND SECRET IMAGE SHARING

Digital Imaging and Computer Vision Series

Series Editor

Rastislav Lukac

Foveon, Inc./Sigma Corporation
San Jose, California, U.S.A.

Visual Cryptography and Secret Image Sharing, *edited by Stelvio Cimato and Ching-Nung Yang*

Computational Photography: Methods and Applications, *edited by Rastislav Lukac*

Super-Resolution Imaging, *edited by Peyman Milanfar*

Digital Imaging for Cultural Heritage Preservation: Analysis, Restoration, and Reconstruction of Ancient Artworks, *by Filippo Stanco, Sebastiano Battiato, and Giovanni Gallo*

Image Processing and Analysis with Graphs: Theory and Practice *by Olivier Lezoray and Leo Grady*

Perceptual Digital Imaging: Methods and Applications *by Rastislav Lukac*

Image Restoration: Fundamentals and Advances *by Bahadir Kursat Gunturk and Xin Li*

VISUAL
CRYPTOGRAPHY
AND SECRET IMAGE SHARING

EDITED BY
STELVIO CIMATO
CHING-NUNG YANG

CRC Press
Taylor & Francis Group
Boca Raton London New York

CRC Press is an imprint of the
Taylor & Francis Group, an **informa** business

CRC Press
Taylor & Francis Group
6000 Broken Sound Parkway NW, Suite 300
Boca Raton, FL 33487-2742

© 2012 by Taylor & Francis Group, LLC
CRC Press is an imprint of Taylor & Francis Group, an Informa business

No claim to original U.S. Government works

Printed in the United States of America on acid-free paper
Version Date: 20110613

International Standard Book Number: 978-1-4398-3721-4 (Hardback)

This book contains information obtained from authentic and highly regarded sources. Reasonable efforts have been made to publish reliable data and information, but the author and publisher cannot assume responsibility for the validity of all materials or the consequences of their use. The authors and publishers have attempted to trace the copyright holders of all material reproduced in this publication and apologize to copyright holders if permission to publish in this form has not been obtained. If any copyright material has not been acknowledged please write and let us know so we may rectify in any future reprint.

Except as permitted under U.S. Copyright Law, no part of this book may be reprinted, reproduced, transmitted, or utilized in any form by any electronic, mechanical, or other means, now known or hereafter invented, including photocopying, microfilming, and recording, or in any information storage or retrieval system, without written permission from the publishers.

For permission to photocopy or use material electronically from this work, please access www.copyright.com (http://www.copyright.com/) or contact the Copyright Clearance Center, Inc. (CCC), 222 Rosewood Drive, Danvers, MA 01923, 978-750-8400. CCC is a not-for-profit organization that provides licenses and registration for a variety of users. For organizations that have been granted a photocopy license by the CCC, a separate system of payment has been arranged.

Trademark Notice: Product or corporate names may be trademarks or registered trademarks, and are used only for identification and explanation without intent to infringe.

Visit the Taylor & Francis Web site at
http://www.taylorandfrancis.com

and the CRC Press Web site at
http://www.crcpress.com

Library
University of Texas
at San Antonio

To the memory of my adviser, C.S. Laih, whose guidance and support have made my achievements possible.—C. N. Yang

To my beloved family, Imma and Silvia, stars shining into my life.
—S. Cimato

Contents

List of Figures

List of Tables

Preface

Secure digital imaging is an important research area combining methods and techniques coming from cryptography and image processing. Visual cryptography and in general secret image sharing techniques enable distributing sensitive visual materials to involved participants through public communication channels, as the generated secure images do not reveal any information if they are not combined in the prescribed way. In visual cryptography, the decoding process is performed directly by the human eyes, while in general, the shared images need some processing to reconstruct the secret image. The increasing number of possibilities to create, publish, and distribute images calls for novel protection methods, offering new sharing and access control mechanisms for the information contained in the published images. Secure image sharing techniques overcome the traditional cryptographic approach, providing new solutions for the development of new and secure imaging applications.

Since both digital imaging and privacy protection are two research themes that involve a continuously growing number of people and applications, the book focuses on secure image sharing techniques, offering an interesting reference point for studying and developing solutions needed in such areas. This book aims to fill the existing literature gap, providing a valid guide for professional people and/or researchers interested in such fields of research. It provides a scientifically and scholarly sound treatment of state-of-the-art techniques to students, researchers, academics, and practitioners who are interested or involved in the study, research, use, design, and development of image sharing techniques. It can serve as a guide for making use of visual cryptographic algorithms and as a basis to develop applications based on image secret sharing solutions.

The book has the form of a contributed volume, where well-known experts address an extensive range of topics related to visual cryptography techniques and solutions for secure image sharing. Besides introducing the relevant issues of such research fields, and providing a coverage of theoretical results, the book illustrates some aspects and interesting results concerning recent research directions. It covers the most prominent topics in the area, such as the possibility of sharing multiple secrets, visual cryptography schemes based on the probabilistic reconstruction of the secret image, possibility to include pictures in the distributed shares, contrast enhancement techniques, visual cryptography schemes based on different logical operations for combining the shared images, color images visual cryptography, cheating prevention, alignment problem for image shares, and the description of some practical applications of visual cryptography. In addition to the above, steganography and authentication are discussed, and different methods of image sharing are presented, including mathematical and probabilistic techniques for generating the image shares.

Each chapter provides the fundamentals for the topics under consideration and the detailed description of the relevant methods, presenting examples and practical applications to demonstrate the effectiveness of the surveyed techniques. Chapters 1–2 introduce visual cryptography and its application to halftone and color images, respectively. The techniques presented in these chapters focus on the problem of providing high quality shares using halftoning, and providing an adaptation of visual cryptography to color images, resolving the problem of darkening caused from the superposition of the shares in the reconstruction phase. Chapter 3 presents an extension of the general technique useful for sharing multiple images using the same set of shares. Chapter 4 shows how visual cryptography schemes can be applied to photograph images, obtaining interesting results for practical applications.

Chapters 5–7 focus on different extensions of basic visual cryptography technique. Probabilistic visual cryptography schemes trade the guarantee of a correct reconstruction of the secret image with a reduced pixel expansion of the generated shares. In Chapter 6 the basic superposition mechanism used for the reconstruction of the secret image is abandoned for a XOR- based reconstruction phase, improving on the contrast of the reconstructed image. Chapter 7 presents random grid techniques, which are derived from visual cryptography, but provide a reduced pixel expansion in the generated shares and do not need a basis matrix in the generation phase of the shares.

Chapters 8–9 focus on the contrast of the reconstructed images. In Chapter 8 the bounds on the contrast of the most known schemes are discussed. In Chapter 9 techniques based on the *reversing* of the images are presented, where black pixels are reversed to white ones and vice versa. Using such an operation, schemes with perfect reconstruction of both white and black pixels can be derived, avoiding any loss of resolution for the shared image.

Chapters 10–12 are more focused on the problems to be faced in the practical applications of visual cryptography schemes. One is cheating prevention, as discussed in Chapter 10, where the problem of malicious participants to the schemes is discussed. The other problem relates to the alignment of the shares in the reconstruction phase. In general, techniques to prevent the alignment of the shares and introduce some degree of tolerance in the reconstruction phase are discussed. Finally, Chapter 12 presents some applications of visual cryptography in the field of securing online transactions and e-voting.

Chapters 13–17 are more focused on general image sharing techniques. Image sharing in halftone images is discussed in Chapter 13, where two general techniques of data hiding are presented and analyzed. An algorithm for color image sharing is presented in Chapter 14, where interpolation is used for hiding secret images. Image sharing techniques based on polynomial secret sharing are discussed in Chapter 15. Chapter 16 discusses steganography and authentication in the context of image sharing, presenting some novel schemes. Finally, Chapter 17 presents a novel technique for image sharing, where the reconstruction phase can be performed both via the superposition of the shares, or via some computation.

In all the chapters, figures, tables, and examples are used to improve the presentation and the analysis of the discussed methods and techniques. Bibliographic references are included in each chapter and provide a good starting point for deeper research and further exploration of the topics covered in this volume.

Contributors

Gonzalo R. Arce
University of Delaware
Newark, Delaware, USA

Oscar C. Au
Hong Kong University of Science
and Technology
Hong Kong, China

Bernd Borchert
Universität Tübingen
Tubingen, Germany

Chin-Chen Chang
Feng Chia University
Taichung, Taiwan

Tse-Shih Chen
National Dong Hwa University
Hualien, Taiwan

Yu-Chi Chen
National Chung Hsing University
Taichung, Taiwan

Yu-Ting Chen
Central Police University
Taoyuan, Taiwan

Stelvio Cimato
Università degli Studi di Milano
Crema (CR), Italy

Chuei-Bang Ciou
National Dong Hwa University
Hualien, Taiwan

Roberto De Prisco
Università di Salerno
Fisciano (SA), Italy

Alfredo De Santis
Università di Salerno
Fisciano (SA), Italy

Giovanni Di Crescenzo
Telcordia Technologies Inc.
Piscataway, New Jersey, USA

Lin Dong
Tsinghua University
Beijing, China

Anna Lisa Ferrara
Università di Salerno
Fisciano (SA), Italy

Yuanfang Guo
Hong Kong University of Science
and Technology
Hong Kong, China

John S. Ho
Hong Kong University of Science
and Technology
Hong Kong, China

Gwoboa Horng
National Chung Hsing University
Taichung, Taiwan

Mohan S. Kankanhalli
National University of Singapore
Singapore, Singapore

Andreas Klein
Ghent University
Ghent, Belgium

Chia-Chen Lin
Providence University
Taichung, Taiwan

Feng Liu
Institute of Software, Chinese
  Academy of Sciences
Beijing, China

Barbara Masucci
Università di Salerno
Fisciano (SA), Italy

Klaus Reinhardt
Universität Tübingen
Tubingen, Germany

Shyong Jian Shyu
Ming Chuan University
Taoyuan, Taiwan

Du-Shiau Tsai
Hsiuping Institute of Technology
Taichung, Taiwan

Daoshun Wang
Tsinghua University
Beijing, China

Shiuh-Jeng Wang
Central Police University
Taoyuan, Taiwan

Zhongmin Wang
University of Delaware
Newark, Delaware, USA

Jonathan Weir
Queen's University Belfast
Belfast, United Kingdom

Yasushi Yamaguchi
The University of Tokyo
Tokyo, Japan

WeiQi Yan
Queen's University Belfast
Belfast, United Kingdom

Chen-Hsing Yang
National Pingtung University of
  Education
Pingtung, Taiwan

Ching-Nung Yang
National Dong Hwa University
Hualien, Taiwan

Zhao-Xia Yin
Anhui University
Anhui, China

1

Visual Cryptography from Halftone Error Diffusion

Gonzalo R. Arce

University of Delaware, USA

Zhongmin Wang

University of Delaware, USA

Giovanni Di Crescenzo

Telcordia Technologies Inc., USA

CONTENTS

1.1   Introduction

Visual cryptography (VC), proposed by Naor and Shamir in [19], is a paradigm for cryptographic schemes that allows the decoding of concealed images without any cryptographic computation. Particularly in a k-out-of-n visual secret sharing scheme (VSS), a secret image is cryptographically encoded into n shares. Each share resembles a random binary pattern. The n shares are then xeroxed onto transparencies respectively and distributed among n participants. The secret images can be visually revealed by stacking together any k or more transparencies of the shares and no cryptographic computation is needed. However, by inspecting less than k shares, one cannot gain any information about the secret image, even if infinite computational power is available. Aside from the obvious applications to information hiding, VC can be applied to access control, copyright protection [10], watermarking [8], visual authentication, and identification [18].

1.1.1   Visual Cryptography

The main instantiation of VC realizes a cryptography protocol called secret sharing (SS). In a conventional SS scheme, a secret image is shared among n participants in such a way that subsets of qualified participants can pull their shares and recover the secret but subsets of forbidden participants can obtain no information about it. Here, both the sharing phase and the reconstruction phase involve algorithms that are run by computers (specially, a dealer runs a distribution algorithm and a set of qualified parties can run a reconstruction algorithm). The surprising novelties of a VSS scheme are in representing data as images and in an elementary realization of the reconstruction phase, consisting of just viewing the image obtained after stacking transparencies. VSS schemes inherit all applications of conventional SS schemes; most notably, access control. As an example, consider a bank vault that must be opened everyday by five tellers, but for security purposes it is desirable not to entrust any two individuals with the combination. Hence, a vault-access system that requires any three of the five tellers may be desirable. This problem can be solved using a 3-out-of-5 threshold scheme. In addition to access control, VSS schemes can be applied to a number of other cryptographic protocols and applications using conventional SS, such as threshold signatures, private multiparty function evaluation, electronic cash, and digital elections.

Another quite intriguing instantiation of VC schemes realizes VSS with innocent-looking images as shares. This version of VSS has applications to a multiparty variant of steganography. In a steganography scheme, a user A sends an innocent-looking image to another user B, in such a way that B can recover some hidden images, but no observer of the communication between A and B even suspects that the communication contains some hidden images. In a multiparty variant of conventional steganography schemes, a user A sends innocent-looking images to users, B_1, ..., B_m in such a way that qualified subsets of the recipients can recover some hidden image, but no observer of the communication between A and B_1, ..., B_m even suspects the existence of hidden images. Interestingly, a version of VSS with innocent-looking images as shares can implement this steganography variant. In practice, however, both regular VSS and VSS with innocent-looking images work by expanding a single pixel of an image into multiple pixels, with consequences on image quality. It is indeed of great interest to design schemes achieving high image quality. Although the literature has paid a significant amount of attention to VSS, some different paradigms of VC have also been studied, giving rise to visual versions of other types of cryptographic protocols.

1.1.2 Halftone Visual Cryptography

Traditional VC constructions are exclusively based on combinational techniques. In the halftoning framework of VC, a secret binary image is encrypted into high-quality halftone images, or halftone shares. In particular, this method applies the rich theory of blue noise halftoning to the construction mechanism used in conventional VSS schemes to generate halftone shares, while the security properties are still maintained. The decoded secret image has uniform contrast. The halftone shares carry significant visual information to the reviewers, such as landscapes, buildings, etc. The visual quality obtained by the new method is significantly better than that attained by any available VSS method known to date. As a result, adversaries, inspecting a halftone share, are less likely to suspect that cryptographic information is hidden. A higher security level is thus achieved [26, 23].

1.1.3 Blue Noise Error Diffusion

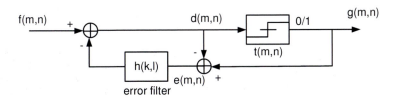

FIGURE 1.1
Block diagram for binary error diffusion. The pixel $f(m, n)$ is passed through a quantizer to obtain the corresponding pixel of the halftone $g(m, n)$. The difference between these two pixels is diffused to the neighboring pixels by means of the filter $h(k, l)$. (Reprinted with permission from IEEE Trans. Inf. Forensics Security, vol. 4, no. 3, pp. 383–396, Sep. 2009 ©IEEE 2009)

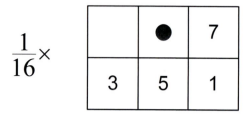

FIGURE 1.2

Floyd-Steinberg error filter. ● indicates the current pixel. The weights are given by: $h(0,1) = 7/16$, $h(1,-1) = 3/16$, $h(1,0) = 5/16$ and $h(1,1) = 1/16$. (Reprinted with permission from IEEE Trans. Inf. Forensics Security, vol. 4, no. 3, pp. 383–396, Sep. 2009 ©IEEE 2009).

Among all blue noise halftoning algorithms, error diffusion is a simple, yet efficient algorithm to halftone a grayscale image. The quantization error at each pixel is filtered and fed back to a set of future input samples. Figure 1.1 shows a binary error diffusion diagram where $f(m,n)$ represents the (m,n)th pixel of the input grayscale image, $d(m,n)$ is the sum of the input pixel value and the "diffused" past errors, and $g(m,n)$ is the output quantized pixel value [12, 16]. Error diffusion consists of two main components. The first component is the thresholding block where the output $g(m,n)$ is given by:

$$g(m,n) = \begin{cases} 1, & \text{if } d(m,n) \geq t(m,n), \\ 0, & \text{otherwise.} \end{cases} \tag{1.1}$$

The threshold $t(m,n)$ can be position-dependent. The second component is the error filter $h(k,l)$ whose input $e(m,n)$ is the difference between $d(m,n)$ and $g(m,n)$. Finally, we can compute $d(m,n)$ as:

$$d(m,n) = f(m,n) - \sum_{k,l} h(k,l)e(m-k,n-l). \tag{1.2}$$

As an example, the widely used Floyd–Steinberg error filter is shown in Figure 1.2 where ● indicates the current pixel. The weights of the filter are given by: $h(0,1) = 7/16$, $h(1,-1) = 3/16$, $h(1,0) = 5/16$ and $h(1,1) = 1/16$.

The recursive structure of the block diagram shown in Figure 1.1 indicates that the quantization error $e(m,n)$ depends not only on the current input and output but also on the entire past history. The error filter is designed in such a way that the low frequency difference between the input and output image is minimized. The error that is diffused away by the error filter is high frequency or "blue noise" in nature, leading to visually pleasing halftone images for human vision [12, 13, 7]. As will be described in this chapter, all the shares in the introduced halftone visual cryptography (HVC) methods are generated by a constrained error diffusion based on the algorithm introduced above.

For the HVC construction methods to be introduced shortly, it is necessary to generate mutually exclusive sets of pixels. To this end, the method of error diffusion is modified so as to produce multitone output pixels where the pixels of each tone are assigned to a pixel set. Multitone error diffusion is obtained by simply replacing the thresholding block by a multilevel quantization block in halftone error diffusion. The number of output levels of the quantization block is the same as the number of tones of the multitone image [15]. Multitone error diffusion can generate multitone images where the pixels of each tone are homogeneously distributed. The multitone error diffusion algorithm proposed in [4] is used here for the generation of mutually exclusive pixel sets. This algorithm jointly optimizes the distribution of multitone pixels by locating the pixels of different tones in a correlated fashion so that the mutual interference between different tones is minimized and multitone pixels are well separated from each other. Refer to [4] for details.

1.2 Visual Secret Sharing

We provide a brief description on visual cryptography where the key concepts will be referenced in subsequent sections. Please refer to [19, 2] for more details on VSS.

1.2.1 Notion and Formal Definitions

To illustrate the principles of VSS, consider a simple 2-out-of-2 VSS scheme shown in Figure 1.3. Each pixel p taken from a secret binary image is encoded into a pair of black and white subpixels in each of the two shares. If p is white/black, one of the first/last two columns tabulated under the white/black pixel in Figure 1.3 is selected. The selection is random such that each column is selected with a 50% probability. Then, the first two subpixels in that column are assigned to share 1 and the following two subpixels are assigned to share 2. Independent of whether p is black or white, p is encoded into two subpixels of black-white or white-black with equal probabilities. Thus, an individual share gives no clue as to whether p is black or white [26, 25]. Now consider the superposition of the two shares as shown in the last row of Figure 1.3. If the pixel p is black, the superposition of the two shares outputs two black subpixels corresponding to a gray level 1. If p is white, it results in one white and one black subpixel, corresponding to a gray level $1/2$. Then by stacking two shares together, we can obtain the full information of the secret image.

Figure 3.24 shows an example of the application of the 2-out-of-2 VSS scheme. Figure 1.4(a) shows a secret binary image SI to be encoded. According to the encoding rule shown in Figure 1.3, each pixel p of SI is split into two subpixels in each of the two shares, as shown in Figure 1.4(b) and Figure 1.4(c). Superimposing the two shares leads to the output secret im-

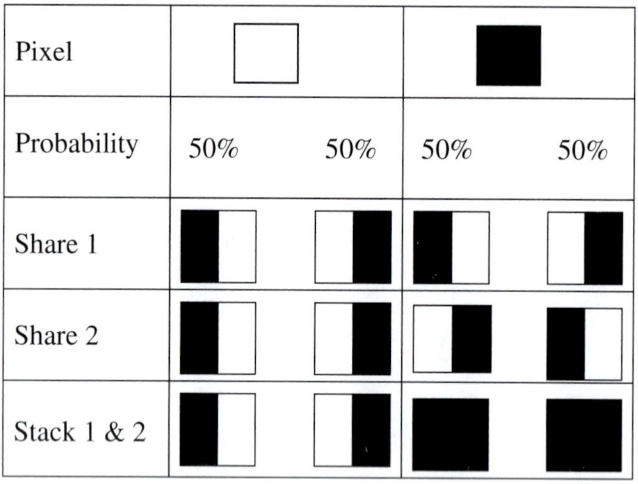

Pixel				
Probability	50%	50%	50%	50%
Share 1				
Share 2				
Stack 1 & 2				

FIGURE 1.3

In a 2-out-of-2 scheme, a secret pixel is encoded into 2 subpixels in each of the two shares. (Reprinted with permission from IEEE Trans. Inf. Forensics Security, vol. 4, no. 3, pp. 383–396, Sep. 2009 ©IEEE 2009).

age shown in Figure 1.4(d). The decoded image is clearly identified, although some contrast loss occurs. The width of the reconstructed image is twice that of the original secret image since each pixel is expanded to two subpixels in each share.

The 2-out-of-2 VSS scheme demonstrated above is a special case of the k-out-of-n VSS scheme [19]. Ateniese et al. designed a more general model for VSS schemes based on general access structures [2]. An access structure is a specification of all the qualified and forbidden subsets of shares. The participants in a qualified subset can recover the secret image while the participants in a forbidden subset cannot. Formal definitions of VSS are presented below.

Let $\mathcal{P} = \{1, \cdots, n\}$ be a set of elements called participants. A visual secret sharing scheme for a set $\mathcal{P}$ of n participants is a method to encode a secret binary image SI into n shadow images called shares, where each participant in $\mathcal{P}$ receives one share. Let $2^{\mathcal{P}}$ denote the set of all subsets of $\mathcal{P}$ and let $\Gamma_{Qual} \subseteq 2^{\mathcal{P}}$ and $\Gamma_{Forb} \subseteq 2^{\mathcal{P}}$, where $\Gamma_{Qual} \cap \Gamma_{Forb} = \emptyset$. We refer to members of Γ_{Qual} as qualified sets and call members of Γ_{Forb} forbidden sets. The pair $(\Gamma_{Qual}, \Gamma_{Forb})$ is called the access structure of the scheme [2].

Any qualified set of participants $X \in \Gamma_{Qual}$ can visually decode SI, but a forbidden set of participants $Y \in \Gamma_{Forb}$ has no information of SI [2, 3]. A visual recovery for a set $X \in \Gamma_{Qual}$ consists of xeroxing the shares given to the participants in X onto transparencies and then stacking them together.

(a) Secret binary image

(b) Share 1

(c) Share 2

(d) Decoded image

FIGURE 1.4

Example of 2-out-of-2 scheme. The secret image (a) is encoded into two shares (b)-(c) showing random patterns. The decoded image (d) shows the secret image with 50% contrast loss. (Reprinted with permission from IEEE Trans. Inf. Forensics Security, vol. 4, no. 3, pp. 383–396, Sep. 2009 ©IEEE 2009).

The participants in X are able to observe the secret image without performing any cryptographic computation. VSS is characterized by two parameters: the pixels expansion γ, which is the number of subpixels on each share that each pixel of the secret image is encoded into, and the contrast α, which is the measurement of the difference of a black pixel and a white pixel in the reconstructed image [5].

For each secret binary pixel p that is encoded into γ subpixels in each of the n shares, these subpixels can be described as an $n \times \gamma$ Boolean matrix $\mathbf{M}$, where a value 0 corresponds to a white subpixel and a value 1 corresponds to a black subpixel. The ith row of $\mathbf{M}$, $\mathbf{r}_i$, contains the subpixel values to be assigned to the ith share. The gray level of the reconstructed pixel p, obtained by superimposing the transparencies in a participant subset $X = \{i_i, i_2, \cdots, i_s\}$, is proportional to the Hamming weight $w(\mathbf{v})$ of the vector $\mathbf{v} = OR(\mathbf{r}_{i_1}, \mathbf{r}_{i_2}, \cdots, \mathbf{r}_{i_s})$, where $\mathbf{r}_{i_1}, \mathbf{r}_{i_2}, \cdots, \mathbf{r}_{i_s}$ are the corresponding rows in the matrix $\mathbf{M}$ [26, 25].

Definition 1 *Let* $(\Gamma_{Qual}, \Gamma_{Forb})$ *be an access structure on a set of n par-*

ticipants. Two collections of $n \times \gamma$ Boolean matrices $\mathcal{C}_0$ and $\mathcal{C}_1$ constitute a VSS scheme if there exists a value $\alpha(\gamma)$ and value t_X for every X in Γ_{Qual} satisfying [3]:

> *1. Contrast condition: any (qualified) subset $X = \{i_1, i_2, \cdots, i_u\} \in \Gamma_{Qual}$ of u participants can recover the secret image by stacking the corresponding transparencies. Formally, for a matrix $\mathbf{M} \in \mathcal{C}_j$, $(j = 0, 1)$ the row vectors $\mathbf{v}_j(X, \mathbf{M}) = OR(\mathbf{r}_{i_1}, \mathbf{r}_{i_2}, \cdots, \mathbf{r}_{i_u})$. It holds that: $w(\mathbf{v}_0(X, \mathbf{M})) \leq t_X - \alpha(\gamma) \cdot \gamma$ for all $\mathbf{M} \in \mathcal{C}_0$ and $w(\mathbf{v}_1(X, \mathbf{M})) \geq t_X$ for all $\mathbf{M} \in \mathcal{C}_1$. $\alpha(\gamma)$ is called the relative difference referred to as the contrast of the decoded image and t_X is the threshold to visually interpret the reconstructed pixel as black or white.*

> *2. Security condition: Any (forbidden) subset $X = \{i_1, i_2, \cdots, i_v\} \in \Gamma_{Forb}$ has no information of the secret image. Formally, the two collections $\mathcal{D}_j (j = 0, 1)$, obtained by extracting rows $i_1, i_2, \cdots, i_v$ from each matrix in $\mathcal{C}_j$, are indistinguishable.*

1.2.2 Construction of VSS Scheme

If the given secret pixel p is black (white), the matrix $\mathbf{M}$ is randomly selected from matrices collections $\mathcal{C}_1$ ($\mathcal{C}_0$). The matrix collections can be obtained by permuting the columns of the corresponding basis matrix $\mathbf{S_0}$ or $\mathbf{S_1}$ in all possible ways [3]. The basis matrices are defined below.

Definition 2 *Two matrices $\mathbf{S}_0$ and $\mathbf{S}_1$ are called basis matrices, if $\mathbf{S}_0$ and $\mathbf{S}_1$ satisfy the following tow conditions [3]:*

> *1. Contrast condition: If $X = \{i_1, i_2, \cdots, i_u\} \in \Gamma_{Qual}$, the row vectors $\mathbf{v}_0$ and $\mathbf{v}_1$, obtained by performing OR operation on rows $i_1, i_2, \cdots, i_u$ of $\mathbf{S}_0$ and $\mathbf{S}_1$ respectively, satisfy $w(\mathbf{v}_0) \leq t_X - \alpha(\gamma) \cdot \gamma$ and $w(\mathbf{v}_1) \geq t_X$.*

> *2. Security condition: If $X = \{i_1, i_2, \cdots, i_v\} \in \Gamma_{Forb}$, one of the two $v \times \gamma$ matrices, formed respectively by extracting rows $i_1, i_2, \cdots, i_v$ from $\mathbf{S}_0$ and $\mathbf{S}_1$, equals to a column permutation of the other.*

The algorithm to construct the basis matrices for a given VSS scheme can be found in [2, 5]. See [5] for the construction algorithm of basis matrices that leads to the best contrast. As an example, the $\mathbf{S}_0$ and $\mathbf{S}_1$ in a 2-out-of-2 scheme are shown below:

$$\mathbf{S}_0 = \begin{bmatrix} 0 & 1 \\ 0 & 1 \end{bmatrix}, \quad \mathbf{S}_1 = \begin{bmatrix} 0 & 1 \\ 1 & 0 \end{bmatrix}. \tag{1.3}$$

$\mathbf{S}_0$ corresponds to the encoding of a white secret pixel and $\mathbf{S}_1$ corresponds to the encoding of a black secret pixel.

1.3   Halftone VSS Construction Using Error Diffusion

The introduced methods for halftone VSS are built upon the fundamental principles of conventional VSS. Given a secret halftone image and multiple grayscale images, halftone shares are generated such that the resultant halftone shares are no longer random patterns, but take meaningful visual images. Without loss of generality, the k-out-of-n scheme is described in the following.

1.3.1   Share Structure

The first step in constructing a halftone VSS scheme is to construct the underlying k-out-of-n VSS scheme where a secret image pixel is encoded into γ pixels in each share. γ is the VSS pixel expansion and only a function of (k, n). Furthermore, in halftone VSS, a share image is divided into nonoverlapping halftone cells of size $q = v_1 \times v_2$ where $q > \gamma$. A secret image pixel is encoded into one halftone cell in each share. Within the q pixels in a halftone cell, only γ pixels called secret information pixels (SIPs) actually carry the secret information. Here γ is exactly the VSS pixel expansion. Since γ SIPs are not designed to carry share visual information, $q \geq 2\gamma$ is desirable for good share image quality.

It is required that when all qualified shares are stacked together, only the secret visual information is revealed. Thus, besides SIPs, auxiliary pixels that are forced to be black (value 1) are also introduced. These pixels are called auxiliary black pixels (ABP). In each halftone cell, there are x ABPs. ABPs are deliberately introduced into the shares so that some ABPs on one share block the visual information of the other shares. Thus, when qualified shares are stacked together, only the secret visual information is revealed on the reconstructed image as a result of the OR operation. In each halftone cell, the remaining $q - \gamma - x$ pixels that are neither SIPs or ABPs are assigned to carry the visual information of the shares.

An example of the halftone cells in a 2-out-of-2 scheme is shown in Figure 1.5 where the 1st and 2nd pixels in each cell are SIPs. The 3rd pixel in share 1 and the 4th pixel in share 2 are ABPs. When stacking two shares together, the result is a white pixel with contrast $1/4$. The 4th pixel in share 1 and the 3rd pixel in share 2 are assigned values to carry visual information of the shares. They can take a value of 0 or 1, which will not affect the decoded image.

There should be a sufficient number of ABPs in the shares so that the visual information of one share is completely blocked by the ABPs on the other shares. Since the ABPs are not designed to carry visual information, the number of ABPs in a share is to be minimized as follows. Let $\mathbf{p}(i, j) = [p_1(i, j), p_2(i, j), \ldots, p_n(i, j)]^T$ be the vector where $p_l(i, j)$ is the (i, j)th pixel

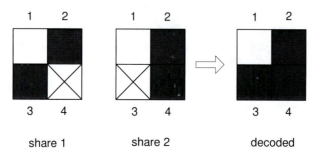

share 1   share 2   decoded

FIGURE 1.5
Example of halftone cells in a 2-out-of-2 scheme using the first method. The 1st and the 2nd pixels in both shares are SIPs. The 3rd pixel in share 1 and the 4th pixel in share 2 are ABPs. Others ("X") are assigned to carry visual information. (Reprinted with permission from IEEE Trans. Inf. Forensics Security, vol. 4, no. 3, pp. 383–396, Sep. 2009 ©IEEE 2009).

of share l. If the (i,j)th pixel is non-SIP, there should be at least $n-k+1$ ABPs in $\mathbf{p}(i,j)$ so that any k entries selected from $\mathbf{p}(i,j)$ contain at least one ABP. In each halftone cell, there are $q-\gamma$ non-SIPs. Thus, the optimal number of ABPs in a halftone cell is:

$$x^* = \lceil \frac{n-k+1}{n}(q-\gamma) \rceil, \tag{1.4}$$

where $\lceil\ \rceil$ returns the smallest integer that is no less than the argument.

To understand how the x ABPs are arranged in a halftone cell, we introduce a configuration matrix $\mathbf{T}$ to describe the $q-\gamma$ non-SIPs. In the matrix $\mathbf{T}$, only the values of ABPs are defined. In a k-out-of-n scheme, $\mathbf{T}$ is of size $(q-\gamma)\times n$ where one row corresponds to one share. There are x ABPs assigned to each row. The ABPs in each row should have minimal overlap with ABPs on the other rows. The construction of the matrix $\mathbf{T}$ is as follows. First, x ABPs are placed on the first row from the first to xth column. Then another x ABPs are placed on the second row from the $(x+1)$th column to the $(2x)$th column. If $2x > q-\gamma$, then the ABPs are placed from the $(x+1)$th column to the last column and then from the first column to the $(2x-q+\gamma)$th column. The x ABPs on other rows are placed sequentially in the same way.

For the 2-out-of-2 example above, the configuration matrix $\mathbf{T}$ for the 2 non-SIPs is:

$$\mathbf{T} = \begin{bmatrix} 1 & \triangle \\ \triangle & 1 \end{bmatrix}, \tag{1.5}$$

where 1 indicates ABP that has value 1 and $\triangle$ indicates an undefined pixel. The corresponding non-SIPs are then assigned values to carry the share visual information through the error diffusion. Each row of $\mathbf{T}$ is assigned to each share. It can be seen that when we stack 2 halftone shares together, all the non-SIPs result in black.

As another example, consider a 2-out-of-3 halftone VSS scheme where a secret image pixel is encoded into a halftone cell of size $q = 12$. The number of SIPs is $\gamma = 6$. By Eqn. 1.4, the optimal number of ABPs is $x = 4$. Thus, the configuration matrix $\mathbf{T}$ of the 6 non-SIPs is given by:

$$\mathbf{T} = \begin{bmatrix} 1 & 1 & 1 & 1 & \triangle & \triangle \\ 1 & 1 & \triangle & \triangle & 1 & 1 \\ \triangle & \triangle & 1 & 1 & 1 & 1 \end{bmatrix}. \tag{1.6}$$

From the content of $\mathbf{T}$, it is concluded that $1/3$ of the pixels on a share will be ABPs.

1.3.2 Distribution of SIPs and ABPs

The locations of the SIPs do not depend on the share images or the secret image, but only on the HVC expansion q and the underlying VSS scheme. Thus, the distribution of SIPs can be generated prior to the generation of halftone shares. For security purposes, the SIPs should be randomly distributed. To achieve good image quality, it is also desirable to distribute the SIPs homogeneously so that one SIP is maximally separated from its neighboring SIPs. Since the SIPs are maximally separated, the quantization error caused by an SIP will be diffused away before the next SIP is encountered leading to visually pleasing halftone shares. Similarly, the distribution of ABPs can also be determined *a priori*. As SIPs, ABPs should be distributed as homogenously as possible and maximally separated from each other. Since there is a strong correlation between the distribution of SIPs and the distribution of ABPs, the distributions of SIPs and ABPs should be optimized jointly to avoid low frequency spectral interference among them [4]. The SIPs and ABPs should also be maximally separated from each other. The jointly optimized distributions of SIPs and ABPs are generated based on a method of blue noise multitoning as follows [4].

We first construct a constant grayscale image with gray level $g = \sum_{i=0}^{w} g_i z_i$, where g_i is a tone arbitrarily chosen between 0 and 1 and $g_i \neq g_j$ for $i \neq j$. z_i is the pixel density for the pixels with tone g_i. The value of z_i, together with w, depends on q, γ and the structure of the configuration matrix $\mathbf{T}$. By using the blue noise multitone error diffusion, an output image with $w+1$ tones is produced. The distribution of pixels with tone g_i indicates a pixel distribution denoted by Z_i. Let $z_0 = \gamma/q$, then Z_0 indicates the distribution of SIPs. The distribution of ABPs in a share is a subset of $\{Z_i\}, i = 1, \ldots, w$.

An example following the previous 2-out-of-3 halftone VSS example is used to illustrate how to set z_i, g_i, and w, where a secret image pixel is encoded into a halftone cell of size $q = 12$. Among the q pixels, $\gamma = 6$ SIPs are characterized by basis matrix $\mathbf{S}_0$ or $\mathbf{S}_1$; $q - \gamma = 6$ non-SIPs are characterized by the configuration matrix $\mathbf{T}$. Thus, the matrix $\mathbf{R}_i$, $i = 0, 1$, is constructed below for the q pixels:

$$\mathbf{R_i} = [\mathbf{S}_i \quad \mathbf{T}]$$

$$= \begin{bmatrix} \square & \square & \square & \square & \square & \square & 1 & 1 & 1 & 1 & \triangle & \triangle \\ \square & \square & \square & \square & \square & \square & \triangle & \triangle & 1 & 1 & 1 & 1 \\ \square & \square & \square & \square & \square & \square & 1 & 1 & \triangle & \triangle & 1 & 1 \end{bmatrix}, \qquad (1.7)$$

$$\underbrace{\qquad\qquad\qquad\qquad}_{Z_0} \quad \underbrace{\qquad}_{Z_1} \quad \underbrace{\qquad}_{Z_2} \quad \underbrace{\qquad}_{Z_3}$$

where each row corresponds to q pixels in one share. The $\square$ indicates the SIPs that are determined by $\mathbf{S}_i$. Columns of $\mathbf{R}_i$ are partitioned into several sets Z_i, where the columns with the same configuration are assigned to the same set. As shown in (1.7), columns are partitioned into 4 sets Z_i, $i = 0, 1, \ldots, 3$. The set Z_0 denotes the distribution of SIPs and contains $1/2$ of all the pixels. The non-SIPs of each share are partitioned between set Z_1, Z_2, and Z_3, where each set contains $1/6$ of all the pixels. Thus, to generate Z_i on the share, we can set the parameters for the multitone error diffusion as follows: $w = 3$, $z_0 = 0.5$, and $z_1 = z_2 = z_3 = 1/6$. The corresponding tone is arbitrarily chosen as: $g_0 = 0$, $g_1 = 0.3$, $g_2 = 0.6$, and $g_3 = 0.9$. By using the algorithm proposed in [4], we obtain homogenous distributions Z_i, $i = 0, 1, \ldots, 3$. The combination of Z_1 and Z_2 is the distribution of ABPs of share 1; the combination of Z_2 and Z_3 is the distribution of ABPs of share 2; and the combination of Z_1 and Z_3 is the distribution of ABPs of share 3, as shown in (1.8)

$$\begin{cases} \text{SIPs of all shares} & \rightarrow & Z_0 \\ \text{ABPs of share 1} & \rightarrow & Z_1 \cup Z_2 \\ \text{ABPs of share 2} & \rightarrow & Z_2 \cup Z_3 \\ \text{ABPs of share 3} & \rightarrow & Z_1 \cup Z_3. \end{cases} \qquad (1.8)$$

As an example, assume the share has size 6×8 and is partitioned into 4 halftone cells of size 3×4, each cell corresponding to one secret image pixel. Within each cell, there are 6 SIPs and 4 ABPs. Suppose the generated distributions $Z_i, i = 0, 1, \ldots, 3$ are shown on the up-left image in Figure 1.6. Then, as also shown in Figure 1.6, the distributions of SIPs and ABPs of each share are determined based on Z_i. Note that without knowing what values the $\triangle$s carry, any two shares can be stacked together to decode the secret image pixels.

Notice that we may not be able to generate an exact number of pixels as desired for sets Z_i, $i = 0, 1, 2, 3$. However, for ABPs, as long as the membership of ABPs to the sets Z_i indicated in (1.8) is maintained, a slight deviation from its desired number of pixels for Z_i, $i = 1, 2, 3$ is allowable. The contrast condition of image decoding is still maintained. Such a point can be clearly illustrated in Figure 1.7 where the composition of each share is shown. It is clear that the relative size of Z_i, $i = 1, 2, 3$ is not important.

However, the distribution of SIPs, denoted by Z_0, needs to be refined to guarantee that there are exactly γ SIPs in each halftone cell. Each halftone cell is checked to find out the number of pixels belonging to Z_0. Assume the

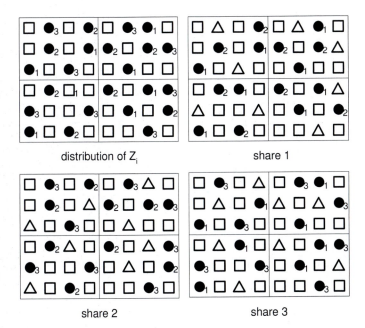

FIGURE 1.6
Upper left: distributions Z_i, $i = 0, 1, \ldots, 3$. $\square$ indicates pixels within Z_0; $\bullet_i$ indicates pixels within Z_i. Upper right: distribution of SIPs and ABPs in share 1. Down left: distribution of SIPs and ABPs in share 2. Down right: distribution of SIPs and ABPs in share 3. $\bullet_i$ indicates ABP and $\triangle$ indicates pixels used to carry share visual information. (Reprinted with permission from IEEE Trans. Inf. Forensics Security, vol. 4, no. 3, pp. 383–396, Sep. 2009 ©IEEE 2009).

number is τ. In most cases, $\tau = \gamma$ and no operation is needed. If $\tau < \gamma$, then $\gamma - \tau$ pixels within the current cell are selected and removed from sets Z_i, $i \neq 0$ and assigned to Z_0. The pixels are added one by one in such a way that each newly added pixel is maximally separated from other pixels belonging to Z_0 within that halftone cell. If $\tau > \gamma$, then within that halftone cell, $\gamma - \tau$ pixels belonging to Z_0 are randomly selected and removed from Z_0. For each removed pixel, a pixel set is randomly selected from $\{Z_i\}$, $i \neq 0$ and the removed pixel is added to that set. After such a procedure, there are exactly γ SIPs in each halftone cell and most of the SIPs are well separated from each other.

The configuration matrix helps to generate the distribution of ABPs. It should be noted that there is no need to find out the exact x ABPs in each halftone cell. The number x only characterizes the local density of the ABPs.

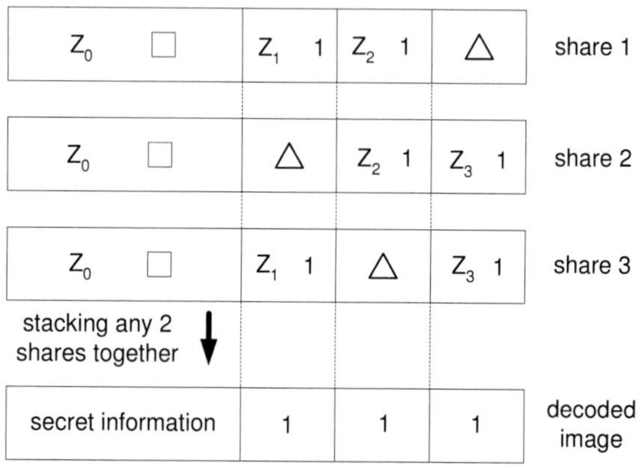

Z_0 □	Z_1 1	Z_2 1	△	share 1
Z_0 □	△	Z_2 1	Z_3 1	share 2
Z_0 □	Z_1 1	△	Z_3 1	share 3

stacking any 2 shares together ↓

secret information	1	1	1	decoded image

FIGURE 1.7
Composition of shares. □ indicates SIPs; 1 indicates ABPs; △ indicates pixels to carry the share visual information. (Reprinted with permission from IEEE Trans. Inf. Forensics Security, vol. 4, no. 3, pp. 383–396, Sep. 2009 ©IEEE 2009).

We also only need to ensure that the contrast condition of the image decoding is satisfied.

1.3.3 Generation of Halftone Shares via Error Diffusion

After the distributions of SIPs and ABPs are generated, the next step is to assign the values to all the SIPs. This procedure only depends on the underlying VSS scheme. Under the k-out-of-n VSS scheme, the basis matrices S_0 and S_1 are constructed first. Then we construct the matrix collections C_0 and C_1 from the basis matrices. For each γ SIPs of a halftone cell, a matrix M is randomly selected from C_0 or C_1 depending on the value of the corresponding secret image pixel p. The SIPs in the ith share are then replaced with the ith row of M. s_{ij}, the jth SIP in share i, is set as: $s_{ij} = M(i, j)$. For the ABPs, they are assigned value 1 (black). Thus, the locations and the values of the SIPs and ABPs are set before halftone shares are actually generated. As it will be described below, the pixels other than ABPs and SIPs are assigned freely to carry the share visual information.

Once the assignments of the SIPs and ABPs are determined, a halftoning algorithm, such as error diffusion and direct binary search (DBS) [1], can be employed to produce the halftone shares from grayscale images. Error diffusion is used in our method as it is a computationally efficient way to generate halftone shares. The basic computation required is linear filtering followed by quantization.

The process of generating halftone shares via error diffusion is shown in

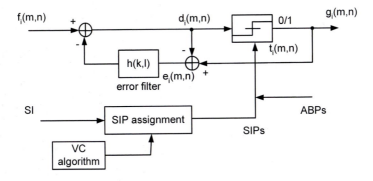

FIGURE 1.8
Block diagram of halftone VSS using the first method. Depending on the secret image and VSS scheme chosen, the SIP assignment block outputs the SIPs. If $g_i(m, n)$ is a SIP or ABP, its value is prefixed. Otherwise, $g_i(m, n)$ is determined by the output of the thresholding block. (Reprinted with permission from IEEE Trans. Inf. Forensics Security, vol. 4, no. 3, pp. 383–396, Sep. 2009 ©IEEE 2009).

Figure 1.8, where the values of the SIPs and ABPs are preset. To produce the halftone share i, a grayscale image is provided. Let $f_i(m, n)$ be the (m, n)th pixel of the grayscale image, then the input to the threshold block is:

$$d_i(m, n) = f_i(m, n) - \alpha \sum_{k,l} h(k, l) e_i(m - k, n - l), \qquad (1.9)$$

where $h(k, l) \in H$ and H is a two dimensional error filter. Here $0 < \alpha \leq 1$ is an error diffusion constant that can be tuned to avoid instability in error diffusion [24]. A $\alpha < 1$ means that not all quantization errors are diffused, which can prevent instability under excessive quantization errors. $e_i(m, n)$ is the quantization error at point (m, n). Assume the threshold for the error diffusion is $t_i(m, n)$, then the output halftone pixel $g_i(m, n)$ is 1, if $d_i(m, n) > t_i(m, n)$; or 0, if $d_i(m, n) \leq t_i(m, n)$. The quantization error $e_i(m, n)$ is given by:

$$e_i(i, j) = g_i(m, n) - d_i(m, n). \qquad (1.10)$$

The above procedure is applied only when $g_i(m, n)$ is not a SIP or an ABP. Otherwise, instead of simple thresholding, the value of the output pixel $g_i(m, m)$ is set equal to the value of the corresponding predetermined SIP or ABP and the error $e_i(m, n)$ is calculated as the difference between the input to the thresholding block and the SIP or ABP value. Since the SIPs and ABPs are separated from each other, the quantization error caused by the introduction of the SIPs and ABPs is diffused away to the neighboring grayscale pixels, as illustrated in Figure 1.8, and will not accumulate to cause visible distortion. In this way, the SIPs and ABPs are seamlessly embedded into the halftone shares generated and the halftone share is structured taking meaningful visual information.

Much like the methods in [26, 25], the above procedure can be extended to an arbitrary access structure $(\Gamma_{Qual}, \Gamma_{Forb})$. The security of the introduced halftone VSS scheme is guaranteed by the properties of the underlying visual secret sharing scheme.

1.4 Halftone VSS Construction Using Parallel Error Diffusion

In the previous method, uniformly distributed auxiliary black pixels are introduced to satisfy the contrast condition of image decoding. The current method exploits the fact that halftoning of the grayscale images alone may generate a sufficient number of black pixels to block the shared visual information from showing on the decoded image.

As in the previous method, in the current method, the shares are also divided into nonoverlapping halftone cells of size $q = v_1 \times v_2$. γ pixels within the halftone cell are SIPs carrying the secret visual information. A method based on error diffusion is used to generate the distribution of SIPs. To this end, a constant-value grayscale image of gray level $z_0 = \gamma/q$, having the same size of a share image is first produced. This grayscale image is then halftoned, producing a distribution of "1"s, denoted by Z_0. Z_0 determines the distribution of the SIPs. To ensure that there are exactly γ "1"s in each halftone cell, the error diffusion is constrained such that the values of some pixels are preset and are not modified by the error diffusion. The error $e(m, n)$, which is the difference between the input to the thresholding block and the resultant pixel value, is still calculated for the preset pixels and then diffused away to neighboring grayscale pixels through the error filter. In this constrained error diffusion mechanism, if the current halftone cell already contains γ "1"s, the rest of the pixels in the halftone cell are prefixed and constrained to be "0"s. The quantization error is accumulated and diffused to grayscale pixels in neighboring cells. If the current halftone cell contains $t < \gamma$ "1"s, and the error diffusion already proceeds to the $(q - \gamma - t)$th pixel of the current cell, then the rest of the $\gamma - t$ pixels in that cell are all constrained to be "1"s. This procedure guarantees that there are exactly γ SIPs in each halftone cell. Since the errors are always diffused to neighboring grayscale pixels, a homogeneous distribution of SIPs is produced, where most of the SIPs are well separated from each other. The assignment of SIPs are predetermined using the same approach as described previously.

Then, the current method halftones the grayscale images in parallel to produce the halftone shares. Within the error diffusion process, all the shares are checked at each non-SIP position to see if a sufficient number of black pixels have been produced. If a sufficient number of black pixels have not yet been generated, black pixels are deliberately inserted at that position. The SIPs are again preserved and not changed.

In a k-out-of-n scheme, if only $\tau < n - k + 1$ black pixels are generated by halftoning at a non-SIP location (i, j), then $\lambda = n - k + 1 - \tau$ shares with the smallest magnitudes of halftone error at (i, j) are selected and black pixels are inserted at (i, j) on these shares. Thus, the contrast condition of image decoding is guaranteed. The quantization error caused by the inserted black pixels will be diffused away to neighboring grayscale pixels and pleasing halftone shares can be obtained. Since far fewer black pixels are deliberately introduced, the second method imposes fewer constrains on the error diffusion and thus it has the potential to achieve better image quality than that of the first method. However, to achieve uniform image quality of the whole share, we need to choose the grayscale images in a selective way.

It is clear that the decision to insert a black pixel or not depends on the image content of the shares. Thus, the inserted black pixels are not evenly distributed. In some regions of the image, the error diffusion mechanism is constrained by the SIPs. In some other regions, error diffusion is constrained not only by the SIPs but also by the inserted black pixels. Therefore, the image quality on some regions in the image may be better than the image quality on some other regions that exhibit more artifacts. The parallel approach may thus generate shares whose image quality is not consistent over the whole image. Such quality discrepancy may cause visible distortions. To mitigate such visible distortion, we need to minimize the number of black pixels inserted. An obvious way to mitigate the distortion is to select grayscale images where the contents of some images tend to be complimentary to those of the others. For example, if there is one bright (white) region on one image, there should be corresponding dark region(s) in some other image(s). Then the halftoning of the grayscale images will generate most of the black pixels needed and the number of inserted black pixels will be greatly reduced, which leads to visually pleasing halftone shares. In a n-out-of-n scheme, if $n \gg 1$, then this approach is especially effective and the visible distortion is less likely to happen.

1.5 Quality of Halftone Shares

In this section, we focus on the quality analysis of the halftone shares for the first method. The analysis also helps to evaluate the share image quality of the second method. With the exception of the SIPs and the ABPs, all pixels in the halftone share produced by the first method are assigned freely to carry the shared visual information. The proportion of these pixels governs the image quality of the resultant halftone shares. The quantity s is called the quality index of the halftone share and is represented as:

$$s = \frac{q - x - \gamma}{q}, \qquad (1.11)$$

where q is the halftone cell size, x is the number of ABPs in a cell, and γ is the number of SIPs in a cell. A large s leads to good image quality of the halftone share. However, s cannot be arbitrarily large and it can be shown that:

$$s \leq (\frac{k-1}{n})(\frac{q-\gamma}{q}) < 1. \qquad (1.12)$$

Thus, the best image quality of a halftone share that can be achieved depends on k, n and the halftone cell size q. If k, n, and s are the design parameters, then q is calculated as:

$$q = \lceil \frac{(k-1)\gamma}{k - ns - 1} \rceil. \qquad (1.13)$$

Consider the 2-out-of-2 scheme. Assume $q = 4$, then it is calculated that $x = 1$ and $s = 0.25$. Since s is small, the image quality of the share is not high. If q is larger, then better image quality can be expected. Furthermore, as $q \to \infty$, s approaches 0.5. However, a larger q leads to worse contrast loss of the reconstructed image. As will be shown later, the contrast loss of the reconstructed image can be improved by filtering.

The quality of each share depends on the quality index s. We can compare the share image with the halftone image generated from the grayscale image without encoding any secret information and then compute the perceived error ϵ between the coded and uncoded halftone image. The perceived error is calculated by employing an appropriate human visual system (HVS) model. See [11, 9] for details.

For the second method, it is difficult to determine the proportion of pixels that carry visual information of the shares. However, it is clear that for $n \gg 1$ in a n-of-of-n scheme, the quantity s approaches:

$$s = \frac{q - \gamma}{q}, \qquad (1.14)$$

which indicates potentially good image quality for a sufficiently large q.

Compared with methods in [26] and [22], the requirement of a complementary pair is removed and all shares generated carry natural images. From (1.12), it is clear that for the first method, the quality index is more correlated to $\{k, n\}$ in the VSS scheme. A visually pleasing halftone image share can be obtained if $n \gg 1$ and $n - k$ is small, and if the HVC expansion q is sufficiently large. If small image quality discrepancy of the share is tolerable, then we should first consider the first method, especially if we have the flexibility to choose the grayscale images. If the grayscale images are carefully chosen, $n \gg 1$, and $n - k$ is small, then the distortion due to image quality discrepancy will be hardly noticeable. Otherwise, only the first method should be considered since it is the only method that guarantees uniform image quality of the shares without using complimentary shares.

1.6 Discussion

1.6.1 Improvement of Image Quality

As stated previously, the quality index s heavily affects the image quality of the halftone shares. A large s leads to visually pleasing halftone shares, but it also introduces higher contrast loss in the reconstructed images. The error filter employed in the error diffusion also affects the image quality of the shares. For example, an error filter with longer weights leads to a sharper contrast in the halftone image [20, 14]. Another factor that affects the image quality is the position-dependent threshold in the thresholding block. To achieve a visually more pleasing halftone image, output-dependent threshold modulation can be used in the error diffusion to spread the minority pixels as homogeneously as possible and suppress some unwanted textures [6]. For various methods that can improve the halftone image by error diffusion, refer to [20, 6, 17, 21], etc.

Error diffusion is employed as the halftoning algorithm to generate halftone shares since error diffusion is able to generate a visually pleasing halftone image with simple computation. However, other halftoning algorithms can also be applied to the second method to generate halftone shares. The DBS algorithm can be used to generate a high quality halftone image but with significant computation [1]. Note that the DBS algorithm for multitoning can also be used to generate the distributions of SIPs and ABPs.

1.6.2 Comparison with Other Methods

To evaluate the performance of the introduced methods, it is illustrative to compare our methods with VC and extended visual cryptography (EVC) [3]. VSS can be treated as a special case in our methods for $s = 0$, which means that no visual information is carried by the share. In EVC, shares carry visual information and there is a tradeoff between the contrast of the reconstructed image and the contrast of the share image. This tradeoff is similar to the tradeoff between the contrast of the reconstructed image and the image quality of the halftone shares in our methods. However, the shares generated by EVC is basically based on pixels expansion. Thus, EVC is unable to generate shares that can show fine details.

1.6.3 Image Decoding

In the methods introduced in this chapter, a large HVC expansion is desirable to make s large. But the contrast loss of the decoded image is severe when the HVC expansion is large. However, a low contrast does not hinder the decoding of the secret image if the decoding can be performed digitally. Initially shares are supposed to be xeroxed on transparencies and decoding of the secret image involves stacking the shares physically. However, both the distribution of the shares and decoding of the secret image can be performed in a digital way where the decoding rule remains the same (OR operation). The human visual system is still the ultimate tool to identify the secret image.

The robustness of the introduced scheme to the contrast loss is attributed to the fact that shares have well-defined local structure. If decoded digitally, we can measure the local intensity of the reconstructed image by filtering the data through a running window. Then we measure the local intensity of the data within the window. We can assign 1 to the current pixel if the local intensity is relatively high and assign 0 to the current pixel if the local intensity is relatively low. By such a simple method, the contrast of the reconstructed image can be enhanced and the content is more readable. The size of the running window should be the same as the size of the halftone cell. However, if the halftone cell size is unknown, a proper window size can be obtained by trying different window sizes until only two different local intensities are most likely to appear.

1.7   Simulation Results

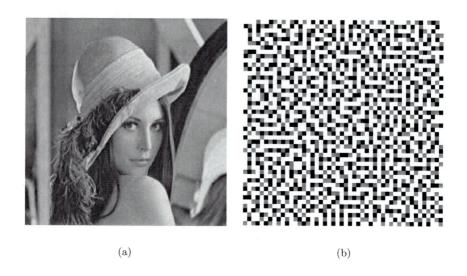

(a)   (b)

FIGURE 1.9
(a) Grayscale image *Lena*. (b) Part of the distribution of SIPs and ABPs. The gray pixels indicate SIPs and the black pixels indicate ABPs. (Reprinted with permission from IEEE Trans. Inf. Forensics Security, vol. 4, no. 3, pp. 383–396, Sep. 2009 ©IEEE 2009).

In this section, examples are provided to illustrate the effectiveness and tradeoff of the introduced methods. Constructions of 3-out-of-3 halftone VSS schemes are illustrated. The secret image to be encoded is the logo of the University of Delaware. Three grayscale natural images, *Lena*, *Earth*, and *Baboon* are provided to generate the shares. The size of the share is 513×513 and the size of the secret image is 171×171. For illustrative purposes, image *Lena* is shown in Figure 1.9(a).

The VSS pixel expansion of the secret pixel is $\gamma = 4$ and the size of the halftone cell is $q = 3 \times 3$. The number of ABPs for each halftone cell is $x = 2$ and the quality index is $s = 1/3$. The distributions of SIPs and ABPs for each share are decided before halftone shares are generated. The jointly optimized distribution of SIPs and ABPs over a local region in one share is shown in Figure 1.9(b) where the gray pixels indicate SIPs and the black pixels indicate ABPs. We use the Floyd-Steinberg error filter shown in Figure 1.2 for the error diffusion. For robust error diffusion, output-dependent threshold modulation is employed and the threshold $t(m, n)$ at the point (m, n) is given by:

$$t(m, n) = 0.25 + 0.33 \times 0.25 \times [g(m, n - 1) + g(m, n - 2) + g(m, n - 3)].$$

The threshold modulation tries to adjust the current threshold by using the information of three preceding halftone pixels. For error diffusion, the error diffusion constant is set to $\alpha = 0.8$ to avoid error diffusion instability.

Three output shares are shown in Figure 1.10(a) to Figure 1.10(c), respectively, where the corresponding perceived errors are also shown. When calculating the perceived error ϵ, an alpha stable human visual system model proposed in [9] is used. All shares show visually pleasing halftone images where the image details can be clearly recognized. The resultant shares do not show either residue image of the encoded image or any residue image of other shares. Figure 1.10(d) shows the reconstructed image when shares 1 to 3 are "stacked" together with the OR operation. The content of the reconstructed image is clearly recognizable with a contrast $\alpha = 1/9$. The decoded image does not bear any residue of any share images.

To illustrate the tradeoff between the share image quality and contrast of the decoded image, we show another set of simulation results of a 3-out-of-3 halftone VSS scheme. Compared with the previous example, the halftone cell size is changed to $q = 4 \times 4$. The sizes of the shared image and the secret image are well adjusted to be 512×512 and 128×128, respectively. The number of ABPs for each halftone cell is $x = 4$ and the quality index increases to $s = 1/2$. The three output shares are shown in Figure 1.11(a) to Figure 1.11(c), respectively, where the corresponding perceived error is also shown. Figure 1.11(d) shows the reconstructed image by stacking three shares together, where the contrast of the decoded image is $\alpha = 1/16$. Shares in the second example have larger halftone cell size, thus larger s and better image quality than that of the shares in the first example, as can be seen from Figure 1.11. However, the

FIGURE 1.10
(a)–(c) Shares of the 3-out-of-3 scheme using the first method, $q = 9$. The perceived errors are 1.73×10^{10}, 8.45×10^9, and 5.46×10^9, respectively. (d) Decoded image by shares (a)-(c), $\alpha = 1/9$. (Reprinted with permission from IEEE Trans. Inf. Forensics Security, vol. 4, no. 3, pp. 383–396, Sep. 2009 ©IEEE 2009).

(a)

(b)

(c)

(d)

FIGURE 1.11
(a)–(c) Shares of the 3-out-of-3 scheme using the first method, $q = 16$. The perceived errors are 7.1×10^9, 3.48×10^9, and 2.27×10^9, respectively. (d) Decoded image by shares (a)-(c), $\alpha = 1/16$. (Reprinted with permission from IEEE Trans. Inf. Forensics Security, vol. 4, no. 3, pp. 383–396, Sep. 2009 ©IEEE 2009).

contrast loss of the decoded image is worse. It is observed that the contrast of Figure 1.11(d) is lower than that of Figure 1.10(d).

(a)

(b)

(c)

(d)

FIGURE 1.12
(a)–(c) Shares of the 3-out-of-3 scheme using the second method, $q = 16$.
(d) Decoded image by shares (a)–(c), $\alpha = 1/16$. (Reprinted with permission from IEEE Trans. Inf. Forensics Security, vol. 4, no. 3, pp. 383–396, Sep. 2009 ©IEEE 2009).

In the following, we present the halftone shares generated by using the second construction method. The performance of the second method is evalu-

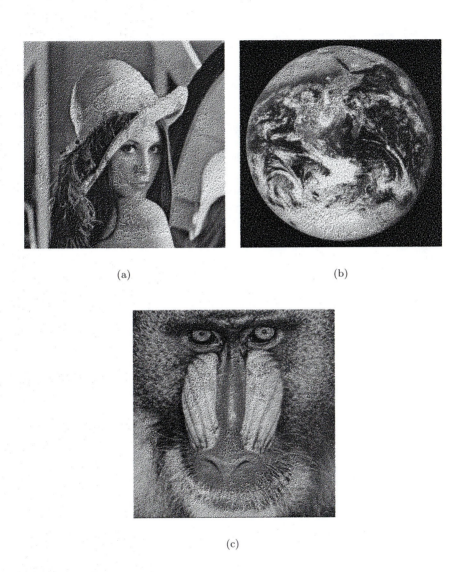

(a)

(b)

(c)

FIGURE 1.13
(a)–(c) Another set of shares generated by the second method, $q = 16$, depicting more pronounced artifacts. (Reprinted with permission from IEEE Trans. Inf. Forensics Security, vol. 4, no. 3, pp. 383–396, Sep. 2009 ©IEEE 2009).

ated by comparing its shares with those of the first method. Figure 1.12(a)–(c) show three shares produced using the second method in a 3-out-of-3 scheme where the size of the image is 512×512 and the size of the halftone cell is 4×4. Here, images *Lena*, *Tank*, and *Baboon* are used. It is obvious that all halftone shares show images with fine details. The reconstructed image is shown in Figure 1.12(d), which preserves all the secret information. Compared with Figure 1.12(a), Figure 1.11(a) is generated using the first method, showing less image details and appearing dark due to the added ABPs.

We should point out that the second method does not necessarily produce visually pleasing halftone shares if the grayscale images are not carefully selected. As an example, Figure 1.13(a)–(c) shows another set of shares generated by *Lena*, *Earth*, and *Baboon* using the second method with the same scheme. Careful inspection shows these shares present more artifacts due to the cross interference between the shares. The reason is that the contents of the three grayscale images are not complementary to each other. Comparing Figure 1.13(a) with Figure 1.11(a) and Figure 1.12(a), local geometric distortion can be observed in Figure 1.13(a). The texture of the distorted region is quite different from that of the neighboring regions, which causes visible artifacts. On the other hand, Figure 1.11(a) and Figure 1.12(a) both show smooth images with uniform image quality. Thus, it is important to choose appropriate grayscale images for the second method.

1.8   Conclusion

In this chapter, HVC construction methods based on error diffusion are introduced, which can generate shares with pleasing visual information. In the introduced methods, the pixels that carry the secret information are preset before a halftone share is generated from a grayscale image. Error diffusion is used to construct the shares so that the noise introduced by the preset pixels is diffused away when halftone shares are generated. The secret information is then naturally embedded into the halftone shares. The homogeneous and isotropic distribution of the preset pixels imposes the least noise in the error diffusion, thus leading to shares with high image quality. Our introduced methods follow the basic principle of VC, thus the security of the construction scheme is guaranteed. The introduced HVC constructions apply not only to VSS but also to VSS used in the context of visual authentication and encryption.

For the first method, by using auxiliary black pixels, the contrast condition of the decoded image is satisfied. Furthermore, the shares do not suffer any interference from other shares. When auxiliary black pixels are employed, blue noise multitone error diffusion is used to generate the distributions of the secret information pixels and black auxiliary pixels. The second method exploits the fact that halftoning of the grayscale images alone can generate most of the black pixels needed. Black pixels are inserted only when a sufficient number of black pixels have not yet been produced. By carefully selecting the grayscale images, the second method can also generate shares with visually pleasing images. For both methods, the decoded image does not suffer any interference from the shared images. It is clear that there is a tradeoff between the shared image quality and the contrast loss of the decoded image. However, by simple linear filtering, the contrast loss can be easily recovered.

Bibliography

[1] M. Analoui and J. P. Allebach. Model-based halftoning using direct binary search. In *Proc. SPIE*, volume 1666, pages 96–108, Feb. 1992.

[2] G. Ateniese, C. Blundo, A. De Santis, and D. R. Stinson. Visual cryptography for general access structures. *Information and Computation*, 129(2):86–106, Sept. 1996.

[3] G. Ateniese, C. Blundo, A. De Santis, and D. R. Stinson. Extended capabilities for visual cryptography. *Theoret. Comput. Sci.*, 250:143–161, 2001.

[4] J. Bacca, G. R. Arce, and D. L. Lau. Blue noise multitone dithering. *IEEE Trans. Image Process.*, 17(8):1368–1382, Sep. 2008.

[5] C. Blundo, P. D'Arco, A. De Santis, and D. R. Stinson. Contrast optimal threshold visual cryptography schemes. *SIAM J. Discrete Math.*, 16(2):224–261, 2003.

[6] R. Eschbach, Z. Fan, K. T. Knox, and G. Marcu. Threshold modulation and stability in error diffusion. *IEEE Signal Process. Mag.*, 20:39–50, Jul. 2003.

[7] F. Faheem, D. L. Lau, and G. R. Arce. Digital multitoning using gray level separation. *J. Imag. Sci. and Tech.*, 46(5):385–397, Sep. 2002.

[8] M. S. Fu and O. C. Au. Joint visual cryptography and watermarking. In *Proc. IEEE Int. Conf. on Multimedia and Expo*, Taipei, Taiwan, Jun. 2004.

[9] A. J. Gonzalez, G. R. Arce, J. Bacca Rodriguez, and D. L. Lau. Human visual alpha-stable models for digital halftoning. In *Proc. 18th Annual Symp. on Electron. Imag. Sci. and Tech.: Human Vision and Electro. Imag. XI*, San Jose, CA, Jan. 2006.

[10] A. Houmansadr and S. Ghaemmaghami. A novel video watermarking method using visual cryptography. In *Proc. IEEE Int. Conf. on Engineering of Intelligent Systems*, Islamabad, Pakistan, Apr. 2006.

[11] S. H. Kim and J. P. Allebach. Impact of HVS models on model-based halftoning. *IEEE Trans. Image Process.*, 11(3):258–269, Mar. 2002.

[12] D. L. Lau and G. R. Arce. *Modern Digital Halftoning*. Marcel Dekker, Inc, 2001.

[13] D. L. Lau, G. R. Arce, and N. C. Gallagher. Green-noise digital halftoning. *Proceedings of the IEEE*, 86(12):2424–2444, Dec. 1998.

[14] D. L. Lau, G. R. Arce, and N. C. Gallagher. Digital halftoning by means of green-noise masks. *J. Opt. Soc. Am. A-Optics Image Sci. and Vis.*, 16(7):1575–1586, Jul. 1999.

[15] D. L. Lau, G. R. Arce, and N. C. Gallagher. Digital color halftoning with generalized error diffusion and multi-channel green-noise masks. *IEEE Trans. Image Processing*, 9(5):923–935, May 2000.

[16] D. L. Lau, R. Ulichney, and G. R. Arce. Blue- and green-noise halftoning models—a review of the spatial and spectral characteristics of halftone textures. *IEEE Signal Processing Magazine*, 10(4):28–38, Jul. 2003.

[17] P. Li and J. P. Allebach. Tone-dependent error diffusion. *IEEE Trans. Image Process.*, 13(2):201–215, Feb. 2004.

[18] M. Naor and B. Pinkas. Visual authentication and identification. *Crypto97, LNCS*, 1294:322–340, 1997.

[19] M. Naor and A. Shamir. Visual cryptography. *Advances in Cryptography: EUROCRYPT'94, LNCS*, 950:1–12, 1995.

[20] R. A. Ulichney. Dithering with blue noise. In *Proc. the IEEE*, volume 76, pages 56–79, Jan. 1988.

[21] N. D. Venkata and B. L. Evans. Adaptive threshold modulation for error diffusion halftoning. *IEEE Trans. Image Process.*, 10(1):104–116, Jan. 2001.

[22] Z. Wang and G. R. Arce. Halftone visual cryptography through error diffusion. In *Proc. IEEE ICIP*, Atlanta, GA, Oct. 2006.

[23] Z. Wang and G. R. Arce. Halftone visual cryptography via error diffusion. *IEEE Trans. Inf. Forensics Security*, 4(3):383–396, Sep. 2009.

[24] C. W. Wu and G. Thompson. Digital watermarking and steganography via overlays of halftone images. In *Proc. SPIE*, volume 5561, pages 152–163, Oct. 2004.

[25] Z. Zhou, G. R. Arce, and G. Di Crescenzo. Halftone visual cryptography. In *Proc. IEEE Int. Conf. on Image Process.*, Barcelona, Spain, Sep. 2003.

[26] Z. Zhou, G. R. Arce, and G. Di Crescenzo. Halftone visual cryptography. *IEEE Trans. Image Process.*, 15(8):2441–2453, Aug. 2006.

2

Visual Cryptography for Color Images

Stelvio Cimato

Università degli Studi di Milano, Italy

Roberto De Prisco

Università di Salerno, Italy

Alfredo De Santis

Università di Salerno, Italy

CONTENTS

2.1   Introduction

The key property used to construct visual cryptography schemes for black and white images is the following: if we superpose transparencies with black and white pixels, the resulting pixel that our eyes see is black if at least one of the superposed pixels is black and is white if all the superposed pixels are white. Such a property can be rephrased as follows: the possible "state" for the pixels can be represented with a bit, using 0 for white and 1 for black, and the human visual systems performs an OR of the input pixels in order to reconstruct the secret pixels.

This key property does not easily extend to colored pixels. With colored pixels the state of each pixel cannot be represented anymore with a single bit and the "reconstruction" operation performed by our eyes is much more complicated than a simple OR.

In this chapter we will first describe the difficulties that arise from the superposition of colored pixels and then we review the work on visual cryptography for colored images.

We assume that the reader is familiar with (at least the basics of) black and white visual cryptography.

2.2   Color Superposition

What happens when we stack together two transparencies so that two pixels get superposed? What is the color that the human eyes see as the result of this superposition? Figure 2.1 illustrates the superposition operation with two examples. In the first one we are using black and white pixels: the superposition of a black pixel with a white pixel yields a black pixel. In the second example, we are using colored pixels: the superposition of a yellow pixel with a magenta pixel yields a red pixels.

Using only black and white images the result of the superposition of pixels printed on transparencies is straightforward: it is black if and only if at least one of the pixels is black.

The answer to the same question gets much more complicated when we use colors. In order to understand what happens when we superpose transparencies with colored pixels we have to talk a bit about light and color theory.

2.2.1   Color Vision and Color Models

Modern understanding of light and color vision is based upon the advances of several great scientists, such as the ones due to Newton. Thanks to them

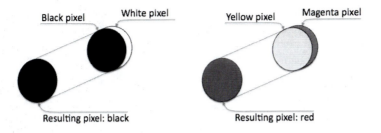

FIGURE 2.1

(See color insert.) Pixels superposition: black and white (left) and colored (right).

today we have a good understanding of light and colors; the topic is quite complex and a rigorous and detailed explanation goes beyond the scope of this chapter. However, we will try to explain the basic properties of light and colors because they are crucial for any visual cryptography scheme that deals with color images.

Roughly speaking, light consists of electromagnetic energy with wavelengths in the approximate range of 350–750 *nm*, as shown in Figure 2.2. The visible range represents only a small fraction of the full electromagnetic spectrum.

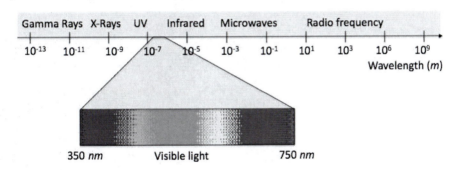

FIGURE 2.2

(See color insert.) Electromagnetic spectrum.

When a particular wavelength in such a range hits the retina in our eyes, it is perceived as a color. In the visible range, shorter wavelengths are perceived as bluish colors, middle wavelengths as greenish colors, and higher wavelengths as reddish colors. When our eyes are hit by several wavelengths we perceive a color that is a sort of "sum" of the wavelengths. If the eyes are hit by all the visible wavelengths, the perceived color is white. That is, a (pure) white light consists of all the visible wavelengths.

The expressions "red light," "yellow light," etc., are technically incorrect, but we will often use them to mean a light whose wavelength is perceived as red, yellow, etc. Since each wavelength corresponds to a color, there are infinite colors, one for each possible wavelength.

An object appears to be of a particular color because when light hits the object some light is absorbed, that is some wavelengths are absorbed; the remaining wavelengths are perceived by our eyes. When light hits an object it can also be reflected (or pass through the object). An object of a particular color χ, has strong absorption properties for the wavelengths that do not correspond to χ while it reflects the light with color χ. For example, an object appears yellow (when hit by a white light) because it reflects the yellow light and absorbs most strongly in the other parts of the spectrum. In the case of a transparency, the light that is not absorbed instead of getting reflected passes through the transparency.

A color model is a formal model that allows us to represent all (or some of) the possible colors. One of the most used color model is the one called the "additive color model." With this method three primary colors (usually red, green, and blue) are mixed to obtain other colors. Figure 2.3 shows this model with the primaries red, green, and blue; the colors yellow, cyan, and magenta are produced when two of these primaries overlap. Varying the "intensity" of each primary in the mixing we can obtain many other colors.

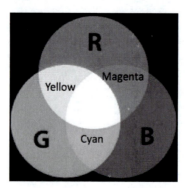

FIGURE 2.3
(See color insert.) Additive color model with primaries red, green, and blue.

The set of all possible colors that we can obtain depends on the three primary colors that we use. Any three colors can be used as primaries; the range of colors that we obtain is the gamut of those primaries. Unfortunately, no three primary colors exist so that their gamut corresponds to the set of all possible colors; however by choosing red, green, and blue as primaries we obtain a very large number of colors in their gamut. This is why these three colors have been chosen for the additive color model and the model is often called Red, Green, and Blue (RGB). Most displays use this model.

Another color model is the "subtractive color model," also called the Cyan, Magenta, and Yellow (CMY) model. In this case the colors are obtained with a subtractive technique that starts from a white light and subtracts wavelengths corresponding to the three colors cyan, magenta, and yellow. Figure 2.4 shows the CMY model.

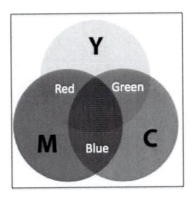

FIGURE 2.4
(See color insert.) Subtractive color model with primaries cyan, magenta, and yellow.

Most modern printers use this model, often exploiting an additional black ink; an additional K in the name CMYK indicates the use of the extra black ink. Notice that we can obtain black by using all three inks (cyan, magenta, yellow) together; however it is more efficient to cover a pixel with just black ink, rather covering it with the three inks cyan, magenta, and yellow.

In the additive model we start from the absence of light, which gives the black color, and we add light to obtain other colors with the extreme case being the white color obtained when we add all possible wavelengths. In the subtractive model we start from a white light and we subtract wavelengths to obtain other colors with the extreme case being the color black obtained when we take out all possible wavelengths.

If we are not very picky, and a discussion about this aspect goes beyond the scope of this chapter, we can say that the RGB and the CMY models are equivalent and complementary. Indeed an ink with the color cyan absorbs the light corresponding to the red color, an ink with the color magenta absorbs the light corresponding to the green color, and an ink with the color yellow absorbs the light corresponding to the blue color. Because of this, for both models, we will formally represent a color χ as a triple (x, y, z), where x, y, and z denote the amount of red, green, and blue, respectively, that χ consists of. The amount of each type of light (red, green, blue) is described by an integer in the range $[0, L]$. With this setting, we can produce $(L+1)^3$ different colors, which, for L sufficiently large, are enough to approximate all colors that the human eyes are able to distinguish. Typically, for computers, we have $L = 255$. To make things easier, throughout this chapter, we use $L = 100$.

Each of the components x, y, and z can be seen as a filter that lets pass through only some light. The color $(0, 0, 0)$, which we will denote also with the symbol "$\bullet$," is the black color: indeed all filters are 0, meaning that there is no light left. The color $(100, 100, 100)$, which we will denote also with the symbol "$\circ$," is white because no light is absorbed by the filters. The colors red, green, and blue are represented, respectively, by $(100, 0, 0), (0, 100, 0)$, and $(0, 0, 100)$; we will refer to these colors also as R, G, and B, respectively. The colors cyan, magenta, and yellow are represented, respectively, by $(0, 100, 100), (100, 0, 100)$, and $(100, 100, 0)$; we will refer to these colors also as C, M, and Y, respectively. The color $(50, 0, 0)$ is also a red, because that is the only component present, but it is darker since some red light has been absorbed. The higher the value of the component, the lighter is the color. If all components are equal, i.e., (x, x, x), then the resulting color is a gray whose intensity depends on x: the smaller is x, the darker is the gray.

Recall that this representation works fine both for the additive model and for the subtractive model. In the additive model we start from $(0, 0, 0)$ and add light while in the subtractive model we start from $(100, 100, 000)$ and subtract light.

In the context of visual cryptography we can think of the transparencies as filters; starting from a white light we "subtract" some light applying filters (the transparencies). The remaining light determines the color that we see when superposing several transparencies. At this point it is worth emphasizing that "white" on a transparency is actually "transparent." We assume to start with a pure white light; if the transparency does not have any ink on it, then the white light just passes through the transparency and we see white.

What is the color of the pixel resulting from the superposition of one or more transparencies?

When we drop some ink on the transparency and hold the transparency to the light we see the color that the ink lets pass through. When more transparencies get stacked together, the color of the resulting pixel depends on the absorption properties of the inks on all of the transparencies.

Let $\chi_1 = (x_1, y_1, z_1)$ and $\chi_2 = (x_2, y_2, z_2)$ be two colors and assume that two pixels of color χ_1 and of color χ_2 are printed on two different transparencies.

The following operator add describes the color superposition operation:

$$\mathtt{add}(\chi_1, \chi_2) = \left(\mathtt{int}\left(\frac{x_1 x_2}{L} \right), \mathtt{int}\left(\frac{y_1 y_2}{L} \right), \mathtt{int}\left(\frac{z_1 z_2}{L} \right) \right).$$

Notice that taking into account only the inks that we have used for each transparency is a simplification: the perception of the final color depends also on the material of the transparencies and the aberrations that the stack of transparencies produces. Moreover, it is likely that the initial light we start with is not a pure white light and that there are also other sources of light in the environment. However, the add operator is a quite good approximation.

The add operation is commutative and thus the order in which we superpose the colors is irrelevant. As expected, it results that $\mathtt{add}(\mathtt{Y}, \mathtt{M}) = \mathtt{R}$, add(R, G) $=$ Y, add(Y, M, C) $=$ •. Figure 2.5 shows some other examples of superposition of colored pixels.

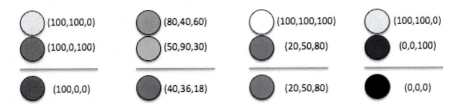

FIGURE 2.5
(See color insert.) Examples of pixels superposition.

The add operator can be easily extended to any number of pixels. Indeed since the operation is commutative it is enough to add any two pixels each time until we get to one pixel. Let $\chi_1 = (x_1, y_1, z_1), \chi_2 = (x_2, y_2, z_2), \ldots, \chi_n = (x_n, y_n, z_n)$ be the colors of the pixels. The color of the pixel that results from the superposition is:

$$\text{add}(\chi_1, \chi_2, \ldots, \chi_n) = (X, Y, Z)$$

where

$$X = \text{int}\left(\frac{x_1 x_2 \ldots x_n}{L^{n-1}}\right), Y = \text{int}\left(\frac{y_1 y_2 \ldots y_n}{L^{n-1}}\right), Z = \text{int}\left(\frac{z_1 z_2 \ldots z_n}{L^{n-1}}\right).$$

Figure 2.6 shows examples of superpositions with 3 pixels.

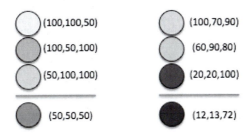

FIGURE 2.6
(See color insert.) More examples of pixels superposition.

2.2.2 Lattices

Some papers (e.g., [7]) use finite lattices to formalize the properties of the superposition of colored pixels. A finite lattice is a partially ordered set for which any two elements of the set have a least upper bound and a greatest lower bound. We can use a lattice to describe a color model.

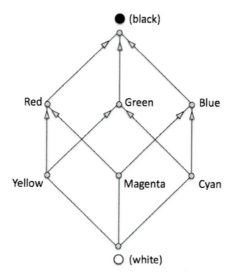

FIGURE 2.7
Lattice for the RGB and CMY color models.

In the additive model the superposition of two colored pixels corresponds to the greatest lower bound while in the subtractive color model the superposition corresponds to the least upper bound. The choice of a particular lattice is equivalent to the choice of a color model. For example, the lattice in Figure 2.7 is equivalent to the color model that uses the following 8 colors: black, white, R, G, B, C, M, and Y. Notice that this particular set of colors is closed under the superposition operation. It is worth emphasizing that this lattice is not equivalent to the RGB and CMY models: it only considers the 8 colors with zero or full intensity while the RGB and CMY models have many more colors.

2.2.3   The Darkening Problem

When we superpose pixels having the same color, unless we have zero or full intensity components, the resulting pixel is a darker version of the original color. This is because each transparency is a filter that absorbs some light, except when the transparency is white, and thus the resulting pixel is darker. Figure 2.8 shows examples of superposition of pixels with the same color, a light grey. As can be noticed in the figure, as we add more pixels the resulting color becomes darker, with the limit being a full black.

We will refer to this problem as the *darkening problem*. Some of the schemes that we will describe later superpose pixels with the same color, but ignore the darkening problem.

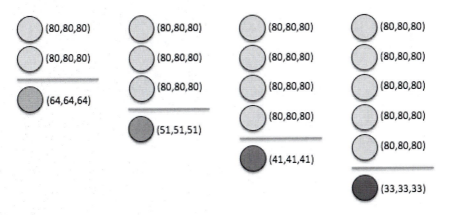

FIGURE 2.8
The darkening problem.

2.2.4 The Annihilator Color

Since for any color χ we have that $\mathtt{add}(\chi, \bullet) = \bullet$, in many visual cryptography schemes for color images the black color is often used to "cover up" other colors so that they don't show up in the reconstructed image. For this reason we call the black color the "annihilator" color. The presence of the annihilator color in the reconstructed image has no meaning and thus the observer has to ignore it. The use of the annihilator color is not a problem from a formal point of view but the visual effect is not good: in many cases the presence of the annihilator color in the reconstructed image is overwhelming (e.g., 90% of the image) and thus it is not reasonable to assume that the observer can recognize the secret image. This is clearly a problem if we want to share images, but it doesn't rule out some applications as we will see in later sections. We remark that the annihilator color, has nothing special: it is just the black color! If the secret image contains black pixels, then we will not be able to distinguish amongst the black pixels in the reconstructed image which ones were originally black and which ones were annihilated.

2.2.5 The Identity Color

Color $\circ$ is the "identity" color, in the sense that for any color χ we have that $\mathtt{add}(\chi, \circ) = \chi$. In some schemes the identity color is used, together with the annihilator color, as a special color. Recall that that in the context of visual cryptography white is actually transparent.

2.3   Formal Models for Colored VCS

In this section, we discuss the formal model for color visual cryptography, or Color-VC for short. We first recall the basic properties of the formal model for black and white visual cryptography, B&W-VC for short, that will be needed also for the case of color images and then we dwell upon the problems that need to be tackled in order to define a formal model for Color-VC.

2.3.1   The Models for B&W-VC

For B&W-VC the formal models used in the literature are all equivalent (with variations on the metrics used for evaluation, like for example the contrast of the scheme). The two key properties, that will be needed also for color images, are:

- *the safety property*, which guarantees that nonqualified sets of participants are not able to reconstruct the secret image;

- *the contrast property*, which guarantees that qualified sets of participants are able to reconstruct the secret image.

To evaluate visual cryptography schemes the most important metric is the *pixel expansion*, that is the number of subpixels used in the reconstructed image for each pixel of the secret image.

Another important measure for the evaluation of B&W-VC schemes is the *contrast* of the reconstructed image that can be defined as a function of the contrast property. Several contrast properties and metrics can be found in literature for B&W-VC. We refer the reader to the relevant papers about the contrast (see for example [5]).

With the exception of the definition of the contrast, the formal model for B&W-VC is pretty standard.

2.3.2   The Models for Color-VC

For color images even the model becomes difficult to define. Do we start with a prespecified palette, perhaps the one used in the secret image, or do we consider all possible colors? What color model do we consider? Do we consider the darkening problem? Is the palette closed under the superposition operation? That is, if we start with a prespecified palette, do we consider the possibility that the reconstructed image contains colors that are not in the original palette? How do we define the contrast property for color images and what is the contrast metric? Do we allow the use of the annihilator color? How do we account for it in the contrast property?

In the following, we discuss all these issues. We start by defining the secret and the shares palettes as follows:

- *Secret palette*: this is the set of colors used in the secret image. This is a finite set of c colors (we can have at most one color per pixel). To make notation easier we will denote these colors simply with the set of integers $\{1, 2, \ldots, c\}$. For the colors white, black, red, green, blue, cyan, magenta, and yellow we will also use the corresponding symbol $(\circ, \bullet, \mathsf{R}, \mathsf{G}, \mathsf{B}, \mathsf{C}, \mathsf{M}, \mathsf{Y})$ instead of the palette index.

- *Shares palette*: this is the set of colors that we can print on the shares or obtain by superposing printed shares. The shares palette might be the same as the original palette, or it might be augmented with some (or even many) other colors. Most of the schemes used in the literature augment the shares palette with the colors $\circ$ and $\bullet$. We denote the colors in the shares palette with the set of integers $\{1, 2, \ldots, d\}$. When the shares palette is a superset of the secret palette (this is the case in almost all of the scheme presented in this chapter) we have that $d \geq c$ and to simplify the notation we assume, without loss of generality, that the first c colors of the shares palette are exactly those in the secret palette.

The secret image consists of a collection of pixels, each one with a color of the secret palette. As for B&W-VC, each pixel of the secret image is encoded in the shares into a certain number m of subpixels. Such an integer m is the pixel expansion of the scheme.

In order to define a scheme we need to specify the qualified and the non-qualified set of participants. There are n participants. For simplicity we consider only the case of threshold schemes: Any set of at least k participants is a qualified set, while any set with less than k participant is a nonqualified set of participants.

In order to share each pixel of the secret image a trusted third party has to create and distribute shares to the n participants. The creation of the shares is defined using *distribution matrices*. These are c collections (multisets) of $n \times m$ matrices $\mathcal{C}^1, \mathcal{C}^2, \ldots, \mathcal{C}^c$, whose elements are in the shares palette.

To share a secret pixel of color i, the dealer randomly chooses one of the matrices in $\mathcal{C}^i$ and distributes row j to participant j. Thus, the chosen matrix defines the m subpixels in each of the n transparencies.

An example of distribution matrix is the following:

$$D = \begin{bmatrix} 1 \bullet 1 \mathsf{M} 1 \\ \mathsf{R} 1 1 \circ 2 \\ 2 1 1 \bullet 3 \\ \circ \mathsf{M} 1 \mathsf{G} \mathsf{B} \end{bmatrix}$$

In this case, there are $n = 4$ participants (number of rows in the distribution matrices) and the pixel expansion of the scheme is $m = 5$ (number of columns in the distribution matrices). If D is the matrix selected for the distribution of the shares then the 5 subpixels in the first share will have colors 1•1M1, while those in the second share will have colors R11∘2.

Given a distribution matrix M and a set of participants X, we denote with $M|X$ the submatrix of M obtained by considering only the rows of M corresponding to the participants in X.

As for the black and white case, the definition of a scheme must satisfy the security and the contrast properties:

Security property: Given a forbidden set X, $|X| < k$, the c collections of $|X| \times m$ matrices, $\mathcal{D}^i$, $i = 1, 2, ..., c$, consisting of $M|X$ for each $M \in \mathcal{C}^i$, contain the same matrices with the same frequencies. This property guarantees that a forbidden set of participants has no information on the secret image.

Contrast property: The contrast property has to guarantee that the secret image will be visible for a qualified set of participants. For B&W-VC this property uses two thresholds ℓ and h, with $\ell < h$, and requires that when the secret pixel is white, the number of black subpixels in the reconstruction is at most ℓ and when the secret pixels is black, the number of black subpixels is at least h. Many papers that deal with color images generalize this definition requiring that in the reconstructed pixel there are at least h subpixels of color i, where i is the color of the secret pixel, and for any other color $j \neq i$ there are at most ℓ subpixels with color j. Notice that this definition can be used only if the shares palette is equal to the secret palette. Moreover, it allows the possibility that the reconstructed pixel is made up of an overwhelming majority of subpixels with a wrong color. For example if $h = 4$, $l = 3$, and $c = 10$ it is possible to have in the reconstructed pixel only 4 subpixels with the right color while other $27 = 3 \cdot 9$ have (mixed) wrong colors. The annihilator color • can be present without any restriction.

Probably a better definition of the contrast property should require that in the reconstructed image there be at least h subpixels with the right color and at most ℓ subpixels with wrong colors. That is, the number of subpixels with the right color should be greater than the number of subpixels with a wrong color (counting all the subpixels with wrong colors).

We will refer to the first property as the *weak* contrast property and to the second one as the *strong* contrast property.

Next, we provide a formalization of such properties. Define the $\mathsf{add}(M)$ for a distribution matrix M to be the vector whose j^{th} component is the add of column j in M and define $w_i(v)$ for a vector v to be the number of elements equal to color i, for $i = 1, 2, \ldots, c$, that is for any color in the secret palette. Moreover we define $\bar{w}_i(v)$ to be the number of elements in v different from color i and from the annihilator color.

Weak contrast property: There must exist h and ℓ, integers $0 \leq \ell < h \leq m$, such that given a qualified set X, $|X| = k$, for any $M \in \mathcal{C}^i$, it holds that $w_i(\mathsf{add}(M|X)) \geq h$ and $w_j(\mathsf{add}(M|X)) \leq \ell$ for any j in the shares palette and $j \neq i$. Note that the annihilator color is not considered, that is, it is allowed that many pixels be •.

Strong contrast property: There must exist h and ℓ, integers $0 \leq \ell < h \leq m$, such that given a qualified set X, $|X| = k$, for any $M \in \mathcal{C}^i$, it holds that $w_i(\text{add}(M|X)) \geq h$ and $\bar{w}_i(\text{add}(M|X)) \leq \ell$. Also in this case the annihilator color can be present without restriction.

In the black and white case, the thresholds ℓ and h, together with the pixel expansion m have been used to define several variants of the contrast metric, such as $\alpha = h - \ell$, $\alpha = (h-\ell)/m$, and $\alpha = (h-\ell)/(h+\ell)$. Similar measures have been used for color schemes and we will specify the definition of the contrast when presenting the schemes. However, for Color-VC schemes we need to account for the presence of the annihilator color in the reconstructed image and this makes the contrast less important. We will evaluate the *annihilator presence* that we can define as $\beta = b/m$, where b is the number of pixels that get annihilated in the reconstruction process.

2.3.3 The SC, ND, and General Models

The schemes that we will review in the rest of the chapter can be classified, based on the formal model that they use, into three classes. In the next paragraph we define three formal models for Color-VC.

The SC (Same Color) model.

The SC model does not allow the superposition of pixels with different colors, with the exception of the identity ($\circ$) and the annihilator ($\bullet$) colors. Hence, the shares have to be constructed in such a way that each column in the distribution matrices have elements taken from the set $\{i, \circ, \bullet\}$, for some color i. Thus, when we superpose several transparencies, we never have a pixel of color i superposed with a pixel of color j.

Moreover the darkening problem is ignored. That is, it is assumed that superposing several pixels with color i, the resulting color is still i.

An example of a distribution matrix for such kinds of schemes is the following (we have used three colors, denoted with the numbers $1, 2$, and 3):

$$
D = \begin{bmatrix}
3\,1\,1\,1\,\bullet\,2\,2\,2\,\circ\,3\,3\,\bullet\,3\,\circ\,\bullet\,\bullet\,\circ\,1\,\bullet\,\circ\,\bullet\,2 \\
3\,1\,1\,\bullet\,1\,2\,2\,\bullet\,2\,\bullet\,3\,\bullet\,3\,\circ\,3\,\bullet\,\bullet\,1\,\bullet\,\circ\,\circ\,2\,\bullet \\
3\,1\,\bullet\,1\,1\,2\,\bullet\,2\,2\,3\,\circ\,3\,\bullet\,\bullet\,3\,\bullet\,1\,\bullet\,\bullet\,\bullet\,2\,\bullet\,\bullet \\
3\,\circ\,1\,1\,1\,\bullet\,2\,2\,2\,3\,\bullet\,\bullet\,3\,3\,\bullet\,1\,\bullet\,\bullet\,\bullet\,2\,\circ\,\circ\,\circ
\end{bmatrix}
$$

As can be noted, in each column, we either have colors $\circ, \bullet$, or pixels with a color $\chi = 1, 2$, or 3. We never have a column that mixes two different colors in the set $\{1, 2, 3\}$.

This restriction and the fact that the darkening problem is ignored avoids the complications that derive from color superposition.

The ND (No Darkening) Model.

The ND model is as the SC model but it considers the darkening problem. Thus, again we cannot superpose pixels with different colors, but if we superpose several pixels with the same color we get a darker version of that color.

The General Model.

In the General model, there are no restrictions about the superposition of pixels and the superposition operation satisfies the real properties of color superposition. This means the darkening problem is considered. Very few schemes have been defined for this model.

2.3.4   Base Matrices

Given a matrix B, we denote by $\mathcal{C}(B)$ the set of matrices obtained by permuting in all possible ways the columns of B. In most schemes, the c collections $\mathcal{C}^i$ are obtained by fixing c matrices B^i, $i = 1, 2, \ldots, c$, and letting $\mathcal{C}^i = \mathcal{C}(B^i)$. The matrices B^i are called the "base matrices." Base matrices constitute an efficient representation of a scheme. Indeed, the dealer has to store only the base matrices and in order to randomly choose a matrix from $\mathcal{C}(B^i)$ it has to randomly choose a permutation of the columns of the base matrix B^i.

Notice that the security property for a base matrices scheme is equivalent to: Given a forbidden set X, the matrices $B^i|X$, for $i = 1, 2, \ldots, c$ are the same up to a permutation of the columns.

2.4   Schemes for the SC Model

In this section, we review the known schemes for the SC model. Verheul and van Tilborg [10] were the first to consider visual cryptography schemes for color images. Their model is equivalent to the SC one; as we will see shortly their model requires a special property, which can be easily implemented using the SC model. The schemes of [10] were improved first by [2] and then by [7, 11]. Paper [3] provides a lower bound on the pixel expansion and also the construction of (n, n)-threshold schemes that achieve the lower bound. It turns out that the (n, n)-threshold schemes of [7, 11] also have optimal pixel expansion, which means that the schemes of [3] and those of [7, 11] are equivalent.

2.4.1   The VV Schemes

The model considered in [10], which we will call the VV model, requires a special property: if we superpose pixels with different colors then the resulting pixel is black. As we have explained earlier, this property is not natural. When we superpose two pixels with different colors, we get a third color that depends on the colors of the two superposed pixels. In some particular cases the resulting color is actually black, but it is not black in most cases.

Verheul and van Tilborg propose a trick that "implements" such a property. The trick works as follows: each pixel is divided into c subpixels, where c is the number of colors in the secret image, subpixels i gets color number i, while all other subpixels get painted with black, as shown in Figure 2.9.

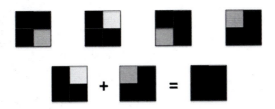

FIGURE 2.9
(See color insert.) The VV trick for the case of 4 colors. Subpixels with different colors are never superposed.

This trick implements the required property and makes the VV model equivalent to the SC model because in the resulting scheme subpixels with different colors are never superposed. However, to implement the trick, we have to pay an extra pixel expansion factor of c and a considerable fraction of the original pixel gets annihilated in the reconstruction.

The schemes of [10] are constructed using finite fields that satisfy certain conditions. We refer the reader to the original paper for a detailed description of the construction.

Assuming that $c > 2$ is a prime power, the construction produces

- (k, k)-threshold schemes with c colors for any k;

- $(k, c - 1)$-threshold schemes with c colors for $k < c$;

- (k, c)-threshold schemes with c colors for $k < c$, if $k - 1$ and $c - 1$ are not relatively prime.

The pixel expansion of the schemes is $m = c^k$; this includes the pixel expansion $m = c^{k-1}$ due to the construction of the scheme and the extra c factor due to implementation of the special property of the VV model.

The reconstruction guarantees that there is at least one pixel of the original color and no pixels with other colors, that is $h = 1$ and $\ell = 0$. The contrast property property considered is the weak one. However, we note that when $\ell =$

0 the weak and the strong contrast property are equivalent. The annihilator presence $\beta = (m-1)/m$, that is only one out of the m pixels is of the original color, while the remaining $m-1$ are annihilated.

As an example, we report the $(3,3)$-threshold 3-color scheme. Here are the three base matrices:

$$B^1 = \begin{bmatrix} 1 & 1 & 1 & 2 & 2 & 2 & 3 & 3 & 3 \\ 1 & 2 & 3 & 1 & 2 & 3 & 1 & 2 & 3 \\ 1 & 2 & 3 & 2 & 3 & 1 & 3 & 1 & 2 \end{bmatrix}$$

$$B^2 = \begin{bmatrix} 1 & 1 & 1 & 2 & 2 & 2 & 3 & 3 & 3 \\ 1 & 2 & 3 & 1 & 2 & 3 & 1 & 2 & 3 \\ 3 & 1 & 2 & 1 & 2 & 3 & 2 & 3 & 1 \end{bmatrix}$$

$$B^3 = \begin{bmatrix} 1 & 1 & 1 & 2 & 2 & 2 & 3 & 3 & 3 \\ 1 & 2 & 3 & 1 & 2 & 3 & 1 & 2 & 3 \\ 2 & 3 & 1 & 3 & 1 & 2 & 1 & 2 & 3 \end{bmatrix}$$

The pixel expansion, that corresponds to the number of columns in the base matrices, is $m = c^{k-1} = 3^2 = 9$. The above base matrices work only in the vv model. Although the annihilator color is not explicitly used, it appears because of the special property. Indeed, implementing the special property using the trick suggested earlier, the base matrices become the following:

$$B^1 = \begin{bmatrix} 1\bullet\bullet1\bullet\bullet1\bullet\bullet\bullet2\bullet\bullet2\bullet\bullet2\bullet\bullet\bullet3\bullet\bullet3\bullet\bullet3 \\ 1\bullet\bullet\bullet2\bullet\bullet\bullet31\bullet\bullet\bullet2\bullet\bullet\bullet31\bullet\bullet\bullet2\bullet\bullet\bullet3 \\ 1\bullet\bullet\bullet2\bullet\bullet\bullet3\bullet2\bullet\bullet\bullet31\bullet\bullet\bullet\bullet31\bullet\bullet\bullet2\bullet \end{bmatrix}$$

$$B^2 = \begin{bmatrix} 1\bullet\bullet1\bullet\bullet1\bullet\bullet\bullet2\bullet\bullet2\bullet\bullet2\bullet\bullet\bullet3\bullet\bullet3\bullet\bullet3 \\ 1\bullet\bullet\bullet2\bullet\bullet\bullet31\bullet\bullet\bullet2\bullet\bullet\bullet31\bullet\bullet\bullet2\bullet\bullet\bullet3 \\ \bullet\bullet31\bullet\bullet\bullet2\bullet1\bullet\bullet\bullet2\bullet\bullet\bullet3\bullet2\bullet\bullet\bullet31\bullet\bullet \end{bmatrix}$$

$$B^3 = \begin{bmatrix} 1\bullet\bullet1\bullet\bullet1\bullet\bullet\bullet2\bullet\bullet2\bullet\bullet2\bullet\bullet\bullet3\bullet\bullet3\bullet\bullet3 \\ 1\bullet\bullet\bullet2\bullet\bullet\bullet31\bullet\bullet\bullet2\bullet\bullet\bullet31\bullet\bullet\bullet2\bullet\bullet\bullet3 \\ \bullet2\bullet\bullet\bullet31\bullet\bullet\bullet\bullet31\bullet\bullet\bullet2\bullet1\bullet\bullet\bullet2\bullet\bullet\bullet3 \end{bmatrix}$$

Hence, the real pixel expansion is $m = c^k = 3^3 = 27$. In this particular case superposing 3 shares we get 26 black pixels out of 27 and just 1 colored pixel. That is, the annihilator presence is $\beta = 26/27$ (about 96%).

This approach doesn't seem practicable for images, but it can be used for other applications, like sharing passwords associating, for example, a digit to each color. For example, as reported in [10], if we use pixels of diameter 0.5 *cm* with 9 colors we can build a $(3,9)$-threshold visual scheme with 9 colors using $9^2 = 81$ pixels for each color of the password; on a standard A4 page there is room for a 90 digit password.

2.4.2 The BDD Schemes

Blundo et al. [2] focus on schemes with maximal contrast. They consider the weak contrast property and define the contrast as $\alpha = (h - \ell)/(h + \ell)$. The following results are provided in [2]:

- A first construction of c-color $(2, n)$-threshold Color-VC schemes with maximal contrast. The construction requires $c > n$.

- A proof that the above condition $c > n$ is necessary to have a maximal contrast scheme. It turns also out that, among the schemes with maximal contrast, the schemes provided by the first construction are also with optimal pixel expansion.

- A second construction of c-color $(2, n)$-threshold Color-VC schemes. Such a construction gives a better pixel expansion with respect to the first one but the schemes are not with maximal contrast.

- A construction of maximal contrast c-color (n, n)-threshold Color-VC schemes with improved pixel expansion with respect to those provided in [10].

We refer the interested reader to [2] for details about the constructions and the lower bound cited in this section.

2.4.3 The KY and YL Schemes

Koga and Yamamoto [7] and independently, Yang and Laih [11] provide (k, n)-threshold c-color schemes that improve on the pixel expansion of the schemes in [10, 2]. Here we report the construction provided in [11], but the one in [7] is equivalent.

Construction 1 *The construction exploits as a building block the base matrices $B_\circ$ and $B_\bullet$ of a scheme for black and white images. In order to obtain the base matrix B_i for color i we can concatenate one modified copy of $B_\circ$ with $c - 1$ modified copies of $B_\bullet$. The required modifications are the following: in $B_\circ$ we substitute $\circ$ with color i while in the $c - 1$ copies of $B_\bullet$ we substitute $\circ$ with the remaining $c - 1$ colors (one color per copy). The pixel expansion of the c-color scheme is c times the pixel expansion of the original black and white scheme.*

As an example, consider the following $(3, 3)$-threshold 3-color scheme. We start with the base matrices of a $(3, 3)$-threshold scheme for black and white images as defined in the paper by Naor and Shamir [8] (however, the construction works with any other choice of the black and white base matrices):

$$
B^\circ = \begin{bmatrix} \circ & \bullet & \circ & \bullet \\ \circ & \bullet & \bullet & \circ \\ \circ & \circ & \bullet & \bullet \end{bmatrix}
\qquad
B^\bullet = \begin{bmatrix} \bullet & \circ & \circ & \bullet \\ \circ & \bullet & \circ & \bullet \\ \circ & \circ & \bullet & \bullet \end{bmatrix}
$$

Then we construct the base matrices for the 3-color scheme as follows:

$$B_1 = [B_\circ^{\circ \to 1} | B_\bullet^{\circ \to 2} | B_\bullet^{\circ \to 3}] = \begin{bmatrix} 1\bullet 1\bullet & \bullet 2\,2\bullet & \bullet 3\,3\bullet \\ 1\bullet\bullet 1 & 2\bullet 2\bullet & 3\bullet 3\bullet \\ 1\,1\bullet\bullet & 2\,2\bullet\bullet & 3\,3\bullet\bullet \end{bmatrix}$$

$$B_2 = [B_\bullet^{\circ \to 1} | B_\circ^{\circ \to 2} | B_\bullet^{\circ \to 3}] = \begin{bmatrix} \bullet 1\,1\bullet & 2\bullet 2\bullet & \bullet 3\,3\bullet \\ 1\bullet 1\bullet & 2\bullet\bullet 2 & 3\bullet 3\bullet \\ 1\,1\bullet\bullet & 2\,2\bullet\bullet & 3\,3\bullet\bullet \end{bmatrix}$$

$$B_3 = [B_\bullet^{\circ \to 1} | B_\bullet^{\circ \to 2} | B_\circ^{\circ \to 3}] = \begin{bmatrix} \bullet 1\,1\bullet & \bullet 2\,2\bullet & 3\bullet 3\bullet \\ 1\bullet 1\bullet & 2\bullet 2\bullet & 3\bullet\bullet 3 \\ 1\,1\bullet\bullet & 2\,2\bullet\bullet & 3\,3\bullet\bullet \end{bmatrix}$$

Using as a building block the original (k, n)-threshold scheme provided in the paper by Naor and Shamir [8], whose pixel expansion is 2^{k-1}, the c-color schemes so obtained have pixel expansion $m = c \times 2^{k-1}$. This greatly improves on the pixel expansion of [2, 10].

Finally, as observed, also in [11], we can delete from the base matrices the columns that have all pixels with color $\bullet$. Using the base matrices provided in the paper by Naor and Shamir [8], for n even we always have one such column in each base matrix, while for n odd we always have $c - 1$ such columns in each base matrix. Hence, the pixel expansion can be further improved to $m = c \times 2^{k-1} - 1$ for n even and to $m = c \times 2^{k-1} - c + 1$ for n odd. This is important as we will see that for $k = n$ this improved pixel expansion matches a lower bound proved in [3].

The contrast property considered is the weak one. The scheme have parameters $h = 1$ and $\ell = 0$ (recall that for the special case of $\ell = 0$ the weak contrast property is equivalent to the strong one). The annihilator presence is $\beta = (m - 1)/m$.

The same idea used for the construction of c-color (k, n)-threshold schemes starting from black and white (k, n)-threshold schemes, can be used also for general access structure schemes. The pixel expansion of the c-color scheme is c times the pixel expansion of the black and white scheme that we start with.

2.4.4   The CDD Schemes and a Lower Bound

Paper [3] defines the contrast as $\alpha = (h-\ell)/m$ and considers the weak contrast property. The following theorems are proved in [3]:

Theorem 1 *In the SC model, the optimal contrast of a c-color (n, n)-threshold scheme is*

$$\alpha_{opt} = \begin{cases} \frac{1}{c \cdot 2^{n-1}-1}, & \text{if } n \text{ is even} \\ \frac{1}{c \cdot 2^{n-1}-c+1}, & \text{if } n \text{ is odd}. \end{cases}$$

Theorem 2 *In the SC model, the pixel expansion of a c-color (n,n)-threshold scheme, for any $c, n \geq 2$, is lower bounded by*

$$m \geq \begin{cases} c \cdot 2^{n-1} - 1, & \text{if } n \text{ is even} \\ c \cdot 2^{n-1} - c + 1, & \text{if } n \text{ is odd.} \end{cases}$$

Note that the above lower bound implies that the schemes of [7, 11] have optimal pixel expansion. In [3] an alternative construction of c-color (n,n)-threshold schemes with optimal pixel expansion is provided. The construction is the following:

Construction 2 *Fix any color i; base matrix C^i consists of the following columns:*

> *1. for $r = 0, 1, \ldots, \lceil n/2 \rceil - 1$ include the $\binom{n}{2r}$ columns, having $2r$ entries equal to $\bullet$ and the remaining ones of color i;*

> *2. for any color $j \neq i$, for $r = 0, 1, \ldots, \lceil \frac{n-1}{2} \rceil - 1$ include the $\binom{n}{2r-1}$ columns having $2r - 1$ entries equal to $\bullet$ and the remaining ones of color j;*

Below is an example for $c = 3$ and $n = 4$. For such a scheme $m = 23$ and $\alpha = 1/23$.

$$C^1 = \begin{bmatrix} 1\,2\,2\,2\,\bullet\,3\,3\,3\,\bullet\,\bullet\,1\,1\,\bullet\,1\,\bullet\,\bullet\,\bullet\,\bullet\,2\,\bullet\,\bullet\,\bullet\,3 \\ 1\,2\,2\,\bullet\,2\,3\,3\,\bullet\,3\,\bullet\,1\,\bullet\,1\,\bullet\,1\,\bullet\,\bullet\,2\,\bullet\,\bullet\,\bullet\,3\,\bullet \\ 1\,2\,\bullet\,2\,2\,3\,\bullet\,3\,3\,1\,\bullet\,1\,\bullet\,\bullet\,1\,\bullet\,2\,\bullet\,\bullet\,\bullet\,3\,\bullet\,\bullet \\ 1\,\bullet\,2\,2\,2\,\bullet\,3\,3\,3\,1\,\bullet\,\bullet\,1\,1\,\bullet\,2\,\bullet\,\bullet\,\bullet\,3\,\bullet\,\bullet\,\bullet \end{bmatrix}$$

$$C^2 = \begin{bmatrix} 2\,1\,1\,1\,\bullet\,3\,3\,3\,\bullet\,\bullet\,2\,2\,\bullet\,2\,\bullet\,\bullet\,\bullet\,\bullet\,1\,\bullet\,\bullet\,\bullet\,3 \\ 2\,1\,1\,\bullet\,1\,3\,3\,\bullet\,3\,\bullet\,2\,\bullet\,2\,\bullet\,2\,\bullet\,\bullet\,1\,\bullet\,\bullet\,\bullet\,3\,\bullet \\ 2\,1\,\bullet\,1\,1\,3\,\bullet\,3\,3\,2\,\bullet\,2\,\bullet\,\bullet\,2\,\bullet\,1\,\bullet\,\bullet\,\bullet\,3\,\bullet\,\bullet \\ 2\,\bullet\,1\,1\,1\,\bullet\,3\,3\,3\,2\,\bullet\,\bullet\,2\,2\,\bullet\,1\,\bullet\,\bullet\,\bullet\,3\,\bullet\,\bullet\,\bullet \end{bmatrix}$$

$$C^3 = \begin{bmatrix} 3\,1\,1\,1\,\bullet\,2\,2\,2\,\bullet\,\bullet\,3\,3\,\bullet\,3\,\bullet\,\bullet\,\bullet\,\bullet\,1\,\bullet\,\bullet\,\bullet\,2 \\ 3\,1\,1\,\bullet\,1\,2\,2\,\bullet\,2\,\bullet\,3\,\bullet\,3\,\bullet\,3\,\bullet\,\bullet\,1\,\bullet\,\bullet\,\bullet\,2\,\bullet \\ 3\,1\,\bullet\,1\,1\,2\,\bullet\,2\,2\,3\,\bullet\,3\,\bullet\,\bullet\,3\,\bullet\,1\,\bullet\,\bullet\,\bullet\,2\,\bullet\,\bullet \\ 3\,\bullet\,1\,1\,1\,\bullet\,2\,2\,2\,3\,\bullet\,\bullet\,3\,3\,\bullet\,1\,\bullet\,\bullet\,\bullet\,2\,\bullet\,\bullet\,\bullet \end{bmatrix}$$

Other results of [3] are

- A characterization of maximal contrast (k, n)-thresholds schemes. The characterization describes the schemes with a linear programming problem.

- A construction of c-color $(2, n)$-threshold schemes with improved pixel expansion with respect to [10, 11].

2.5   Schemes for the ND Model

In this section we describe schemes that work for the ND model. This model has been considered only in [4] where a construction for c-color (k, n)-threshold schemes is presented.

In order to have pixels with exactly the same color as the original one the schemes of [4] have the property that in any shares superposition at most one pixel is colored; all other pixels have one of the two special colors $\circ$ or $\bullet$.

The construction uses as a building block a black and white $(k-1, k-1)$-threshold scheme.

Construction 3 *Let S_{k-1}° and $S_{k-1}^\bullet$ be the basis matrices of a $(k-1, k-1)$-threshold scheme and let m' be the pixel expansion of such a scheme. Denote the rows of S_{k-1}° and $S_{k-1}^\bullet$ with w_i and b_i, respectively:*

$$S_{k-1}^\circ = \begin{bmatrix} w_1 \\ w_2 \\ \dots \\ \dots \\ w_{k-1} \end{bmatrix}, \qquad S_{k-1}^\bullet = \begin{bmatrix} b_1 \\ b_2 \\ \dots \\ \dots \\ b_{k-1} \end{bmatrix}.$$

Let $S_1^\bullet = [\bullet]$ and $S_1^\circ = [\circ]$. Then let $F_{k,n}(i, S_{k-1}^\phi)$, where $i \in \{1, 2, ..., c\}$ and $\phi \in \{\circ, \bullet\}$ be the $n \times \binom{n}{k} m'$ matrix constructed by $\binom{n}{k}$ submatrices, called "blocks," with dimension $n \times m'$ each consisting of the following rows: $n - k$ ("black") rows of m' elements $\bullet$; Each block differs from the others in the choice of the $n - k$ "black" rows; The remaining k rows are filled with one row of elements equal to i followed in order by the $k - 1$ rows of S_{k-1}^ϕ.

Base matrix for color i, for $i \in \{1, 2, ..., c\}$, is given by

$$\begin{aligned} B^i \;=\; & F_{k,n}(1, S_{k-1}^\bullet) + \dots + F_{k,n}(i-1, S_{k-1}^\bullet) + F_{k,n}(i, S_{k-1}^\circ) + \\ & F_{k,n}(i+1, S_{k-1}^\bullet) + \dots + F_{k,n}(c, S_{k-1}^\bullet) \end{aligned}$$

where $+$ denotes the concatenation of the matrices.

An example will clarify the above construction. Let $k = 3$ and $n = 4$ and consider the matrices S_{k-1}° and $S_{k-1}^\bullet$ given by the Naor and Shamir $(2, 2)$-threshold scheme [8], that is,

$$S_2^\circ = \begin{bmatrix} \circ & \bullet \\ \circ & \bullet \end{bmatrix}, \qquad\qquad S_2^\bullet = \begin{bmatrix} \circ & \bullet \\ \bullet & \circ \end{bmatrix}.$$

The F matrices will have $\binom{n}{k} = 4$ blocks, since we have to place 1 black row in each of 4 possible positions. Hence, we have

$$F_{3,4}(i, S_2^o) = \begin{bmatrix} i\,i & i\,i & i\,i & \bullet\,\bullet \\ \circ\,\bullet & \circ\,\bullet & \bullet\,\bullet & i\,i \\ \circ\,\bullet & \bullet\,\bullet & \circ\,\bullet & \circ\,\bullet \\ \bullet\,\bullet & \circ\,\bullet & \circ\,\bullet & \circ\,\bullet \end{bmatrix}, \quad F_{3,4}(i, S_2^\bullet) = \begin{bmatrix} i\,i & i\,i & i\,i & \bullet\,\bullet \\ \circ\,\bullet & \circ\,\bullet & \bullet\,\bullet & i\,i \\ \bullet\,\circ & \bullet\,\bullet & \circ\,\bullet & \circ\,\bullet \\ \bullet\,\bullet & \bullet\,\circ & \bullet\,\circ & \bullet\,\circ \end{bmatrix}.$$

The vertical bars identify the 4 blocks. As can be seen each block is given by 1 black row, and the remaining rows filled, in this order, by one row of i's and the rows of S_2^o (or $S_2^\bullet$), from the first to the last. Using the above F matrices we can build the following 3-color $(3,4)$-threshold scheme.

$$B^1 = \begin{bmatrix} 1\,1\,1\,1\,1\,1\,\bullet\,\bullet & 2\,2\,2\,2\,2\,2\,\bullet\,\bullet & 3\,3\,3\,3\,3\,3\,\bullet\,\bullet \\ \circ\,\bullet\,\circ\,\bullet\,\bullet\,\bullet\,1\,1 & \circ\,\bullet\,\circ\,\bullet\,\bullet\,\bullet\,2\,2 & \circ\,\bullet\,\circ\,\bullet\,\bullet\,\bullet\,3\,3 \\ \circ\,\bullet\,\bullet\,\bullet\,\circ\,\bullet\,\circ\,\bullet & \bullet\,\circ\,\bullet\,\bullet\,\circ\,\bullet\,\circ\,\bullet & \bullet\,\circ\,\bullet\,\bullet\,\circ\,\bullet\,\circ\,\bullet \\ \bullet\,\bullet\,\circ\,\bullet\,\circ\,\bullet\,\circ\,\bullet & \bullet\,\bullet\,\bullet\,\circ\,\bullet\,\circ\,\bullet\,\circ & \bullet\,\bullet\,\bullet\,\circ\,\bullet\,\circ\,\bullet\,\circ \end{bmatrix},$$

$$B^2 = \begin{bmatrix} 2\,2\,2\,2\,2\,2\,\bullet\,\bullet & 1\,1\,1\,1\,1\,1\,\bullet\,\bullet & 3\,3\,3\,3\,3\,3\,\bullet\,\bullet \\ \circ\,\bullet\,\circ\,\bullet\,\bullet\,\bullet\,2\,2 & \circ\,\bullet\,\circ\,\bullet\,\bullet\,\bullet\,1\,1 & \circ\,\bullet\,\circ\,\bullet\,\bullet\,\bullet\,3\,3 \\ \circ\,\bullet\,\bullet\,\bullet\,\circ\,\bullet\,\circ\,\bullet & \bullet\,\circ\,\bullet\,\bullet\,\circ\,\bullet\,\circ\,\bullet & \bullet\,\circ\,\bullet\,\bullet\,\circ\,\bullet\,\circ\,\bullet \\ \bullet\,\bullet\,\circ\,\bullet\,\circ\,\bullet\,\circ\,\bullet & \bullet\,\bullet\,\bullet\,\circ\,\bullet\,\circ\,\bullet\,\circ & \bullet\,\bullet\,\bullet\,\circ\,\bullet\,\circ\,\bullet\,\circ \end{bmatrix},$$

$$B^3 = \begin{bmatrix} 3\,3\,3\,3\,3\,3\,\bullet\,\bullet & 1\,1\,1\,1\,1\,1\,\bullet\,\bullet & 2\,2\,2\,2\,2\,2\,\bullet\,\bullet \\ \circ\,\bullet\,\circ\,\bullet\,\bullet\,\bullet\,3\,3 & \circ\,\bullet\,\circ\,\bullet\,\bullet\,\bullet\,1\,1 & \circ\,\bullet\,\circ\,\bullet\,\bullet\,\bullet\,2\,2 \\ \circ\,\bullet\,\bullet\,\bullet\,\circ\,\bullet\,\circ\,\bullet & \bullet\,\circ\,\bullet\,\bullet\,\circ\,\bullet\,\circ\,\bullet & \bullet\,\circ\,\bullet\,\bullet\,\circ\,\bullet\,\circ\,\bullet \\ \bullet\,\bullet\,\circ\,\bullet\,\circ\,\bullet\,\circ\,\bullet & \bullet\,\bullet\,\bullet\,\circ\,\bullet\,\circ\,\bullet\,\circ & \bullet\,\bullet\,\bullet\,\circ\,\bullet\,\circ\,\bullet\,\circ \end{bmatrix}.$$

Construction 3 builds a c-color (k,n)-threshold scheme with pixel expansion $m = c\binom{n}{k}m'$, where m' is the pixel expansion of the black and white scheme used as building block. The thresholds ℓ and h depend on the b&w scheme used as building block, If such a scheme is with perfect reconstruction of black pixels the resulting scheme has $\ell = 0$, $h \geq 1$. Notice that the contrast property satisfied is the weak one.

Using as a building block the best, with respect to the pixel expansion, b&w $(k-1, k-1)$-threshold scheme, provided in [8], whose pixel expansion is $m' = 2^{k-2}$, the resulting scheme has pixel expansion

$$m = c\binom{n}{k}2^{k-2}.$$

For $k = n$ the pixels expansion is $m = c2^{n-2}$. The model assumes the weak contrast property. The parameters h and ℓ are $h = 1$ and $\ell = 0$ and the annihilator presence is $\beta = (m-1)/m$.

We remark that the schemes of [4] are constructed with the restriction that the shares have only one colored pixel. This is not a restriction on the model but just on the kind of schemes that can be constructed. Although this limits the search space for good schemes, it guarantees that the reconstructed pixels are exactly of the same original color (and not a darker version of it).

If we consider a model that requires this special property the c-color (n, n)-threshold schemes of [4] are optimal with respect to the pixel expansion:

Theorem 3 *[4] If the shares are restricted to be such that for any superposition it is possible to have at most one colored pixel, any c-color (n, n)-threshold scheme has pixel expansion $m \geq c2^{n-2}$.*

Other results presented in [4]:

- A construction of c-color $(2, n)$-threshold with pixel expansion $m = c(n-1)$.

- A matching lower bound $m \geq c(n-1)$.

- A construction of c-color $(2, n)$-threshold with contrast $\alpha = \frac{2}{cn}$. The contrast is defined as $\alpha = (h - \ell)/m$ and the thresholds h and ℓ satisfy the weak contrast property.

- An upper bound on the contrast $\alpha \leq \frac{k}{cn}$. This matches the construction for $k = 2$.

2.6   Schemes for the General Model

In this last section we finally describe schemes for Color-VC that consider the General model, that is we consider schemes that superimpose pixels with different colors. In the rest of the section, we present several $(2, 2)$-threshold schemes from [7, 1] and a construction for $(2, n)$-threshold schemes from [1].

2.6.1 $(2, 2)$-Threshold Schemes

In this section, we present schemes for the particular case of $k = n = 2$.

Scheme 1 *[7] The secret palette is $\{Y, C, G\}$ while the shares palette is $\{Y, C, G, \circ, \bullet\}$. The base matrices are:*

$$S^Y = \begin{bmatrix} Y \circ \bullet C \\ \circ Y C \bullet \end{bmatrix} \qquad S^C = \begin{bmatrix} C \circ \bullet Y \\ \circ C Y \bullet \end{bmatrix} \qquad S^G = \begin{bmatrix} Y C \circ \bullet \\ C Y \bullet \circ \end{bmatrix}$$

It is easy to see that for this scheme the pixel expansion is $m = 4$ and we have $h = 2$, $\ell = 0$. The annihilator presence is $\beta = 1/2$ because 2 out of 4 pixels are annihilated.

Scheme 2 *[7] Both the secret palette and the shares palette are* $\{\circ, Y, M, C, R, G, B, \bullet\}$. *The base matrices are:*

$$S^{\circ} = \begin{bmatrix} \circ\, Y\, M\, C\, \bullet\, \bullet\, \bullet\, \bullet \\ \circ\, \bullet\, \bullet\, \bullet\, Y\, M\, C\, \bullet \end{bmatrix} \qquad S^{Y} = \begin{bmatrix} Y\, \circ\, M\, C\, \bullet\, \bullet\, \bullet\, \bullet \\ \circ\, Y\, \bullet\, \bullet\, M\, C\, \bullet\, \bullet \end{bmatrix}$$

$$S^{M} = \begin{bmatrix} M\, \circ\, C\, Y\, \bullet\, \bullet\, \bullet\, \bullet \\ \circ\, M\, \bullet\, \bullet\, C\, Y\, \bullet\, \bullet \end{bmatrix} \qquad S^{C} = \begin{bmatrix} C\, \circ\, Y\, M\, \bullet\, \bullet\, \bullet\, \bullet \\ \circ\, C\, \bullet\, \bullet\, Y\, M\, \bullet\, \bullet \end{bmatrix}$$

$$S^{R} = \begin{bmatrix} Y\, M\, C\, \circ\, \bullet\, \bullet\, \bullet\, \bullet \\ M\, Y\, \bullet\, \bullet\, C\, \circ\, \bullet\, \bullet \end{bmatrix} \qquad S^{G} = \begin{bmatrix} C\, Y\, M\, \circ\, \bullet\, \bullet\, \bullet\, \bullet \\ Y\, C\, \bullet\, \bullet\, M\, \circ\, \bullet\, \bullet \end{bmatrix}$$

$$S^{B} = \begin{bmatrix} M\, C\, Y\, \circ\, \bullet\, \bullet\, \bullet\, \bullet \\ C\, M\, \bullet\, \bullet\, Y\, \circ\, \bullet\, \bullet \end{bmatrix} \qquad S^{\bullet} = \begin{bmatrix} Y\, M\, C\, \circ\, \bullet\, \bullet\, \bullet\, \bullet \\ \bullet\, \bullet\, \bullet\, \bullet\, Y\, M\, C\, \circ \end{bmatrix}$$

For this scheme the pixel expansion is $m = 8$ and we have $h = 1$, $\ell = 0$. The annihilator presence is $\beta = 7/8$ because in most cases 6 out of 8 pixels are annihilated and for the color white 7 out of 8 pixels are annihilated. Because of this, if we restrict the secret palette to $\{Y, M, C, R, G, B, \bullet\}$ and add $\circ$ for the shares palette the resulting scheme has $h = 2$ improving the contrast.

Scheme 3 *[7] Both the secret palette and the shares color palette are* $\{\circ, Y, M, C, R, G, B, \bullet\}$. *The base matrices are:*

$$S^{\circ} = \begin{bmatrix} \circ\, Y\, M\, C\, \bullet \\ \circ\, B\, G\, R\, \bullet \end{bmatrix} \qquad S^{Y} = \begin{bmatrix} Y\, M\, C\, \bullet\, \circ \\ \circ\, G\, R\, B\, \bullet \end{bmatrix}$$

$$S^{M} = \begin{bmatrix} M\, C\, Y\, \bullet\, \circ \\ \circ\, R\, B\, G\, \bullet \end{bmatrix} \qquad S^{C} = \begin{bmatrix} C\, Y\, M\, \bullet\, \circ \\ \circ\, B\, G\, R\, \bullet \end{bmatrix}$$

$$S^{R} = \begin{bmatrix} \circ\, Y\, M\, C\, \bullet \\ R\, B\, G\, \bullet\, \circ \end{bmatrix} \qquad S^{G} = \begin{bmatrix} \circ\, C\, Y\, M\, \bullet \\ G\, R\, B\, \bullet\, \circ \end{bmatrix}$$

$$S^{B} = \begin{bmatrix} \circ\, M\, C\, Y\, \bullet \\ B\, G\, R\, \bullet\, \circ \end{bmatrix} \qquad S^{\bullet} = \begin{bmatrix} \bullet\, \circ\, Y\, M\, C \\ \circ\, \bullet\, B\, G\, R \end{bmatrix}$$

It is easy to see that for this scheme the pixel expansion is $m = 5$ and we have $h = 1$, $\ell = 0$. The annihilator presence $\beta = 4/5$ because 4 out of 5 pixels are annihilated.

Scheme 4 *[1] The secret and shares palette are* $\{R, G, B, C, M, Y\}$. *The base matrices are:*

$$S^{R} = \begin{bmatrix} Y\, M\, C\, \bullet\, \bullet\, \circ \\ M\, Y\, \bullet\, C\, \circ\, \bullet \end{bmatrix} \quad S^{G} = \begin{bmatrix} Y\, C\, M\, \bullet\, \bullet\, \circ \\ C\, Y\, \bullet\, M\, \circ\, \bullet \end{bmatrix} \quad S^{B} = \begin{bmatrix} M\, C\, Y\, \bullet\, \bullet\, \circ \\ C\, M\, \bullet\, Y\, \circ\, \bullet \end{bmatrix}$$

$$S^{C} = \begin{bmatrix} C\, \circ\, M\, Y\, \bullet\, \bullet \\ \circ\, C\, \bullet\, \bullet\, M\, Y \end{bmatrix} \quad S^{M} = \begin{bmatrix} M\, \circ\, Y\, C\, \bullet\, \bullet \\ \circ\, M\, \bullet\, \bullet\, Y\, C \end{bmatrix} \quad S^{Y} = \begin{bmatrix} Y\, \circ\, C\, M\, \bullet\, \bullet \\ \circ\, Y\, \bullet\, \bullet\, C\, M \end{bmatrix}$$

It is easy to see that for this scheme the pixel expansion is $m = 6$ and we have $h = 2$, $\ell = 0$. The annihilator presence $\beta = 2/3$ because 4 out of 6 pixels are annihilated.

2.6.2   The $(2,n)$-Threshold AS Schemes

In [1] a constructions of $(2,n)$-threshold schemes is provided. The construction uses, as a building block, the base matrix $S^\bullet$ for the black color of the $(2,n)$-threshold scheme for black and white images defined in [2]. Matrix $S^\bullet$ is defined as all the binary column-vector with weight $\binom{n}{\lfloor n/2 \rfloor}$, with the substitutions $1 \leftrightarrow \bullet$ and $0 \leftrightarrow \circ$. For example, for $n = 4$, we have

$$
S_4^\bullet =
\begin{bmatrix}
1\,0\,0\,1\,0\,1 \\
1\,1\,0\,0\,1\,0 \\
0\,1\,1\,1\,0\,0 \\
0\,0\,1\,0\,1\,1
\end{bmatrix}
=
\begin{bmatrix}
\bullet\,\circ\,\circ\,\bullet\,\circ\,\bullet \\
\bullet\,\bullet\,\circ\,\circ\,\bullet\,\circ \\
\circ\,\bullet\,\bullet\,\bullet\,\circ\,\circ \\
\circ\,\circ\,\bullet\,\circ\,\bullet\,\bullet
\end{bmatrix}
$$

Then, to obtain the color scheme, the black and white pixels are substituted with the rows of a specific $(2,2)$-threshold color scheme. For example, using the KY scheme for the set of colors $\{C, Y, G\}$ with $m = 4$ provided in the previous section and substituting $\bullet$ with the first row of the base matrix for a given color and $\circ$ with the second row of the base matrix we get the base matrix for that color. For example, to get the base matrix for color Y for the $(2,4)$-threshold scheme, we substitute in S_4° the symbol $\bullet$ with Yo$\bullet$C and the symbol $\circ$ with $\circ$YC$\bullet$.

The scheme that we obtain is:

$$
S^C =
\begin{bmatrix}
\circ\,C\bullet Y\,C\circ Y\bullet C\circ Y\bullet\circ C\bullet Y\,C\circ Y\bullet\circ C\bullet Y \\
\circ\,C\bullet Y\circ C\bullet Y\,C\circ Y\bullet C\circ Y\bullet\circ C\,Y\bullet\circ C\bullet Y \\
C\circ Y\bullet\circ C\bullet Y\circ C\bullet Y\circ C\bullet Y\,C\circ Y\bullet C\circ Y\bullet \\
C\circ Y\bullet C\circ Y\bullet\circ C\bullet Y\,C\circ Y\bullet\circ C\bullet Y\circ C\bullet Y
\end{bmatrix}
$$

$$
S^Y =
\begin{bmatrix}
\circ\,Y\bullet C\,Y\circ C\bullet Y\circ C\bullet\circ Y\bullet C\,Y\circ C\bullet\circ Y\bullet C \\
\circ\,Y\bullet C\circ Y\bullet C\,Y\circ C\bullet Y\circ C\bullet\circ Y\,C\bullet\circ Y\bullet C \\
Y\circ C\bullet\circ Y\bullet C\circ Y\bullet C\circ Y\bullet C\,Y\circ C\bullet Y\circ C\bullet \\
Y\circ C\bullet Y\circ C\bullet\circ Y\bullet C\,Y\circ C\bullet\circ Y\bullet C\circ Y\bullet C
\end{bmatrix}
$$

$$
S^Y =
\begin{bmatrix}
\circ\,Y\bullet C\,Y\circ C\bullet Y\circ C\bullet\circ Y\bullet C\,Y\circ C\bullet\circ Y\bullet C \\
\circ\,Y\bullet C\circ Y\bullet C\,Y\circ C\bullet Y\circ C\bullet\circ Y\,C\bullet\circ Y\bullet C \\
Y\circ C\bullet\circ Y\bullet C\circ Y\bullet C\circ Y\bullet C\,Y\circ C\bullet Y\circ C\bullet \\
Y\circ C\bullet Y\circ C\bullet\circ Y\bullet C\,Y\circ C\bullet\circ Y\bullet C\circ Y\bullet C
\end{bmatrix}
$$

2.7   Other Schemes

In [9] Shyu proposes a construction that is very similar to the one used in [7, 11]. However, the model, although for many aspects equal to the SC model, has a crucial difference: the author assumes that the color perceived by the human eyes is an "average" of the colors present in the subpixels of the reconstructed pixels. For example, if a given surface is evenly covered with red and green we should see yellow as a result. Although this is in principle true, in practice it works only if the pixels are so tiny and evenly distributed that our eyes are not able to distinguish the single pixels and perceives an average color mixing the two primary colors. What really happens is that our eyes perceive the mixture of red and green. However, this does not mean that we cannot use this model. We have to accept the fact that a secret color (yellow, for example) is reconstructed as a mixture of other colors (red and green, for example). This model allows building schemes with a better pixel expansion, namely $m = \lceil \log c \rceil \times 2^{n-1}$. The contrast properties that we have used throughout this paper are not applicable to this model.

In [6] Hou proposes a method that first splits the secret image into the cyan, magenta, and yellow components and then uses ad-hoc $(2,2)$-threshold schemes to share those components. Although the paper claims that this method is easily extensible to the (k,n)-threshold scheme it is not clear how to use the ad-hoc $(2,2)$-threshold schemes for the general case of the (k,n)-threshold scheme. A proof of the security property is also missing.

2.8 Conclusions

Stepping from visual cryptography for black and white images to visual cryptography for color images is not immediate. The color model poses some tricky questions that arise from the complex behavior of colors superposition. Many visual cryptography schemes for color images avoid the problem by not superposing pixels with different colors. Very few known schemes do actually exploit color superposition. In this chapter we have first emphasized the difficulties that arise from the superposition of colored pixels; then we have provided a survey of the models of visual cryptography for color images that have been considered in the literature and a survey of the schemes that have been proposed for such models.

Visual cryptography for black and white images has been thoroughly studied. The case of color images is still pretty much unexplored. A first direction of research concerns the definition of a reference model. We believe that the General model is the one that best represents the real world. All the models proposed in the literature lack a well-defined notion of contrast, which is a very important measure for the evaluation of the schemes. A second direction of research concerns the search for schemes that do use the properties of color superposition. The construction of schemes for color images seems to be much more difficult than for black and white images.

Bibliography

[1] A. Adhikari and S. Sikdar. A new (2,n)-visual threshold scheme for color images. In *Proceedings of the Indocrypt 2003*, volume 2904, pages 148–161. Springer Verlag, LNCS, 2005.

[2] C. Blundo, A. De Bonis, and A. De Santis. Improved schemes for visual cryptography. *Designs, Codes and Cryptography*, 24:255–278, 2001.

[3] S. Cimato, R. De Prisco, and A. De Santis. Optimal colored threshold visual cryptography schemes. *Designs, Codes and Cryptography*, 35(3):311–335, 2005.

[4] S. Cimato, R. De Prisco, and A. De Santis. Colored visual cryptography without color darkening. *Theoretical Computer Science*, 374(1-3):261–276, 2007.

[5] P.A. Eisen and D.R. Stinson. Threshold visual cryptography schemes with specified whitness levels of reconstructed pixels. *Designs, Codes and Cryptography*, 25:15–61, 2002.

[6] Y.-C. Hou. Visual cryptography for color images. *Pattern Recognition*, 36:1619–1629, 2003.

[7] H. Koga and H. Yamamoto. Proposal of a lattice-based visual secret sharing scheme for color and gray-scale images. *IEICE Trans. on Fundamentals of Electronics, Communication and Computer Sciences*, 81-A(6):1262–1269, 1998.

[8] M. Naor and A. Shamir. Visual cryptography. In *Proceedings of the Eurocrypt 1995*, volume 950, pages 1–12. Springer Verlag, LNCS, 1995.

[9] S. J. Shyu. Efficient visual secret sharing scheme for color images. *Pattern Recognition*, 35:866–880, 2006.

[10] E.R. Verheul and H.C.A. van Tilborg. Constructions and properties of k out of n visual secret sharing schemes. *Designs, Codes and Cryptography*, 11(2):179–196, 1997.

[11] C.-N. Yang and C.-A. Laih. New colored visual secret sharing schemes. *Designs, Codes and Cryptography*, 20:325–335, 2000.

3

Visual Cryptography for Multiple Secrets

Shyong Jian Shyu

Ming Chuan University, Taiwan

CONTENTS

3.1 Introduction

Visual cryptography proposed by Naor and Shamir [7] discloses the possibility for using human visual ability to perform the decryption process. Specifically, one secret image is encoded into two shares that are seemingly random pictures. By xeroxing them onto transparencies, the dealer distributes the two random transparencies to two participants (one share for each participant).

Each participant cannot tell the secret from his own transparency, but when the two participants superimpose their transparencies pixel by pixel, they recognize the secret from the superimposed result by their visual system. Neither computational devices nor cryptographic knowledge is required for the decryption process.

With such an interesting characteristic that the decryption process is by the human visual system only, instead of any computational device, visual cryptography attracts much attention from researchers. In particular, it is much useful in situations where computing devices are not available or not possible to use. Naor and Shamir [7] first presented *k out of n visual secret-sharing schemes*, which ensure that the secret is concealed from groups of less than k participants, while it can be seen by groups of at least k participants when they stack their shares altogether. Since this pioneer research, many theoretical results on the *construction* or *contrast* (the relative difference between the reconstructed white and black pixels in the superimposed image) of visual secret sharing schemes for binary images have been proposed in the literature [1, 10, 11, 2]. Some studies [4, 5, 9] focused on the practical realization of visual cryptographic schemes for gray-level or color images. So far, the above-mentioned results concern sharing "one" secret in a visual sense.

Wu and Chen [12] might be the first researchers to consider the problem of sharing two secret images in two shares in visual cryptography. They concealed two secret binary images into two seemingly random shares, namely S_1 and S_2, such that the first secret can be seen by stacking the two shares, denoted by $S_1 \otimes S_2$, and the second secret can be obtained by $S_1^\theta \otimes S_2$ where θ denotes the superimposition operation and S_1^θ is the result of rotating S_1 θ counterclockwise. S_1 and S_2 are in the shape of squares of the same size. In order to align the encoded pixels on S_1 and S_2 as well as on S_1^θ and S_2, they designed the rotation angle θ to be $90°$. Nevertheless, it is easy to obtain one that can be $180°$ or $270°$.

Wu and Chang [13] refined the idea of Wu and Chen [12] by consciously designing the encoded shares to be circles so that the restrictions to the rotating angles ($\theta = 90°, 180°,$ or $270°$) can be removed. Let the two encoded circle shares in their approach be denoted as A and B. The first secret is revealed by $A \otimes B$, while the second secret is obtained by $A^{-\theta} \otimes B$ where θ may be any angle within $(0°, 360°)$ and $A^{-\theta}$ is the result of rotating A θ clockwise. Both of the studies successfully share two secrets in two shares in a visual sense.

Shyu et al. [8] devised an elegant algorithm to achieve the sharing of multiple sercte images using two circle shares. It is the first true approach that shares any general number of multiple secrets in two shares. Consider x secret images. Consider a set of $x \geq 2$ secret images. Their scheme produces two circle shares A and B such that none of them individually leaks any secret and the x secrets can be obtained one by one by $A \otimes B$, $A^\theta \otimes B$, $A^{2\theta} \otimes B, \ldots, A^{(x-1)\theta} \otimes B$ where $A^{(i-1)\theta}$ is the result of rotating $A^{(i-1)\theta}$ counterclockwise for $1 \leq i \leq x$ for $\theta = 360°/x$ and $A^{0°} = A$. This is the first true result of sharing multiple secrets in visual cryptography for any $x \geq 2$ secrets in two shares. Their pixel expansion is $2x$, which is the best result so far in the model that each pixel would be expanded.

Based on a different set of encoding patterns, Feng et al. [3] developed another scheme to achieve the same goal using two cylinder shares. The pixel expansion needed is $3x$.

In this chapter, we introduce these interesting algorithms. The rest of the paper is organized as follows. In Section 3.2, we briefly review the visual one-secret sharing scheme in two shares proposed by Naor and Shamir [7]. The visual two-secret sharing scheme by Wu and Chen [12] and Wu and Chang [13] are discussed in Sections 3.3.1 and 3.3.2, respectively. The schemes for visual multisecret are examined in Section 3.4 in which the experimental results and discussions are also presented. Section 3.5 gives some concluding remarks.

3.2 Naor and Shamir's Basic Visual Secret Sharing Scheme

The basic idea of Naor and Shamir's encoding scheme [7] for sharing a single pixel, say p, in a binary image P into two shares s_1 and s_2 is illustrated in Table 3.1. If p is white, the dealer randomly chooses one of the first two rows of Table 3.1 to encode s_1 and s_2. If p is black, the dealer randomly chooses one of the last two rows in Table 3.1 to encode s_1 and s_2. The possibilities of the two encoding cases are equally likely to occur, independently of whether the original pixel is black or white. Thus, neither s_1 nor s_2 exposes any clue about the binary color of p. When these two shares are stacked together, i.e., $s_1 \otimes s_2$, two black subpixels appear if p is black, while one black subpixel and one white subpixel appear if p is white as indicated in the rightmost column in Table 3.1. Based upon the contrast between these two kinds of *reconstructed* pixels, our visual system can tell whether p is black or white by observing $s_1 \otimes s_2$.

Note that s_1 (or s_2) in Table 3.1 is not a single pixel, but two subpixels. We call s_1 (or s_2) an *extended block* and the pair (s_1, s_2) *the pair of two extended blocks* with respect to p. The number of the subpixels in each of the two extended blocks (s_1, s_2) for encoding p is referred to as the pixel expansion. In Table 3.1, the pixel expansion is 2. In realistic implementations, it may be chosen as 4 $(= 2 \times 2)$ in order to retain the aspect ratio of the original secret image. Since there are six possible patterns for a 2×2 extended block, all pairs of two extended blocks (s_1, s_2)'s for encoding a specific binary pixel p (visual one-secret sharing) are summarized in Table 3.2.

When p is white (black), the dealer randomly chooses one of the first (last) six rows of Table 3.2 to encode p into s_1 and s_2. It is seen from the last column of Table 3.2 that the reconstructed pixel $r = s_1 \otimes s_2$ may contain two white

TABLE 3.1

Encoding a binary pixel p into two shares s_1 and s_2.

p	probability	s_1	s_2	$s_1 \otimes s_2$
☐	1/2			
	1/2			
■	1/2			
	1/2			

TABLE 3.2

Implementing the visual one secret sharing scheme with a pixel expansion of 4.

p	Probability	s_1	s_2	$r = s_1 \otimes s_2$
	1/6			
	1/6			
	1/6			
☐	1/6			
	1/6			
	1/6			
	1/6			
	1/6			
	1/6			
■	1/6			
	1/6			
	1/6			

and two black subpixels if p is white, or all four black subpixels when p is black. When all pixels in p are encoded in this way, where each p's random choice for encoding alternatives is independent, the encoded shares S_1 (containing all s_1's) and S_2 (containing all s_2's) are indeed random pictures, respectively. When S_1 and S_2 are superimposed, all of the four subpixels are black in the reconstructed blocks corresponding to each black pixel in P, while two subpixels are white and the other two are black corresponding to each white pixel in P. Based upon such a difference, our visual system recognizes the white and black pixels in P from $S_1 \otimes S_2$. We say that the reconstructed image $S1 \otimes S_2$ *recovers* P.

Figure 3.1 shows the implementation results of the encoding scheme in Table 3.2. Figure 3.1(a) is a secret binary image P, Figures 3.1(b) and (c) are the two encoded shares S_1 and S_2, which are random pictures revealing no information about P and Figure 3.1(d) illustrates the reconstructed image $S_1 \otimes S_2$ that recovers P visually.

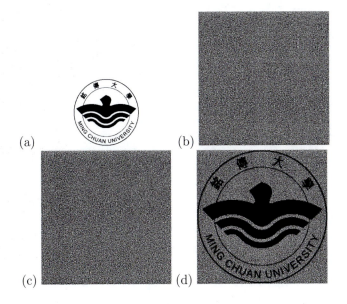

FIGURE 3.1
Implementation results of visual one-secret sharing in two shares: (a) P, (b) S_1, (c) S_2, (d) $S_1 \otimes S_2$.

3.3   Visual Two-Secret Sharing Schemes

3.3.1   Wu and Chen's Scheme

Following the research of Naor and Shamir, Wu and Chen [12] developed a visual secret sharing scheme that encrypts two secrets into two shares. Given two $N \times N$ (square) secret binary images P_1 and P_2, their scheme produces two shares, namely S_1 and S_2, which reveal no information about P_1 or P_2 individually. Yet when stacking S_1 and S_2, we obtain P_1 visually; moreover, when stacking $S_1^{90°}$ and S_2, we see P_2.

Consider a pair of pixels $p_1 = P_1[i, j]$ and $p_2 = P_2[u, v]$ in P_1 and P_2, respectively. We refer to (p_1, p_2) as the *corresponding pixels* of P_1 and P_2 if and only if $i = u$ and $j = v$. Given a set of corresponding pixels (p_1, p_2),Wu and Chen's encoding scheme for visual two-secret sharing in two shares is summarized in Table 3.3.

It is seen from Table 3.3 that each pair of corresponding pixels (p_1, p_2) of (P_1, P_2) is encoded into extended blocks s_1 (as well as $s_1^{90°}$) and s_2 in which the pixel expansion is $m = 4$. Note that $s_1^{90°}$ is exactly the result of rotating s_1 $90°$ counterclockwise. We explain how Wu and Chen's encoding scheme works by using a simple example. Assume that the two secret images P_1 and P_2 are composed in a square of 12×12 pixels. Then, the two encoded shares S_1 and S_2 are composed in a square of 48×48 ($48 = 12 \times 4$) pixels. They first decompose S_1 into four triangle-like areas with an equal size as shown in Figure 3.2(a). All of the four areas are composed of an equal amount of extended blocks (2×2 pixels each), which are indexed as shown in Figure 3.2(b) where each triangle-like area contains 36 blocks. Let block j in area k be denoted as b_j^k for $1 \leq k \leq 4$ and $1 \leq j \leq 36$. The extended blocks in area I, b_j^1, are randomly selected out of those in Figure 3.2(c). Each block, say b_j^t, in area II, III, IV is assigned to be the same as b_j^1 in area I, that is, $b_j^t = b_j^1$ for $t = 2, 3, 4$, and $1 \leq j \leq 36$.

Let us pay attention to the four pixels at the top-right, top-left, bottom-left, and bottom-right corners in sequence (counterclockwise) in P_1 and P_2. Assume that those pixels in P_1 (P_2) are □, □, ■, ■ (□, ■, □, ■) as shown in Figure 3.3(a) and (Figure 3.3(b)). Assume that corresponding block b_{26} at S_1 is randomly determined as ■, then as mentioned b_{26}^t is ■ for $2 \leq t \leq 4$ (see Figure 3.3(c)). The above-mentioned pixels in P_1 and P_2 constitute four sets of corresponding pixels: (□, □), (□, ■), (■, □), and (■, ■). Since b_{26}^k in S_1 is ■ for $1 \leq k \leq 4$, according to Table 3.3 the four blocks $b_{26}^1, b_{26}^2, b_{26}^3$ and b_{26}^4, in S_2 with respect to the four sets of the corresponding pixels are ■, ■, □ and ■, respectively (see the 2nd, 6th, 10th, and 14th rows in column s_2 of Table 3.3). Figure 3.3(d) illustrates the encoding result of S_2. As expected,

TABLE 3.3
Wu and Chen's encoding scheme for visual two-secret sharing in two shares.

p_1	p_2	Probability	s_1	$s_1^{90°}$	s_2	$s_1 \otimes s_2$	$s_1^{90°} \otimes s_2$
		1/4					
		1/4					
□	□	1/4					
		1/4					
		1/4					
		1/4					
□	■	1/4					
		1/4					
		1/4					
		1/4					
■	□	1/4					
		1/4					
		1/4					
		1/4					
■	■	1/4					
		1/4					

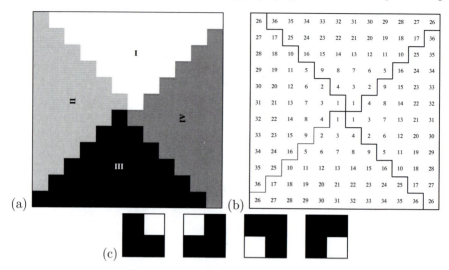

(a)   (b)

(c)

FIGURE 3.2
Encoding S_1 in Wu and Chen's scheme: (a) Four triangle-like areas. (b) Indexing the blocks in each of the four areas. (c) Blocks to be assigned.

the four corners in the above-mentioned order in $S_1 \otimes S_2$ reveal $\square, \square, \blacksquare, \blacksquare$, respectively (see Figure 3.3(e)) to our visual system. When S_1 is rotated as $S_1^{90°}$ as indicated in Figure 3.3(f), where all blocks are in the form of $s_1^{90°}$, the four corresponding corners in $S_1^{90°} \otimes S_2$ recover $\square, \blacksquare, \square, \blacksquare$, respectively (see Figure 3.3(g)). It is not hard to see that by encoding all pixels in S_1 and S_2 with respect to the corresponding pixels in P_1 and P_2 according to Table 3.3, P_1 and P_2 can be recovered by $S_1 \otimes S_2$ and $S_1^{90°} \otimes S_2$, respectively.

Note that S_1 and S_2 are in the shape of squares of the same size. $S_1 \otimes S_2$ reveals P_1, while $S_1^\theta \otimes S_2$ reveals P_2. Wu and Chen set θ to be 90°. It is easy to extend their idea to design θ as one of 90°, 180°, or 270°, but the other degrees are infeasible. This is because the rotated S_1 (S_1^θ) cannot be aligned to S_2 pixel by pixel when $\theta \neq 0°, 90°, 180°$, or 270°. Except for the fact that it is restricted, there is another pitfall in their scheme: since the encoded pixels in each of areas I, II, III, and IV in S_1 are exactly the same, S_1 is not a random picture. In fact, only $1/4$ shares of S_1 are purely random pictures.

3.3.2 Wu and Chang's Scheme

Based upon the idea of Wu and Chen [12], Wu and Chang [13] devised another visual two-secret sharing scheme that allows the rotation angle to be an arbitrary one between 0° and 360° by adopting circle shares. Given an angle and two secret images P_1 and P_2, their approach produces two circle shares A and B such that any single A or B is a seemingly random picture that leaks

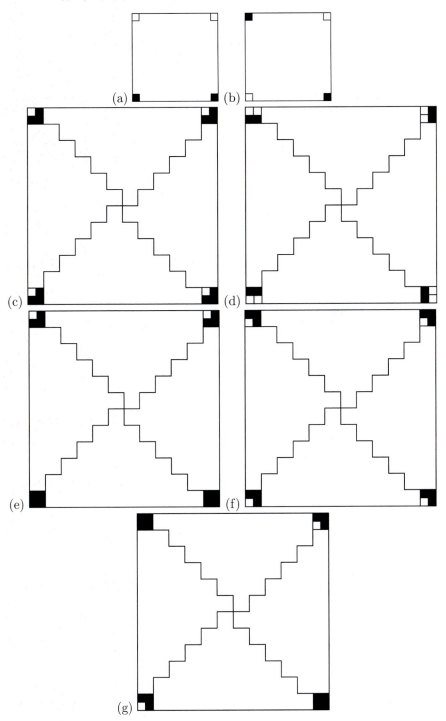

FIGURE 3.3
Example for illustrating the idea of Wu and Chen [12]: (a) P_1, (b) P_2, (c) S_1, (d) S_2, (e) $S_1 \otimes S_2$, (f) $S_1^{90°}$, (g) $S_1^{90°} \otimes S_2$.

nothing about P_1 or P_2; while $A \otimes B$ reconstructs P_1 and $A^{-\theta} \otimes B$ recovers P_2 where $A^{-\theta}$ denotes the result of rotating A θ clockwise. Note that $A^{-\theta} \otimes B$ is equivalent to $A \otimes B^{\theta}$. Intuitively, it is reasonable to choose circles as the encoded shares since they ease the correct alignments between A and B as well as $A^{-\theta}$ and B pixel by pixel where $0° < \theta < 360°$.

They deliberately decomposed circle share A into $360°/\theta$ areas where each area contains an equal amount of 2×2 *sector blocks*. Figure 3.4(a) shows the four typical patterns for sector blocks, namely s_1^1, s_1^2, s_1^3, and s_1^4, used in their approach. That is, the whole circle share A is composed by all these four sector blocks. Note that s_1^1 (s_1^2, s_1^3, s_1^4) can be consciously regarded as the result of rotating s_1^2 (s_1^3, s_1^4, s_1^1, respectively) 90° counterclockwise (or s_1^2 can be consciously regarded as the result of rotating 90° clockwise). We say that s_1^1's (s_1^2's, s_1^3's, s_1^4's) *previous* sector block is s_1^4 (s_1^1, s_1^2, s_1^3, respectively) and its *next* sector block is s_1^2 (s_1^3, s_1^4, s_1^1, respectively) as summarized in Figure 3.4(b).

(a)

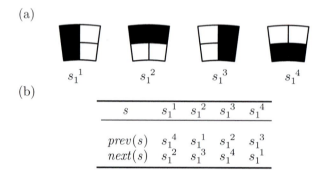

(b)

s	s_1^1	s_1^2	s_1^3	s_1^4
$prev(s)$	s_1^4	s_1^1	s_1^2	s_1^3
$next(s)$	s_1^2	s_1^3	s_1^4	s_1^1

FIGURE 3.4

2×2 sector blocks for A in Wu and Chang's approach: (a) 2×2 sector blocks for A, (b) $prev(s)$ and $next(s)$ of sector block s.

Let the number of areas in circle share A be $\alpha(= 360°/\theta)$ and the number of sector blocks in each area be β. These α areas are indexed clockwise. Let a_j^k be the jth sector block in area k in A, $1 \leq j \leq \beta$ and $1 \leq k \leq \alpha$. At first, the β sector blocks in the first area are randomly selected out of those in Figure 3.4(a). Then, sector blocks in area t are defined according to those in area $t-1$ by assigning a_j^t as the next sector block of a_j^{t-1}, i.e., $a_j^t = next(a_j^{t-1})$ (or $a_j^{t-1} = prev(a_j^t)$) for $1 \leq j \leq \beta$ and $2 \leq t \leq \alpha$.

Given a pair of corresponding pixels p_1 and p_2 in P_1 and P_2, respectively, each sector block b_j^k in B is determined by p_1, p_2, and the corresponding block a_j^k in A for $1 \leq j \leq \beta$ and $1 \leq k \leq \alpha$. Table 3.4 summarizes such an encoding scheme.

Note that in Wu and Chen's scheme each extended block s_2 in S_2 would be superimposed with s_1 and $s_1^{90°}$ when s_1 is rotated 0° (or fixed) and 90° counterclockwise, respectively. In Wu and Chang's scheme, each sector block

TABLE 3.4

Wu and Chang's encoding scheme for visual two-secret sharing in two shares.

p_1	p_2	Probability	s_1	$s_1^{-\theta}$	s_2	$s_1 \otimes s_2$	$s_1^{-\theta} \otimes s_2$
		1/4					
□	□	1/4					
		1/4					
		1/4					
		1/4					
□	■	1/4					
		1/4					
		1/4					
		1/4					
■	□	1/4					
		1/4					
		1/4					
		1/4					
■	■	1/4					
		1/4					
		1/4					

b_j^k in B is superimposed with a_j^k in area k and a_j^{k-1} in k's previous area $k-1$ when A is rotated $0°$ and θ clockwise where $a_j^{k-1} = prev(a_j^k)$ (or $a_j^k = next(a_j^{k-1})$). Note that a_j^{k-1} is the result of rotating a_j^k $90°$ counterclockwise (or a_j^k is the result of rotating a_j^{k-1} $90°$ clockwise; see Figure 3.4). That means the result of $A^{-\theta} \otimes B$ in Wu and Chang's scheme emulates that of $S_1^{90°} \otimes S_2$ in Wu and Chen's scheme. There is no restriction for θ to be $90°$, $180°$, or $270°$ merely. Yet there exist some inconsistent situations in some of the areas in $A^{-\theta} \otimes B$ when $\alpha = 360°/\theta > 4$. Interested readers refer to Ref. [13] for details.

As mentioned above, square share S_1 in Wu and Chen's scheme is not a totally random image. Strictly speaking, neither is circle share A in Wu and Chang's scheme due to the reason that sector block a_j^t is assigned as $next(a_j^{t-1})$ (i.e., the result of rotating a_j^{t-1} $90°$ clockwise) for $2 \leq t \leq \alpha$; that is, only sector block a_j^1 in the first area is randomly determined from those in Figure 3.4(a) for $1 \leq j \leq \beta$, while the other areas are not. Furthermore, sector blocks in the first area of circle share A (see Figure 3.4(a)) in Wu and Chang's scheme (or extended blocks in the first area of square share S_1 (see Figure 3.2(c)) contain only four patterns, instead of six, which is the number of all possible combinations for four subpixels with two white and two black subpixels (see Table 3.2).

3.4   Visual Multiple-Secret Sharing Schemes

Both of the above-mentioned schemes accomplish visual secret sharing for only two secrets in two shares. In this section we discuss two more generalized visual secret sharing scheme for $x \geq 2$ secrets in two shares by Shyu et al. [8] and Feng et al. [3] in Sections 3.4.1 and 3.4.2, respectively.

3.4.1   Shyu et al.'s Scheme

3.4.1.1   Informal Description

Let us start by using a simple example. Assume that the number of secret images to be shared is $x = 3$. Let P_1, P_2, and P_3 be the three binary secret images with the same size $h \times w$. Let p_1, p_2, and p_3 denote the corresponding pixels in P_1, P_2, and P_3, respectively. Let A and B denote the two circle shares encoded by the scheme. Our aim is to assure $A \otimes B$ recovers P_1, $A^{120°} \otimes B$ recovers P_2, and $A^{240°} \otimes B$ recovers P_3.

Since there are three secrets, we decompose circle share A and B into three $(x = 3)$ *chord-areas* (*chords* for short), respectively, in which the angle of each chord extends up to $120°(= 360°/x = 360°/3)$. Each chord is divided into a set of 2×3 ($2 \times x$) chord blocks. Let the number of 2×3 blocks in each chord be β. Let a_j^k and b_j^k denote block j of chord k in A and B, respectively, $1 \leq j \leq \beta$ and $1 \leq k \leq 3(= x)$. The chords are indexed clockwise and the divided blocks in A and B are indexed as shown in Figures 3.5(a) and (b), respectively. We call a_j^k and b_j^k the *corresponding blocks* in A and B.

3.4.1.2 Encoding Circle Share A

We first define three 2×3 *elementary blocks*, namely s_A^1, s_A^2, and s_A^3, for circle share A as shown in Figure 3.6. That is, these elementary blocks are the basic constituents of A and there are one white and five black subpixels in each of the elementary blocks.

In order to guarantee the randomness when using s_A^k as a constituent of A for $1 \leq k \leq 3$, we permute the subpixels within s_A^k before assigning s_A^k as a constituent block in A. Let $\Sigma = (\sigma_1, \sigma_2, \sigma_3, \sigma_4, \sigma_5, \sigma_6)$ be a permutation of $1, 2, 3, 4, 5, 6$ (in which $6 = 2x = 2 \times 3$). We define a function *permute*(s, Σ) re-arranging the subpixels in elementary block s by permutation Σ. Figure 3.7(a) shows a certain typical ordering of the subpixels in a 2×3 elementary block s and Figure 3.7(b) shows the result of *permute*(s, Σ) with $\Sigma = (3, 5, 1, 6, 2, 4)$. Note that the order of the subpixels in the elementary block s can be defined arbitrarily.

We call the set of three blocks (a_j^1, a_j^2, a_j^3) the *related blocks* of the three chords in A for $1 \leq j \leq \beta$. Obviously, there are totally β sets of the related blocks in A. For a certain set of related blocks (a_j^1, a_j^2, a_j^3), we generate one permutation, denoted as Σ_j, and assign a_j^k to be *permute*(s_A^k, Σ_j) for $1 \leq k \leq 3$ and $1 \leq j \leq \beta$. That is,

$$(a_j^1, a_j^2, a_j^3) = (permute(s_A^1, \Sigma_j), permute(s_A^s, \Sigma_j), permute(s_A^3, \Sigma_j)) \quad (3.1)$$

for $1 \leq j \leq \beta$.

For the purpose of illustration, we show how the first set of the related blocks (a_1^1, a_1^2, a_1^3) in A is encoded. Assume that $\Sigma_1 = (1, 2, 3, 4, 5, 6)$. Figure 3.8(a) exposes the results of encoding (a_j^1, a_j^2, a_j^3) in A. Note that for this particular Σ_1 *permute*$(s_A^k, \Sigma_1) = s_A^k$ for $1 \leq k \leq 3$. In real implementation, a new random permutation Σ_j is adopted when encoding (a_j^1, a_j^2, a_j^3) in A for each $j, 1 \leq j \leq \beta$. Figures 3.8(b) and (c) show the results of $A^{120°}$ and $A^{240°}$, respectively.

Let $[k, j]$ denote the absolute location with respect to block j of chord k in a circle share (see Figure 3.9) and $A^\theta[k, j]$ denote the content of block $[k, j]$ in A^θ (i.e., the results of rotating A θ counterclockwise) where $1 \leq k \leq 3$ $(= x)$, $1 \leq j \leq \beta$ and $\theta = 120°(= 360°/3)(A^{0°}[k, j] = A[k, j])$. The relationship among the related blocks is easily seen from Figures 3.8 and 3.9:

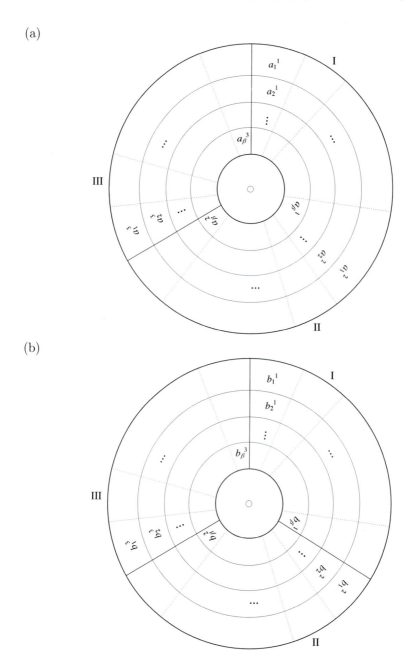

FIGURE 3.5
Decomposing circle shares A and B into chords, which are further divided into blocks: (a) A, (b) B.

FIGURE 3.6
Elementary blocks for circle share A for sharing 3 secrets: (a) S_A^1, (b) S_A^2, (c) S_A^3.

3	2	1
4	5	6

1	5	3
6	2	4

(a) (b)

FIGURE 3.7
Subpixels in 2×3 elementary block s and $permute(s, \Sigma)$: (a) a certain ordering of the subpixels in s, (b) the ordering of those in $permute(s, \Sigma)$ where $\Sigma = (3, 5, 1, 6, 2, 4)$.

$$A[1,j] = a_j^1, \quad A^{120°}[1,j] = a_j^2, \quad A^{240°}[1,j] = a_j^3$$
$$A[2,j] = a_j^2, \quad A^{120°}[2,j] = a_j^3, \quad A^{240°}[2,j] = a_j^1$$
$$A[3,j] = a_j^3, \quad A^{120°}[3,j] = a_j^1, \quad A^{240°}[3,j] = a_j^2 \tag{3.2}$$

for $1 \le j \le \beta$.

3.4.1.3 Encoding Circle Share B

First of all, with regard to the three given $h \times w$ secret images P_1, P_2, and P_3, we divide P_i evenly into $\beta = h \times (w/3) = h \times (w/x)strips$. Let $(p_i)_j^k$ denote the jth pixel of strip k in P_i and $(p_1, p_2, p_3)_j^k = ((p_1)_j^k, (p_2)_j^k, (p_3)_j^k)$ be the jth corresponding pixels of strip k for (P_1, P_2, P_3) where $1 \le j \le \beta$ and $1 \le i$, $k \le 3$. Each block, say b_j^k, in B is determined according to the related blocks (a_j^1, a_j^2, a_j^3) in A and the corresponding pixels $(p_1, p_2, p_3)_j^k$ in (P_1, P_2, P_3) for $1 \le j \le \beta$ and $1 \le k \le 3$.

Consider a particular block b_j^1 in the first chord of B. Note that the corresponding block a_j^1 (a_j^2 and a_j^3) in A ($A^{120°}$ and $A^{240°}$, respectively) is a permuted result of elementary block s_A^1 (s_A^2 and s_A^3, respectively) according to Σ_j, which is decided in run time. The encoding scheme for b_j^1 before run time is essentially based upon (s_A^1, s_A^2, s_A^3) and $(p_1, p_2, p_3)_j^1$ as summarized in Table 3.5.

It is observed from Table 3.5 that given a set of corresponding pixels (p_1, p_2, p_3), the results of $s_A^1 \otimes s_B$ ($s_A^2 \otimes s_B$, $s_A^3 \otimes s_B$) reveals p_1 (p_2, p_3, respectively) to our visual system. For instance, consider the third row of Table 3.5 where $(p_1, p_2, p_3) = (\square, \blacksquare, \square)$. When the related blocks in A are in

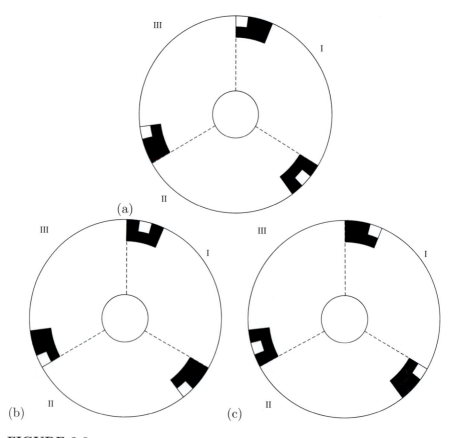

FIGURE 3.8
Encoding the first three blocks in each of the three chords by Σ_1 in A: (a) A,
(b) $A^{120°}$, (c) $A^{240°}$.

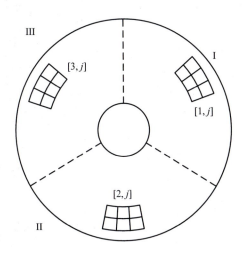

FIGURE 3.9
Absolute location of block $[1, j]$, $[2, j]$, and $[3, j]$.

TABLE 3.5
Encoding a set of corresponding pixels $(p_1, p_2, p_3)_j^1$ into a_j^1 (a_j^2 and a_j^3) and b_j^1 in terms of s_A^1 (s_A^2, s_A^3, respectively) and s_B in the first chords of A and B, respectively for visual 3-secret sharing.

p_1	p_2	p_3	s_A^1	s_A^2	s_A^3	s_B	$s_A^1 \otimes s_B$	$s_A^2 \otimes s_B$	$s_A^3 \otimes s_B$
□	□	□							
□	□	■							
□	■	□							
□	■	■							
■	□	□							
■	□	■							
■	■	□							
■	■	■							

their elementary forms s_A^1, s_A^2, and s_A^3, we design s_B to be 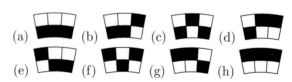 so that both $s_A^1 \otimes s_B$ and $s_A^3 \otimes s_B$ reveal one white and five black subpixels, while $s_A^2 \otimes s_B$ show six black subpixels. Our eyes recognize $s_A^1 \otimes s_B$ and $s_A^3 \otimes s_B$ as white, while $s_A^2 \otimes s_B$ as black. That means (p_1, p_2, p_3) is recovered by $(s_A^1 \otimes s_B, s_A^2 \otimes s_B, s_A^3 \otimes s_B)$ in a visual sense.

In actual implementation, the set of three related blocks (a_j^1, a_j^2, a_j^3) in A is deliberately assigned as $(permute(s_A^1, \Sigma_j), permute(s_A^2, \Sigma_j), permute(s_A^3, \Sigma_j))$ so that we only need to assign b_j^1 to be $permute(s_B, \Sigma_j)$ to preserve the superimposition results designed in Table 3.5. Then, when we superimpose a_j^1 and b_j^1, we identify $(p_1)_j^1$ form $a_j^1 \otimes b_j^1$. When we rotate A $120°$ counterclockwise, b_j^1's corresponding block in $A^{120°}$ turns out to be a_j^2 (i.e., $A^{120°}[1, j]$, see formula (3.2) and Figure 3.9) and $a_j^2 \otimes b_j^1$ reveals $(p_2)_j^1$ in a visual sense. Likewise, when rotating A $240°$ counterclockwise, b_j^1's corresponding block in $A^{240°}$ is a_j^3 and $a_j^3 \otimes b_j^1$ reveals $(p_3)_j^1$. In general, when rotating A $(i-1)\theta$ counterclockwise we recognize $a_j^i \otimes b_j^1$ as $(p_i)_j^1$ in the first chord of $A^{(i-1)\theta} \otimes B$ by our visual system for $1 \leq i \leq 3 (= x)$ and $\theta = 120°(= 360°/x)$.

We call the blocks in column s_B of Table 3.5 the *elementary blocks circle share* B for sharing 3 secrets, which consists of three white and three black subpixels. They are named by $s_B^0, s_B^1, \ldots, s_B^7$ in sequence as indicated in Figure 3.10. When we denote □ as 0 and ■ as 1, the superscript l of s_B^l is equal to the code formed by $p_1 p_2 p_3$ in binary, i.e. $l = btod(p_1 p_2 p_3)$ where $btod(b)$ is a function that returns the decimal representation of a binary number b. It means that based upon Table 3.5, once (a_j^1, a_j^2, a_j^3) is assigned to be (s_A^1, s_A^2, s_A^3) and b_j^1 is encoded to be $s_B^{btod(p_1 p_2 p_3)}$ (specifically, $(a_j^1, a_j^2, a_j^3) = (permute(s_A^1, \Sigma_j), permute(s_A^2, \Sigma_j), permute(s_A^3, \Sigma_j))$ and $b_j^1 = permute(s_B^{btod(p_1 p_2 p_3)}, \Sigma_j)$ in practical implementation with respect to give $(p_1, p_2, p_3)_j^1$, $(a_j^1 \otimes b_j^1, a_j^2 \otimes b_j^1, a_j^3 \otimes b_j^1)$ recovers $(p_1, p_2, p_3)_j^1$.

FIGURE 3.10
Elementary blocks of share B for sharing 3 secrets: (a) s_B^0, (b) s_B^1, (c) s_B^2, (d) s_B^3, (e) s_B^4, (f) s_B^5, (g) s_B^6, (h) s_B^7.

Now, we take the instances in Figure 3.11, in which the first three pixels of the three divided strips in P_i are depicted for $1 \leq i \leq 3$, as an example to show how the corresponding blocks in B are encoded. From Figure 3.11, we have $(p_1, p_2, p_3)_1^1 = (\square, \blacksquare, \square)$. According to Table 3.5, the elementary block for b_1^1 is chosen to be $s_B^{btod(p_1 p_2 p_3)} = s_B^{btod(010)} = s_B^2$. Since in practical implementation, b_1^1's corresponding block a_1^1 (a_1^2, a_1^3) in A ($A^{120°}$, $A^{240°}$, respectively) has been encoded as $permute(s_A^1, \Sigma_1)$ $(permute(s_A^2, \Sigma_1), permute(s_A^3, \Sigma_1)$, respectively), the same permutation Σ_1 should be adopted for encoding b_1^1 to preserve the superimposition result of $s_A^1 \otimes s_B$ $(s_A^2 \otimes s_B$ and $s_A^3 \otimes s_B$, respectively). Let us simply set $\Sigma_1 = \{1, 2, 3, 4, 5, 6\}$. Therefore, b_1^1 is encoded as $permute(s_B^2, \Sigma_1)$ (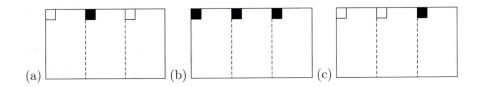) as show in Figure 3.12. It is easily seen that

$$
\begin{aligned}
a_1^1 \otimes b_1^1 &= permute(s_A^1, \Sigma_1) \otimes permute(s_B^2, \Sigma_1) \\
&= s_A^1 \otimes s_B^2 = \blacksquare \otimes \blacksquare = \blacksquare \\
a_1^2 \otimes b_1^1 &= permute(s_A^2, \Sigma_1) \otimes permute(s_B^2, \Sigma_1) \\
&= s_A^2 \otimes s_B^2 = \blacksquare \otimes \blacksquare = \blacksquare \\
a_1^3 \otimes b_1^1 &= permute(s_A^3, \Sigma_1) \otimes permute(s_B^2, \Sigma_1) \\
&= s_A^3 \otimes s_B^2 = \blacksquare \otimes \blacksquare = \blacksquare
\end{aligned}
$$

FIGURE 3.11
Instances of the first three pixels of the three strips in (a) P_1, (b) P_2, and (c) P_3.

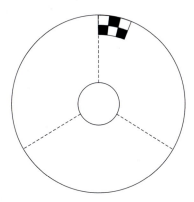

FIGURE 3.12
Encoding b_1^1 in B.

Therefore, the first blocks in the first chords of $A \otimes B$, $A^{120°} \otimes B$, and $A^{240°} \otimes B$ reconstruct $(p_1)_1^1$ ($\square$), $(p_2)_1^1$ ($\blacksquare$), and $(p_3)_1^1$ ($\square$), respectively.

In summary, given a certain $(p_1, p_2, p_3)_j^1$ in the first strips of (P_1, P_2, P_3), b_j^i

is encoded as $permute(s_B^{btod(p_1p_2p_3)}, \Sigma_j)$ where Σ_j is a random permutation for $1 \leq j \leq \beta$. It is noted that for a specific black b_j^1 in the first chord of B, when A is rotated $0°$, $120°$, and $240°$ counterclockwise, the blocks that are superimposed onto b_j^1 are a_j^1, a_j^2, and a_j^3, respectively where $a_j^k = permute(s_A^k, \Sigma_j)$ for $1 \leq j \leq \beta$ and $1 \leq k \leq 3$.

Now, consider a certain block b_j^2 in the second chord of B. When A is rotated $0°$, $120°$, and $240°$ counterclockwise, the blocks that are superimposed onto b_j^2 are a_j^2, a_j^3, and a_j^1 accordingly (see Figure 3.9 and formula (3.2)). Thus, to recover a given set of $(p_1, p_2, p_3)_j^2$, we should assure that $s_A^2 \otimes s_B$, $s_A^3 \otimes s_B$, and $s_A^1 \otimes s_B$ (or more precisely $permute(s_A^2, \Sigma_j) \otimes permute(s_B, \Sigma_j), permute(s_A^3, \Sigma_j) \otimes permute(s_B, \Sigma_j)$ and $permute(s_A^1, \Sigma_j) \otimes permute(s_B, \Sigma_j))$ resconstruct $(p_1)_j^2$, $(p_2)_j^2$ and $(p_3)_j^2$, respectively. Table 3.6 is designed for this principle.

TABLE 3.6

Encoding a set of corresponding pixels $(p_1, p_2, p_3)_j^2$ into a_j^2 (a_j^3 and a_j^1) and b_j^2 in terms of s_A^2 (s_A^3, s_A^1, respectively) and s_B in the first chords of A and B, respectively for visual 3-secret sharing.

p_1	p_2	p_3	s_A^2	s_A^3	s_A^1	s_B	$s_A^2 \otimes s_B$	$s_A^3 \otimes s_B$	$s_A^1 \otimes s_B$
□	□	□							
□	□	■							
□	■	□							
□	■	■							
■	□	□							
■	□	■							
■	■	□							
■	■	■							

In fact, we can rearrange Table 3.6 to make the columns 4–10 exactly the same as those in Table 3.5. Table 3.7 is such a consequence. Note that Tables 3.5 and 3.7 are the same except for the headings of columns 1–3. From Table 3.7, we observe that given a set of corresponding pixels $(p_1, p_2, p_3)_j^2$, the elementary block of b_j^2 can be easily determined by $s_B^{btod(p_3p_1p_2)}$.

Following the above example shown in Figure 3.11, we have $(p_1, p_2, p_3)_1^2 = (■, ■, □)$ and $\Sigma_1 = (1, 2, 3, 4, 5, 6,)$. Since $btod(p_3p_1p_2) = btod(011) = 3$ b_1^2 is encoded as $permute(s_B^3, \Sigma_1)$ as show in Figure 3.13. It is easily seen that

TABLE 3.7

Encoding scheme equivalent to Table 6 for the second chords of A and B for visual 3-secret sharing.

p_3	p_1	p_2	s_A^1	s_A^2	s_A^3	s_B	$s_A^1 \otimes s_B$	$s_A^2 \otimes s_B$	$s_A^3 \otimes s_B$
□	□	□							
□	□	■							
□	■	□							
□	■	■							
■	□	□							
■	□	■							
■	■	□							
■	■	■							

$$
\begin{aligned}
a_1^2 \otimes b_1^2 &= permute(s_A^2, \Sigma_1) \otimes permute(s_B^3, \Sigma_1) \\
&= s_A^2 \otimes s_B^3 = \blacksquare \otimes \blacksquare = \blacksquare \\
a_1^3 \otimes b_1^2 &= permute(s_A^3, \Sigma_1) \otimes permute(s_B^3, \Sigma_1) \\
&= s_A^3 \otimes s_B^3 = \blacksquare \otimes \blacksquare = \blacksquare \\
a_1^1 \otimes b_1^2 &= permute(s_A^1, \Sigma_1) \otimes permute(s_B^3, \Sigma_1) \\
&= s_A^1 \otimes s_B^3 = \blacksquare \otimes \blacksquare = \blacksquare
\end{aligned}
$$

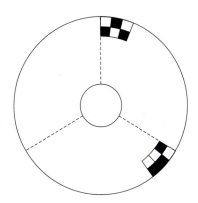

FIGURE 3.13

Encoding b_1^2 in B.

That is, $a_1^2 \otimes b_1^2$, $a_1^3 \otimes b_1^2$, and $a_1^1 \otimes b_1^2$ (the first blocks in the second chords of $A \otimes B$, $A^{120°} \otimes B$, and $A^{240°} \otimes B$) reconstruct $(p_1)_1^2$ (■), $(p_2)_1^2$ (■), and $(p_3)_1^2$ (□), respectively.

Based upon the experience above, Table 3.8 summarizes the encoding scheme for the blocks in the third chord of B for sharing 3 secrets. We can see from Table 3.8 that given a set of corresponding pixels $(p_1, p_2, p_3)_j^3$, the elementary block of b_j^3 is chosen to be $s_B^{btod(p_2 p_3 p_1)}$.

TABLE 3.8

Encoding a set of corresponding pixels $(p_1, p_2, p_3)_j^3$ into a_j^3 (a_j^1 and a_j^2) and b_j^3 in terms of s_A^3 (s_A^1, s_A^2, respectively) and s_B in the first chords of A and B, respectively for visual 3-secret sharing.

p_2	p_3	p_1	s_A^1	s_A^2	s_A^3	s_B	$s_A^1 \otimes s_B$	$s_A^2 \otimes s_B$	$s_A^3 \otimes s_B$
□	□	□				▨	▰	▰	▰
□	□	■				▨	▰	▰	▰
□	■	□				▨	▰	▰	▰
□	■	■	▰	▰	▰	▨	▰	▰	▰
■	□	□				▨	▰	▰	▰
■	□	■				▨	▰	▰	▰
■	■	□				▨	▰	▰	▰
■	■	■				▨	▰	▰	▰

Following the previous example in Figure 3.11, consider the particular case $(p_1, p_2, p_3)_1^3 = (□, ■, ■)$. Since $btod(p_2 p_3 p_1) = btod(110) = 6$, we encode b_1^3 as $permute(s_B^6, \Sigma_1)$ (▰) as show in Figure 3.14. It is clearly seen that

$$
\begin{aligned}
a_1^3 \otimes b_1^2 &= permute(s_A^3, \Sigma_1) \otimes permute(s_B^6, \Sigma_1) \\
&= s_A^3 \otimes s_B^6 = ▰ \otimes ▰ = ▰ \\
a_1^1 \otimes b_1^2 &= permute(s_A^1, \Sigma_1) \otimes permute(s_B^6, \Sigma_1) \\
&= s_A^1 \otimes s_B^6 = ▰ \otimes ▰ = ▰ \\
a_1^2 \otimes b_1^2 &= permute(s_A^2, \Sigma_1) \otimes permute(s_B^6, \Sigma_1) \\
&= s_A^2 \otimes s_B^6 = ▰ \otimes ▰ = ▰
\end{aligned}
$$

We have that $a_1^3 \otimes b_1^3$, $a_1^1 \otimes b_1^3$, and $a_1^2 \otimes b_1^3$ (the first blocks in the third chords of $A \otimes B$, $A^{120°} \otimes B$, and $A^{240°} \otimes B$) recover $(p_1)_1^3$ (□), $(p_2)_1^3$ (■), and $(p_3)_1^3$ (■), respectively.

Figure 3.15 depicts the results of the first three blocks in the three chords of $A \otimes B$, $A^{120°} \otimes B$ and $A^{240°} \otimes B$, which reconstruct $(p_1)_1^k$ in P_1 (□, ■, □) (Figure 3.15(a) vs. Figure 3.11(a)), $(p_2)_1^k$ in P_2 (■, ■, ■) (Figure 3.15(b) vs.

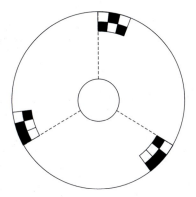

FIGURE 3.14
Encoding b_1^3 in B.

Figure 3.11(b)) and $(p_3)_1^k$ in P_3 ($\square, \square, \blacksquare$) (Figure 3.15(c) vs. Figure 3.11(c)), respectively for $1 \le k \le 3$.

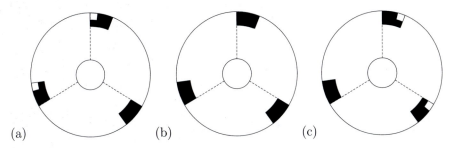

(a)   (b)   (c)

FIGURE 3.15
Results of (a) $A \otimes B$, (b) $A^{120°} \otimes B$, and (c) $A^{240°} \otimes B$.

Based upon the above discussions, we obtain $b_j^k = permute(s_B^l, \Sigma_j)$ where

$$l = \begin{cases} btod(p_1 p_2 p_3) & \text{if } k = 1; \\ btod(p_3 p_1 p_2) & \text{if } k = 2; \\ btod(p_2 p_3 p_1) & \text{otherwise}, \end{cases}$$

for the given set of corresponding pixels $(p_1, p_2, p_3)_j^k$ with respect to pixel j of strip k in (P_1, P_2, P_3) where $1 \le j \le \beta$ and $1 \le k \le 3$.

Let $rotate(r_1 r_2 \ldots r_x, d)$ denote a function that rotates $r_1 r_2 \ldots r_x$ right d bits where $r_i \in \{0, 1\}$, $1 \le i \le x$ and $0 \le d \le x - 1$; that is,

$$rotate(r_1 r_2 \ldots r_x, d) = \begin{cases} r_1 r_2 \ldots r_x \\ \quad \text{if } d = 0; \\ r_{x-d+1} r_{x-d+2} \ldots r_x r_1 r_2 \ldots r_{x-d} \\ \quad \text{otherwise } (1 \le d \le x - 1). \end{cases} \quad (3.3)$$

Then the above formula can be simplified as

$$b_j^k = permute(s_B^{btod(rotate(p_1p_2p_3,k-1))}, \Sigma_j) \qquad (3.4)$$

with respect to the set of corresponding pixels $(p_1, p_2, p_3)_j^k$ where $1 \le j \le \beta$ and $1 \le k \le 3$.

Since the number of subpixels in both of the elementary blocks of A and B is $6(= 2x)$ in the above case, the *pixel expansion* (i.e., the number of subpixels in the shares needed to encode a set of corresponding pixels in the secret images) in this algorithm is 6 for $x = 3$. The visual x-secret sharing scheme for any general number $x \ge 2$ will be formalized in the following section.

3.4.1.4   General Algorithm

The definitions of the elementary blocks for circle shares A (formula (3.1)) and B (formula (3.4)) and the encoding scheme in Tables 3.5, 3.7, and 3.8 for visual 3-secret sharing can be generalized to accomplish the visual multisecret sharing for $x \ge 1$ (including $x = 1$) secrets. Furthermore, there is no need to store any codebook like Tables 3.5, 3.7, and 3.8. Thus, this scheme formally presented in the following is not only general but also efficient for physical implementation.

Assume that there are x secrets to be shared by two participants. The two circle shares A and B are evenly decomposed into x chords, respectively. Let θ denote the degree expanded in each chord of A and B. It is computed as

$$\theta = 360^\circ / x.$$

We refer to the elementary block of the x secrets as a block with $2x$ ordered subpixels as shown in Figure 3.16. It is noted that the pixel expansion in the scheme is $2x$ when x secrets are shared. The width and height of the elementary block can be any combination as long as their multiplication is $2x$ (or even any number larger than $2x$ for some special purposes, such as retaining aspect ratios, to ease the production of the circle shares, and so on). The order of the $2x$ subpixels in the elementary block can also be arbitrarily defined. In the following discussions, we follow the shape and order of the elementary block as shown in Figure 3.16.

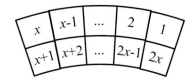

FIGURE 3.16
Elementary block for x secrets.

We define the set of the elementary blocks for share A as follows:

$$E_A^x = \{s_A^k | 1 \le k \le x\},$$

where s_A^k is an elementary block consisting of one white and $2x - 1$ black subpixels in which the jth subpixel, denoted as $s_A^k[j]$, is defined by

$$s_A^k[j] = \begin{cases} 0 & \text{if } j = x + 1 - k; \\ 1 & \text{otherwise,} \end{cases} \tag{3.5}$$

for $1 \leq j \leq 2x$ and $1 \leq k \leq x$.

Figure 3.17 shows the elementary blocks of A for encoding $x = 4$ secrets. As an example, we show how the subpixels in s_A^2 are computed by formula (3.5). Since $k = 2$ and $x + 1 - k = 4 + 1 - 2 = 3$, thus $s_A^k[3] = 0$ and $s_A^k[j'] = 1$ for $1 \leq j' \neq 3 \leq 8 (= 2x)$ as shown in Figure 3.17(b).

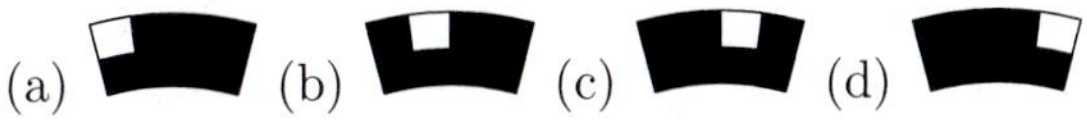

(a) ■ (b) ■ (c) ■ (d) ■

FIGURE 3.17
Elementary blocks in E_A^4: (a) s_A^1, (b) s_A^2, (c) s_A^3, (d) s_A^4.

We define the set of the elementary blocks for share B as follows:

$$E_B^x = \{s_B^\gamma | 0 \leq \gamma \leq 2^x - 1\},$$

where s_B^γ is also an elementary block containing x white and x black subpixels in which the jth subpixel, denoted as $s_B^\gamma[j]$, is defined by

$$s_B^\gamma[j] = \begin{cases} r_j & 1 \leq j \leq x; \\ \overline{r}_{2x+1-j} & \text{otherwise,} \end{cases} \tag{3.6}$$

where $\gamma = btod(r_x r_{x-1} \ldots r_2 r_1)$, r_t is the tth least significant bit of γ is represented in binary (x-bit) in which $1 \leq t \leq x$ and $0 \leq \gamma \leq 2^x - 1$ for $1 \leq j \leq 2x$ and $\overline{r}_t$ is the inverse of r_t.

Figure 3.18 illustrates the elementary blocks of B for $x = 4$. Consider s_B^4. Since $\gamma = 4 = btod(r_4 r_3 r_2 r_1) = btod(0100)_2$, we have $(s_B^4[1], s_B^4[2], s_B^4[3], s_B^4[4]) = (r_1, r_2, r_3, r_4) = (0, 0, 1, 0)$ and $(s_B^4[5], s_B^4[6], s_B^4[7], s_B^4[8]) = (\overline{r}_{2\times4+1-5}, \overline{r}_{2\times4+1-6}, \overline{r}_{2\times4+1-7}, \overline{r}_{2\times4+1-8}) = (\overline{r}_4, \overline{r}_3, \overline{r}_2, \overline{r}_1) = (1, 0, 1, 1)$. Thus, s_B^4 is as shown in Figure 3.18(e).

Formulae (3.1) and (3.4) about the encoding of blocks in A and B, respectively, for $x = 3$ can now be formulated in a more generalized form as follows. The blocks in A are encoded by

$$(a_j^1, a_j^2, \ldots, a_j^x) = (permute(s_A^1, \Sigma_j), permute(s_A^2, \Sigma_j), \ldots, permute(s_A^x, \Sigma_j)), \tag{3.7}$$

where Σ_j is a random permutation of $\{1, 2, \ldots, 2x\}$ for $1 \leq j \leq \beta$.

Given a set of corresponding pixels $(p_1, p_2, \ldots, p_x)_j^k$ in block j of strip k in $(P_1, P_2, \ldots, P_x)$, b_j^k (i.e., block j of chord k in B) is encoded by

$$b_j^k = permute(s_B^{btod(rotate(p_1 p_2 \ldots p_x, k-1))}, \Sigma_j) \tag{3.8}$$

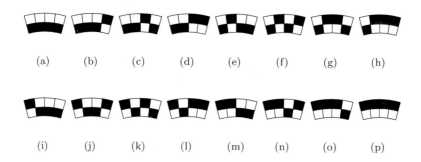

FIGURE 3.18
Elementary blocks in E_B^4: (a) s_B^0, (b) s_B^1, (c) s_B^2, (d) s_B^3, (e) s_B^4, (f) s_B^5, (g) s_B^6, (h) s_B^7, (i) s_B^8, (j) s_B^9, (k) s_B^{10}, (l) s_B^{11}, (m) s_B^{12}, (n) s_B^{13}, (o) s_B^{14}, (p) s_B^{15}.

for $1 \leq j \leq \beta$ and $1 \leq k \leq x$ where function $ratote(p_1 p_2 \ldots p_x, k-1)$ is defined by formula (3.3).

Based upon the above definitions and formulae (3.5)–(3.8), our visual multi-secrets sharing scheme is formally presented in Algorithm 1.

Algorithm 1. *Encoding x secret images into two circle shares*

Input: x $h \times w$ binary secret image $P_1, P_2, \ldots, P_x$
Output: two circle shares A and B such that any single A or B leaks no information about any one of the secret images, while $A^{(i-1)\theta} \otimes B$ recovers P_i for $1 \leq i \leq x$ in the human visual system where $\theta = 360°/x$ and $A^{0°}$

1. Create A and B as circle shares, which are decomposed into x chords where each chord is composed by $\beta = h \times (w/x)$ chord-shaped blocks referred to as a_j^k and b_j^k, $1 \leq k \leq x$ and $1 \leq j \leq \beta$, respectively and each block contains $2x$ subpixels.

2. Generate E_A^x and E_B^x according to formulae (5) and (6), respectively.

3. **for** (each block j, $1 \leq j \leq \beta$) **do**

3.1 {   Determine $\Sigma_j = (\sigma_1, \sigma_2, \ldots, \sigma_{2x})$, a random permutation of $\{1, 2, \ldots, 2x\}$

3.2   **for** (each chord k, $1 \leq k \leq x$) **do**
    {   // all β blocks in x chords of A and B adopt the same permutation Σ_j

3.2.1 $a_j^k = permute(s_A^k, \Sigma_j)$

3.2.2 **for** (each secret image i, $1 \le i \le x$) **do** $q_i = (p_i)_j^k$

3.2.3 $\gamma = btod(rotate(q_1 q_2 \ldots q_x, k-1))$

3.2.4 $b_j^k = permute(s_B^\gamma, \Sigma_j)$ }

   }

4. **output**(A, B) // A and B are composed by all a_j^k's and b_j^k's, respectively

Note that a new permutation Σ_j is determined to permute the subpixels in each pair of a_j^k and b_j^k for $1 \le k \le x$ to ensure the entire randomness that the subpixels in a_j^k and b_j^k can provide. Further, a_j^k and b_j^k are encoded by using the same permutation Σ_j so that the numbers of the white and black subpixels in $a_j^k \otimes b_j^k$ and $s_A^k \otimes b_B^\gamma$ are exactly the same for $1 \le k \le x$.

The pixel expansion of Shyu et al.'s scheme is $2x$ when x secrets are shared. In the case of $x = 2$, the pixel expansion is $2x = 4$ which is the same as that of Wu and Chen [12] as well as Wu and Chang [13]. The number of all possible patterns in an extended block in S_1 [12] (see Figure 3.2(c)) or any sector block in A [13] (see Figure 3.4(a)) is 4, which contains two white and two black subpixels, while that in b_j^k of the scheme is $(4!)/(2! \times 2!) = 6$. The randomness of this scheme, in the case of $x = 2$, is surely better than that of Wu and Chen as well as Wu and Chang.

It is seen from Algorithm 1 that we do not physically store any information about Tables 3.5, 3.7, and 3.8 in memory. The elementary blocks s_A^k's and s_B^γ's are generated in the run time (Step 2 in Algorithm 1) according to formulae (3.5) and (3.6). The encoding process is guaranteed by formulae (3.7) and (3.8) (Step 3 in Algorithm 1).

3.4.2 Feng et al.'s Scheme

Regarding Feng et al.'s $(2, 2, x)$ scheme [3], x secret images $P_1, P_2, \ldots, P_x$ are encoded into two shares A and B. Each set of x corresponding pixels (in x secret images) is encoded into two blocks, namely $s_A \in A$ and $s_B \in B$, each of which consists of x rows containing 3 pixels each. The pixel expansion is thus $3x$. The rotation relationship for revealing each of the x secrets is similar to Shyu et al.'s scheme where the ith secret is revealed by $A \otimes B^{(i-1)\theta}$ for $1 \le i \le x$. One special design in their scheme is that the encoded shares are in the form of cylinders to avoid the distortion of the revealed secrets.

They chose four types of 3-pixel patterns, referred to as the *effective block* $B_e = \blacksquare\blacksquare\square$, *ineffective block* $B_i = \blacksquare\square\square$, *white block* $B_w = \square\blacksquare\blacksquare$ and *black block* $B_b = \square\blacksquare\blacksquare$ to construct s_A and s_B. It is evident that $B_i \otimes B_w = B_i \otimes B_b = \blacksquare\blacksquare\blacksquare$, $B_e \otimes B_w = \blacksquare\blacksquare\blacksquare$, and $B_e \otimes B_b = \blacksquare\blacksquare\blacksquare$.

Figure 3.19 specifically illustrated the stacked results of these chosen patterns in their scheme.

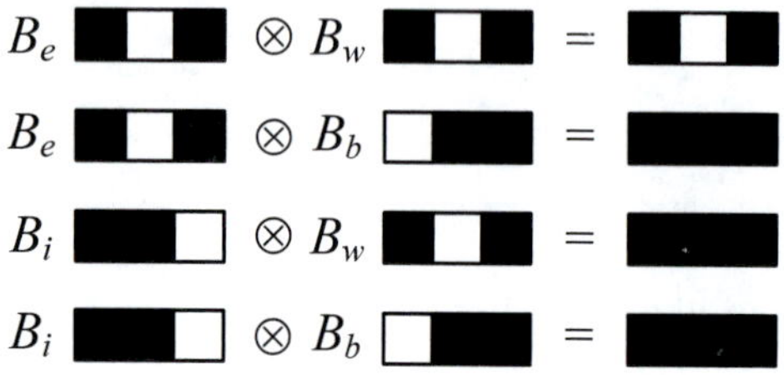

FIGURE 3.19
Stacking results of the chosen visual patterns for Feng et al.'s scheme.

Table 3.9 lists a possible set of encoded patterns for s_A^1, s_A^2, s_A^3, and s_B; and their stacked results for $x = 3$. The reason that p_i is reconstructed by $s_A^i \otimes s^B$ is precisely explained in this table. Surely, each set of the encoded pattern could be permuted correspondingly to accomplish the randomness for the whole shares.

TABLE 3.9
Possible set of encoded patterns for s_A^1, s_A^2, s_A^3, and s_B; and their stacked results for $x = 3$.

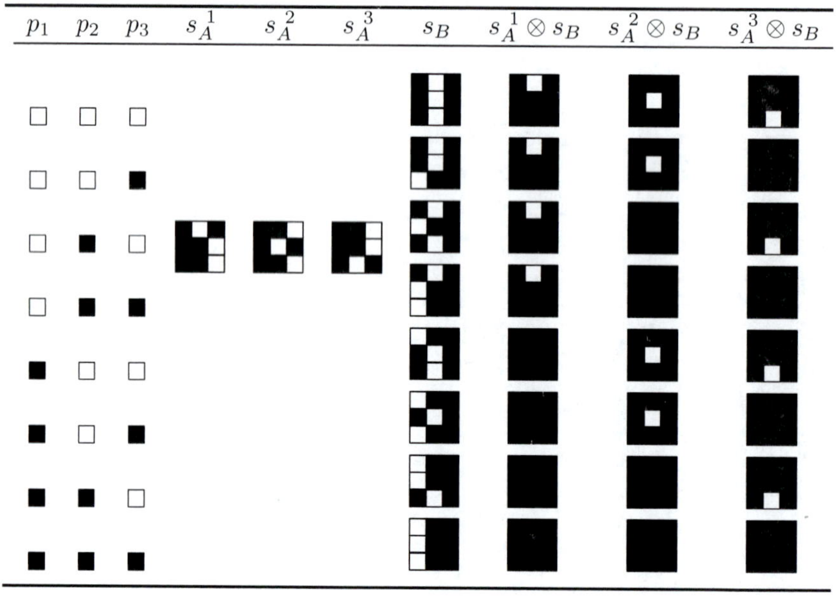

3.4.3 Experimental Results

Here, we implement Shyu et al.'s visual multisecret sharing scheme due to its generality on the abstraction and the superiority on the pixel expansion (over Feng et al.'s scheme).

We coded the program by using Borland C++ Builder (BCB) in a personal computer running MS Windows. Since the blocks are in the shape of chords, we called the embedded functions in BCB such as circle drawing, line drawing, flood-filling a closed area, and so on, to build the chord-shaped blocks in the scheme.

Four experiments were designed to explore the feasibility and applicability of the visual multisecret sharing scheme. Experiment 1 verifies the correctness of the scheme for $x = 3$ where the starting position for encoding on the circle shares are fixed as above-mentioned. Experiment 2 demonstrates that the scheme can be easily extended in such a way that the starting position for encoding can be arbitrarily assigned. This increases the secrecy of the proposed scheme. Experiment 3 gives the implementation results of the visual 4-secret sharing scheme. Experiment 4 presents implementation results of encoding the shares using cylinder (instead of circle) shares.

Experiment 1: Figure 3.20 illustrates the results of a computer implementation of the proposed scheme for sharing three secret images. Figures 3.20(a)–(c) are the three secrets to be shared, namely P_1, P_2, and P_3, respectively. Figures 3.20(d) and (e) show the circle shares A and B encoded by Algorithm 1, which expose no information about P_1, P_2, and P_3 individually. Figures 3.20(f)–(h) reveal the superimposed results of $A \otimes B$, $A^{120°} \otimes B$, and $A^{240°} \otimes B$, which reconstruct P_1, P_2, and P_3 in our visual system, respectively. Figure 3.20(i) gives another superimposed result, $A^{85°} \otimes B$ that leaks no information about any of the three secrets. In fact, any result of $A^\theta \otimes B$, for $\theta = 0°, 120°, 240°$, is merely a seemingly random picture.

Experiment 2: The encoding processes of A and B in the algorithm start from the 0° position and move on in a clockwise direction (see Figure 3.5). However, the starting position for encoding in A (or B) can be predefined arbitrarily.

Figure 3.21 shows the implementation results of using the same example as in Experiment 1 with a different starting starting position in B; that is, we encoded B by starting from the 85° position (85° counterclockwise to the 0° position) while we encoded A by starting from the 0° position as mentioned. The three secret images are the same as those in Figures 3.20(a)–(c). Figures 3.21(a) and (b) are the circle shares A' and B' encoded by Algorithm 1. Figure 3.21(c) shows the result of $A' \otimes B'$, which reveals nothing about the secrets, while Figures 3.21(d)–(f) display the superimposed results of $(A')^{85°} \otimes B'$, $(A')^{205°} \otimes B'$ and $(A')^{325°} \otimes B'$ that reconstruct P_1, P_2, and P_3, respectively, in our visual system. Note that both $A \otimes B$ (Figure 3.20(f)) and $(A')^{85°} \otimes B'$

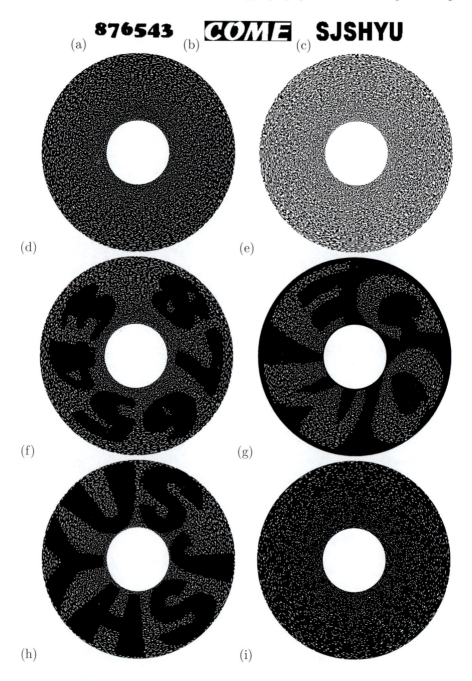

FIGURE 3.20

Implementation results for the proposed visual 3-secret sharing scheme: (a) P_1, (b) P_2, (c) P_3, (d) A, (e) B, (f) $A \otimes B$, (g) $A^{120°} \otimes B$, (h) $A^{240°} \otimes B$, (i) $A^{85°} \otimes B$.

is 85° counterclockwise away from that in $A \otimes B$.

Experiment 3: Figure 3.22 gives the implementation results of the proposed scheme for sharing four secrets. Figures 3.22(a)–(d) are the four secrets to be shared, namely P_1, P_2, P_3, and P_4, respectively. Figures 3.22(e) and (f) are the encoded circle shares A and B. Figures 3.22(g)–(j) show the superimposed results of $A \otimes B$, $A^{90°} \otimes B$, $A^{180°} \otimes B$, and $A^{270°} \otimes B$ that recover P_1, P_2, P_3, and P_4 in our visual system, respectively.

Experiment 4: One disadvantage in applying circle shares is that the reconstructed secrets might be distorted. This shortcoming could be easily refined by introducing *cylinder* shares.

Suppose that we encode each set of x pixels into square blocks (instead of chord blocks) in Shyu et al.'s scheme. The encoded shares evolve into the shape of rectangles. Each of the two rectangle shares can be easily rolled up into a cylinder by aligning the rightmost column next to the leftmost one. Figure 3.23 shows an example of applying cylinder shares to reveal the corresponding distorted secrets where (a) and (b) are distorted reconstructed secrets (which are the same as those in Figure 3.22(g) and (j), respectively) using circular shares, while (c) and (d) are the corresponding counterparts using cylinder shares which avoid any distortion when exposing the secrets.

The results in Experiments 1–4, as expected, demonstrate the feasibility and applicability of Shyu et al.'s visual multisecret sharing scheme. We compare the performances of the aforementioned schemes in terms of the capability of sharing secrets, pixel expansion, contrast, and shape of shares in the next subsection.

3.4.4 Comparison and Discussions

When we deal with x secrets, the pixel expansion of Shyu et al.'s scheme [8] is $2x$ and the contrast (i.e., the relative difference between the reconstructed white and black pixels in the superimposed image) of the scheme is $1/(2x)$ since all $2x$ subpixels in a reconstructed black pixel are black, while those in a reconstructed white pixel are $2x$. Suppose that Feng et al.'s scheme is applied. The pixel expansion becomes $3x$ and the contrast is $1/(3x)$. Note that when $x = 2$, the pixel expansions (contrasts) in Wu and Chen's [12], Wu and Chang's [13], and Shyu et al's schemes are all 4 (1/4); while in Feng et al.'s scheme is 6 (1/6).

Table 3.10 summarizes the numbers of secrets shared (denoted as x), pixel expansions(denoted as m), contrasts, and the shapes of shares in these visual multiple-secret sharing schemes for the comparison purpose.

The pixel expansion of Shyu et al.'s scheme [8] is $2x$ when x secrets are shared. It would be challenging to prove whether or not it is optimal. Is there any algorithm that improves the contrast in the scheme? It is surely worthy of further study. How to extend Shyu et al.'s scheme such that multiple secrets

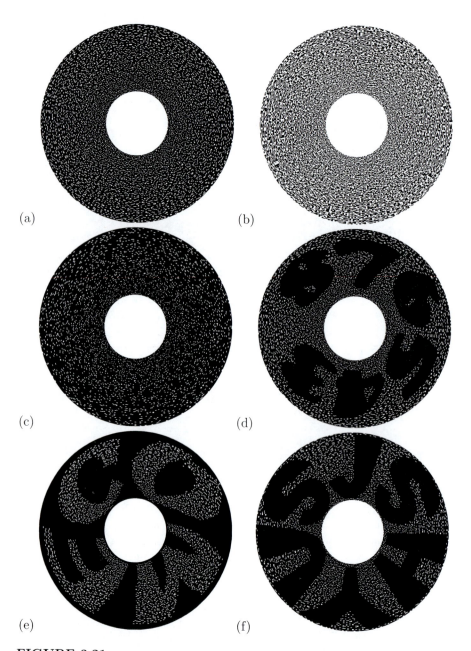

FIGURE 3.21

Implementation results for the proposed visual 3-secret sharing scheme with a different starting encoding position: (a) A', (b) B', (c) $A' \otimes B'$, (d) $(A')^{85°} \otimes B'$, (e) $(A')^{205°} \otimes B'$, (f) $(A')^{325°} \otimes B'$.

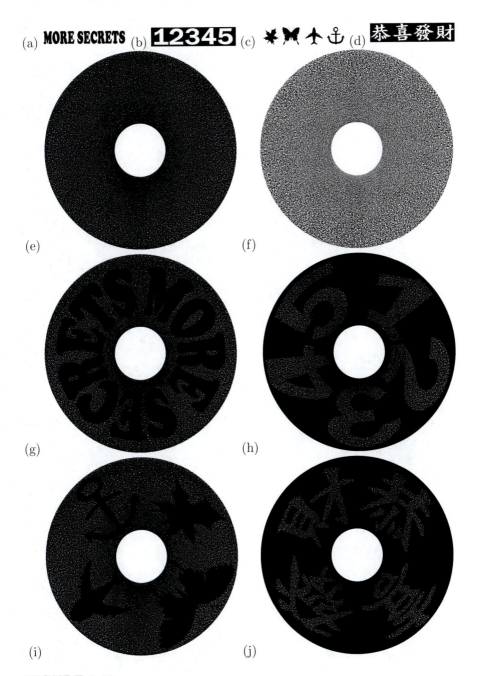

FIGURE 3.22

Results of computer implementation for 4-secret sharing: (a) P_1, (b) P_2, (c) P_3, (d) P_4, (e) A, (f) B, (g) $A \otimes B$, (h) $A^{90°} \otimes B$, (i) $A^{180°} \otimes B$ (j) $A^{270°} \otimes B$.

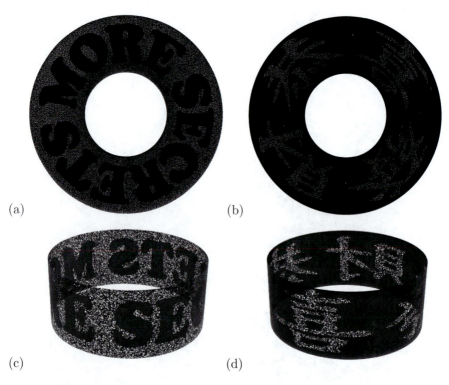

(a)   (b)

(c)   (d)

FIGURE 3.23
Transforming circle shares (a) and (b) into cylinder counterparts (c) and (d), respectively.

TABLE 3.10
Comparison of visual multiple secrets sharing schemes.

schemes	x	m	contrast	shares' shape	
Wu and Chen [12]	2	4	1/4	square	
Wu and Chang [13]	2	4	1/4	square	
Feng et al.[3]	≥ 2	$3x$	$1/3x$	cylnder	
Shyu et al.[8]	≥ 2	$2x$	$1/2x$	circle	or
				cylinder	

can be shared by more than two shares is also an interesting topic. Potentially, sharing multiple secrets may have more flexibilities and applications than sharing only one secret. Visual identification and visual authentication are some typical applications in visual cryptography [6]. It would be of much significance to reexamine these topics from a viewpoint of sharing multiple secrets.

As a matter of fact, the sharing of multiple secrets visually brings forth new problems to be considered. For instance, with regard to the "starting position for encoding" in A or/and B in Experiment 2, we may design such a concern to be some kind of *private key*, which is only accessible between the dealer and authorized participant(s). Without the correct starting positions in A or/and B, the alignment of A and B cannot recover the secret yet. In addition, the second secret of the three secrets in Experiment 1 might be designed to be fake for the purpose of diffusion. That is to say whether the whole secret message is "Help is never on its way" or "Help is on its way" may be treated to be another private key between the dealer and authorized participant(s). Mainly, the number of secrets, the degree of the starting position for encoding, the combination of the true or fake reconstructed secrets, and so on, can be designed as private keys to increase the level of security in the visual multi-secret sharing system.

3.5 Concluding Remarks

By adopting circle or cylinder shares, we discuss general visual secret sharing schemes for $x \geq 1$ (indeed, these schemes work well for $x = 1$) secrets in two shares in this chapter. The previous studies considered sharing only two secrets in two shares [12, 13]. Shyu et al.'s scheme can be implemented easily and it takes only some constant working space. All encoding information can be determined in run time. By introducing an independent random permutation (i.e., Σ_j, see formulae (3.1) and (3.4)) when encoding each pair of the corresponding blocks (i.e., a_j^k and b_j^k, see Step 3.2.1 and 3.2.4 in Algorithm 1), the scheme ensures the maximum randomness that the subpixels in an encoded block may possibly provide. For the transmitter, one machine capable of running the encoding scheme is needed, while for the receivers, no computing device is required and the decryption process is simply by the human visual system. The proposed scheme can be easily extended to gray-level images by adopting the halftone technology [4] or even color images by exploiting color decomposition [4] or color composition [9].

In traditional visual secret-sharing schemes, rectangle shares are encoded to conceal one shared secret. They are easily superimposed by aligning the rectangular corners. As compared to the rectangle shares, the circle or cylinder shares are relatively hard to superimpose since there are no reference points to align with. Basically, the usage of the circle or cylinder shares increases the complexity in decoding the secrets. In practical applications, the dealer might add additional information in the circle shares, such as some supplementary points, lines, or markers, to ease the superimposition (decoding process) for the participants. Figure 3.24 shows one possible arrangement in the case of sharing three secrets as in Experiment 1. Note that circle share A has three markers (see Figure 3.24(a)), while B has only one (see Figure 3.24(b)) based upon which A, $A^{120°}$, and $A^{240°}$ can be superimposed with B easily. Or, the dealer can deliberately organize such information as private key(s) such that only the legal receivers are informed how to obtain the key(s). The same reasoning could be applied if the cylinder shares are adopted.

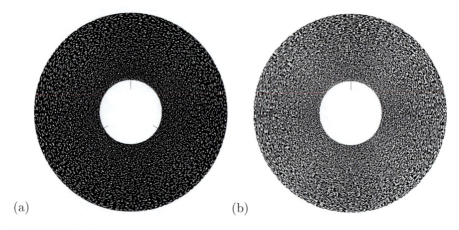

(a)   (b)

FIGURE 3.24
Shares (based upon Experiment 1) with supplementary lines to ease the alignments: (a) A with three markers, (b) B with one marker.

Generally speaking, the use of circle or cylinder shares to convey several secrets discloses some new issues that have not yet been considered in traditional visual cryptography, such as "How many secrets are there?," "How to superimpose the shares (where to align with or in what rotation angles)?," "Is there any fake secret(s) for diffusion?," and so on. These concerns can be designed as a set of private keys. The consideration and distribution of these private keys can be further discussed.

Bibliography

[1] G. Ateniese, C. Blundo, A. De Santis, and D.R. Stinson. Visual cryptography for general access structures. *Inf. Comput.*, 129:86–106, 1996.

[2] C. Blundo, A. De Santis, and D.R. Stinson. On the contrast in visual cryptography schemes. *J. Cryptogr.*, 12:261–289, 1999.

[3] J. B. Feng, H. C. Wu, C. S. Tsai, Y. F. Chang, and Y. P. Chu. Visual secret sharing for multiple secrets. *Pattern Recognition*, 41:3572–3581, 2008.

[4] Y.-C. Hou. Visual cryptography for color images. *Pattern Recognition*, 36:1619–1629, 2003.

[5] C.-C. Lin and W.-H. Tsai. Visual cryptography for grey-level images by dithering techniques. *Pattern Recognition Lett.*, 24:349–358, 2003.

[6] M. Naor and B. Pinkas. Visual authentication and identification. *in: B.S. Kaliski Jr. (Ed.), Advances in Cryptology: CRYPTO97, Lecture Notes in Computer Science*, 1294:322–336, 1997.

[7] M. Naor and A. Shamir. Visual cryptography. *in: A. De Santis (Ed.), Advances in Cryptology: Eurpocrypt'94, Lecture Notes in Computer Science*, 950:1–12, 1995.

[8] S. J. Shyu, S.-Y. Huang, Y.-K. Lee, R.-Z. Wang, and K. Chen. Sharing multiple secrets in visual cryptography. *Pattern Recognition*, 40:3633–3651, 2007.

[9] S.J. Shyu. Efficient visual secret sharing scheme for color images. *Pattern Recognition*, 39:866–880, 2006.

[10] D.R. Stinson. An introduction to visual cryptography. *Public Key Solution '97*, pages 28–30, April 1997.

[11] E.R. Verheul and H.C.A. Van Tilborg. Constructions and properties of k out of n visual secret sharing schemes. *Designs Codes Cryptography*, 11:179–196, 1997.

[12] C.C. Wu and L.H. Chen. *A study on visual cryptography, Master Thesis*. PhD thesis, Institute of Computer and Information Science, National Chiao Tung University, Taiwan, R.O.C., 1998.

[13] H.-C. Wu and C.-C. Chang. Sharing visual multi-secrets using circle shares. *Comput. Stand. Interfaces*, 134 (28):123–135, 2005.

4

Extended Visual Cryptography for Photograph Images

Yasushi Yamaguchi

The University of Tokyo, Japan
Japan Science and Technology Agency / CREST, Japan

CONTENTS

4.1 Introduction

Visual cryptography is a kind of cryptography that can be decoded directly by the human visual system without any computation for decryption. It usually prints certain images on transparencies and the secret image is reconstructed by simply stacking the transparencies together. Extended visual cryptography allows the printing of meaningful images on transparencies so that it can conceal the very existence of "secret" in the transparencies. There have been a lot of studies to incorporate photograph images into extended visual cryptography. This chapter attempts to survey the studies on extended visual cryptography for photograph images.

4.2 Basic Visual Cryptography Schemes

In order to determine basic terminology in this chapter, this section explains basic concepts of visual cryptography, namely, k out of n *Visual Secret Sharing Scheme* ((k, n) VSSS), an *Extended Visual Cryptography Scheme* (EVCS), and *Random Grids*.

4.2.1 (k, n) Visual Secret Sharing Schemes

This scheme was proposed by Naor and Shamir in 1994 [31]. It generates n transparencies from an original *secret image*. The transparencies are usually shared by n participants so that each participant is expected to keep one transparency. Thus, a secret image is sometimes called a *shared image*. The secret image can be observed if any k or more of them are stacked together. However, the secret image is totally invisible if fewer than k transparencies are stacked. The images on transparencies are called *shadow images*.

Each pixel of a shadow image is generated separately in the conventional VSSS. An original secret pixel will be transformed to n patterns of pixels for shadow images. These pixels on shadow images are called *shares*. A share consists of m black and white subpixels. The human visual system observes the average of subpixels, because they exist in close proximity. This structure is usually described by an $n \times m$ Boolean matrix $M = [m_{ij}]$. Here $m_{ij} = 0$ or 1 if the jth subpixel in the ith shadow is white or black, respectively. If transparencies of r shadows $i_1, i_2, \cdots, i_r$ out of n are stacked in a way that properly aligns the subpixels, each combined share can be represented by the Boolean "OR" of the corresponding rows $i_1, i_2, \cdots, i_r$ in the Boolean matrix

FIGURE 4.1
Six possible patterns of subpixel arrangements with 50% gray. Each pattern is represented as $[0\ 0\ 1\ 1]$, $[1\ 1\ 0\ 0]$, $[0\ 1\ 0\ 1]$, $[1\ 0\ 1\ 0]$, $[0\ 1\ 1\ 0]$, $[1\ 0\ 0\ 1]$ from left to right.

M. Let M_r denote the m-D vector obtained by taking the Boolean "OR" of r row vectors. The gray level of a pixel combined by r shares is obtained by the Hamming weight $H(M_r)$ of the "OR"ed m-D vector M_r. Users interprets this gray level as black if $H(M_r) \geq t$ and as white if $H(M_r) < t - \alpha m$. Here, $t \in \{1, \cdots, m\}$ is called *threshold*, while the value $\alpha > 0$ and the number $\alpha m \geq 1$ are called *relative difference* and *contrast*, respectively.

The (k, n) VSSS consists of two collections of $n \times m$ Boolean matrices $\mathcal{C}_w$ and $\mathcal{C}_b$ where any matrix in $\mathcal{C}_w$ generates a white pixel with k or more of shares (rows) while a matrix in $\mathcal{C}_b$ generates a black pixel. The scheme is valid if it fulfills the following three conditions:

1. For any M in $\mathcal{C}_w$, the "OR"ed vector M_k of any k rows of M satisfies $H(M_k) < t - \alpha m$.

2. For any M in $\mathcal{C}_b$, the "OR"ed vector M_k of any k rows of M satisfies $H(M_k) \geq t$.

3. For any subset $\{i_1, i_2, \cdots, i_q\}$ of $\{1, 2, \cdots, n\}$ with $q < k$, the two collections of $q \times m$ matrices $\mathcal{D}_w$ and $\mathcal{D}_b$ obtained by extracting rows $i_1, i_2, \cdots, i_q$ from $n \times m$ matrices in $\mathcal{C}_w$ and $\mathcal{C}_b$ are indistinguishable so that the collections contain the same matrices with the same frequencies.

The above first two conditions are *contrast* conditions that k or more of shadow images can recover the secret image with contrast αm. The recovered secret image is usually called the *reconstructed image*. The last condition is related to *security*, which implies that none can gain any information on the secret image by investigating fewer than k shadow images.

Here, two parameters m and α are very important to this discussion. The parameter m indicates the number of subpixels in a share, which is called *pixel expansion*. Each pixel of the original secret image is represented by m subpixels so that the reconstructed image as well as the shadow images will be m times large as the original image. People would like m to be as small as possible. The parameter α indicates the relative difference between combined shares of an originally white pixel and an originally black pixel. Since it means the loss of contrast of the reconstructed image, people would like α to be as large as possible.

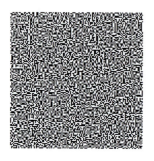  

FIGURE 4.2
An example of visual secret sharing scheme (VSSS). Two shadow images of random patterns (left and middle) and reconstructed secret image (right).

Let us consider a special case of $(2,2)$ VSSS. Each share consists of 4 subpixels of a 2×2 array in a physical implementation, where two of them are white and the rest two are black. The Boolean matrix of this scheme is 2×4 where each row consisting of two 0's and two 1's represents an arrangement of subpixels in a share. For instance, six possible patterns of shares having 50% gray as shown in Figure 4.1 are represented as $\{[0\ 0\ 1\ 1], [1\ 1\ 0\ 0], [0\ 1\ 0\ 1], [1\ 0\ 1\ 0], [0\ 1\ 1\ 0], [1\ 0\ 0\ 1]\}$. The scheme is accomplished by the following two collections:

$$
\begin{aligned}
\mathcal{C}_w &= \{\text{matrices obtained by permuting the columns of } \mathcal{S}_w\}, \\
\mathcal{C}_b &= \{\text{matrices obtained by permuting the columns of } \mathcal{S}_b\},
\end{aligned}
$$

where $\mathcal{S}_w$ and $\mathcal{S}_b$ are given as below:

$$
\mathcal{S}_w = \begin{bmatrix} 0 & 0 & 1 & 1 \\ 0 & 0 & 1 & 1 \end{bmatrix}, \quad
\mathcal{S}_b = \begin{bmatrix} 0 & 0 & 1 & 1 \\ 1 & 1 & 0 & 0 \end{bmatrix}.
$$

The above matrices $\mathcal{S}_w$ and $\mathcal{S}_b$ are called *basis matrices*. Because the collections are obtained by permutation of subpixels, each share may have randomly arranged two white and two black subpixels, which looks 50% gray. A pair of shares from $\mathcal{C}_w$ has the same arrangement of subpixels. The combined result is the same pattern, which looks 50% gray. A pair of shares from $\mathcal{C}_b$ has the complementary arrangement of subpixels. The combined result consists of four black subpixels, which looks completely black. Figure 4.2 shows an example of resulting shadow images and a reconstructed secret image. The size of all images are 128×128 pixels, because the original secret image has 64×64 pixels.

The original scheme proposed by Naor and Shamir is *uniform*, such that any combined shares from $q < k$ shadow images yield "OR"ed m-D vector M_q with $H(M_q) = f(q)$ with uniform probability distribution, regardless of if the matrices were taken from $\mathcal{C}_w$ or $\mathcal{C}_b$. Suppose the case of $q = 1$, the above-mentioned combined share is a single share of each shadow image. It means all the shadow images consist of uniformly random pattern of black and white subpixels. Naor and Shamir pointed out an extension of this scheme for concealing the very existence of the secret image.

4.2.2 Extended Visual Cryptography Scheme

Ateniese et al. extended the VSSS in the sense of a *General Access Structure* (GAS) [1] and extended capability. [2] A General Access Structure controls the qualified set of transparencies with which one can recover the secret image, while any k or more transparencies can reconstruct the secret image in (k, n) VSSS. An extended capability is able to introduce a meaningful image as a shadow image which Naor and Shamir pointed out in their very first paper [31]. An innocent-looking image of a house, dog, or something else would be much less suspicious than a random-dotted image as a shadow image.

In the *Extended Visual Cryptography Scheme* (EVCS), for an *access structure* $(\Gamma_{\mathsf{Qual}}, \Gamma_{\mathsf{Forb}})$ on a set of n participants, the shared (secret) image can be recovered by any *qualified set* $X \in \Gamma_{\mathsf{Qual}}$ with no trace of the shadow images, but any *forbidden set* $X \in \Gamma_{\mathsf{Forb}}$ has no information on the secret image. Moreover, the shadow images are meaningful so that each participant can recognize the image on one's transparency.

Similar to the (k, n) VSSS, an EVCS can be constructed in a pixel-wise manner. Since n participants share one secret image and have their own images in the n shadow images, we have to consider $n+1$ colors, $c, c_1, \cdots, c_n \in \{w, b\}$ where w and b stands for white and black, respectively. The value c denotes the color of the secret image pixel and c_i denotes the color of the original image pixel for i-th participant's shadow image. In order to realize an EVCS that obtains a c pixel when transparencies associated to a set $X \in \Gamma_{\mathsf{Qual}}$, we need 2^n pairs of collections of $n \times m$ Boolean matrices, $(\mathcal{C}_w^{c_1 \cdots c_n}, \mathcal{C}_b^{c_1 \cdots c_n})$, one for each possible combination of white and black pixels in the n original images for the shadow images.

An EVCS for an access structure $(\Gamma_{\mathsf{Qual}}, \Gamma_{\mathsf{Forb}})$ for n participants is valid if it fulfills the following conditions.

1. For any $X \in \Gamma_{\mathsf{Qual}}$ and for any $c_1, \cdots, c_n \in \{b, w\}$, the threshold t_X and the relative difference α_R exist, which satisfy $H(M_X) \le t_X - \alpha_R m$ for any $M \in \mathcal{C}_w^{c_1 \cdots c_n}$ and $H(M_X) \ge t_X$ for any $M \in \mathcal{C}_b^{c_1 \cdots c_n}$. Here M_X denotes the m-D vector obtained by taking Boolean "OR" of the row vectors of M corresponding to the participants in X and $H(M_X)$ denotes the Hamming weight of the vector M_X.

2. For any $X = \{i_1, \cdots, i_q\} \in \Gamma_{\mathsf{Forb}}$ and for any $c_1, \cdots, c_n \in \{b, w\}$, the two collections of $q \times m$ matrices, $\mathcal{D}_w^{c_1 \cdots c_n}$ and $\mathcal{D}_b^{c_1 \cdots c_n}$, obtained by extracting rows $i_1, \cdots, i_q$ from each $n \times m$ matrix in $\mathcal{C}_w^{c_1 \cdots c_n}$ and $\mathcal{C}_b^{c_1 \cdots c_n}$, respectively, are indistinguishable so that the collections contain the same matrices with the same frequencies.

3. For any $i \in \{1, 2, \cdots, n\}$ and any $c_1, \cdots, c_{i-1}, c_{i+1}, \cdots, c_n \in \{b, w\}$,

it results that

$$\min_{M \in \mathcal{M}_b} H(M_i) - \max_{M \in \mathcal{M}_w} H(M_i) \geq \alpha_S m,$$

where

$$\begin{aligned}
\mathcal{M}_b &= \mathcal{C}_b^{c_1 \cdots c_{i-1} b c_{i+1} \cdots c_n} \cup \mathcal{C}_w^{c_1 \cdots c_{i-1} b c_{i+1} \cdots c_n}, \\
\mathcal{M}_w &= \mathcal{C}_b^{c_1 \cdots c_{i-1} w c_{i+1} \cdots c_n} \cup \mathcal{C}_w^{c_1 \cdots c_{i-1} w c_{i+1} \cdots c_n},
\end{aligned}$$

and $H(M_i)$ denotes the Hamming weight of the i-th row vector M_i of a matrix M.

The values $\alpha_R > 0$ and $\alpha_S > 0$ are referred to as the *relative difference of the reconstructed image* and *relative difference of shadow images*, respectively. The number $\alpha_R m \geq 1$ and $\alpha_S m \geq 1$ are *contrasts* of the reconstructed image and the shadow images. People would like both α_R and α_S to be as large as possible.

The first condition is the *contrast* condition that indicates any qualified set $X \in \Gamma_{\mathsf{Qual}}$ can recover the secret image. The secret image can be recovered by stacking the transparencies of a qualified set, belonging to Γ_{Qual}. The second condition is the *security* condition that states any forbidden set $X = \{i_1, \cdots, i_q\} \in \Gamma_{\mathsf{Forb}}$ has no information on the secret image. People cannot get any information on the secret image by inspecting the shadow images of a forbidden set. The third condition is the *extended* condition that implies that the shadows images are still meaningful after the original images are encoded. Any participant can recognize the shadow image on one's transparency. Although the collection $\mathcal{M}_b$ is obtained by combining two collections $\mathcal{C}_b^{c_1 \cdots c_{i-1} b c_{i+1} \cdots c_n}$ and $\mathcal{C}_w^{c_1 \cdots c_{i-1} b c_{i+1} \cdots c_n}$, we have the same set of $\{M_i\}$ only with one of the collections, because $\{M_i : M \in \mathcal{C}_w^{c_1 \cdots c_n}\} \equiv \{M_i : M \in \mathcal{C}_b^{c_1 \cdots c_n}\}$ for any $c_1, \cdots, c_n \in \{b, w\}$ and any $i \in \{1, \cdots, n\}$ due to the second condition.

Here we show how to accomplish a 2 out of 2 EVCS. Each share consists of 4 subpixels like $(2, 2)$ VSSS. However, it contains either two 1's or three 1's depending on the colors of pixels of the corresponding original image, white or black, respectively. The scheme is given by the 4 pairs of collections $(\mathcal{C}_w^{c_1 c_2}, \mathcal{C}_b^{c_1 c_2})$, namely 8 collections $\mathcal{C}_c^{c_1 c_2}$, where $c, c_1, c_2 \in \{b, w\}$. The collections are obtained by permuting the columns of the following 8 basic matrices, $\mathcal{S}_c^{c_1 c_2}$:

$$\mathcal{S}_w^{ww} = \begin{bmatrix} 0 & 0 & 1 & 1 \\ 0 & 1 & 0 & 1 \end{bmatrix}, \quad \mathcal{S}_b^{ww} = \begin{bmatrix} 0 & 0 & 1 & 1 \\ 1 & 1 & 0 & 0 \end{bmatrix},$$

$$\mathcal{S}_w^{wb} = \begin{bmatrix} 0 & 0 & 1 & 1 \\ 0 & 1 & 1 & 1 \end{bmatrix}, \quad \mathcal{S}_b^{wb} = \begin{bmatrix} 0 & 0 & 1 & 1 \\ 1 & 1 & 1 & 0 \end{bmatrix},$$

$$\mathcal{S}_w^{bw} = \begin{bmatrix} 0 & 1 & 1 & 1 \\ 0 & 1 & 0 & 1 \end{bmatrix}, \quad \mathcal{S}_b^{bw} = \begin{bmatrix} 0 & 1 & 1 & 1 \\ 1 & 1 & 0 & 0 \end{bmatrix},$$

$$\mathcal{S}_w^{bb} = \begin{bmatrix} 0 & 1 & 1 & 1 \\ 0 & 1 & 1 & 1 \end{bmatrix}, \quad \mathcal{S}_b^{bb} = \begin{bmatrix} 0 & 1 & 1 & 1 \\ 1 & 1 & 1 & 0 \end{bmatrix}.$$

FIGURE 4.3
An example of extended visual cryptography scheme (EVCS). Two resulting shadow images (left and middle) and reconstructed secret image (right).

The reconstructed pixel has 3 or 4 black subpixels if the original secret pixel is white or black, respectively. In this scheme, the relative contrasts are given as $\alpha_R = \alpha_S = \frac{1}{4}$. Figure 4.3 shows an example of resulting shadow images and reconstructed secret image. The size of all images are 128×128 pixels, because all the original shadow and secret images have 64×64 pixels.

Ateniese et al. also pointed out some of the most important aspects of the extended capability [2]. One is related to the contrasts of images. A trade-off between two relative differences exists, α_R and α_S, in any (k, k) EVCS as below:

$$2^{k-1}\alpha_R + \frac{k}{k-1}\alpha_S \leq 1.$$

This means we cannot increase both contrasts of a reconstructed image and shadow images, $\alpha_R m$ and $\alpha_S m$, simultaneously. They also specified the lower bound of the pixel expansion m in (k, k) EVCS as below:

$$m \geq 2^{k-1} + 2.$$

This means we need more pixels to obtain EVCS. Although people would like contrasts to be as large as possible and pixel expansion as small as possible, there exist certain limits of them.

4.2.3   Random Grids

Random Grids (RG) give a very different approach to visual cryptography, which can keep the size of resulting shadow images to be the same as that of the original image. In other words, the pixel expansion of this method is $m = 1$ and no more expansion problems exist. The method is first introduced by Kafri and Keren in 1987 [17] and reinvestigated by Shyu in 2007 [37]. A random grid R is defined as a two-dimensional array of pixels. Each pixel is either transparent (white) or opaque (black) by a coin-flip procedure. The numbers of transparent pixels and opaque pixels are probabilistically same

FIGURE 4.4
An example of random grid (RG). Two random grids (left and middle) and reconstructed secret image (right).

and the *average opacity*[1] of a random grid is 50%:

$$\mathcal{O}(R) = \frac{1}{2}.$$

Let $R(\mathbf{p})$ denote a pixel value of the random grid R at the position $\mathbf{p}$ and $\overline{R(\mathbf{p})}$ denote its *inverse*.

$$R(\mathbf{p}) = \begin{cases} 0 & \text{if } R(\mathbf{p}) \text{ is transparent (white)} \\ 1 & \text{if } R(\mathbf{p}) \text{ is opaque (black)} \end{cases},$$

$$\overline{R(\mathbf{p})} = \begin{cases} 0 & \text{if } R(\mathbf{p}) = 1 \\ 1 & \text{if } R(\mathbf{p}) = 0 \end{cases}.$$

We must note that the inverse of a random grid is also a random grid and its opacity is 50 %, $\mathcal{O}(\overline{R}) = \frac{1}{2}$. The *superimposition* of two random grids, R_1 and R_2, pixel by pixel is computed by taking Boolean "OR" operation of their *corresponding pixels*, $R_1(\mathbf{p})$ and $R_2(\mathbf{p})$, as VSSS and EVCS:

$$(R_1 + R_2)(\mathbf{p}) = R_1(\mathbf{p}) + R_2(\mathbf{p}).$$

It is obvious that the superimposition of the same random grids results in the original random grid. The superimposition of a random grid and its inverse is a grid whose pixels are all opaque. Thus, the average opacity will be as below:

$$\mathcal{O}(R + R) = \mathcal{O}(R) = \frac{1}{2}, \quad \mathcal{O}(R + \overline{R}) = 1.$$

The encryption algorithm for a binary image B, which generates a pair of random grids R_1 and R_2 that can achieve the highest contrast is as follows.

---

[1]Originally a concept of *average transmission* was used both in [17] and [37] instead of average opacity. However, it is slightly confusing because people usually use 0 for a transparent (white) pixel and one for an opaque (black) pixel in visual cryptography studies. Thus, here we use the term average opacity.

Algorithm

1. Generate a random grid R_1 with the same size as B.
2. For each pixel $B(\mathbf{p})$, a grid R_2 is determined as below:

$$R_2(\mathbf{p}) = \begin{cases} R_1(\mathbf{p}) & \text{if } B(\mathbf{p}) = 0 \\ \overline{R_1(\mathbf{p})} & \text{otherwise} \end{cases}.$$

Let T and O denote transparent and opaque regions of the original binary image B so that $B(\mathbf{p}) = 0|_{\mathbf{p} \in T}$ and $B(\mathbf{p}) = 1|_{\mathbf{p} \in O}$. The regions T and O fulfills the following constraints:

$$\Omega = T \cup O \quad \text{and} \quad \phi = T \cap O,$$

where Ω stands for the entire region of the original image while ϕ represents the null region, because B is a binary image. Due to the definition of the above algorithm, random grids R_1 and R_2 satisfies the following relations:

$$R_1(\mathbf{p}) = R_2(\mathbf{p})|_{\mathbf{p} \in T} \quad \text{and} \quad \overline{R_1(\mathbf{p})} = R_2(\mathbf{p})|_{\mathbf{p} \in O}.$$

Therefore, R_2 as well as R_1 is a random grid and the average opacity of their superimposition depends on the regions T and O of the input binary image:

$$\mathcal{O}\left((R_1 + R_2)(\mathbf{p})\right)|_{\mathbf{p} \in T} = \frac{1}{2} \quad \text{and} \quad \mathcal{O}\left((R_1 + R_2)(\mathbf{p})\right)|_{\mathbf{p} \in O} = 1.$$

The difference of the average opacities of region T and O corresponds to relative difference α in VSSS. Figure 4.4 shows an example of random grids and a reconstructed secret image. The size of all images are 64×64 pixels, which is the same as that of the original secret images, because Random Grids are free from pixel expansion, $m = 1$.

4.3 Fundamentals of Photograph Visual Cryptography

Digital cameras have become very popular and people can easily obtain continuous-tone digital image data. However, all the schemes explained in the last section accept binary images as input. Thus, a photograph image must be converted to a binary image that can be observed similar to the original image by the human visual system. The algorithm that can achieve such a conversion is referred as *digital halftoning* or *halftoning* in short [38, 18].

4.3.1 Digital Halftoning

There are several approaches to digital halftoning, namely, noise-encoding, ordered dither, error diffusion, iterative and search-based methods, etc. Here we explain some of the approaches.

6	7	8	9
5	0	1	10
4	3	2	11
15	14	13	12

0	8	2	10
12	4	14	6
3	11	1	9
15	7	13	5

FIGURE 4.5

Samples of ordered dither matrices. Clustered-dot matrix (left) and dispersed-dot matrix (right).

Density Pattern

It is obvious to achieve $(l + 1)$ gray levels, i.e., tones, with l binary pixels. *The density pattern* method uses l subpixels for representing each pixel value. This is similar to the visual cryptography in the sense that a resulting image requires more pixels than the original image. If we adopt this type of halftoning method, the resulting shadow images and reconstructed secret image are lm times larger than the original continuous-tone images. Of course, this is inappropriate since people want to make a resulting image as small as possible.

Noise-Encoding

The easiest way to obtain a binary image from a continuous-tone image of the same size is *thresholding*, which assigns 0 to a pixel of the resulting binary image if the original pixel value is smaller than a threshold value. Otherwise it assigns 1 to the binary pixel. However, the resulting binary image usually suffers from pseudo-contours. *Noise-encoding* is a key concept for improving image quality. In the early stage of digital halftoning studies, random noise, i.e., white noise, is used for this purpose. A binary image obtained by adding random noise followed by thresholding yields better quality than a simply-thresholded image, because it reduces pseudo-contours. *Random dither* is a simple extension of noise-encoding. It uses a random threshold array and thresholds each pixel with a random number instead of using a constant threshold after adding a random number to the pixel value.

Ordered Dither

Noises need not be random and a threshold matrix can be generated with a certain order. *Ordered dither* generates a binary image by comparing a pixel of an original continuous-tone image with a threshold value of the periodic ordered matrix. The methods based on ordered dither are classified into two categories, *clustered-dot ordered dither*, and *dispersed-dot ordered dither* depending on the nature of generated dots.

Clustered-dot ordered dither turns adjacent pixels *on* which form a cluster in the matrix. The period length of dots is determined by that of the matrix. The tone level of a region is modulated by the area size of clustered-dot. Thus,

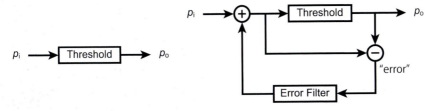

FIGURE 4.6
Diagrams of point process and error diffusion.

this method is categorized as an *amplitude modulation* (AM) technique. A spiral-dot screen, whose sample threshold matrix is shown in Figure 4.5 left, is a kind of cluster-dot ordered dither.

Dispersed-dot ordered dither turns scattered pixels *on* without making any clusters. The result may have a high-frequency fidelity and better appearance in constant gray regions. Thus, people prefer dispersed-dot ordered dither to clustered-dot ordered dither. The tone level of a region is modulated by the density (or frequency) of dots. Thus, this method is categorized as a *frequency modulation* (FM) technique. A famous Bayer's matrix shown in Figure 4.5 right is an example of dispersed-dot ordered dither.

Error Diffusion

The above-mentioned approaches are said to be *point process* in a sense that each point (or pixel) is processed independently. In other words, an output value of a pixel depends only on a value of the input pixel as shown in Figure 4.6 left. *Error diffusion*, a commonly-used halftoning algorithm, takes a neighborhood into account so that it is no more a point process. It is an adaptive algorithm that uses the threshold error feedback to produce patterns having different spatial frequency content. A single pass is carried out over the input image each pixel of which is processed sequentially. A single pixel process consists of a binary thresholding of the input pixel and an error computation caused by the binarization. This error is distributed to the neighboring pixels that have not been processed according to an error filter (or error matrix). Famous error filters proposed by Floyd and Steinberg [8] and Jarvis et al. [16] are shown in Figure 4.7 where X indicates the current pixel. In other words, the values of neighboring pixels are corrected to keep the total tone of a local region. The schematic diagram of the algorithm is illustrated in Figure 4.6 right.

Iterative and Search-Based Methods

Iterative and search-based methods attempt to obtain the optimum solution as a halftoned image by iteration and/or search manner. Since the ultimate goal of digital halftoning is to accomplish an illusion of a binary image that is

	X	7/16
3/16	5/16	1/16

		X	7/48	5/48
3/48	5/48	7/48	5/48	3/48
1/48	3/48	5/48	3/48	1/48

FIGURE 4.7
Samples of error filters for error diffusion proposed by Floyd and Steinberg [8] (left) and Jarvis et al. [16] (right).

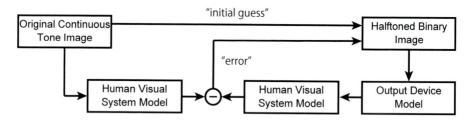

FIGURE 4.8
A diagram of iterative and search-based method.

observed completely same as the original continuous-tone image, the method tries to minimize the perceived difference between the binary image output and the original image. The perceived error is estimated by a spatial filter that simulates a human visual system as well as the output device model. Direct binary search is one of the popular methods of this kind. The schematic diagram of this method is illustrated in Figure 4.8.

4.3.2 Image Quality and Related Parameters

Let us discuss quality of resulting images in terms of its parameters in this section.

Tone levels (l) Due to the nature of the visual cryptography scheme, the input images, namely, the original images as a secret image and shadow images, must be binary. This is a big limitation in the sense of image quality, especially for photograph images. Halftoning allows us to convert a continuous-tone image to a binary image that can be observed similar to the original image by human eyes. However, the halftoned image is not completely the same as the original image. It may lose a certain quality of photograph images. On the contrary, some types of images, such as images of logos or text, are still meaningful even if the tone levels are limited to two or three.

Pixel expansion (m) Pixel expansion is also an important parameter that affects quality of images as well as its data size. The resulting image requires m times more subpixels, which means that subpixels must be m

times smaller than the original pixel if the image size is fixed to that of the original image. This results in difficulty of alignment so that transparencies with smaller subpixels are more difficult to be stacked properly. Misaligned transparencies cannot reconstruct the secret image. This cannot be neglected because one of the most important characteristics of visual cryptography is the capability that the secret information is revealed by simply stacking transparencies without any computation.

Relative difference (α) It is obvious that contrast is also one of the most important parameters related to image quality. An image with low contrast is obscure and difficult to see its details. Furthermore, there exists a certain tradeoff between contrasts of shadow and secret images in case of extended visual cryptography. It is impossible to increase both contrasts of a secret image and shadow images simultaneously.

4.3.3   Photograph Visual Cryptography with Basic Schemes

The straightforward way to incorporate photograph images into visual cryptography is as below:

1. Convert photograph (continuous-tone) images to binary images by halftoning.

2. Encrypt a secret image by one of the schemes explained in Section 4.2.

Of course the quality of resulting images may be changed by the halftoning algorithm. But here we would like to focus on the differences among the encryption schemes.

Table 4.1 summarizes the characteristics of visual cryptography with photograph images according to the encryption schemes, i.e., Visual Secret Sharing Scheme (VSSS), Extended Visual Cryptography Scheme (EVCS), and Random Grids (RG), in the case of $(2, 2)$. Since all three schemes assume that shadow images are printed on transparencies and stacked together, the superimposition (stacking operation) can be seen as Boolean "OR" in mathematical sense. The basic properties of VSSS and RG are very similar except for the pixel expansions. The pixel expansion of $(2, 2)$ VSSS is $m = 2$ (it would be 4 if one wants to preserve the aspect ratio of a image), while RG's pixel expansion is $m = 1$. VSSS and RG have no extended capability. They cannot incorporate photograph images into shadow images in order to conceal the existence of "secret," which means the original shadow images are simple monotone images ("mono.") and encrypted results are random-dot binary images ("rand."). In other words, the relative difference of shadow images is $\alpha_S = 0$. However, they can accept a continuous-tone image ("cont.") as a secret image. A reconstructed secret image is a halftoned binary image ("half.") whose relative difference is $\alpha_R = \frac{1}{2}$. Only EVCS among the three basic schemes can incorporate continuous-tone images into shadow images. The resulting shadow images

TABLE 4.1
Comparison of visual cryptography schemes, namely, VSSS, EVCS, and RG, with photograph images in the case of $(2,2)$. Here, m denotes the pixel expansion of each scheme. α_S and α_R stand for relative differences of shadow images and reconstructed secret image, respectively. $l_S^{(O)}$ and $l_R^{(O)}$ denote the tone levels of original images to be processed, while l_S and l_R denote the tone levels of the resulting (encrypted/decrypted) images. Strictly speaking, l_S and l_R should be 2, because every resulting image consists of white and black subpixels. However, for human visual system, an image can be observed as a gray-scale image because of halftoning. In this sense, we specify "$l_S > 2$" for EVCS and "$l_R > 2$" for all three schemes. "mono.," "cont.," "half.," and "rand." mean a monotone image, continuous-tone image, halftoned binary image, and random-dot binary image, respectively.

scheme	m	shadow image			secret image		
		α_S	$l_S^{(O)}$	l_S	α_R	$l_R^{(O)}$	l_R
VSSS	2 (4)	0	1 (mono.)	1 (rand.)	1/2	∞ (cont.)	>2 (half.)
EVCS	4	1/4	∞ (cont.)	>2 (half.)	1/4	∞ (cont.)	>2 (half.)
RG	1	0	1 (mono.)	1 (rand.)	1/2	∞ (cont.)	>2 (half.)

to be printed on transparencies and reconstructed secret image by stacking shadow images are binary images converted by halftoning. The pixel expansion of $(2,2)$ EVCS is $m = 4$. A tradeoff between relative differences of shadow image and reconstructed image exists. If we restrict both relative differences to be the same, the maximum relative differences are $\alpha_S = \alpha_R = \frac{1}{4}$.

4.4 Variations of Photograph Visual Cryptography

Although extended capability is a crucial aspect for visual cryptography with a photograph image, contrast of resulting images becomes very low ($\frac{1}{4}$ at maximum) due to the tradeoff between relative differences of the shadow image and reconstructed image. Contrast tends to much lower if the number of shadow images increases. In order to enhance contrast of encrypted images, several researches assumed other kinds of operations for superimposition, namely, "Cover," [32] "XOR," [3, 23] "NOT," etc. [40] Sometimes even a certain computation is required for decryption [5, 20, 24]. Most of the approaches are no longer realized by stacking transparencies.[2] Those schemes may not be categorized as visual cryptography, because the most important characteristics of visual cryptography is the capability of visual decryption without any computation. Therefore, we will not discuss these type of approaches assuming other operations than Boolean "OR." Instead our discussion will mainly focus on $(2, 2)$ schemes. There have been a lot of studies that aim at incorporating color into visual cryptography [27, 35, 39, 22, 44, 12, 13, 7, 30, 36, 46, 19]. These approaches are strongly related to the techniques for handling continuous-tone images. However, this chapter will not discuss those color studies because this book contains a special chapter dedicated to color visual cryptography.

4.4.1 Approaches to Photograph Visual Cryptography

The main issue for incorporating photographs into visual cryptography is the quality of resulting images, i.e., pixel expansion, relative differences, tone levels, as we discussed in Section 4.3.2. A lot of approaches to photograph visual cryptography intend to improve the image quality by introducing a certain limitation and/or by exploiting image processing techniques. Table 4.2 summarizes those approaches. For instance, [4, 15, 37, 14, 6] limit their shadow image to a random-dot binary image ("rand."). Actually they do not intend extended visual cryptography [33, 34, 21, 41, 9, 47]. Generate very similar images ("sim.") or a positive/negative pair of images ("p/n") as encrypted shadow images. Sometimes only logo-like images with the trace of shadows ("logo tr.") can be reconstructed as a secret image [33, 34, 21, 41, 9, 10, 45]. [10, 25, 42] utilize halftoning techniques to make pixel expansion $m = 1$. [28, 42, 45, 25, 26] adjust tone, i.e., dynamic range, of images for improving image quality. Some studies introduce continuous-tone subpixels into encrypted shadow images to obtain continuous-tone results [45, 29]. The rest of this section explains those approaches.

---

[2]Only "Cover" can be physically realized with transparencies and opaque sheets. Physical implementation of "XOR" is possible by exploiting polarization.

TABLE 4.2

Approaches to photograph visual cryptography of $(2,2)$ scheme. "pair" represents constraints on a pair of shadow images such as random-dot binary image ("rand."), similar images ("sim."), positive/negative pair ("p/n"), and different pair, i.e., no constraint ("diff."). "tone" indicates types of images to be reconstructed. "logo" stands for a logo-like image, "logo tr." stands for a logo-like image with traces of shadows, and "gray" stands for a gray-scale image. "GAS," "RG," "not ext.," and "cont. tone" mean General Access Structure, Random Grids, not extended, and continuous-tone, respectively.

approaches	m	shadow image			secret image			features
		pair	α_S	l_S	tone	α_R	l_R	
[4, 15]	>1	rand.	0	1	gray	$1/2$	$<m$	GAS, not ext.
[37]	1	rand.	0	1	gray	$1/2$	>2	RG, not ext.
[14, 6]	1	rand.	0	1	gray	$1/2$	>2	(k,n), not ext.
[33]	>1	sim.	$\simeq 1$	$m+1$	logo tr.	≥ 0	$-$	density pattern
[34, 21, 41]	1	sim.	$\simeq 1$	>2	logo tr.	≥ 0	$-$	ordered dither
[9]	1	sim.	$\simeq 1$	>2	logo tr.	≥ 0	$-$	error diffusion
[47]	1&4	p/n	$\leq 1/2$	>2	logo	$1/4$	2	GAS, w/o trace
[10]	1	diff.	≤ 1	>2	logo tr.	≥ 0	$-$	error diffusion
[45]	4	diff.	$1/4$	∞	logo tr.	$1/4$	$-$	GAS, cont. tone
[28]	>1	diff.	$\geq 1/4$	$<m$	gray	$\geq 1/4$	$<m$	density pattern
[42]	1	diff.	$\geq 1/4$	>2	gray	$\geq 1/4$	>2	iterative search
[25, 26]	1	diff.	$\geq 1/4$	>2	gray	$\geq 1/4$	>2	error diffusion
[29]	2	diff.	$\geq 1/4$	∞	gray	$\geq 1/4$	∞	cont. tone

4.4.2 Random-Dot Shadow Images

Due to the severe tradeoff between relative differences of the shadow image and reconstructed image, some studies focused on the quality of the reconstructed image by giving up extended capability. For instance, in 2000, Blundo et al. examined a secret sharing scheme with an access structure that can reconstruct a gray-scale image with g gray levels and specified the upper limit of relative differences $\alpha_1, \cdots, \alpha_{g-1}$ as well as the lower limit of pixel expansion m as below [4]:

$$\min\{\alpha_1, \cdots, \alpha_{g-1}\} \leq \frac{1}{(g-1)2^{k-1}}, \quad m \geq (g-1)2^{k-1}.$$

Iwamoto and Yamamoto precisely discussed an (n, n) secret sharing scheme for a gray-scale image in 2002 [15].

Another important concern is pixel expansion. Both studies explained above accomplish gray levels by controlling the number of white/black subpixels, which means pixel expansion is inevitable. The research for reducing the pixel expansion in VSSS is mostly based on a probabilistic approach such as Random Grids revisited by Shyu [37]. Since Random Grids have been already explained in Section 4.2.3, here we discuss another kind of probabilistic approach. Ito et al. proposed a secret sharing scheme with $m = 1$ by introducing randomness into a conventional (k, n) VSSS in 1999 [14]. Their scheme first determines a basis matrix according to the value of a secret pixel, black or white. Then it randomly selects one of a column of the basis matrix, which stands for values of corresponding subpixels in shadow images, and uses them as those of shadow pixels. Chen et al. proposed the similar scheme in 2007 [6]. They also proposed to use histogram equalization for enhancing contrast. We will explain this contrast enhancement technique later in Section 4.4.7.

4.4.3 Similar Shadow Images

Another approach is entirely opposite to the studies explained in the previous section. It limits the relative difference of secret image to nearly zero and attempts to enhance the relative difference of shadow images as much as possible. Some research has succeeded in achieving a full relative difference for shadow images, namely, $\alpha_S \simeq 1$, by using very similar shadow images. The key of this approach is the reconstructed secret image. The secret image can be observed with the trace of a shadow image. In other words, by overlapping very similar shadow images, one can observe obscure dark logos or text within the shadow image.

This type of scheme was first proposed by Oka et al. in 1996 as a watermarking technique [33]. One can conceal one's signature within a halftoned image so that one can claim one's copyright on the image. It uses multiple subpixels, i.e., density patterns, to represent a gray pixel of the original shadow image. The second shadow image is generated by rearranging dots, i.e., black

0	8	2	10
12	4	14	6
3	11	1	9
15	7	13	5

8	0	10	2
4	12	6	14
11	3	9	1
7	15	5	13

FIGURE 4.9
Dither matrices for similar shadow scheme proposed in [34].

subpixels, of the first shadow image if the corresponding original secret pixel is black. On the contrary, the same arrangement is used if the corresponding secret pixel is white. The secret image appears when the shadow images are stacked together. It is easy to rearrange subpixels if the gray level of the original pixel is approximately 50% gray. But it is difficult to control resulting darkness in the nearly white or black region of the shadow image. Thus, it is impossible to guarantee the quality of the reconstructed secret image.

[34, 21, 41] are modifications of this approach without pixel expansion, namely, $m = 1$, by using ordered dither. For instance, [34] specified sample dither matrices based on a Bayer's matrix as shown in Figure 4.9. Shadow images are halftoned by using the same dither matrix if the corresponding secret region is white, while they are halftoned with the different dither matrices if the corresponding secret region is black. Fu and Au proposed a variation of this scheme in 2001 [9]. It uses the error diffusion technique to get halftoned results instead of ordered dither.

4.4.4   Positive and Negative Shadow Images

The underlying concept of this approach is quite similar to the previous approach. It can achieve a large relative difference for shadow images by using a positive and negative pair of shadow images instead of a similar pair. It also reconstructs a logo or text image as a secret image, i.e., binary-tone image, without any trace of shadow images.

This scheme was proposed by Zhou et al. in 2006 [47]. It is not limited to $(2, 2)$ VSSS and can handle an access structure. However, here we explain a basic algorithm to establish a $(2, 2)$ secret sharing scheme due to space limitations.

1. The positive shadow image is generated by halftoning the original shadow image. The negative shadow image is obtained by reversing the positive one. Thus, the overlapping result is entirely black at this moment.

2. A secret pixel is encrypted into a square region of halftoned pixels, $Q_1 \times Q_2$. A pair of black and white pixels, referred to as *secret*

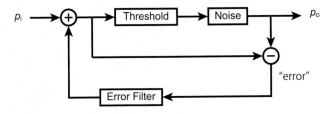

FIGURE 4.10
The conjugate error diffusion algorithm proposed in [10].

information pixels, are selected in each region. The secret information pixels are at the same positions in the shadow images. Therefore, corresponding pairs are complementary or a reversal of each other. One of the secret information pixels are swapped to obtain a brighter result if the corresponding secret pixel is white.

The shadow image is free from pixel expansion, $m = 1$, while the pixel expansion of a secret image is $m = Q_1Q_2$. It uses the void and cluster algorithm to select secret information pixels to keep better image quality. However, it requires a black and white pair within a square region to encrypt a secret image, the relative difference of shadow image is a matter of the size of a square region. The shadow images must be grayish if the square region is relatively small.

4.4.5 Error and Secret Diffusion

This approach considers secret information as extra noise to the shadow image and takes into account binarizing error. It can realize a $(2,2)$ EVCS with completely different shadow images. This approach was first proposed by Fu and Au in 2003 [10]. The basic algorithm is as below:

1. The first shadow image is halftoned by the error diffusion algorithm.

2. The second shadow image is also halftoned by the conjugate error diffusion algorithm shown in Figure 4.10. The "Noise" is added according to the corresponding pixels of the secret image and the first halftoned shadow image. This noise causes a pixel-wise distortion to the second shadow image, which is controlled by some appropriate threshold T_1. A large T_1 allows more pixels to hide the secret image but results in a large distortion of the second shadow image.

This approach is free from pixel expansion, $m = 1$. However, [10] originally was applied to a logo image as a secret image in order to avoid a huge distortion in the second shadow image. It could reconstruct only a faint logo with the traces of both shadow images. One can recognize a logo as well as shadow images at the same time.

Myodo et al. extended this approach to be able to handle a photograph as a secret image [25]. It can reconstruct a secret photograph image with very little or no trace of shadow images by properly adjusting tones of images. This aspect of tone adjustment will be discussed in Section 4.4.7.

4.4.6 Simultaneous Iterative Search

This type of method was proposed by Wu et al. in 2004 [42]. It can handle three different photographs as two shadow images and a secret image without pixel expansion, $m = 1$. This method consists of two major steps, tone adjustment and simultaneous iterative search. It first adjusts the tones of three input images to satisfy a condition on relative differences of shadow and secret images. Then it simultaneously searches three halftoned images. Since [42] contains very little explanation about the two steps, it is hard to know the exact algorithm. But it would take a certain amount of time to obtain a result if it process is images in a brute-force manner.

4.4.7 Tone Adjustment

This type of approach attempts to improve image quality with image processing technique. It tries to enhance contrasts, namely, dynamic ranges, of the resulting images as much as possible. The conventional visual cryptography studies consider relative differences, α_S and α_R, which represent a limitation of possible pixel values. However, the pixel values of shadow and secret images are actually limited by lower and upper limits, *dynamic ranges*, and there is a certain interaction among them.

Nakajima and Yamaguchi precisely examined the interactions of pixel values in $(2, 2)$ EVCS [28]. There exist constraints among the pixel values of three corresponding pixels in shadow and secret images. Let us call the three corresponding pixels a *triplet*. The constraints among values of a triplet are represented as below:

$$o_R \in [\max(o_1, o_2), \min(o_1 + o_2, 1)], \qquad (4.1)$$

where o_R denotes pixel opacity of reconstructed secret image and o_1 and o_2 denote pixel opacities of resulting shadow images.[3] This expression indicates that any reconstructed pixel must be equal to or more opaque than the most opaque corresponding shadow pixel, $\max(o_1, o_2)$. It also indicates that the reconstructed pixel must be equal to or less opaque than the sum of opacities of corresponding shadow pixels, $o_1 + o_2$.

---

[3]In [28], Nakajima and Yamaguchi discussed pixel transparency instead of opacity. However, as we already indicated, people usually use 0 for a transparent (white) pixel and 1 for an opaque (black) pixel in visual cryptography studies. Thus, here we consider pixel opacity rather than pixel transparency.

This approach adjusts the tones of given images to make the dynamic ranges as large as possible while every triplet fulfills the constraints given by Equation (4.1). Affine transformation or piece-wise linear transformation is most commonly used for tone adjustment [28, 42, 45, 25, 26]. Wu et al. suggested to calculate optimum parameters [42]. However, [42] does not explain any details how to obtain optimum parameters. Myodo et al. [25] proposed a method that can determine optimum parameters at once [26]. They claimed that their method can enhance relative differences to 0.28 on average without any violation. Their method can control the relative differences independently by specifying the weights.

Another approach uses a contrast enhancement technique called *histogram equalization* or *histogram linearization* transformation [11], which is very well-known for improving the contrast of images. The *histogram* of an image can be seen as a function $h(i)$ that returns a frequency or probability density of pixels having an intensity level i, namely a transparency. Histogram equalization equalizes or flattens a histogram. This means that frequencies of tone levels are totally uniform and the resulting image may have a high contrast. Chen et al. [6] as well as Wu et al. [43] suggested a way to improve image quality by applying histogram equalization to the input images before encryption.

4.4.8 Continuous-Tone Subpixel

Image quality can be improved by increasing tone levels as discussed in Section 4.3.2. There have been some studies improving image quality by introducing continuous-tone subpixels into encrypted shadow images [29, 45].

Yang and Chen [45] introduced continuous tone into the resulting shadow images. They extended usual EVCS explained in Section 4.3.3 by substituting a black subpixel by a gray subpixel having the same gray value as the original shadow pixel. This approach can be applied to EVCS with an access structure. The drawback of this approach is the trace of shadow images. The reconstructed secret image can be observed in superimposed shadow images. Thus, a secret image should be a logo or text image. One can recognize a logo as well as shadow images at the same time when shadow images are overlapped.

Nakajima and Yamaguchi [29] proposed a very unique approach for improving image quality by introducing continuous-tone subpixels. Their method also deals with a misalignment problem caused by pixel expansion. We will discuss this method in the next section.

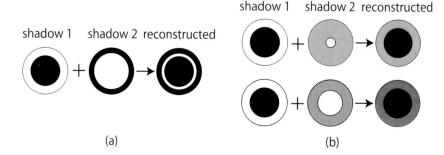

(a)   (b)

FIGURE 4.11
Examples of subpixel arrangements with enhanced misalignment tolerance.
If the two subpixels take only binary values as in (a), there is no way to
control the secret pixel transparency. The reconstructed transparency can be
controlled by allowing a continuous value for one of the four subpixels as in
(b).

4.5   Misalignment-Tolerant Photograph Visual Cryptography

Nakajima and Yamaguchi [29] attempt to improve the image quality while
maintaining the misalignment tolerance. Their method generates encrypted
shadow images that are robust to the misalignment error while the resulting
images may have continuous-tone levels. In this section, we use transparency
t rather than opacity o in order to simplify the equations. One should note
that transparency and opacity are simply inverse, $t = 1 - o$.

4.5.1   Theory and Implementation

The basic idea of their method is to use two concentric regions of variable area.
In other words, it virtually uses two concentric subpixels whose relative sizes
may change. Here, we treat a pixel as a unit circle instead of a square to make
the explanation simple. The most enhanced misalignment tolerance between
the corresponding pixels would be achieved by this concentric arrangement of
subpixels as shown in Figure 4.11 (a). However, such a subpixel arrangement
with binary values allow only one degree of freedom even if the area changes,
i.e., the radius of the inner circle, for each shadow pixel. This is enough for
realizing the transparency of each shadow pixel but is insufficient to control
the reconstructed secret pixel transparency.

To increase the degree of freedom to control the reconstructed transparency, they extended the halftoning technique by introducing continuous
(gray-scale) values rather than just binary values to the concentric subpixels.

(a)   (b)   (c)

FIGURE 4.12
Variation of subpixel arrangement having the same transparency. To maintain an average transparency shown in (a), the inner circle must be more opaque than the original transparency if the outer ring is completely transparent as in (b). Similarly, if the outer ring is completely opaque, the inner circle must become more transparent than the original transparency as in (c).

There are four subpixels in the corresponding shadow pixels: one inner circle and one outer ring for both pixels. A gray-scale value is assigned to one of the four subpixels, and binary values to other three subpixels. Figure 4.11 (b) shows an example of assigning a gray value to the outer ring of shadow 2. A variety of reconstructed transparencies are obtained by changing the gray value and the radius of the inner white circle simultaneously.

Let $i_j, o_j \in [0, 1]$ be the transparencies of the inner circle and the outer ring, respectively, of shadow $j \in \{1, 2\}$, and $r_j \in [0, 1)$ be the radii of the inner circles, respectively. The transparency of shadow j is written as

$$t_j = (1 - r_j{}^2)o_j + r_j{}^2 i_j.$$

If we suppose $r_1 \geq r_2$, the reconstructed transparency t_R becomes

$$t_R = (1 - r_1{}^2)o_1 o_2 + (r_1{}^2 - r_2{}^2)i_1 o_2 + r_2{}^2 i_1 i_2,$$

because the superimposed transparency is calculated by taking the product of transparencies.

The encryption corresponds to calculating $i_1, i_2, o_1, o_2, r_1, r_2$ for given t_1, t_2, t_R. Here, only one of i_1, i_2, o_1 and o_2 takes a value within the interval $[0, 1]$, while the remaining three are either 0 or 1. If all four subpixels take binary values, i.e., $i_1, i_2, o_1, o_2 \in \{0, 1\}$, there are four possible cases shown as below:

(a) $i_1 = 0,$ $o_1 = 1,$ $i_2 = 0,$ $o_2 = 1,$
(b) $i_1 = 0,$ $o_1 = 1,$ $i_2 = 1,$ $o_2 = 0,$
(c) $i_1 = 1,$ $o_1 = 0,$ $i_2 = 0,$ $o_2 = 1,$
(d) $i_1 = 1,$ $o_1 = 0,$ $i_2 = 1,$ $o_2 = 0.$

Now consider the other cases where either one of i_1, i_2, o_1, o_2 takes a continuous value. For example, suppose the inner circle of shadow 1 is gray, namely, $i_1 \in (0, 1)$. When the outer ring is completely transparent as shown in Figure 4.12 (b), the gray level of the inner circle, i_1, must be more opaque than the

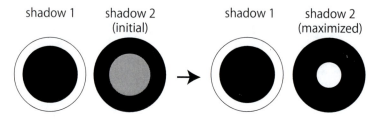

shadow 1   shadow 2   shadow 1   shadow 2
            (initial)   (maximized)

FIGURE 4.13

An example of pattern change for maximizing the difference of inner circles'
radii, $r_1 - r_2$.

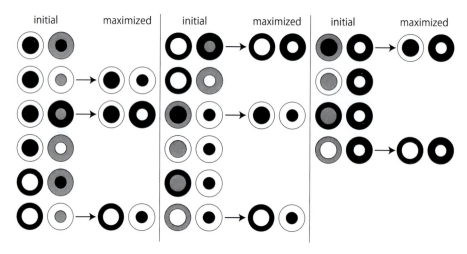

FIGURE 4.14

All the possible pattern combinations of subpixel arrangements and the results
of maximizing the difference, $r_1 - r_2$.

original transparency t_1 to preserve the average transparency of shadow 1.
Otherwise, the average pixel transparency exceeds t_1. Similarly, i_1 must become more transparent than t_1 if the outer ring is completely opaque as shown
in Figure 4.12 (c). Therefore, when $i_1 \in (0, 1)$, there are four possibilities as
below:

$$
\begin{array}{llll}
\text{(i)} & 0 < i_1 \leq t_1, & o_1 = 1, & i_2 = 0, & o_2 = 1, \\
\text{(ii)} & 0 < i_1 \leq t_1, & o_1 = 1, & i_2 = 1, & o_2 = 0, \\
\text{(iii)} & t_1 \leq i_1 < 1, & o_1 = 0, & i_2 = 0, & o_2 = 1, \\
\text{(iv)} & t_1 \leq i_1 < 1, & o_1 = 0, & i_2 = 1, & o_2 = 0.
\end{array}
$$

There are $16(= 4 \times 4)$ possibilities, if either one of $i_1, i_2, o_1,$ and o_2 exclusively can take a value of $(0, 1)$. One may hypothesize that the difference
of the radii of two inner circles determines the misalignment tolerance. The
misalignment tolerance is enhanced by maximizing the difference of the radii,

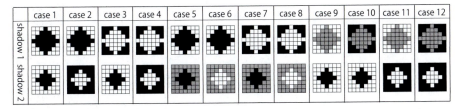

FIGURE 4.15
Physical implementation of the concentric subpixel arrangements using square patterns.

$r_1 - r_2$. For example, Figure 4.13 illustrates a case where the inner circle of shadow 2 becomes smaller for maximizing the difference by being completely transparent. Half of the sixteen cases amount to the first four binary cases by this difference maximization as illustrated in Figure 4.14. Thus, the number of possible cases after the maximization is 12 $(= 4 + 16 - 8)$.

The case whose difference, $r_1 - r_2$, is the largest among the 12 cases should be adopted in order to maximize the misalignment tolerance. However, using such cases results in patches of the same patterns in the shadow images. In such situations, the patch boundaries are noticeable. Pattern changes may be caused by the other two pixels of the pixel's corresponding triplet. Thus, the boundary may imply the information of the secret image as well as the other shadow image. To avoid this unpleasant visual effect, a case should be chosen in a weighted random manner using the difference, $r_1 - r_2$, as the weight.

In the actual procedure, a pixel is implemented by tiny subpixels aligned in a square. The inner circle is approximated by subpixels aligned in a square tilted by 45 degrees as shown in Figure 4.15. This is because the human eye's sensitivity to artifacts produced by a periodic pattern is the least when the periodicity axis makes an angle of about $45°$ or $-45°$ with the horizontal direction. The number of subpixels forming the inner square is determined by the area of the inner circle, r_j^2.

The entire process of encryption is as below:

1. Take three input gray-scale images and adjust their tones to alleviate the constraint condition on triplet value (4.1).

2. Generate two encrypted shadow images by processing each pixel triplet:

   (a) choose one of the cases shown in Figure 4.15 in a weighted random manner and

   (b) generate the actual subpixel arrangements according to the fixed parameters, i_j, o_j, and r_j $(j \in \{1, 2\})$.

FIGURE 4.16
Input images. From left to right, original shadow 1, shadow 2, and secret images, respectively.

4.5.2 Results

Figure 4.16 shows the three input images. Each image corresponds to the shadow 1, shadow 2, and secret image, respectively, from left to right. Figure 4.17 gives the encrypted shadow 2 (above) and the secret image reconstructed by superimposing the output shadows (below). They are generated with 15×15 physical subpixels per pixel. For comparison, the shadow 2 (above) and reconstructed secret images (below) generated with the straightforward density pattern are depicted in Figure 4.18. The images in Figure 4.18 contain 3×3 subpixels per pixel, a practical number of subpixels considering the superimposition by human hands.

The method is possible to generate quite pleasant results of high image quality, especially with smoother shading of the petals and the background stems, or the whiskers and stripes of the cat. This is because the method can express at least $226 = (15 \times 15 + 1)$ gray levels, which means that the resulting images can almost fully express the gray levels of the input images. Moreover, the encrypted shadow images can be superimposed by human hands with little difficulty, as they allow more misalignment tolerance.

4.6 Conclusions

This chapter explained extended visual cryptography handling photograph images. First, we overlooked the three basic visual cryptography schemes for binary images, namely, the Visual Secret Sharing Scheme, the Extended Visual Cryptography Scheme, and Random Grids. Some fundamental concepts for incorporating photographs into visual cryptography, such as halftoning techniques, some parameters related to image quality, and issues for handling

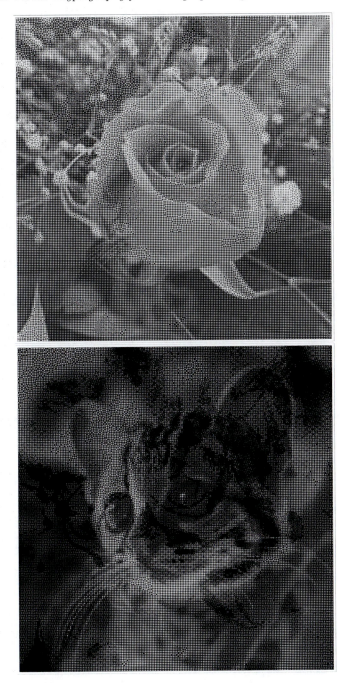

FIGURE 4.17
Examples of resulting images. The upper image is the encrypted shadow 2
and the lower image is the secret image reconstructed by superimposing two
shadows by computer simulation.

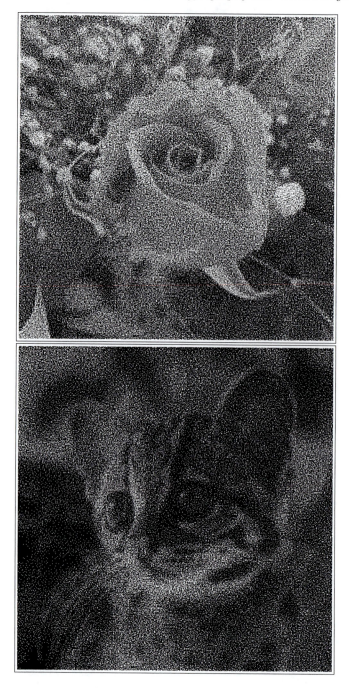

FIGURE 4.18
Examples of the output with density pattern using 3×3. The upper image is the encrypted shadow 2 and the lower image is the secret image reconstructed by superimposing the two shadows by computer simulation.

photographs are observed. Then we surveyed approaches for handling photograph images, most of which are aimed at improving image quality. Finally, we have discussed one the most unique approaches proposed by Nakajima and Yamaguchi [29]. Although the method does not guarantee perfect security, it can generate very pleasant results of high image quality while maintaining misalignment tolerance.

Bibliography

[1] G. Ateniese, C. Blundo, A. De Santis, and D.R. Stinson. Visual cryptography for general access structure. *Information and Computation*, 129:86–106, 1996.

[2] G. Ateniese, C. Blundo, A. De Santis, and D.R. Stinson. Extended capabilities for visual cryptography. *Theoretical Computer Science*, 250:143–161, 2001.

[3] E. Biham and A. Itzkovitz. Visual cryptography with polarization. In *The Rump Session of EUROCRYPTf98*. Available at http://www.cs.technion.ac.il/~biham/Reports/visual.ps.gz, 1998.

[4] C. Blundo, A. De Santis, and M. Naor. Visual cryptography for grey level images. *Information Processing Letters*, 75:255–259, 2000.

[5] C.-C. Chang and T.-X. Yu. Sharing a secret gray image in multiple images. In *International Symposium on Cyber Worlds*, pages 230–237. 2002.

[6] Y.-F. Chen, Y.-K. Chan, C.-C. Huang, M.-H. Tsai, and Y.-P. Chu. A multiple-level visual secret-sharing scheme without image size expansion. *Information Sciences*, 177(21):4696–4710, 2007.

[7] S. Cimato, R. De Prisco, and A. De Santis. Colored visual cryptography without color darkening. In *Lecture Notes in Computer Science 3352 (Intl. Conference on Security in Communication Networks)*, pages 235–248, 2005.

[8] R.W. Floyd and L. Steinberg. An adaptive algorithm for spatial grayscale. In *Society for Information Display*, 17(2):75–77, 1976.

[9] M.S. Fu and O.C. Au. Data hiding in halftone images by stochastic error diffusion. In *Intl. Conference on Acoustics, Speech, and Signal Processing*, pages 1965–1968, 2001.

[10] M.S. Fu and O.C. Au. A novel method to embed watermark in different halftone images: Data hiding by conjugate error diffusion (DHCED). In *Intl Conference on Acoustics, Speech, and Signal Processing*, volume III, pages 529–532, 2003.

[11] R.C. Gonzalez and R.E. Woods. *Digital Image Processing, Third Edition*. Pearson Education International, 2010.

[12] Y.-C. Hou. Visual cryptography for color images. *Pattern Recognition*, 36(7):1619–1629, 2003.

[13] T. Ishihara and H. Koga. A visual secret sharing scheme for color images based on meanvalue-color mixing. *IEICE Trans. on Fundamentals*, E86-A(1):194–197, 2003.

[14] R. Ito, H. Kuwakado, and H. Tanaka. Image size invariant visual cryptography. *IEICE Trans. on Fundamentals*, E82-A(10):2127–2177, 1999.

[15] M. Iwamoto and H. Yamamoto. The optimal n-out-of-n visual secret sharing scheme for gray-scale images. *IEICE Trans. on Fundamentals*, E85-A(10):2238–2247, 2002.

[16] J.F. Jarvis, C.N. Judice, and W.H. Ninke. A survey of techniques for the display of continuous tone pictures on bilevel displays. *Computer Graphics and Image Processing*, 5(1):13–40, 1976.

[17] O. Kafri and E. Keren. Image encryption by multiple random grids. *Optics Letters*, 12(6):377–379, 1987.

[18] H.R. Kang. *Digital Color Halftoning*. SPIE Optical Engineering Press, IEEE Press, 1999.

[19] I. Kang, G.R. Arce, and H.-K. Lee. Color extended visual cryptography using error diffusion. In *Intl. Conference on Acoustics, Speech and Signal Processing*, pages 1473–1476, 2009.

[20] H.J. Kim and Y. Choi. A new visual cryptography using natural images. In *Intl. Symposium on Circuits and Systems*, pages 5537–5540, 2005.

[21] K.T. Knox. *Digital Watermarking Using Stochastic Screen Patterns*. US Patent No. 5,734,752, 1998.

[22] H. Koga and H. Yamamoto. Proposal of a lattice-based visual secret sharing scheme for color and gray-scale images. *IEICE Trans. on Fundamentals*, E81-A(6):1262–1269, 1998.

[23] S.-S. Lee, J.-C. Na, S.-W. Sohn, C. Park, D.-H. Seo, and S.-J. Kim. Visual cryptography based on an interferometric encryption technique. *ETRI Journal*, 24(5):373–380, 2002.

[24] R. Lukac and K.N. Plataniotis. Bit-level based secret sharing for image encryption. *Pattern Recognition*, 38(5):767–772, 2005.

[25] E. Myodo, K. Takagi, S. Miyaji, and Y. Takishima. Halftone visual cryptography embedding a natural grayscale image based on error diffusion technique. In *Intl. Conference on Multimedia and Expo*, pages 2114–2117, 2007.

[26] E. Myodo, K. Takagi, and A. Yoneyama. Auto tone mapping method to generate and decode high contrast halftone images in visual cryptography. *IEICE Technical Report*, IE2009-53, 109(149):25–30, 2009. (in Japanese)

[27] D. Naccache. Colorful Cryptography—a purely physical secret sharing scheme based on chromatic filters. In *The French-Israeli Workshop on Coding and Information Integrity*. Available at http://www.gemplus.com/smart/rd/publications/ps/Nac94col.ps. 1994.

[28] M. Nakajima and Y. Yamaguchi. Extended visual cryptography for natural images. *Journal of WSCG*, 10(2):303–310, 2002.

[29] M. Nakajima and Y. Yamaguchi. Enhancing registration tolerance of extended visual cryptography. *Journal of Electronic Imaging*, 13(3):654–662, 2004.

[30] M. Nakajima and Y. Yamaguchi. Device-dependent color neutralization method. In *Color Imaging X*, SPIE volume 5667, pages 581–588, 2005.

[31] M. Naor and A. Shamir. Visual cryptography. In *Lecture Notes in Computer Science 950 (Advances in Cryptology- EUROCRYPTf94)*, pages 1–12, 1994.

[32] M. Naor and A. Shamir. Visual cryptography II: Improving the contrast via the cover base. In *Lecture Notes in Computer Science 1189 (Security Protocols)*, pages 197–202, 1997.

[33] K. Oka, Y. Nakamura, and K. Matsui. Embedding signature into a hardcopy image using micro-patterns. *IEICE Trans. on Information and Systems*, J79-D-II(9):1624–1626, 1996. (in Japanese)

[34] K. Oka, Y. Nakamura, and K. Matsui. Embedding signature into a hardcopy image of dithered image. *IEICE Trans. on Information and Systems*, J80-D-II(3):820–823, 1997. (in Japanese)

[35] V. Rijmen and B. Preneel. Efficient colour visual encryption or "Shared colors of Benetton." In *The Rump Session of EUROCRYPTf96*. Available at http://www.iacr.org/conferences/ec96/rump/preneel.ps.gz, 1996.

[36] S.J. Shyu. Efficient visual secret sharing scheme for color images. *Pattern Recognition*, 39(5):866–880, 2006.

[37] S.J. Shyu. Image encryption by random grids. *Pattern Recognition*, 40(3):1014–1031, 2007.

[38] R. Ulichney. *Digital Halftoning*. The MIT Press, 1987.

[39] E.R. Verheul and H.C.A. van Tilborg. Constructions and properties of k out of n visual secret sharing schemes. *Designs, Codes and Cryptography*, 11(2):179–196, 1997.

[40] D.Q. Viet and K. Kurosawa. Almost ideal contrast visual cryptography with reversing. In *Lecture Notes in Computer Science 2964 (Topics in Cryptography)*, pages 353–365, 2004.

[41] S.-G. Wang. *Digital Watermarking Using Conjugate Halftone Screens*. US Patent No. 5,790,703, 1998.

[42] C.W. Wu, G. Thompson, and M. Stanich. *Digital watermarking and steganography via overlays of halftone images*. Electrical Engineering IBM Research Report, RC23267 (W0407-013), pages 1–12. 2004.

[43] X. Wu, D.S. Wong, and Q. Li. Threshold visual cryptography scheme for color images with no pixel expansion. In *Intl. Symposium on Computer Science and Computational Technology*, pages 310–315, 2009.

[44] C.-N. Yang and C.-S. Laih. New colored visual secret sharing schemes. *Designs, Codes and Cryptography*, 20(3):325–335, 2000.

[45] C.-N. Yang and T.-S. Chen. Extended visual secret sharing schemes: Improving the shadow image quality. *International Journal of Pattern Recognition and Artificial Intelligence*, 21:879–898, 2007.

[46] C.-N. Yang and T.-S. Chen. Colored visual cryptography scheme based on additive color mixing. *Pattern Recognition*, 41(10):3114–3129, 2008.

[47] Z. Zhou, G. R. Arce, and G. Di Crescenzo. Halftone visual cryptography. *IEEE Trans. on Image Processing*, 15(8):2441–2453, 2006.

5

Probabilistic Visual Cryptography Schemes

Stelvio Cimato

Università degli Studi di Milano, Italy

Roberto De Prisco

Università di Salerno, Italy

Alfredo De Santis

Università di Salerno, Italy

CONTENTS

5.1   Introduction

Visual cryptography schemes allow the encoding of a secret image, consisting of black or white pixels, into n shares that are distributed to the set $\mathcal{P}$ of n

participants. The shares are such that only qualified subsets of participants can "visually" recover the secret image. The secret pixels are shared with techniques based on subdividing each secret pixel into a certain number m, $m \geq 2$ of subpixels. Such a parameter m is called the *pixel expansion*, since the reconstructed shared image becomes m times bigger than the original. This cryptographic paradigm was introduced by Naor and Shamir [16]. They analyzed the case of (k, n)-threshold visual cryptography schemes, in which a black and white secret image is visible if and only if any k transparencies are stacked together.

The pixel expansion has a number of drawbacks, affecting the quality of the reconstructed image and the complexity of the visual cryptography scheme (VCS). In some cases, the pixel expansion is exponential, and this limits the applicability of the VCS. In general, the "quality" of the reconstructed image depends both on the pixel expansion and on the contrast, which is another measure of the goodness of the scheme. A number of papers studying the best pixel expansion and the best contrast have appeared in the literature. A partial list of such papers include [2, 4, 5, 6, 7, 12, 14, 15]. Some other papers have focused on different models or properties. For example, in [1], visual cryptography schemes for general access structures (where the qualified set of participants are arbitrary and not defined by a threshold of participants) have been studied. Schemes where the shares show meaningful pictures (not related to the secret) are studied in [3]. In [24] the problem of not distorting the original image is considered. Some research has also considered the case of colored images (see for example [10, 9, 19, 25]).

To deal with the pixel expansion, Yang [22, 23] has introduced a new model of visual cryptography in which the reconstruction of the secret image is probabilistic, but the shares have the same size of the secret image, i.e., the schemes have no pixel expansion. To be fair, a first attempt to provide VCS without pixel expansion has been done by Ito et al. in [13]. In both Ito and Yang models, each pixel is reconstructed "OR"ing the corresponding single pixel contained in the shares. Such models are called *probabilistic*, because they give no absolute guarantee on the correct reconstruction of the original pixel: in some cases, the reconstructed pixel is wrong. This differs from the traditional VCS, which are now called *deterministic*, where the reconstruction of an "approximation" of the secret pixel is guaranteed. Here the approximation means that a white (black) pixel can be, in some cases, replaced in the reconstructed image by a set of subpixels having a given set of whiteness (blackness). Since in probabilistic models the secret pixel is correctly reconstructed with some probability, the quality of the reconstructed images depends on how big is the probability of correctly reconstructing the secret pixels.

Between deterministic schemes and probabilistic schemes it is possible to set a trade-off. In a deterministic scheme a certain pixel expansion is paid for the guarantee of a correct reconstruction. In a probabilistic scheme a reconstruction with no pixel expansion is paid with a (small) probability of making mistakes in reconstructing the secret image. In some cases it is possible to sacrifice some pixel expansion in order to improve the probabilistic reconstruction of the secret image or vice versa. Yang's model has been generalized in Cimato et al. [11] showing how it is possible to trade pixel expansion for the probability of a good reconstruction. Such a model can be seen as a generalization of both the classical deterministic model and the probabilistic model introduced by Yang [22]. Moreover, there exists a one-to-one mapping between probabilistic schemes with no expansion and deterministic schemes; such a mapping trades the contrast of the deterministic scheme with the probability factor of the probabilistic scheme. Other proposals in literature have been introduced to deal with non-OR-based vcs and to extend the approach to color and grayscale images.

5.2 Visual Cryptography Schemes

A formal definition of the probabilistic model has been given in [11], generalizing Yang's approach and extending the traditional definition of VCS. In the next subsection we review the notions related to traditional VCS, before introducing the definition of probabilistic visual cryptography schemes in subsection 5.2.2

5.2.1 The Deterministic Model

The secret image consists of black and white, where usually white color is interpreted as transparent, so that the superposition of white pixels, let the color of the pixel contained in the other shares pass. In order to share each pixel of the secret image the owner of the secret, usually called the *dealer*, provides each participant with a *share*, which is an enlarged version of the secret pixel consisting of a certain number m of subpixels. So the shared version of the original secret pixel will consists of m pixels, which are called *subpixels* because all together they represent the original secret pixel.

The shares can be conveniently represented with $n \times m$ matrices where each row represents one share, i.e., m subpixels, and each element is either 0, for a white subpixel, or 1 for a black subpixel. A matrix representing the shares is called the *distribution matrix*. Physically, the shares are given out in the form of printed transparencies. Given a distribution matrix M and a set Q of participants, the notation M^Q refers to the submatrix of M consisting of only the rows corresponding to participants in Q.

To reconstruct the secret image a group of participants stacks together the shares. Since each secret pixel is represented by m pixels in the shares, the reconstructed image will be bigger than the original (depending on m and on the actual positions of the pixels, the image can also be distorted; a perfect square is a good choice for m because it avoids distortion). Depending on the stacked shares, each secret pixel will be reconstructed with a certain number of black and white subpixels. A reconstructed pixel is considered white if the number of white pixels in its reconstruction is big enough, i.e., the number of black subpixel is less than or equal to a given threshold ℓ, and is considered black if the number of black subpixels is big enough, i.e., greater or equal a given threshold h. Obviously one has to require that $\ell < h$. 1 ℓ and h are called the *contrast thresholds* of the scheme. In some other papers, the contrast thresholds are used slightly differently: $m-h$ is an upper bound on the number of black subpixels in a white pixel and $m - \ell$ is a lower bound on the number of black subpixels in a black pixel.

Since a reconstructed pixel has to be either black or white, we consider only schemes such that in the reconstructed image each reconstructed pixel has a number of black pixels, which is either $\leq \ell$ or $\geq h$. An easy way to resolve ambiguities in the reconstruction is to assume $\ell = h - 1$.

We consider threshold schemes where a qualified set of participants consists of k or more participants. For these schemes, a nonqualified set of participants, i.e., a set of less than k participants, will not have any information about the secret image from the shares. Instead, a qualified set of participants, i.e., a set of at least k participants will be able to reconstruct the secret image. The quality of the reconstructed image depends on the scheme.

In a deterministic scheme the quality of the reconstructed image depends on the so-called *contrast* that is a function of the pixel expansion m, and the contrast thresholds ℓ and h. The contrast of a scheme is defined as $\gamma = (h - \ell)/m$.

In a deterministic scheme it is guaranteed that, for any qualified set of participants, the pixel is reconstructed correctly; that is, if the secret pixel is white then the number of black subpixels in the reconstructed image, corresponding to that secret pixel, is at most ℓ, whereas if the secret pixel is black, the number of black subpixels in the reconstructed pixel is at least h.

In order to provide shares to the participants the dealer chooses uniformly at random a distribution matrix from a collection of matrices $\mathcal{C}_B$, if the secret pixel is black, or from a collection of matrices $\mathcal{C}_W$, if the secret pixel is white. Hence, for a deterministic scheme it holds that for any distribution matrix M of the set $\mathcal{C}_B$, the reconstruction of a pixel obtained by M^Q for any qualified set Q, gives at least h black subpixels, whereas for any distribution matrix M of the set $\mathcal{C}_W$ the reconstruction of a pixel obtained by M^Q for any qualified set Q, gives at most ℓ black subpixels. Let us report here the formal definition of a deterministic VCS:

Definition 1 *Let* $(\Gamma_{\mathsf{Qual}}, \Gamma_{\mathsf{Forb}})$ *be an access structure on a set of n participants. Two collections (multisets) of $n \times m$ boolean matrices $\mathcal{C}_W$ and $\mathcal{C}_B$ constitute a visual cryptography scheme* $(\Gamma_{\mathsf{Qual}}, \Gamma_{\mathsf{Forb}}, m)$-*VCS if there exist the integers ℓ and h, $\ell < h$, such that:*

1. Any (qualified) set $Q = \{i_1, i_2, \ldots, i_p\} \in \Gamma_{\mathsf{Qual}}$ can recover the shared image by stacking their transparencies.

Formally, for any $M \in \mathcal{C}_W$, the "or" V of rows $i_1, i_2, \ldots, i_p$ satisfies $w(V) \leq \ell$; whereas, for any $M \in \mathcal{C}_B$ it results that $w(V) \geq h$.

2. Any (forbidden) set $X = \{i_1, i_2, \ldots, i_p\} \in \Gamma_{\mathsf{Forb}}$ has no information on the shared image.

Formally, the two collections of $p \times m$ matrices $\mathcal{D}_t$, with $t \in \{B, W\}$, obtained by restricting each $n \times m$ matrix in $\mathcal{C}_X$ to rows $i_1, i_2, \ldots, i_p$ are indistinguishable in the sense that they contain the same matrices with the same frequencies.

In many schemes, the collection $\mathcal{C}_W$ (resp. $\mathcal{C}_B$) consists of all the matrices that can be obtained by permuting all the columns of a matrix M_W (resp. M_B). For such schemes, the matrices M_W and M_B are called the *base matrices* of the scheme. Base matrices constitute an efficient representation of the scheme. Indeed, the dealer has to store only the base matrices and in order to randomly choose a matrix from $\mathcal{C}_X$ he has to randomly choose a permutation of the columns of the base matrix M_X.

A scheme is characterized by several parameters: the number of participants n, the threshold k that determines whether a set of participants is qualified to reconstruct the image, the pixel expansion m, and the contrast thresholds ℓ and h, which determine whether a reconstructed pixel is considered white or black.

5.2.2 The Probabilistic Model

In a probabilistic scheme the reconstruction property is no more guaranteed, but each pixel can be correctly reconstructed only with a probability given as a parameter of the schema. This means that the distribution matrices must be carefully selected in order to satisfy the above properties. For a probabilistic scheme, as done in [22], it is possible to define the probabilities of (un)correctly reconstructing a (black)white pixel, given a qualified set of participants Q. With $p_{i|j}$ is denoted the probability of having a reconstructed pixel i, given that the corresponding pixel in the secret image was j, where $i, j \in \{b, w\}$. Then $p_{w|w}(Q)$, denotes the probability of correctly reconstructing a white pixel when superimposing the shares of Q, and $p_{b|w}(Q)$ as the probability of incorrectly reconstructing a white pixel. Notice that $p_{w|w}(Q) = \frac{z}{|\mathcal{C}_W|}$ where z is the number of distribution matrices M in $\mathcal{C}_W$ for which M^Q reconstructs a pixel with at most ℓ black subpixels and $p_{b|w}(Q) = \frac{z}{|\mathcal{C}_W|}$ where z is the number of distribution matrices M in $\mathcal{C}_W$ for which M^Q reconstructs a pixel with at least h black subpixels. In a similar way $p_{b|b}(Q)$ and $p_{w|b}(Q)$ can be defined.

The quantities

$$p_{b|b}(Q) - p_{b|w}(Q)$$

and

$$p_{w|w}(Q) - p_{w|b}(Q)$$

return then a measure of the goodness of the scheme for the given qualified set Q. The bigger the above differences the better is the scheme. If both quantities are equal to 1 for all qualified sets, then the scheme is again a deterministic one, with no possible error in the reconstruction.

If there exists a positive constant β such that for any qualified set Q it holds

$$p_{b|b}(Q) - p_{b|w}(Q) \geq \beta$$

and

$$p_{w|w}(Q) - p_{w|b}(Q) \geq \beta.$$

then the scheme is called β-probabilistic, meaning that β denotes the possible error in the reconstruction. Notice that when every reconstructed pixel is either white or black, that is, the reconstructed pixel has at most ℓ black subpixels or at least h black subpixels, we have that for any Q,

$$p_{w|w}(Q) = 1 - p_{b|w}(Q)$$

and that

$$p_{b|b}(Q) = 1 - p_{w|b}(Q)$$

and thus,

$$p_{w|w}(Q) - p_{w|b}(Q) = p_{b|b}(Q) - p_{b|w}(Q).$$

The value ℓ and h are the thresholds that define the number of subpixels needed to correctly distinguish between a white and black pixel in the reconstructed image. For some scheme, for some qualified set Q, it could be possible to obtain a secret pixel with a number of black subpixels strictly greater than ℓ and strictly less than h, where the reconstructed pixel is neither white nor black. In this case the value $p_{b|b}(Q) - p_{b|w}(Q)$ might be different from $p_{w|w}(Q) - p_{w|b}(Q)$. To avoid such kind of ambiguous situations, it is possible to fix the value $\ell = h - 1$, so that the equation $p_{w|w}(Q) - p_{w|b}(Q) = p_{b|b}(Q) - p_{b|w}(Q)$ always holds.

Probabilistic schemes are then characterized by a further parameter: the probabilistic factor β. In the following a probabilistic scheme will be described by all these parameters (that is, β, k, n, ℓ, h, and m) which are referred to as the *characteristic parameters* of the scheme.

It is now possible to provide the formal definition of a β-probabilistic threshold visual cryptography scheme with characteristic parameters (k, n, ℓ, h, m), for short β-probabilistic (k, n, ℓ, h, m)-VCS. The definition naturally extends also to general access structures.

Definition 2 *A β-probabilistic (k, n, ℓ, h, m)-VCS consists of two collections of $n \times m$ binary matrices, $\mathcal{C}_W$ and $\mathcal{C}_B$, satisfying the following properties:*

1. There exists $\beta > 0$, such that for any set Q of exact k participants $p_{b|b}(Q) - p_{b|w}(Q) \geq \beta$ and $p_{w|w}(Q) - p_{w|b}(Q) \geq \beta$

2. For any set Q of strictly less than k participants, the sets $\mathcal{C}_W^Q = \{M^Q | M \in \mathcal{C}_W\}$ and $\mathcal{C}_B^Q = \{M^Q | M \in \mathcal{C}_B\}$ are equal.

Notice that the definition of the scheme requires the reconstruction of a secret pixel (Property 1) to be well defined for qualified sets of exact k participants. This is without loss of generality because if a qualified set has more than k participants one can simply choose k of the shares and use only those k shares to reconstruct the image. Hence in the rest of the chapter a qualified set is assumed to consist of exact k participants.

The goodness of a scheme is measured by the pixel expansion m, the contrast $\gamma = (h - \ell)/m$ and by the probabilistic factor β. Notice that for $m = 1$ the above definition is equivalent to the one provided by Yang [22]. Whereas for a big enough m, it is possible to construct schemes with $\beta = 1$, and in such a case the above definition is equivalent to the classical definition of a visual cryptography scheme.

If the pixel expansion m is assumed to be a parameter of a scheme, then on one extreme, when $m = 1$ one gets the probabilistic model with no pixel expansion, and on the other extreme, when m is big enough, one obtains the deterministic model. In between such two extremes it is possible to consider probabilistic models with a given pixel expansion, trading the probability of a good reconstruction with the number of subpixels required to reconstruct each secret pixel.

5.3   Canonical Probabilistic Schemes

By restricting the attention on a particular class of probabilistic schemes, it is possible to show that results valid for all the other schemes can be obtained, without loss of generality. The trick is to prove that for a given scheme it is possible to define a similar scheme satisfying well-defined additional properties, without modification on the parameters of the scheme. For this reason *canonical β-probabilistic (k, n, ℓ, h, m)-VCS* schemes are defined as those schemes that satisfy the following properties:

1.   The cardinality of the collections $\mathcal{C}_W$ and $\mathcal{C}_B$ are equal and

2.   For any two qualified sets Q_1 and Q_2 of participants, we have that $p_{x|y}(Q_1) = p_{x|y}(Q_2)$, for $x \in \{w, b\}$ and $y \in \{w, b\}$.

The first lemma says that given a probabilistic scheme S a new scheme S' can be constructed such that the cardinality of $\mathcal{C}_W(S')$ is the same as that of $\mathcal{C}_B(S')$ and such that S' has the same characteristic parameters as S.

The following lemma is similar to the analogous result proved in Section 2.1 of [1] for deterministic schemes.

Lemma 1 *Given a β-probabilistic (k, n, ℓ, h, m)-VCS S, there exists a β-probabilistic (k, n, ℓ, h, m)-VCS S' such that $|\mathcal{C}_W(S')| = |\mathcal{C}_B(S')|$.*

Proof *Fix a β-probabilistic (k, n, ℓ, h, m)-VCS S, let $r_1 = |\mathcal{C}_W(S)|$ and $r_2 = |\mathcal{C}_B(S)|$ and assume that $r_1 \neq r_2$ (otherwise we can choose $S' = S$ and we are done). Let $r = r_1 \cdot r_2$ and construct S' by letting $\mathcal{C}_W(S')$ (resp. $\mathcal{C}_B(S')$) consists of all the matrices of $\mathcal{C}_W(S)$ (resp. $\mathcal{C}_B(S)$) each one repeated r_2 (resp. r_1) times. It is obvious that $|\mathcal{C}_W(S')| = |\mathcal{C}_B(S')| = r$.*

Since we have only repeated matrices of the collections $\mathcal{C}$ it is easy to see that k, n, and m remain the same. Keeping the same ℓ and h also β stays the same: indeed $\mathcal{C}_W(S')$ is obtained by replicating the same number of times each matrix of $\mathcal{C}_W(S)$ and thus the probabilities $p_{w|w}(Q)$ and $p_{b|w}(Q)$ do not change for any Q and similarly for $p_{w|b}(Q)$ and $p_{b|b}(Q)$ since $\mathcal{C}_B(S')$ is obtained by replicating the same number of times each matrix of $\mathcal{C}_B(S)$.

The following lemma shows that schemes can be built where the number of black subpixels in a reconstructed pixel does not depend on the particular qualified set of participants chosen to reconstruct the secret pixel and the characteristic parameters remain the same.

Lemma 2 *Given a β-probabilistic (k, n, ℓ, h, m)-VCS S, there exists a β-probabilistic (k, n, ℓ, h, m)-VCS S' such that for any two qualified sets Q_1 and Q_2 of participants, we have that $p_{x|y}(Q_1) = p_{x|y}(Q_2)$, for $x \in \{w, b\}$ and $y \in \{w, b\}$.*

Proof *Fix a β-probabilistic (k, n, ℓ, h, m)-VCS S. Consider first $\mathcal{C}_W(S)$ and assume that the desired property does not hold. Then we build a new scheme S' where $\mathcal{C}_W(S')$ is obtained from $\mathcal{C}_W(S)$ in the following way: for each matrix M of $\mathcal{C}_W(S)$ insert into $\mathcal{C}_W(S')$ all the matrices that we can build from M by permuting in all possible $n!$ ways its rows. The collection $\mathcal{C}_B(S')$ is obtained in the same way from $\mathcal{C}_B(S)$.*

Let Q_1 and Q_2 be two qualified sets. We have that $\mathcal{C}_W^{Q_1}(S') = \mathcal{C}_W^{Q_2}(S')$; indeed, by construction both sets contain, for any qualified set Q, all the matrices in $\mathcal{C}_W^Q(S)$, with the same multiplicity (the multiplicity is due to the fact that when restricting the attention to the rows in Q, the other $n - k$ rows can be taken in any order). Hence, we have that $p_{w|w}(Q_1) = p_{w|w}(Q_2)$, and $p_{b|w}(Q_1) = p_{b|w}(Q_2)$. For the same reason we have $\mathcal{C}_B^{Q_1}(S') = \mathcal{C}_B^{Q_2}(S')$ and thus $p_{b|b}(Q_1) = p_{b|b}(Q_2)$, and $p_{w|b}(Q_1) = p_{w|b}(Q_2)$.

By Lemmas 1 and 2, it follows that considering only canonical schemes is without loss of generality, because for any scheme S an equivalent canonical scheme S' having the same characteristic parameters of scheme S can be provided.

In the following only canonical schemes will be considered, remembering that for a canonical scheme $p_{x|y}(Q_1) = p_{x|y}(Q_2)$, for $x \in \{w, b\}$ and $y \in \{w, b\}$ and thus we will just write $p_{x|y}$, without specifying the qualified set.

5.4 Probabilistic Schemes with No Pixel Expansion

Probabilistic threshold schemes give the possibility to construct a VCS scheme with no pixel expansion, that is having $m = 1$. In [22] it has been proved that a deterministic scheme S with contrast $\gamma(S)$ can be transformed into a β-probabilistic scheme S' with $\beta(S') = \gamma(S)$ and no pixel expansion. In [11] a complementary result has been proven showing that there is a one-to-one correspondence between the probabilistic model with no pixel expansion and the deterministic one where the contrast is traded for the probabilistic factor. Indeed a β-probabilistic scheme S with no pixel expansion can be transformed into a deterministic scheme S' with contrast $\gamma(S') = \beta(S)$. An immediate consequence is that any bound on the contrast of a deterministic scheme is also a bound on the probabilistic factor of probabilistic schemes with no pixel expansion. In the following we report the lemma proving the correspondence between probabilistic schemes with no pixel expansion and deterministic schemes.

The following lemma has been proved in [22] and applies to VCS with basis matrices.

Lemma 3 *[22] Let S be a deterministic (k, n, ℓ, h, m)-VCS with base matrices M_B and M_W. Then, there exists a canonical β-probabilistic $(k, n, \ell', h', 1)$-VCS scheme with $\beta = \gamma(S)$.*

Proof *Let S be a deterministic (k, n, ℓ, h, m)-VCS. Construct a probabilistic scheme S', by letting $\mathcal{C}_B(S')$ (resp. $\mathcal{C}_W(S')$) consists of all the $n \times 1$ vectors that appear in the matrices M_B (resp. M_W).*

We need to prove that S' is a β-probabilistic (k, n, ℓ', h', m')-VCS scheme with $\ell' = \ell, h' = h, m' = 1$ and $\beta = \gamma(S) = (h - \ell)/m$. Obviously S' has pixel expansion $m' = 1$.

By the properties of S, we know that when the secret pixel is black the reconstruction in S gives at least h black subpixels, that is $p_{b|b} \geq h/m$. Obviously this gives $p_{w|b} \leq (m - h)/m$. Similarly, we have $p_{w|w} \geq (m - \ell)/m$ and $p_{b|w} \leq \ell/m$.

Hence, in S' we have that

$$p_{w|w} - p_{w|b} \geq \frac{m - \ell}{m} - \frac{m - h}{m} = \frac{h - \ell}{m}$$

and

$$p_{b|b} - p_{b|w} \geq \frac{h}{m} - \frac{\ell}{m} = \frac{h - \ell}{m}$$

and thus we can set $\beta = \frac{h-\ell}{m}$. *The security property is immediate.*

Example 1 *Consider as starting point the deterministic scheme* $(3,4)$ *having basis matrices* M_W *and* M_B *reported below, whose parameters are* $m = 6$, $h = 5$, $l = 4$, $\alpha = 1/6$:

$$
M_W = \begin{bmatrix} 0 & 0 & 1 & 1 & 1 & 0 \\ 0 & 0 & 1 & 1 & 0 & 1 \\ 0 & 0 & 1 & 0 & 1 & 1 \\ 0 & 0 & 0 & 1 & 1 & 1 \end{bmatrix}
\qquad
M_B = \begin{bmatrix} 1 & 1 & 0 & 0 & 0 & 1 \\ 1 & 1 & 0 & 0 & 1 & 0 \\ 1 & 1 & 0 & 1 & 0 & 0 \\ 1 & 1 & 1 & 0 & 0 & 0 \end{bmatrix}.
$$

To illustrate the construction, consider the following $\frac{1}{6}$-*probabilistic* $(3, 4, 4, 5, 1)$-*VCS* S.

$$
\mathcal{C}_W = \left\{ \begin{bmatrix} 0 \\ 0 \\ 0 \\ 0 \end{bmatrix}, \begin{bmatrix} 0 \\ 0 \\ 0 \\ 0 \end{bmatrix}, \begin{bmatrix} 1 \\ 1 \\ 1 \\ 0 \end{bmatrix}, \begin{bmatrix} 1 \\ 1 \\ 0 \\ 1 \end{bmatrix}, \begin{bmatrix} 1 \\ 0 \\ 1 \\ 1 \end{bmatrix}, \begin{bmatrix} 0 \\ 1 \\ 1 \\ 1 \end{bmatrix} \right\}
$$

$$
\mathcal{C}_B = \left\{ \begin{bmatrix} 1 \\ 1 \\ 1 \\ 1 \end{bmatrix}, \begin{bmatrix} 1 \\ 1 \\ 1 \\ 1 \end{bmatrix}, \begin{bmatrix} 0 \\ 0 \\ 0 \\ 1 \end{bmatrix}, \begin{bmatrix} 0 \\ 0 \\ 1 \\ 0 \end{bmatrix}, \begin{bmatrix} 0 \\ 1 \\ 0 \\ 0 \end{bmatrix}, \begin{bmatrix} 1 \\ 0 \\ 0 \\ 0 \end{bmatrix} \right\}.
$$

Let S *be the* $(n, n, 0, 0, 1)$-*VCS obtained applying Lemma 3 to scheme* S_D. *For scheme* S *we have that* $p_{b|b} = 5/6$, $p_{w|b} = 1/6$, $p_{w|w} = 1/3$, $p_{b|w} = 2/3$, $\beta = \gamma(S) = 1/6$.

Lemma 3 shows how starting from a deterministic scheme with base matrices it is possible to obtain a corresponding probabilistic scheme. A more general version of the above lemma we report below has been given in [11] showing the transformation from a general (also with no base matrices) deterministic scheme to a probabilistic one.

Lemma 4 *Let* S *be a deterministic* (k, n, ℓ, h, m)-*VCS. Then, there exists a canonical* β-*probabilistic* $(k, n, 0, 1, 1)$-*VCS scheme with* $\beta = \gamma(S)$.

Proof *Let* S *be a deterministic* (k, n, ℓ, h, m)-*VCS. Construct a probabilistic scheme* S', *by letting* $\mathcal{C}_B(S')$ *(resp.* $\mathcal{C}_W(S')$*) consists of all the* $n \times 1$ *vectors that appear in all the matrices of* $\mathcal{C}_B(S)$ *(resp.* $\mathcal{C}_W(S)$*).*

We need to prove that S' *is a* β-*probabilistic* (k, n, ℓ', h', m')-*VCS scheme with* $\ell' = 0, h' = 1, m' = 1$ *and* $\beta = \gamma(S) = (h - \ell)/m$. *Obviously* S' *has pixel expansion* $m' = 1$. *We set* $\ell' = 0$ *and* $h' = 1$ *and we have to prove that this results in a* β-*probabilistic scheme.*

Let $r = |\mathcal{C}_B(S)|$. *By the properties of* S, *we know that when the secret* *pixel is black the reconstruction in S gives at least h black subpixels, that each matrix of $\mathcal{C}_B(S)$ has at least h columns, which reconstruct black. Hence, in S' we have at least $r \cdot h$ columns that reconstruct black. Since $|\mathcal{C}_B(S')| = r \cdot m$ we have that $p_{b|b} \geq h/m$. Obviously this gives $p_{w|b} \leq (m - h)/m$. Similarly, we have $p_{w|w} \geq (m - \ell)/m$ and $p_{b|w} \leq \ell/m$.*

Hence, in S' we have that $p_{w|w} - p_{w|b} \geq \frac{h-\ell}{m}$ and $p_{b|b} - p_{b|w} \geq \frac{h-\ell}{m}$ and thus we can set $\beta = \frac{h-\ell}{m}$. The security property is immediate.

For a canonical scheme it is possible to define the *characteristic vectors* as:

$$c_B = (c_B^0, c_B^1, \ldots, c_B^m)$$

and

$$c_W = (c_W^0, c_W^1, \ldots, c_W^m)$$

where c_X^i is the number of matrices of $\mathcal{C}_X$ that provide a reconstruction, by any qualified set, of the secret pixel with exactly i black subpixels. The characteristic vectors are well defined because, by Property 2 of canonical schemes, each c_X^i does not depend on the particular qualified set chosen for reconstructing the secret pixel.

The following lemma complements the results presented above, showing how it is possible to transform a probabilistic (canonical) scheme into a deterministic scheme.

Lemma 5 *Let S be a β-probabilistic canonical $(k, n, 0, 1, 1)$-VCS and let $c_B = (c_B^0, c_B^1)$ and $c_W = (c_W^0, c_W^1)$ be its characteristic vectors. Then, there exists a deterministic (k, n, ℓ', h', m')-VCS S' with $\ell' = c_W^1$, $h' = c_B^1$, $m' = |\mathcal{C}_W(S)|$ and contrast $\gamma(S') \geq \beta(S)$.*

Proof *Let S be a β-probabilistic $(k, n, 0, 1, 1)$-VCS satisfying the hypothesis of the lemma. Scheme S' is constructed by letting its base matrix M_B (resp. M_W) consist of all the vectors of $\mathcal{C}_B(S)$ (resp. $\mathcal{C}_W(S)$). Fix $\ell' = c_W^1$ and $h' = c_B^1$. We have to prove that S' is a deterministic (k, n, ℓ', h', m')-VCS with $m' = r$, where $r = |\mathcal{C}_W(S)| = |\mathcal{C}_B(S)|$.*

Both $\mathcal{C}_B(S)$ and $\mathcal{C}_W(S)$ contain r matrices of dimension $n \times 1$; hence, the dimension of both M_B and M_W is $n \times r$; thus, $m' = r$.

Since S is a β-probabilistic scheme, we have that $p_{b|b} - p_{b|w} \geq \beta > 0$ hence, $p_{b|b} > p_{b|w}$, and since $p_{b|b} = c_B^1/r$ and $p_{b|w} = c_W^1/r$, we have $c_B^1 > c_W^1$. Thus, $\ell' < h'$. Now we need prove that the scheme is deterministic. Scheme S' is a basis matrices scheme, so all the matrices of the collections $\mathcal{C}_W(S')$ and $\mathcal{C}_B(S')$ are equal up to a permutation of the columns; moreover it is easy to see that a reconstructed black pixel always has h' black subpixels and that a reconstructed white pixel always has ℓ' black subpixels. Hence, we have $p'_{b|b} = p'_{w|w} = 1$ and $p'_{w|b} = p'_{b|w} = 0$. Hence, scheme S' is deterministic.

Let us consider the security property. Fix a nonqualified set of participants and consider the corresponding $k' < k$ rows of the basis matrices. Each of these rows can be seen as the concatenation of shares of S (one per each matrix in the collections $C_B(S)$ and $C_W(S)$). By the security property of S we have that the k' rows of M_B are equal to the k' rows of M_W, except for a permutation of the columns. Hence, S' satisfies the security property.

Finally, the contrast of S' is $\gamma(S') = \frac{h'-\ell'}{m'} = \frac{c_B^1 - c_W^1}{r} = p_{b|b} - p_{b|w} \geq \beta(S)$.

As already said, the correspondence between deterministic schemes and probabilistic schemes with no pixel expansion allows the reuse of bounds found on the contrast as bounds on the corresponding probabilistic factor. For example, in [14] it has been proved that the contrast of any deterministic $(2, n)$-VCS is upper bounded by

$$\gamma \leq \gamma^* = 4^{-(k-1)} \frac{n^k}{n(n-1)\cdots(n-(k-1))}. \tag{5.1}$$

As an immediate, the following lemma, which is a corollary of Lemma 5, also holds:

Corollary 6 *For any β-probabilistic $(2, n, 0, 1, 1)$-VCS we have that*

$$\beta \leq \beta^* = 4^{-(k-1)} \frac{n^k}{n(n-1)\cdots(n-(k-1))}.$$

Proof *Assume by contradiction that a probabilistic scheme with $\beta > \beta^*$ exists. By Lemma 5, we can construct a scheme with contrast $\gamma \geq \beta$. Since $\beta > \beta^* = \gamma^*$ we get that $\gamma > \gamma^*$, contradicting Equation (5.1).*

5.5   Trading Pixel Expansion with Probabilities

In the above section, the correspondence between deterministic and probabilistic schemes has been stated in Lemma 3 and Lemma 5, showing a trade-off between pixel expansion and probability factor. It shows that by using a large enough pixel expansion we can transform a probabilistic scheme into a deterministic one. Indeed, on one extreme it is possible to have schemes with no pixel expansion, where the reconstruction relies entirely on the probability factor of the scheme. On the other extreme, it is possible to have deterministic schemes, where the reconstruction is guaranteed but a certain pixel expansion is required. In the next section we show how it is possible to stay in between the two extremes, realizing schemes for which the probability factor is traded for the pixel expansion.

5.5.1   Probabilistic Schemes with Given Pixel Expansion

Lemma 5 shows the basic technique for the construction of a deterministic scheme with pixel expansion equal to the cardinality r of the collections C_B

and $\mathcal{C}_W$ of the probabilistic scheme with no pixel expansion taken as a starting point. Extending such a technique, it is possible to obtain schemes with arbitrary pixel expansion m', with $1 < m' \leq r$. The new collections of matrices $\mathcal{C}'_B$ and $\mathcal{C}'_W$ for the resulting scheme is built by constructing in all possible ways matrices with m' columns from the vectors of the collections $\mathcal{C}_B$ and $\mathcal{C}_W$ of the starting probabilistic scheme (in the particular case of $m' = r$, the resulting scheme is deterministic). We remark that Construction 1 does not allow repetition of the same column. It is possible to build schemes by also allowing repetitions of the columns; however, the resulting schemes have a worst probabilistic factor. Notice that it is useless to construct probabilistic schemes with $m' > r$, since if $m' = r$ then a deterministic scheme can be realized.

Construction 1 *Let S be a canonical β-probabilistic $(k, n, 0, 1, 1)$-VCS. Fix $1 < m' \leq r$, where $r = |\mathcal{C}_B(S)| = |\mathcal{C}_W(S)|$. Construct a scheme S' whose collection $\mathcal{C}_B(S')$ (resp. $\mathcal{C}_W(S')$) consists of all the matrices of dimension $n \times m'$ that we can build by choosing m' vectors of $\mathcal{C}_B(S)$ (resp. $\mathcal{C}_W(S)$).*

Notice that we also need to fix the contrast thresholds ℓ' and h' of the new scheme S'. There can be several valid choices.

To illustrate the construction, consider the following $1/3$-probabilistic $(2, 3, 0, 1, 1)$-VCS S.

$$
\mathcal{C}_B = \left\{ \begin{bmatrix} 1 \\ 0 \\ 0 \end{bmatrix}, \begin{bmatrix} 0 \\ 1 \\ 0 \end{bmatrix}, \begin{bmatrix} 0 \\ 0 \\ 1 \end{bmatrix} \right\} \qquad \mathcal{C}_W = \left\{ \begin{bmatrix} 1 \\ 1 \\ 1 \end{bmatrix}, \begin{bmatrix} 0 \\ 0 \\ 0 \end{bmatrix}, \begin{bmatrix} 0 \\ 0 \\ 0 \end{bmatrix} \right\}.
$$

For such a scheme, the parameters are $p_{b|b} = 2/3$ and $p_{b|w} = 1/3$.

If $m' = 2$ is fixed, a $\frac{1}{3}$-probabilistic $(2, 3, \cdot, \cdot, 2)$-VCS is obtained by applying Construction 1 to scheme S:

$$
\mathcal{C}_B = \left\{ \begin{bmatrix} 1 & 0 \\ 0 & 1 \\ 0 & 0 \end{bmatrix}, \begin{bmatrix} 1 & 0 \\ 0 & 0 \\ 0 & 1 \end{bmatrix}, \begin{bmatrix} 0 & 0 \\ 1 & 0 \\ 0 & 1 \end{bmatrix}, \begin{bmatrix} 0 & 1 \\ 1 & 0 \\ 0 & 0 \end{bmatrix}, \begin{bmatrix} 0 & 1 \\ 0 & 0 \\ 1 & 0 \end{bmatrix}, \begin{bmatrix} 0 & 0 \\ 0 & 1 \\ 1 & 0 \end{bmatrix} \right\}
$$

$$
\mathcal{C}_W = \left\{ \begin{bmatrix} 1 & 0 \\ 1 & 0 \\ 1 & 0 \end{bmatrix}, \begin{bmatrix} 1 & 0 \\ 1 & 0 \\ 1 & 0 \end{bmatrix}, \begin{bmatrix} 0 & 0 \\ 0 & 0 \\ 0 & 0 \end{bmatrix}, \begin{bmatrix} 0 & 1 \\ 0 & 1 \\ 0 & 1 \end{bmatrix}, \begin{bmatrix} 0 & 1 \\ 0 & 1 \\ 0 & 1 \end{bmatrix}, \begin{bmatrix} 0 & 0 \\ 0 & 0 \\ 0 & 0 \end{bmatrix} \right\}.
$$

The thresholds of S' can be selected in different ways returning in every case a $\frac{1}{3}$-probabilistic scheme. If values $\ell' = 0$ and $h' = 1$ are selected, the resulting probabilities for the scheme S' are $p'_{b|b} = 1$ and $p'_{b|w} = 2/3$. For

$\ell' = 1$ and $h' = 2$, scheme S' has $p'_{b|b} = 1/3$ and $p'_{b|w} = 0$. As already said, here we are considering values satisfying the condition $\ell' = h' - 1$. Notice that for the values the case $\ell' = 0$ and $h' = 2$, some reconstructed pixels cannot be classified as either black or white. In such a scheme the probabilities of correctly reconstructing secret pixels are smaller (since some of the matrices are wasted with a reconstruction that is neither black nor white). We also remark that one could try to eliminate distribution matrices for which the reconstruction gives a number of black subpixels in the gap between ℓ' and h'; however it is not clear whether this can be done without violating the security property. In the particular case above, $\ell' = 0$ and $h' = 2$ is clearly not possible (indeed we would have a perfect reconstruction of both white and black). One could also extend the formal model to allow this possibility; but then the contrast should be redefined in order to account for unclassified reconstructed pixels.

The matrices M_B and M_W consisting of all the columns of $\mathcal{C}_B$ and $\mathcal{C}_W$ can be used to represent the new scheme S', since they give an efficient representation of the collections $\mathcal{C}'_B$ and $\mathcal{C}'_W$; clearly together with M_B and M_W, the pixel expansion m' and the thresholds ℓ' and h' need to be specified.

Construction 1 starts from probabilistic schemes with no pixel expansion. Using Lemma 3 or Lemma 4 a probabilistic scheme with no pixel expansion can be obtained starting from a deterministic scheme. Hence, probabilistic schemes with pixel expansion can be constructed by starting from a deterministic scheme, applying first Lemma 3 and then Construction 1.

The probabilistic schemes obtained with Construction 1 satisfy the security property. Indeed, consider a nonqualified set Q of participants and let $\mathcal{C}_W^Q(S)$ (resp. $\mathcal{C}_B^Q(S)$) be the vectors of $\mathcal{C}_W(S)$ (resp. $\mathcal{C}_B(S)$) restricted to the rows corresponding to participants in Q. By the security property of S, $\mathcal{C}_W^Q(S)$ and $\mathcal{C}_B^Q(S)$ are the same collection of vectors. Since the way in which $\mathcal{C}_W(S')$ and $\mathcal{C}_B(S')$ are constructed is the same (except that for the former we start from $\mathcal{C}_W(S)$ and for the latter from $\mathcal{C}_B(S)$), also $\mathcal{C}_W^Q(S')$ and $\mathcal{C}_B^Q(S')$ are the same collection of matrices. Hence, the security property for S' holds.

In this section, a formula for the probabilities of the scheme built with Construction 1 as a function of the probabilities of the starting scheme is provided. Let S be a canonical probabilistic scheme with no pixel expansion and let $p_{b|b}, p_{b|w}, p_{w|b}$ and $p_{w|w}$ be the probabilities of S.

Fix an m' and build a probabilistic scheme with pixel expansion m' using Construction 1. Fix also a threshold ℓ', $0 \le \ell' < m'$. Fixing ℓ' also gives $h' = \ell' + 1$. Notice that, even with this restriction, not all choices of ℓ' will result in valid schemes. Let r be the cardinality of the collections $\mathcal{C}_W$ and $\mathcal{C}_B$ of S.

The cardinality of the collections $\mathcal{C}'_W$ and $\mathcal{C}'_B$, of scheme S', is $r' = r(r - 1) \cdot \ldots \cdot (r - m' + 1)$ because the first column can be selected in r ways, the second in $r - 1$ ways, and so on until the last column for which $r - m' + 1$

choices are possible. It is easy to see that $r' = m'!\binom{r}{m'}$. An explanation of this expression is that the binomial coefficient gives the number of possible ways of choosing m' vectors among the r of the collections, and the factorial coefficient accounts for all the possible permutations for each choice.

Fix a qualified set Q of participants. To compute $p'_{b|b}(Q)$, the number of matrices of $\mathcal{C}^Q_B(S)$ that yield a black reconstructed pixel is needed. Notice that, for a qualified set, the number of matrices (vectors) $\mathcal{C}^Q_B(S)$ that would give the correct reconstruction of a black subpixel (i.e., a black pixel) is $r \cdot p_{b|b}$, while the remaining $r - r \cdot p_{b|b}$, would give a wrong reconstruction of a black secret pixel (i.e., a white pixel). Recall that since S is canonical, $p_{b|b}$ does not depend on Q. Hence, the number of matrices in $\mathcal{C}_B(S')$ that will have a certain number z of black subpixels, is given by

$$m'!\binom{r \cdot p_{b|b}}{z}\binom{r - r \cdot p_{b|b}}{m' - z}$$

because z vectors can be selected from the $r \cdot p_{b|b}$ vectors that yield a reconstructed black subpixel, and $m' - z$ vectors from the $r - r \cdot p_{b|b}$ ones that yield a reconstructed white subpixel. Hence, z is constrained by $0 \le z \le r \cdot p_{b|b}$ and $0 \le m' - z \le r - r \cdot p_{b|b}$, which implies that binomial coefficients are defined. (By definition $\binom{0}{0} = 1$ and that $\binom{a}{b} = 0$ when $b > a$.) Hence, the values for the probabilities of the resulting scheme can be expressed as follows:

$$p'_{b|b}(Q) = \frac{\sum_{z=h'}^{m'} \binom{r \cdot p_{b|b}}{z}\binom{r - r \cdot p_{b|b}}{m' - z}}{\binom{r}{m'}} \tag{5.2}$$

and similarly

$$p'_{b|w}(Q) = \frac{\sum_{z=h'}^{m'} \binom{r \cdot p_{b|w}}{z}\binom{r - r \cdot p_{b|w}}{m' - z}}{\binom{r}{m'}} \tag{5.3}$$

$$p'_{w|w}(Q) = \frac{\sum_{z=0}^{\ell'} \binom{r - r \cdot p_{w|w}}{z}\binom{r \cdot p_{w|w}}{m' - z}}{\binom{r}{m'}} \tag{5.4}$$

$$p'_{w|b}(Q) = \frac{\sum_{z=0}^{\ell'} \binom{r - r \cdot p_{w|b}}{z}\binom{r \cdot p_{w|b}}{m' - z}}{\binom{r}{m'}}. \tag{5.5}$$

Since from Equations 5.2–5.5, probabilities of S' do not depend on the particular qualified set Q, it is possible to conclude that S' is canonical.

5.6   Constructing Probabilistic Schemes

The construction previously described can be applied for the construction of novel probabilistic schemes. The starting point however is the selection of a particular deterministic scheme that determines the parameters of the resulting scheme. Obviously, by changing the selected scheme, different probabilistic schemes are obtained. In the next sections we will describe some probabilistic schemes for the (n, n) and the $(2, n)$ cases starting from two selected deterministic schemes. In the first case a simple formula for the probabilistic factor can be obtained, relating β to the pixel expansion of the resulting scheme, while in the second case, the computation of the parameters results is more complicated and depends on the selected pixel expansion of the new scheme.

5.6.1 (n, n)-Threshold Probabilistic Schemes with Any Pixel Expansion

In this section a (n, n)-threshold probabilistic schemes with any pixel expansion is built, for $n \geq 2$ starting from deterministic scheme S_D, which is the (n, n)-threshold deterministic scheme of Naor and Shamir [16]. S_D has pixel expansion $m = 2^{n-1}$, and thresholds $\ell = m - 1$ and $h = m$. Moreover, the scheme consists of a white base matrix containing all vectors with an even (including 0) number of black subpixels and a black base matrix containing all vectors with an odd number of black subpixels. For example, for $n = 4$ the scheme S_D is given by:

$$
M_W = \begin{bmatrix} 0 & 1 & 1 & 1 & 0 & 0 & 0 & 1 \\ 0 & 1 & 0 & 0 & 1 & 1 & 0 & 1 \\ 0 & 0 & 1 & 0 & 1 & 0 & 1 & 1 \\ 0 & 0 & 0 & 1 & 0 & 1 & 1 & 1 \end{bmatrix} \quad M_B = \begin{bmatrix} 0 & 0 & 0 & 1 & 1 & 1 & 1 & 0 \\ 0 & 0 & 1 & 0 & 1 & 1 & 0 & 1 \\ 0 & 1 & 0 & 0 & 1 & 0 & 1 & 1 \\ 1 & 0 & 0 & 0 & 0 & 1 & 1 & 1 \end{bmatrix}.
$$

Let S be the $(n, n, 0, 0, 1)$-VCS obtained applying Lemma 3 to scheme S_D. For scheme S we have that the cardinality of the collections $\mathcal{C}_B$ and $\mathcal{C}_W$ is $r = 2^{n-1}$, and that $p_{b|b} = 1$, $p_{w|b} = 0$, $p_{w|w} = 1/2^{n-1}$, $p_{b|w} = 1 - 1/2^{n-1}$.

$$
\mathcal{C}_B = \left\{ \begin{bmatrix} 0 \\ 0 \\ 0 \\ 0 \end{bmatrix}, \begin{bmatrix} 1 \\ 1 \\ 0 \\ 0 \end{bmatrix}, \begin{bmatrix} 1 \\ 0 \\ 1 \\ 0 \end{bmatrix}, \begin{bmatrix} 1 \\ 0 \\ 0 \\ 1 \end{bmatrix}, \begin{bmatrix} 0 \\ 1 \\ 1 \\ 0 \end{bmatrix}, \begin{bmatrix} 0 \\ 1 \\ 0 \\ 1 \end{bmatrix}, \begin{bmatrix} 0 \\ 0 \\ 1 \\ 1 \end{bmatrix}, \begin{bmatrix} 1 \\ 1 \\ 1 \\ 1 \end{bmatrix} \right\}
$$

$$
\mathcal{C}_W = \left\{ \begin{bmatrix} 0 \\ 0 \\ 0 \\ 1 \end{bmatrix}, \begin{bmatrix} 0 \\ 0 \\ 1 \\ 0 \end{bmatrix}, \begin{bmatrix} 0 \\ 1 \\ 0 \\ 0 \end{bmatrix}, \begin{bmatrix} 1 \\ 0 \\ 0 \\ 0 \end{bmatrix}, \begin{bmatrix} 1 \\ 1 \\ 1 \\ 0 \end{bmatrix}, \begin{bmatrix} 1 \\ 1 \\ 0 \\ 1 \end{bmatrix}, \begin{bmatrix} 1 \\ 0 \\ 1 \\ 1 \end{bmatrix}, \begin{bmatrix} 0 \\ 1 \\ 1 \\ 1 \end{bmatrix} \right\}.
$$

Fix an m', $2 \le m' \le r$, and let S' be the probabilistic scheme with pixel expansion m' obtained using Construction 1 from scheme S. The threshold ℓ' for S' also needs to be selected, remembering also that the threshold h' will be obtained if the assumption $h' = \ell' + 1$ is satisfied. Observe that ℓ' must be $m' - 1$ because there is only one column with all white pixels and thus it would be never possible to get more than 1 white subpixel in a reconstructed pixel (equivalently always at least $m' - 1$ black subpixels in a reconstructed pixel will be obtained). This implies that if by selecting $\ell' < m' - 1$ all reconstructed pixels will always be considered black and thus no valid scheme can be obtained. To construct schemes, values $\ell' = m' - 1$ and $h' = m'$ must be selected.

To compute the probabilistic factor of scheme S' it is necessary to compute $p'_{b|b}$ and $p'_{b|w}$. All the columns of the black base matrix M_B of the deterministic scheme S_D have at least a 1; hence, all columns of $\mathcal{C}_B(S')$ have at least a 1. Thus, m' black subpixels when reconstructing a black secret pixel are always returned. This means that $p'_{b|b} = 1$ and $p'_{w|b} = 0$. To compute $p'_{b|w}$ consider the white base matrix M_W of scheme S_D, which determines $\mathcal{C}_W(S')$. Since M_W has one column with all 0s there will be some matrices of $\mathcal{C}_W(S')$ that include such a column. In order for a reconstructed pixel to be considered black it must have $h' = m'$ black subpixels. Among the $\binom{r}{m'}$ possible distribution matrices of $\mathcal{C}_W(S')$, exactly $\binom{r-1}{m'}$ do not include the unique column with all 0s. Hence,

$$
p'_{b|w} = \binom{r-1}{m'} \Big/ \binom{r}{m'} = 1 - m'/r
$$

and

$$
p'_{w|w} = m'/r
$$

The formula expressing the probability factor of S' is then the following:

$$
\beta = \frac{m'}{2^{n-1}}.
$$

The formula shows that there is a linear relation between the pixel expansion of the scheme and the probability factor. A probabilistic scheme with no pixel expansion implies a small probability factor ($\beta = 1/r$) and then a reconstructed image with many errors; the probability factor of the scheme can be increased by increasing the pixel expansion to obtain a better reconstructed image. Clearly for $m' = 2^{n-1}$ one would get $\beta = 1$, that is, a deterministic scheme.

5.6.2 $(2, n)$-Threshold Probabilistic Schemes with Any Pixel Expansion

In this section a $(2, n)$-threshold probabilistic scheme with pixel expansion, for $n \geq 3$ is provided, starting from the $(2, n)$-threshold deterministic scheme S_D described in [7]. Recall that S_D has pixel expansion $m = \binom{n}{\lfloor \frac{n}{2} \rfloor}$, and thresholds $\ell = \binom{n-1}{\lfloor \frac{n}{2} \rfloor - 1}$ and $h = \binom{n}{\lfloor \frac{n}{2} \rfloor} - \binom{n-2}{\lfloor \frac{n}{2} \rfloor}$. The scheme has a white base matrix consisting of ℓ columns of all 1s and $m - \ell$ columns with all 0s, and a black base matrix consisting of all binary vectors with exactly $\lfloor n/2 \rfloor$ elements equal to 1.

For example, for $n = 4$, scheme S_D has $m = 6$, $\ell = 3$, and $h = 5$ and is given by:

$$
M_W = \begin{bmatrix} 1 & 1 & 1 & 0 & 0 & 0 \\ 1 & 1 & 1 & 0 & 0 & 0 \\ 1 & 1 & 1 & 0 & 0 & 0 \\ 1 & 1 & 1 & 0 & 0 & 0 \end{bmatrix} \qquad M_B = \begin{bmatrix} 1 & 1 & 1 & 0 & 0 & 0 \\ 1 & 0 & 0 & 1 & 1 & 0 \\ 0 & 1 & 0 & 1 & 0 & 1 \\ 0 & 0 & 1 & 0 & 1 & 1 \end{bmatrix}.
$$

Let S be the $(2, n, 0, 0, 1)$-VCS obtained applying Lemma 3 to scheme S_D. Scheme S is a canonical scheme, and the cardinality of the collections $\mathcal{C}_B$ and $\mathcal{C}_W$ is $r = m$. The values of the probabilities are $p_{b|b} = h/m$, $p_{w|b} = 1 - h/m$, $p_{b|w} = \ell/m$, $p_{w|w} = 1 - \ell/m$. Since S is canonical the scheme S' obtained with Construction 1 is canonical. The value of the pixel expansion m' can be arbitrarily fixed between 1 and r.

Consider the case of $m' = 2$. For such a value of m' there are only two possible choices for the thresholds: $\ell' = 0, h' = 1$ or $\ell' = 1, h' = 2$. It is easy to see from the structure of M_W and M_B that the following two properties hold:

- PW: For any set Q of two participants, the matrix M_W^Q consists of ℓ columns of 2 ones and $m - \ell$ columns of 2 zeroes.

- PB: For any set Q of two participants, the matrix M_B^Q consists of $s = \binom{n-2}{\lfloor \frac{n}{2} \rfloor}$ columns of 2 zeroes and $m - s$ columns with at least a 1 (this is because for a qualified set Q of 2 participants matrix M_B^Q reconstructs a pixel with exactly h black subpixels, hence the remaining $m - h = s$ are white).

Let us now determine the parameters of the resulting scheme in the two considered cases.

Case $\ell' = 0, h' = 1$.

To compute $p'_{b|b}$, consider that in this case a reconstructed pixel is black if at least one subpixel is black (and white otherwise). If a qualified set Q is fixed, by property PB, it is easy to see that the matrices of the collection

$C_B^Q(S')$ that consists of all zeroes are exactly $s(s-1)$. Thus, $p'_{w|b} = \frac{s(s-1)}{m(m-1)}$ and consequently $p'_{b|b} = 1 - \frac{s(s-1)}{m(m-1)}$.

To compute $p'_{b|w}$, consider a qualified set Q. By property PW, the matrices of the collection $C_W^Q(S')$ that consists of all zeroes are exactly $(m-\ell)(m-\ell-1)$. Hence, $p'_{w|w} = \frac{(m-\ell)(m-\ell-1)}{m(m-1)}$, which gives $p'_{b|w} = 1 - \frac{(m-\ell)(m-\ell-1)}{m(m-1)}$.

Hence, for the $(2,n)$-threshold scheme obtained by choosing $\ell' = 0$ the probabilistic factor is:

$$\beta_0 = \frac{(m-\ell)(m-\ell-1) - s(s-1)}{m(m-1)}.$$

Case $\ell' = 1, h' = 2$.

To compute $p'_{b|b}$, consider that in this case a reconstructed pixel is black if both subpixels are black (and white otherwise). If a qualified set Q is fixed, by property PB, the matrices of the collection $C_B^Q(S')$ that have at least a 1 in each column are exactly $(m-s)(m-s-1)$. Thus, $p'_{b|b} = \frac{(m-s)(m-s-1)}{m(m-1)}$.

To compute $p'_{b|w}$, fix any qualified set Q and consider that by property PW, the matrices of the collection $C_B^Q(S')$ that have at least a 1 in each column are exactly $\ell(\ell-1)$. Thus, $p'_{b|w} = \frac{\ell(\ell-1)}{m(m-1)}$.

Hence, for the $(2,n)$-threshold scheme obtained by choosing $\ell' = 1$ the probabilistic factor is:

$$\beta_1 = \frac{(m-s)(m-s-1) - \ell(\ell-1)}{m(m-1)}.$$

Compare now the value of β for the above two cases trying to figure out whether one case is better than the other, i.e., the probabilistic factor is bigger and a better quality image is reconstructed. It is possible to express the difference $\beta_1 - \beta_0$ as

$$\beta_1 - \beta_0 = \frac{\psi_m(s) - \psi_m(\ell)}{m(m-1)}$$

where $\psi_m(x) = (m-x)(m-x-1) + x(x-1)$. The first and second derivatives of ψ with respect to x are $\frac{\partial \psi_m}{\partial x} = 4x - 2m$ and $\frac{\partial^2 \psi_m}{\partial x^2} = 4$, respectively. Hence, function $\psi_m(x)$ is a convex $\cup$ function of x, with a minimum at $x = m/2$. A simple algebra shows that, for n even, $s = \frac{m}{4} \cdot \frac{n-2}{n-1}$ and $\ell = \frac{m}{2}$ and that for n odd, $s = \frac{m}{4} \cdot \frac{n+1}{n}$ and $\ell = \frac{m}{2} \cdot \frac{n-1}{n}$. Since in both cases $s < \ell \le m/2$, it is possible to conclude that $\psi_m(s) > \psi_m(\ell)$ and then $\beta_1 > \beta_0$. A simple analysis shows that the limit of β_1 as n approaches infinity is $5/16 \simeq 0.31$, while the limit of β_0 as n approaches infinity is $3/16 \simeq 0.18$.

In the case of any m' Equations (5.2)–(5.5) can be used to compute the probabilities of probabilistic schemes, over all choices of ℓ' and h'. Table 5.1 gives the resulting values of the probabilistic factor β of S' over all possible choices of ℓ' and m', for the case of n. Since for $n = 3$ scheme S_D has $m = 3$ clearly for $m' = 3$ we get a deterministic scheme; such a scheme is obtained for $\ell' = 1$, $h' = 2$. It is worth to notice that for other choices of ℓ' the probabilistic factor of the scheme is 0 meaning that no probabilistic scheme with pixel expansion $m' = 3$ can be constructed with Construction 1. For the case of $m' = 2$ $\ell' = 1$ and $\ell' = 2$ are valid choices and they both yield a $\frac{1}{3}$-probabilistic scheme.

TABLE 5.1

Values of β for scheme S' for $n = 3$. The max over each row is in boldface.

$m' \backslash \ell', h'$	0,1	1,2	2,3
1	**1/3**	–	–
2	1/3	**1/3**	–
3	0	**1**	0

TABLE 5.2

Values of β for $n = 4$. The max over each row is in boldface.

$m' \backslash \ell', h'$	0,1	1,2	2,3	3,4	4,5	5,6
1	**1/3** 0.333	––	–	–	–	–
2	1/5 0.200	**7/15** 0.467	–	–	–	–
3	1/20 0.050	**1/2** 0.500	9/20 0.450	–	–	–
4	0	1/5 0.050	**4/5** 0.800	1/3 0.333	–	–
5	0	0	1/2 0.500	**1**	1/6 0.167	–
6	0	0	0	**1**	**1**	0

Table 5.2 gives the resulting values of the probabilistic factor β of S' over all possible choices of ℓ' and m', for the case of $n = 4$. Notice that for $n = 4$ scheme S_D has $m = 6$, hence, for $m' = 6$ deterministic schemes can be obtained (in this case both $\ell' = 3$ and $\ell' = 4$ yield a deterministic scheme). As reported in the table, also for $m' = 5$ a deterministic scheme can be obtained by choosing $\ell' = 3$. For the other possible choices of m', the table shows in boldface the biggest probabilistic factor β found.

Finally, Table 5.3 gives the resulting values of the probabilistic factor β of S' over all possible choices of ℓ' and m', for the case of $n = 5$. For $n = 5$ scheme S_D has $m = 10$, hence, m' can range from 2 to 10. As reported in

TABLE 5.3

Values of β for $n = 5$. The max over each row is in boldface and ϵ_1 and ϵ_2 denote very small values.

$m'\backslash \ell', h'$	0,1	1,2	2,3	3,4	4,5	5,6	6,7	7,8	8,9	9,10
1	**3/10**	–	–	–	–	–	–	–	–	–
2	4/15	**1/3**	–	–	–	–	–	–	–	–
3	19/120	**29/60**	31/120	–	–	–	–	–	–	–
4	1/14	44/105	**23/42**	17/105	–	–	–	–	–	–
5	1/42	11/42	**55/84**	10/21	1/12	–	–	–	–	–
6	ϵ_1	5/42	23/42	**16/21**	1/3	1/30	–	–	–	–
7	0	1/30	1/3	**5/6**	17/24	11/60	ϵ_2	–	–	–
8	0	0	2/15	2/3	**1**	8/15	1/15	0	–	–
9	0	0	0	2/5	**1**	**1**	3/10	0	0	–
10	0	0	0	0	**1**	**1**	**1**	0	0	0

the table, a deterministic schemes is obtained from $m' = 8$ (in such a case, by choosing $\ell' = 4$). Again the best probabilistic factor found is reported in boldface.

5.7 Probabilistic Schemes with Boolean Operations

One of the peculiar characteristics of VCS is the fact that the reconstruction of the secret image is done via the human visual system. This means that representing a black pixel as '1' and the white pixel as '0,' the reconstruction is performed via an "OR" operation of the superposed pixels contained in the shares. Relaxing such an assumption, it is possible to obtain different classes of VCS where the secret pixel is reconstructed after performing different boolean operations on the subpixels contained in the shares. For example XOR-based VCS have been proposed in [17]. As regards probabilistic VCS, several proposals have been done [20, 18, 8], where the basic operation for the reconstruction is no more based on the OR of the shared subpixels, and both the distribution and the reconstruction phases are consequently modified.

5.7.1 (2,n) Scheme for Binary Images (Wang)

In [20], Wang et al. proposed a probabilistic $(2, n)$-VCS based on binary XOR and AND operations, denoted with $\oplus$ and $\&$, respectively. The construction is given in Figure 5.3.

Let A be the binary matrix representing the secret image.

Distribution phase.
The dealer:

- generates $n+1$ random matrices $B_1, \ldots, B_{n+1}$; compute intermediate matrices $C_1, \ldots, C_n$ where $C_i = B_i \& A$, for $i = 1, \ldots, n$;

- distributes the n shares $A_1, \ldots, A_n$ where $A_I = B_{n+1} \oplus C_i$ for $i = 1, \ldots, n$.

Reconstruction phase. A qualified set $\{i_j, i_k\}$ of two participants reconstructs the secret image as follows:

- superimpose their shares executing binary XOR operation to obtain $A' = A_j \oplus A_k$.

FIGURE 5.1
A construction for (2,n) Boolean probabilistic VCS.

In Wang's scheme, a reverse coding of the pixel is adopted, assigning '0' to black pixel and '1' to white pixels. In this scheme it is easy to see that a black pixel in the secret image will be always mapped to a black pixel in each of the computed shares A_i because of the AND operation performed in the computation of the C_j (for a '0' pixel, $C_j = B_j \& 0 = 0$) and the XOR operation in the reconstruction phase (the contribution for $C_j = 0$ means that $A_j = B_{n+1} \oplus C_j = B_{n+1}$ and then $A' = A_j \oplus A_k = B_{n+1} \oplus B_{n+1} = 0$). The probabilities of reconstructing a black pixel are then $p_{b|b} = 1$ and $p_{w|b} = 0$. A white pixel will be reconstructed on the basis of the result of the XOR operation performed on the two matrices B_i and B_k, which are randomly selected. Then $p_{w|w} = p_{b|w} = 1/2$. Hence, the construction returns a $\beta = 1/2$ probabilistic scheme with contrast $1/2$.

The above presented scheme has been modified by Ulutas et al. in [18] and Chang et al. in [8]. In the first work the construction is modified by adding a mechanism to augment the probability of selecting a white (black) pixel given that a white (black) pixel was present in the secret image. Basically, matrices B_i are no more randomly generated as in Wang's scheme, but are selected in two different sets, one for white pixels and the other for black pixels, in order to maximize the probability of reconstructing a pixel of the right color. Then a random matrix R of the same size of the secret image is generated and used to compute the share. The reconstruction phase is not modified. The modified scheme is given in Figure 5.2.

In Chang et al's construction, a voting strategy is adopted to apply the Wang method to grayscale images [8]. In such a method, for each original pixel, the binary representation of its grayscale value is considered, and for each bit composing the representation, the Wang scheme is executed and repeated $2m + 1$ times in the generation phase. In the reconstruction phase, the voting method is used to recover the best result out of the shared bits in order to restore the original grayscale value of the pixel.

Let A be the binary matrix representing the secret image.

Distribution phase.
The dealer:

- generates a white image (D) and n shares $B_1, \ldots, B_n$ with D as secret image; compute intermediate matrices $C_1, \ldots, C_n$ where $C_i = B_i \& A$, for $i = 1, \ldots, n$;

- generates a random matrix R;

- distributes the n shares $A_1, \ldots, A_n$ where $A_I = R \oplus C_i$ for $i = 1, \ldots, n$

FIGURE 5.2
Ulutas et al. [18] construction for (2,n) Boolean probabilistic VCS.

5.7.2 (n,n) Scheme for Binary Images (Wang)

A construction similar to the one above described, was proposed by Wang also for a (n,n) probabilistic VCS. The construction is given in Figure 9.6

Let A be the binary matrix representing the secret image.

Distribution phase.
The dealer:

- generates $n - 1$ random matrices $B_1, \ldots, B_{n-1}$;
- distributes the n shares $A_1, \ldots, A_n$ where $A_1 = B_1$, $A_i = B_{i-1} \oplus B_i$, and $A_n = B_{n-1} \oplus A$.

Reconstruction phase. The n participants reconstruct the secret image as follows:

- superimpose their shares executing binary XOR operation to obtain $A' = A_1 \oplus A_2 \cdots \oplus A_n$.

FIGURE 5.3
A construction for (n, n) Boolean probabilistic VCS.

In such a scheme it is easy to see that a black pixel in the secret image will be always mapped to a black pixel in each of the computed shares A_i because of the and operation performed in the computation of the C_j (for a '0' pixel, $C_j = B_j$ & $0 = 0$) and the XOR operation in the reconstruction phase (the contribution for $C_j = 0$ means that $A_j = B_{n+1} \oplus C_j = B_{n+1}$ and then $A' = A_j \oplus A_k = B_{n+1} \oplus B_{n+1} = 0$. Then $p_{b|b} = 1$ and $p_{w|b} = 0$. A white pixel will be reconstructed according to the pixel returned after the XOR between the two matrices B_i and B_k, which are randomly selected. Then $p_{w|w} = p_{b|w} = 1/2$. Hence, the construction returns a $\beta = 1/2$ probabilistic scheme with contrast $1/2$.

The same construction has been extended by Wang et al. in [21] to deal with grayscale images. Basically each pixel is coded with a binary string $g - 1$ bits long, where g denotes the gray level number, and having a number of "1"s equivalent to the 1s present in the binary string obtained by the grey level g (without respecting the order). A pixel having grey level k in a range from 0 to $g - 1$, is represented by a binary string having $g - k$ '0's and $k - 1$ '1's. The construction returns a probabilistic scheme with $m = g - 1$.

5.8   Conclusions and Open Problems

The probabilistic model for VCS schemes has been first described in Yang [22], where the reconstruction of the secret pixel has been given in probabilistic terms, no more guaranteeing the correctness property of the traditional VCS schemes. A generalization of the model has been given then in [11], where probabilistic schemes with pixel expansion have been described, showing that there is a one-to-one correspondence between probabilistic schemes with no pixel expansion and deterministic schemes and that such a one-to-one mapping trades the probabilistic nature of the scheme with the contrast of the deterministic scheme. Probabilistic schemes with pixel expansion can be obtained from deterministic schemes and their probabilistic factors can be studied. While for (n, n)-threshold schemes it has been proved that there is a linear relation between the pixel expansion and the probabilistic factor, for $(2, n)$-threshold schemes no closed expression for the probabilistic factor has been found. Finally, a probabilistic model has been extended and alternative operations (instead of OR) have been considered for the distribution of the shares and the reconstruction of the secret images.

Bibliography

[1] Ateniese G., Blundo C., De Santis A., and Stinson D. R. (1996). Visual cryptography for general access structures. *Information and Computation*, 129, 86–106.

[2] Ateniese G., Blundo C., De Santis A., and Stinson D. R. (1996). Constructions and bounds for visual cryptography. Proceedings of ICALP 1996, Paderbon, Germany, 8–12 July, pp. 416–428, Springer Verlag LNCS 1099.

[3] Ateniese G., Blundo C., De Santis A., and Stinson D. R. (2001). Extended schemes for visual cryptography. *Theoretical Computer Science*, 250, 143–161.

[4] Blundo C., D'Arco P., De Santis A., and Stinson D. R. (2003). Contrast optimal threshold visual cryptography schemes. *SIAM J. on Discrete Mathematics*, 16, 224–261.

[5] Blundo C., De Bonis A., and De Santis A. (2001). Improved schemes for visual cryptography. *Designs, Codes, and Cryptography*, 24, 255–278.

[6] Blundo C. and De Santis A. (1998). Visual cryptography schemes with perfect reconstruction of black pixels. *Journal for Computers & Graphics*, 22, 449–455.

[7] Blundo C., De Santis A., and Stinson D. R. (1999). On the contrast in visual cryptography schemes. *Journal of Cryptology*, 12, 261–289.

[8] Chang C., Lin C., Le T.H., and Le H.B. (2008). A probabilistic visual secret sharing scheme for grayscale images with voting strategy. Proceedings of the International Symposium on Electronic Commerce and Security, 2008, pp. 184–188.

[9] Cimato S., De Prisco R., and De Santis A. (2004). Colored visual cryptography without color darkening. In Proceedings of the 4th Conference on Security in Communication Networks, Amalfi, Italy, 8–10 September, pp. 236–251, Springer Verlag LNCS 3352.

[10] Cimato S., De Prisco R. and De Santis A. (2005), Optimal colored threshold visual cryptography schemes. *Designs, Codes and Cryptography*, 35, 311–335.

[11] Cimato S., De Prisco R., and De Santis A., (2006). Probabilistic visual cryptography schemes. *The Computer Journal*, 49(1), 97-107.

[12] Hofmeister T., Krause M. and Simon H. U. (2000). Contrast-optimal *k* out of *n* secret sharing schemes in visual cryptography. *Theoretical Computer Science*, 240, 471–485.

[13] Ito R., Kuwakado H., and Tanaka H. (1999). Image size invariant visual cryptography. *IEICE Transactions on Fundamentals of Electronics, Communications and Computer Sciences*, E82A(10), 2172-2177.

[14] Krause M. and Simon H. U. (2003). Determining the optimal contrast for secret sharing schemes in visual cryptography. *Combinatorics, Probability and Computing*, 12, 285–299.

[15] Kuhlmann C. and Simon H. U. (2000). Construction of visual secret sharing schemes with almost optimal contrast. Proceedings of the 11th ACM-SIAM Symposium on Discrete Algorithms, San Francisco, USA, 9–11 January, pp. 262-272.

[16] Naor M. and Shamir A. (1994). Visual cryptography. Proceedings of Eurocrypt 94, Perugia, Italy, May 9–12, pp. 1–12, Springer Verlag LNCS 950.

[17] Tuyls P., Hollmann H.D.L., van Lint J.H., and Tolhuizen L. (2005). XOR-based visual cryptography schemes. *Designs, Codes, and Cryptography*, 37, 169–186.

[18] Ulutas M., Nabiyev V.V. Ulutas G., and Shamir A. (2009). A PVSS scheme based on Boolean operations with improved contrast. Proceedings of International Conference on Network and Service Security, Paris, 2009, pp. 1–5.

[19] Verheul E. R. and van Tilborg H. C. A. (1997). Constructions and properties of *k* out of *n* visual secret: Sharing schemes. *Designs, Codes, and Cryptography*, 11, 179–196.

[20] Wang D., Zhang L., Ma N., and Li X. (2006). Two secret sharing schemes based on Boolean operations. *Pattern Recognition* 40(10), 2776–2785

[21] Wang D., Li X., Yi F. Daoshun Wang, Xiaobo Li, and Feng Yi. (2007). Probabilistic (n, n) visual secret sharing scheme for grayscale images. In *Information Security and Cryptology, Lecture Notes in Computer Science*, Vol. 4990. Springer-Verlag, Berlin, pp.192–200

[22] Yang C.-N. (2004). New visual secret sharing schemes using probabilistic method. *Pattern Recognition Letters*, 25, 481–494.

[23] Yang C.-N. and Chen T.-S. (2005). Size-adjustable visual secret sharing schemes. *IEEE Trans. Fundamentals*, 88, 2471–2474.

[24] Yang C.-N. and Chen T.-S. (2005). Aspect ratio invariant visual secret sharing schemes with minimum pixel expansion. *Pattern Recognition Letters*, 26, 193–206.

[25] Yang C.-N. and Laih C.-S. (2000). New colored visual secret sharing schemes. *Designs, Codes, and Cryptography*, 20, 325–335.

6

XOR-Based Visual Cryptography

Daoshun Wang

Tsinghua University, China

Lin Dong

Tsinghua University, China

CONTENTS

6.1   Introduction

A (k, n) visual cryptography (VC) scheme [16] is a type of secret sharing scheme with the special property that a secret image can be recovered visually by the human eye and does not require any calculation on a computer. However, the recovered secret image has low quality. In this case, some researchers attempt to consider other different approaches to improve the quality (contrast) of the recovered image.

Lee et al. [12] presented a VC scheme using an XOR process to share a binary image. Since phase masks were placed on any path of the Mach-Zehnder interferometer for reconstruction, the method is impractical and expensive

[17]. Biham and Itzkovitz [2] investigated a VC system based on passive light polarization. This is more flexible than the Naor and Shamir schemes but cannot be modeled by an XOR [17]. Tuyls et al. [17, 18] gave a VC system that uses the polarization of light. The operation of the VC system is mathematically described by an XOR operation. In Section 6.3, we will show that a (n, n)-VC scheme with optimal resolution and contrast exist, and that $(2, n)$-VC schemes are equivalent to binary codes. Three explicit constructions for general k out of n schemes are also introduced [17, 13].

Viet and Kurosawa [21] noticed the phenomenon that most copy machines nowadays can change a black image into a white one and vice versa, then first gave a VC scheme with reversing for binary image. In the VC scheme with reversing, a dealer needs run the distribution phase of a VC scheme c times (with c arbitrary constant), hence requiring each participant to held c shadows. The almost ideal contrast of a recovered secret image is almost obtained for a large number of runs c. Cimato et al. [7] presented two elegant construction methods to improve the contrast and pixel expansion of the scheme in Reference [21]. In order to reduce run times of two schemes [21, 7], Yang et al. [25, 26] overcame the weakness of reversing only a based perfect VC scheme and first introduced a nonperfect black VC scheme, this approach uses a Boolean XOR operation for decoding. Reducing the stacking and reversing operations and minimizing number held by each participant, Hu and Tzeng [10] proposed an ideal contrast VC scheme with less reversing and stacking operations in only two runs. The scheme needs to perform XOR operations to decode the secret image. In section 6.4, we will introduce Yang et al.'s scheme and Hu and Tzeng's [25, 26] scheme with ideal contrast.

The construction of the above schemes is all based on basis matrices, so they may suffer pixel expansion and loss of contrast. Probabilistic VC schemes are proposed in [11, 24, 6] with no pixel expansion. However, the recovered secret image has low contrast. Wang et al. [23] proposed two secret sharing schemes based on a Boolean operation and the recovering operation is XOR. One scheme is $(2, n)$ for the binary image, and the other is (n, n) for the grayscale image. Both have no pixel expansion. Chao and Lin [5] improved Wang et al.'s [23] scheme in order to obtain a (k, n) scheme, which is fast and with a reasonable pixel expansion rate. In section 6.5, we introduce the $(2, n)$, (n, n), and (k, n) schemes.

6.2 Preliminaries

In a black and white (k, n)-visual cryptography (VC) scheme [16], the secret image consists of a collection of black and white pixels and each pixel is subdivided into a collection of m black and white subpixels in each of the n shares. These subpixels are printed in close proximity to each other so that the human visual system averages their individual black and white contributions. The collection of subpixels can be represented by an $n \times m$ Boolean matrix $S = [s_{ij}]$, where the element s_{ij} represents the j-th subpixel in the i-th share. A white subpixel is represented as a 0, and a black subpixel is represented as a 1. The white subpixels are let through the light while black subpixels stop it. $s_{ij} = 1$ if and only if the j-th subpixel in the i-th share is black. The gray level of the combined share obtained by stacking shares $i_1, \cdots, i_\gamma$ is proportional to the Hamming weight (the number of 1's in the vector V) $H(V)$ of the OR-ed ("OR" operation) m-vector $V = OR(i_1, \cdots, i_\gamma)$. Based on the definition of Naor and Shamir [16], Verheul and van Tilborg [20] gave a more general definition. Following the notation from [16, 20], a definition of k out of n XOR-based visual cryptography scheme is given by Tuyls in Reference [17]. A (k, n) VC scheme $S = (C_0, C_1)$ consists of two collections of $n \times m$ binary matrices C_0 and C_1. To share a white (black) pixel, the dealer randomly chooses one of the matrices in $C_0(C_1)$ and distributes its rows as shares among the n participants of the system. The following is the definition.

Definition 1 *[20] Let k, n, n, m, l be positive integers satisfying $1 \leq k \leq n$ and $m \geq h \geq l$. Let $Z(v)$ be the number of 0's in the vector v. A $[(k, n); m, h, l]$ visual cryptography (VC) scheme consists of two collections of $n \times m$ Boolean matrices C_0 and C_1 such that:*

1. *For any $s \in C_0$, the XOR v of any k of the n rows of s satisfies $Z(v) \geq h$.*

2. *For any $s \in C_1$, the XOR v of any k of the n rows of s satisfies $Z(v) \leq l$.*

3. *For any $i_1 < i_2 < \cdots < i_t$ in $\{1, 2, \cdots, n\}$ with $t < k$ the two collections of $t \times m$ matrices D_j for $j \in \{0, 1\}$, obtained by restricting each $n \times m$ matrix in C_j, to rows $i_1, i_2, \cdots, i_t$ are indistinguishable in the sense that they contain the same matrices with the same frequencies.*

The number h and l are called the white level and black level, respectively, of the scheme. The parameter m is called the block length or pixel expansion and determines the resolution of the scheme. The contrast α is defined as $\alpha = \frac{(h-l)}{h+l}$. Note that $\alpha \in [0, 1]$ and α is maximal when $l = 0$. A scheme with $l = 0$ are called maximal contrast schemes. When $\alpha = 1$, the contrast is defined as an ideal contrast.

The following symmetry property follows very easily and is therefore stated without proof.

Proposition 1 *[17] Let $S = (C_0, C_1)$ be a $[(k, n); m, h, l]$ VC scheme. Let $\hat{C}_i$ be obtained from C_i by replacing zeroes by ones and vice versa. If k is even, then the scheme $(\hat{C}_0, \hat{C}_1)$ is a $[(k, n); m, h, l]$ scheme as well. If k is odd, then $(\hat{C}_0, \hat{C}_1)$ is a $[(k, n); m, m - l, m - h]$ scheme with contrast α given by $\hat{\alpha} = (h - l)/(2m - l - h)$. It follows that $\hat{\alpha} > \alpha$ whenever $l + h > m$.*

6.3   Visual Cryptography Scheme Using the Polarization of Light

In this section, we will introduce XOR-based visual cryptography scheme. They are constructed using basis matrices and the secret image is recovered by an XOR certain number of shares. We will show how to construct $(2, n)$, (n, n), and (k, n) XOR-based visual cryptography schemes as follows.

6.3.1   (2, n) Scheme

In this section, we show how to construct a $(2, n)$ XOR-based visual cryptography scheme. The $(2, n)$ scheme is equivalent to binary error-correcting codes. By a (m, n, d) code, that is, a binary code of length m consists of n words and a minimum Hamming distance of at least d.

Theorem 1 *[17] Let m, l and h be positive integers such that $l < h \leq m$. The three following statements are equivalent.*

(i)   *$A [(2, n); m, m, l]$ VC scheme exists.*

(ii) *$A [(2, n); m, h, l]$ VC scheme exists.*

(iii) *A binary $(m, n, m - l)$ code exists.*

Proof: It is clear that (i) implies (ii).

In order to show that (ii) implies (iii), let $S = (C_0, C_1)$ be a $[(2, n); m, h, l]$ VC scheme. Take a matrix $A_1 \in C_1$ and let C consist of the rows from A_1. As S is a $[(2, n); m, h, l]$ VC scheme, the Hamming distance between two words from C is at least $m - l$. Consequently, C is a $(m, n, m - l)$ code. Finally, to show (iii) implies (i), let C be a binary $(m, n, m - l)$ code. For $c \in C$, let $A(c)$ denote the $n \times m$ matrix for which each row equals c. Moreover, let B be an $n \times m$ matrix containing each word from C as a row, and for $0 \leq i \leq n - 1$, let $B(i)$ be the matrix obtained by a cyclic shift of the rows of B over i positions. We claim that $(C_0, C_1) = (\{A(c)|c \in C\}, \{B(0), B(1), \cdots, B(n - 1)\})$ is a $[(2, n); m, m, l]$ scheme. It is clear that both collections contain n matrices, and that in each row, each word from C occurs in one matrix from C_0 and in one matrix from C_1, showing the indistinguishability. The sum of any two rows from a matrix in C_0 equals 0. Finally, the Hamming distance between any two rows of a matrix from C_1 is at least $m - l$, showing that the XOR of these two rows contains at most $m - (m - l) = l$ zeros.   ■

An example is given as follows.

Example 1 *Let C be a binary $(3, 3, 2)$ code containing words $100, 010$ and 001. Construct C_0 and C_1 as follows:*

$$C_0 = \left\{ \begin{bmatrix} 1 & 0 & 0 \\ 1 & 0 & 0 \\ 1 & 0 & 0 \end{bmatrix}, \begin{bmatrix} 0 & 1 & 0 \\ 0 & 1 & 0 \\ 0 & 1 & 0 \end{bmatrix}, \begin{bmatrix} 0 & 0 & 1 \\ 0 & 0 & 1 \\ 0 & 0 & 1 \end{bmatrix} \right\},$$

$$C_1 = \left\{ \begin{bmatrix} 1 & 0 & 0 \\ 0 & 1 & 0 \\ 0 & 0 & 1 \end{bmatrix}, \begin{bmatrix} 0 & 1 & 0 \\ 0 & 0 & 1 \\ 1 & 0 & 0 \end{bmatrix}, \begin{bmatrix} 0 & 0 & 1 \\ 1 & 0 & 0 \\ 0 & 1 & 0 \end{bmatrix} \right\}.$$

(C_0, C_1) is a $[(2,3); 3, 3, 1]$ scheme. The contrast $\alpha = \frac{(h-l)}{h+l} = \frac{2}{4} = \frac{1}{2}$.

6.3.2 (n, n) Scheme

In this section, we show how to construct a (n, n) XOR-based visual cryptography scheme.

Proposition 2 *[17] Let C_0 and C_1 be the set of all binary vectors of length n with even, odd number of ones, respectively. Then (C_0, C_1) is an $[(n, n); 1, 1, 0]$ scheme.*

An example is given as follows.

Example 2 *Let C_0 and C_1 be*

$$C_0 = \left\{ \begin{bmatrix} 0 \\ 0 \\ 0 \end{bmatrix}, \begin{bmatrix} 1 \\ 1 \\ 0 \end{bmatrix}, \begin{bmatrix} 1 \\ 0 \\ 1 \end{bmatrix}, \begin{bmatrix} 0 \\ 1 \\ 1 \end{bmatrix} \right\}, C_1 = \left\{ \begin{bmatrix} 1 \\ 0 \\ 0 \end{bmatrix}, \begin{bmatrix} 0 \\ 1 \\ 0 \end{bmatrix}, \begin{bmatrix} 0 \\ 0 \\ 1 \end{bmatrix}, \begin{bmatrix} 1 \\ 1 \\ 1 \end{bmatrix} \right\}$$

(C_0, C_1) is a $[(3,3); 1, 1, 0]$ scheme with contrast $\alpha = \frac{(h-l)}{h+l} = 1$.

6.3.3 (k, n) Scheme

In this section we will introduce three (k, n) XOR-based visual cryptography schemes. Construction 1 and Construction 2 are given by Tuyls (see more detail in Reference [17]). Construction 3 comes from Droste's OR-based visual cryptography [9] and Liu et al. [13] shows that it is also a (k, n) XOR-based visual cryptography scheme.

Before introducing Construction 1 and Construction 2, some definitions and theorems will be given, which are used in both constructions.

For describing the constructions, we use the following notation. If A is a binary matrix, then $P(A)$ is the multi set of matrices obtained by permuting the columns of A. Moreover, we use the concept of (k, n) pairs, defined as follows.

Definition 2 *[17]A pair (A, B) of binary $n \times m$ matrices is called a (k, n) pair if there exist numbers $a_1, \cdots, a_k$ and $b_1, \cdots, b_k$ such that*

1. *For each i with $1 \le i \le k$, the weight of the sum of any i rows from A equals a_i and the weight of the sum of any i rows from B equals b_i, and*

2. *$a_i = b_i$ for $1 \le i < k$, and $a_k \ne b_k$.*

The importance of this definition stems from the following theorem.

Theorem 2 *[17] If (A, B) is a (k, n)-pair of $n \times m$ matrices, then $(P(A), P(B))$ is a $[(k, n); m, h, l]$ VC scheme with $h = \max(m - a_k, m - b_k)$ and $l = \min(m - a_k, m - b_k)$. Here, a_k and b_k denote the weight of the sum of any k rows from A and B, respectively.*

Some notations are given here for later use. For a binary vector $\mathbf{v}$ of length m, we define $Z(\mathbf{v})$ as the number of zeros in $\mathbf{v}$, and $H(v)$ as its number of ones. Moreover, we define the unbalance $\delta(\mathbf{v})$ of v by $\delta(\mathbf{v}) = Z(\mathbf{v}) - H(\mathbf{v})$. Note that $\delta(\mathbf{v}) = m - 2H(v)$. For later use, we also observe that $\delta(\mathbf{v}) = \sum_{j=1}^{m} (-1)^{v_j}$.

With each binary $n \times m$ matrices A we associate two vectors $\delta(A)$ and $N(A)$ of length 2^n, with the components indexed by binary vectors of length n. For each binary vector $\mathbf{x}$ of length n, the $\mathbf{x}$-th component $\delta_{\mathbf{x}}(A)$ of $\delta(A)$ is defined as $\delta_{\mathbf{x}}(A) = \delta(\mathbf{x}^T A)$, the unbalance of the sum of the rows in A whose index i satisfies $x_i = 1$; also, the $\mathbf{x}$-th component $N_{\mathbf{x}}(A)$ of $N(A)$ is defined as the number of columns of A that are equal to $\mathbf{x}$. Lemma 1 will show that the vectors $\delta(A)$ and $N(A)$ can be computed from each other. To make this precise, define the $2^n \times 2^n$ matrix H by $H(\mathbf{x}, \mathbf{y}) = (-1)^{(\mathbf{x}, \mathbf{y})}$, where $\mathbf{x}$, $\mathbf{y}$ are binary vectors of length n and $(\mathbf{x}, \mathbf{y}) = \sum_{i=1}^{n} x_i y_i$ denotes the inner product of $\mathbf{x}$ and $\mathbf{y}$. Then we have the following lemma:

Lemma 1 *[17]*

1. The matrix H is a Hadamard matrix, that is, $HH^T = 2^n I$.

2. The vectors $\delta(A)$ and $N(A)$ are related by $\delta(A) = HN(A)$.

We are now in the position to prove a generalization of Theorem 2. Before stating it, we need one more notation: if A is a $n \times m$ matrix, then $w(A_I)$ denotes the weight of the sum of the rows of A indexed by I. Note that $w(A_I) = \frac{1}{2}(m - \delta_{\chi(I)}(A))$, where $\chi(I)$ is the characteristic vector of the set I.

Theorem 3 *[17] Let A and B be $n \times m$ matrices such that for each $I \subset \{1, 2, \cdots, n\}$ of size at most $k - 1$, $w(A_I) = w(B_I)$. Assume moreover that there exist integers h and l such that $h > l$ and that for each $I \subset \{1, 2, \cdots, n\}$ of size k, $m - w(A_I) \geq h$ and $m - w(B_I) \leq l$. Then $(P(A), P(B))$ is a $[(k, n); m, h, l]$ VC scheme.*

Proof The only nontrivial thing we have to prove is the indistinguishability. Let $I = \{i_1, \cdots, i_t\}$ be a subset of $\{1, 2, \cdots, n\}$ of size $t < k$. Let $\overline{A}$ and $\overline{B}$ denote the restrictions of A and B to the t rows indexed by I. Let $\mathbf{x} = \{x_1, \cdots, x_t\}$ be a binary vector of length t, and let $\tilde{\mathbf{x}}$ be the binary vector of length n for which entry i_j is equal to x_j for $j = 1, 2, \cdots, t$, and whose other entries equal zero. It is clear that $\delta_{\mathbf{x}}(\overline{A}) = \delta_{\tilde{\mathbf{x}}}(\overline{A})$. As $\tilde{\mathbf{x}}$ has weight at most t, we find, using properties of A and B, that $\delta_{\mathbf{x}}(\overline{A}) = \delta_{\mathbf{x}}(\overline{B})$.

Hence, by Lemma 1, the number of columns $N_y(\overline{A})$ in $\overline{A}$ and the number of columns $N_y(\overline{B})$ in $\overline{B}$ of type $\mathbf{y} = \{y_0, \cdots, y_{t-1}\}$ are equal for all binary vectors $\mathbf{y}$ of length t. As a consequence, the matrices $\overline{A}$ and $\overline{B}$ are equal up to a column permutation, which readily implies indistinguishability in the rows under consideration in the multisets $P(A)$ and $P(B)$. ∎

Construction 1

Now an explicit construction of (k, n) VC schemes for all n and all k with $1 \leq k \leq n$ will be given below. According to Theorem 3, it is sufficient to construct (k, n)-pairs for all such n and k. We will obtain such pairs by concatenation of matrices from a fixed collection of building blocks.

For each n and w with $0 \leq w \leq n$, we let the $n \times \binom{n}{w}$-matrix C_w consist of all the $\binom{n}{w}$ different $0-1$ column vectors of weight w, in any order.

In the sequel, we will need an explicit expression for the weight of the sum of any j rows from C_w. In Lemma 2, we state the result. Here and in what follows, we will use the standard convention that $\binom{n}{k} = 0$ whenever $k < 0$ or $k > n$.

Lemma 2 *[17] The weight of the sum of any j rows, $1 \leq j \leq n$, from C_w does not depend on the choice of the rows and is equal to $M_{j,w}$, where*

$$M_{j,w} = \sum_{i \ odd} \binom{j}{i} \binom{n-j}{w-i}.$$

Let $\lambda = (\lambda_0, \cdots, \lambda_n)^T$ be a vector of length $n + 1$ with nonnegative integer entries. We define the matrix $C(\lambda)$ to be the matrix consisting of the concatenation of λ_0 copies of C_0, λ_1 copies of C_1, $\cdots$, λ_n copies of the matrix C_n. It is clear that $c_0(\lambda)$, the number of columns of $C(\lambda)$, satisfies $C_0(\lambda) = \sum_w \lambda_w \binom{n}{w}$.

According to Lemma 2, the weight of the sum of any j rows of $C(\lambda)$ equals $c_j(\lambda)$, where

$$c_j(\lambda) = \sum_w \lambda_w \sum_{i \ odd} \binom{j}{i} \binom{n-j}{w-i}.$$

Hence, if we define $c(\lambda) = (c_0(\lambda), \cdots, c_n(\lambda))^T$, then the above can also be written as $c(\lambda) = M\lambda$, where M is the $(n+1) \times (n+1)$ matrix with entries the $M_{j,w}$ as defined in Lemma 2 for $j \geq 1$ and with $M_{0,w} = \binom{n}{w}$. Now suppose that λ and μ are nonnegative integer vectors such that $M\lambda$ and $M\mu$ agree in position $0, 1, \cdots, k-1$, but differ in position k. Then $(C(\lambda), C(\mu))$ is a (k, n) pair of matrices to which we can apply Theorem 2. The way to find such vectors λ and μ is described in Theorem 5, which uses Lemma 3, Corollary 4, and Lemma 4 below.

Lemma 3 *[17] For $1 \leq j \leq n$, $1 \leq w \leq n$, we have that $M_{j,w} = \sum_{k \geq 1} (-2)^{k-1} \binom{j}{k} \binom{n-k}{w-k}$.*

The following corollary is a direct consequence of Lemma 3.

Corollary 4 *Let R and L be the matrices defined as $R_{k,w} = \binom{n-k}{w-k}$ for $0 \leq k$, $w \leq n$ and $L_{0,0} = 1$, $L_{i,0} = L_{0,i} = 0$ if $i > 0$, and $L_{j,k} = (-2)^{k-1} \binom{j}{k}$ if $1 \leq j, k \leq n$. Then $M = LR$.*

We define the $(n+1) \times (n+1)$ matrix S by $S_{i,j} = (-1)^{i+j} \binom{n-i}{j-i}$.

Lemma 4 *[17]The matrices R and S are inverses of each other.*

Theorem 5 *[17] Let $1 \leq k \leq n-1$. Let $\theta = (\theta_0, \theta_1, \cdots, \theta_n)$ be an integer-valued vector such that $\theta_j = 0$ if $0 \leq j \leq k-1$, and $\theta_k \neq 0$, and let $\phi := S\theta$. For $0 \leq j \leq n$, we define*

$$\lambda_j = \max(0, \phi_j) \text{ and } \mu_j = -\min(0, \phi_j).$$

Then λ and μ are vectors with nonnegative integer entries, and $(C(\lambda), C(\mu))$ is a (k, n) pair. The parameters of the corresponding $[(k, n); m, h, l]$ VC scheme satisfy the following equations:

$$m = \frac{1}{2} \sum_{w=0}^{n} |\phi_w| \binom{n}{w}, h - l = 2^{k-1} |\theta_k|,$$

and $h + l = m + \frac{1}{2} \sum_{w=0}^{n} |\phi_w| \sum_i (-1)^i \binom{k}{i} \binom{n-k}{w-i}.$

Proof As S has integer entries, ϕ has integer entries, hence λ and μ have nonnegative integer entries. Using Corollary 4, Lemma 4, and the fact that $\phi = \lambda - \mu$, we find that

$$c(\lambda) - c(\mu) = M\lambda - M\mu = M(\lambda - \mu) = M\phi = LRS\theta = L\theta.$$

As L is a lower triangular matrix, and $\theta_j = 0$ if $j < k$, it follows that $c_j(\lambda) - c_j(\mu)$ if $0 \leq j \leq k-1$, and that

$$h - l = |c_k(\lambda) - c_k(\mu)| = |L_{k,k}\theta_k| = 2^{k-1}|\theta_k|.$$

Moreover, $2m = c_0(\lambda) + c_0(\mu) = c_0(\lambda + \mu) = c_0(|\phi|)$, and similarly $(m - h) + (m - l) = c_k(\lambda + \mu) = c_k(|\phi|)$. ∎

We will give an example to illustrate the construction.

Example 3 *Take* $\theta = (0, \cdots, 0, 1, 2, \cdots, n - k, n - k + 1)^T$. *As* $\phi = S\theta$, *we have by definition*

$$\phi_i = \sum_{v=k-i-1}^{n-i} (-1)^v \binom{n-i}{v} (v - k + i + 1).$$

If $k = 3$, *then* $\phi = (2 - n, 1, 0, 0, \cdots, -1, n - 2)$. *Consequently,* $\lambda = (0, 1, 0, \cdots, 0, n - 2)$ *and* $\mu = (n - 2, 0, \cdots, 0, 1, 0)$. *That is to say,* $A = C(\lambda)$ *consists of the* $n \times n$ *identity matrix and* $n - 2$ *columns of weight* n, *while* $B = C(\mu)$ *consists of* $n - 2$ *all-zero columns and furthermore contains each column of weight* $n - 1$ *once. It is clear that* A *and* B *both contain* $2n - 2$ *columns. Straightforward computations show that* $a_1 = b_1 = n - 1$, $a_2 = b_2 = 2$, $a_3 = n + 1$, $b_3 = n - 3$. *As a consequence, we obtain a* $[(3, n); 2n - 2, n + 1, n - 3]$ *VC scheme with* $\alpha = 2/(n - 1)$.

Construction 2

In general, it seems hard to give manageable expression for the physical parameters of the schemes obtained with Construction 1. Tuyls [17] gave another explicit construction of a (k, n) VC scheme that has the virtue that the physical parameters of these schemes can readily be computed in Reference [17]. For constructing these matrices, they used maximum distance separable (MDS) codes over $GF(q)$, the finite field with q elements. An $[n, k]$ MDS code over $GF(q)$ consists of q^k vectors of length n with entries from $GF(q)$ such that any two codewords have a Hamming distance of at least $n - k + 1$. It is known that such a code exists whenever $q + 1 \geq n$. Therefore, we choose $q \geq n - 1$.

Lemma 5 *[17] Let* C *be an* $[n, k]$ *MDS code over* $GF(q)$. *In any set of* k *positions, each of the* q^k *possible patterns occurs in exactly one of the words of* C.

Proof Fix k positions, two codewords that agree in these positions, differ in at most $n - k$ positions. We conclude that each of the q^k patterns agrees with at most one codeword. As the number of patterns equals the number of codewords, each pattern agrees with exactly one codeword in the given positions. ∎

Let C be an $[n, k]$ MDS code over $GF(q)$. Let $U(C)$ be an $n \times q^k$ matrix over $GF(q)$ in which each word from C occurs as a column once, and let $A(C)$ be the binary $n \times q^k$ matrix obtained from $U(C)$ by replacing each nonzero symbol in $GF(q)$ by a '1,' and the zero symbol in $GF(q)$ by a '0.'

Proposition 3 *[17] Let* $1 \leq j \leq k$, *the sum of any* j *rows of* $A(C)$ *has weight* $\frac{1}{2} q^{k-j} [q^j - (2 - q)^j]$.

Proof Consider j rows from $U(C)$. Lemma 5 implies that each of the q^j possible patterns occurs in these j positions in q^{k-j} words from C. The number of patterns with w nonzero elements equals $\binom{j}{w}(q-1)^w$. Each such pattern yields a column in $A(C)$ with exactly w ones in the j prescribed rows. Consequently, the number of ones in the sum of the j rows from $A(C)$ under consideration equals

$$q^{k-j}\sum_{w \text{ odd}}\binom{j}{w}(q-1)^w = \frac{1}{2}q^{k-j}((q-1+1)^j - (-1)^j(q-1-1)^j).$$

∎

Proposition 4 *[17] The sum of any $k+1$ rows of $A(C)$ has a weight $\frac{1}{2q}(q^{k+1} - (2-q)^{k+1}) - \frac{q-1}{q}2^k$.*

Proof Consider $k+1$ positions in C. Any two distinct words from C differ in at least two of these positions. That is, restricted to these $k+1$ positions, C is a $[k+1,k,2]$ code over $GF(q)$. The number of words of weight w in such a code equals

$$b_w = \binom{k+1}{w}(q-1)\sum_{j=0}^{w-2}(-1)^j\binom{w-1}{j}q^{w-2-j}$$

$$= \binom{k+1}{w}\left(\frac{(q-1)^{w-1}-(-1)^{w-1}}{q}\right).$$

The weight of the sum of the rows of $A(C)$ corresponding to the $k+1$ chosen positions is obtained by summing the above expressions of b_w over all odd w.

∎

Theorem 6 *[17] Let $2 \le k \le n-1$, and let q be a prime power not smaller than $n-1$. There exists a $[(k,n);q^k,\frac{1}{2}(q^k+(-1)^k(q-2)^k+(q-1)2^k),\frac{1}{2}(q^k+(-1)^k(q-2)^k)]$ VC scheme with contrast $(q-1)2^{k-1}/[q^k+(-1)^k(q-2)^k+(q-1)2^{k-1}]$.*

Proof Let C be an $[n,k]$ MDS code over $GF(q)$, and let D be an $[n,k-1]$ MDS code over the same field. In the notation of this section, let A equal $A(C)$, and let B equal the concatenation of q copies of $A(D)$. By combining the above results, (A,B) is a (k,n) pair, and $(P(A),P(B))$ is a VC scheme with parameters as claimed in the theorem.

∎

Bounds on the parameters m, h and l

Tuyls provided bounds on the parameters of a (k,n)-VC scheme. We start by proving that maximal contrast schemes ($l = 0$) do not exist. We note that for OR-based schemes maximal contrast schemes can always be constructed.

Proposition 5 *[17] Let $3 \le k < n$. There exists neither a $[(k,n);m,h,0]$ VC scheme, nor a $[(k,n);m,m,l]$ VC scheme.*

Proof Let $S = (C_0, C_1)$ be a $[(k,n); m, h, 0]$ VC scheme and let $B \in C_1$. Denote by σ^1, σ^2 two arbitrary rows in B. Since $n - 2 \geq k - 1$, B contains (at least) $k - 1$ more rows. We denote these rows by $\sigma^3, \cdots, \sigma^{k+1}$. Since S is a scheme with $l = 0$, the XOR of $\sigma^1, \sigma^3, \cdots, \sigma^{k+1}$ is the all-one vector, as is the XOR of $\sigma^2, \sigma^3, \cdots, \sigma^{k+1}$. If follows that $\sigma^1 = \sigma^2$, so all rows of B are equal.

Next, let $A \in C_0$ and consider row i and j of A. As $k \geq 3$, the indistinguishability property of Definition 1 implies that there is a $B \in C_1$ that agrees with A in these rows. As all rows of B are equal, the i-th and j-th row of A are equal. Since i and j are arbitrary, all rows of A are equal, so $A = B$, a contradiction.

The second statement follows from an analogous reasoning. ∎

The next two propositions show that XOR-based VC schemes with odd and even k fundamentally differ.

Proposition 6 *[17] Let k be odd, and let $k < n$. For each $\epsilon > 0$, there are integers m, h, and l such that $l/m < \epsilon$ and a $[(k,n); m, h, l]$ VC scheme exists.*

Proposition 7 *[17] Let k be even, and let $k < n$. If a $[(k,n); m, h, l]$ VC scheme exists, then $l/m \geq 1/(k+1)$.*

Corollary 7 *[17] For even $k < n$, the contrast of a k out of n VC scheme is at most $k/(k+2)$.*

Proof Let S be a $[(k,n); m, h, l]$ scheme. By definition, the contrast α is equal to $(h-l)/(h+l)$. It is clear that α is increasing in h, and so $\alpha \leq (m-l)/(m+l)$. As $(m-l)/(m+l)$ is decreasing in l, we obtain an upper bound on α by plugging in the upper bound for l from Proposition 7. ∎

Lemma 6 *[17] Let k be an even integer. Let B be a binary matrix with n rows such that the sum of any k rows from B differs from $\mathbf{0}$. Then B has at least $n - k + 2$ distinct rows.*

Lemma 7 *[17] Let (C_0, C_1) be a $[(k,n); m, h, l]$ VC scheme with $k \geq 3$, and let c_1 and c_2 be two rows of a matrix in C_0 and hence also two rows of some matrix in C_1. Then, the Hamming distance between c_1 and c_2 satisfies*

$$d(c_1, c_2) \leq \min\{2l, 2(m - h)\}.$$

Proposition 8 *[17] Let k be even, $k \geq 4$. If a $[(k,n); m, h, l]$ VC scheme exists, then $n - k + 1 \leq \sum_{i=0}^{\min(l, 2(m-h))} \binom{m}{i}$.*

Proof Let k be even, $k \geq 4$, and let $S = (C_0, C_1)$ be a $[(k,n); m, h, l]$ scheme. Let B be a matrix in C_1. As $l \neq m$, no k rows of B add to the all-zero word. Lemma 6 implies that B has at least $n - k + 2$ distinct rows. Since according to Lemma 7, all rows from B have a Hamming distance at most $2(m - h)$ to its top row, we obtain $n - k + 2 \leq \sum_{i=0}^{2(m-h)} \binom{m}{i}$.

Now we assume without loss of generality that the top $n - k + 1$ rows of B are distinct. Let $\mathbf{c}$ be the sum of the $k - 1$ bottom rows of B. For $1 \leq i \leq n - k + 1$, the sum of $\mathbf{c}$ and the i-th row of B contains at most l ones; that is to say, the i-th row of B has a Hamming distance at most l to the complement of $\mathbf{c}$. As the $n - k + 1$ top rows of B are distinct, $n - k + 1$ is at most the number of vectors at distance at most l from the complement of $\mathbf{c}$, so $n - k + 1 \leq \sum_{i=0}^{l} \binom{m}{i}$. ∎

Construction 3

Droste [9] proposed an algorithm to construct a (k, n)-VC scheme under the OR operation. Liu et al. [13] proved that the basis matrices constructed by that algorithm are also the basis matrices of a (k, n)-VC scheme under the XOR operation. Droste's algorithm can be described as follows.

For easy specification, we call a column of a Boolean matrix with an even number of 1's even and otherwise odd. If B is a Boolean matrix, we define $P(B)$ as the multiset of matrices obtained by permuting the columns of B, i.e., each permutation corresponds exactly to one element of $P(B)$.

The following lemma will be needed later.

Lemma 8 *[9] Let B_0 and B_1 be two $n \times m$ Boolean matrices so that there exist $m - 2^{k-1}$ column vectors $v_1, \cdots, v_{m-2^{k-1}} \in \{0,1\}^k$ with the following property: for every $\{i_1, \cdots, i_k\} \subseteq \{1, \cdots, n\}$ the restriction of B_0 (resp. B_1) to the rows $i_1, \cdots, i_k$ contains every even (resp. odd) column of length k exactly once and all columns $v_1, \cdots, v_{m-2^{k-1}}$. Then $P(B_0)$ and $P(B_1)$ are a k out of n secret sharing scheme with relative contrast $1/m$; here the contrast is defined as $\alpha = \frac{h-l}{m}$.*

The main idea for construction is to start with an empty matrix (which has no columns) and, for various $q \in \{0, \cdots, n\}$ and all $\binom{n}{q}$ columns, which have exactly q 1's. Because of the symmetry of this construction with respect to rows, all restrictions of such a matrix of k rows contain the same columns. And one can exactly determine which columns they contain.

Lemma 9 *[9] For $q \in \{0, \cdots, n\}$ let B be an $n \times \binom{n}{q}$ Boolean matrix that contains every column with q 1's exactly once. Then every restriction of B to k rows (with $k \leq n$) contains every column with p 1's exactly $\binom{n-k}{q-p}$-times (where $p \in \max\{0, q - (n-k), \cdots, \min(q,k)\}$).*

Lemma 9 shows a possibility to expand a matrix, if we add to all its restrictions every column with p 1's exactly once: just add all columns with $q = p$ or $q = p + n - k$ 1's to the entire matrix. So a subroutine ADD(p, B) adds to each restriction of k rows of a matrix B every column with p 1's by adding columns to the entire matrix.

ADD(p,B)

Step 1: If $p \leq k - p$, add all the columns with $q = p$ 1's to B.
Step 2: If $p \geq k - p$, add all the columns with $q = p + n - k$ 1's to B.

The subroutine makes it easy to construct basis matrices B_0 (resp. B_1) whose restrictions always contain every even (resp. odd) column. But besides these columns, every restriction of B_0 and B_1 can contain remaining columns (which are the same for all restrictions of one matrix because of the construction principle). To be appropriate for a k out of n scheme these remaining columns have to be the same for B_0 and B_1 (see Lemma 8). So the remaining columns of every restriction of B_0, which are not remaining columns of every restriction of B_1, called the rest of B_0, have to be added to every restriction of B_1 and vice versa. In most cases, these added columns will create new rests that cause new columns to be added.

Droste's algorithm

Step 1: For all even $p \in \{0, \cdots, k\}$, call ADD(p, B_0).
Step 2: For all odd $p \in \{0, \cdots, k\}$, call ADD(p, B_1).
Step 3: While the rests of B_0 and B_1 are not empty:

   (a) Add to B_0 all columns adjusting the rest of B_1 by calling ADD.

   (b) Add to B_1 all columns adjusting the rest of B_0 by calling ADD.

Theorem 8 shows that Droste's algorithm also generates a (k,n)-VC scheme under the XOR operation, by Liu et al. [13].

Theorem 8 *[13] Droste's algorithm generates the basis matrices of a (k,n)-VCS, B_0 and B_1, under the XOR operation.*

Proof We need to prove that the basis matrices B_0 and B_1 satisfy the contrast and security conditions of Definition 1.

First, for the contrast condition, we need to prove that the Hamming weight of the stacking (XOR operation) of any k out of n rows of B_0 is less than that of B_1. Denote B_0^k (resp. B_1^k) as the submatrix generated by restricting to arbitrary k rows of B_0 (resp. B_1). According to the Steps 1 and 2 in Droste's algorithm, it is clear that all the even (resp. odd) columns appear in B_0^k (resp. odd). Denote I_0^k (resp. I_1^k) as the matrix whose columns are all the even (resp. odd) columns of length k. Because Droste's algorithm terminates when the rests of B_0 and B_1 are empty, it implies that the remaining columns of B_0 and B_1 are the same, that is, $B_0^k \backslash I_0^k = B_1^k \backslash I_1^k$. Denote R as the remaining columns of B_0 and B_1, we have $B_0^k = I_0^k \cup R, B_1^k = I_1^k \cup R$. Because the XOR (operation) of the entries of an even (resp. odd) column is 0 (resp. 1), we have the result that the Hamming weight of the stacking (XOR operation) of the rows of B_0^k is less than that of B_1^k. Hence, the contrast condition is satisfied.

Second, for the security condition, we need to prove that the submatrices of any less than k rows of B_0 and B_1 have the same columns, and only in such a case, all the column permutations of the two submatrices will generate the same collection, that is, the security condition is satisfied. Denote B_0^t (resp. B_1^t) as the submatrix generated by restricting to arbitrary t rows of B_0 (resp. B_1), where $t < k$. Denote B_0^k (resp. B_1^k) as the submatrix generated by concatenating B_0^t (resp. B_1^t) and arbitrary $k - t$ rows chosen from the remaining rows of B_0 (resp. B_1) (other than the rows in B_0^t and B_1^t). As discussed above, we have $B_0^k = I_0^k \cup R$, $B_1^k = I_1^k \cup R$, where I_0^k (resp. I_1^k) is the matrix that contains all the even (resp. odd) columns of length k. Note that I_0^k and I_1^k are the basis matrices of a (k, k)-VC scheme proposed in References [16, 3]. We have the result that the submatrices generated by restricting to any t rows of B_0 and B_1 have the same columns. Hence, the submatrices generated by restricting to any t rows of B_0 and B_1 have the same columns, that is, the security condition is satisfied. ∎

Besides the above schemes, Liu et al. [15] proposed a step construction to construct XOR-based visual cryptography for general access structure by applying $(2, 2)$-VCS recursively, where a participant may receive multiple share images. Readers can refer to [15] for more detail. An approach to construct extended XOR-based visual cryptography was proposed in Reference [14].

6.4 Visual Cryptography Scheme with Reversing

First we introduce the following notations that will be used throughout the section. Let $A||B$ denote the concatenation of two matrices A and B of the same number of rows. Let $|X|$ be the number of elements in set X. The symbol "$\oplus$" denotes an XOR operation. Let $GRAY(P)$ be the gray level of a pixel P and defined as $GRAY(P) = |black\ pixel|/m$.

6.4.1 (k, n) VC Scheme Using Cyclic-Shift Operation

Let the shadow image $s = [s_{ijk}]$, and the element s_{ijk} is the secret pixel s_{ij} in a $(W \times H)$-pixel secret image replaced by m subpixels $(s_{ij1}, s_{ij2}, \cdots, s_{ijm})$, where $i \in [1, W]$, $j \in [1, H]$, and $k \in [1, m]$. The cyclic-shift operation is $\Gamma([s_{ijk}]) = [\gamma(s_{ijk})]$, where $\gamma(\cdot)$ is a 1-bit cyclical right shift function, i.e.,

$$\gamma(s_{ij1}, s_{ij2}, \cdots, s_{ijm}) = (s_{ijm}, s_{ij1}, \cdots, s_{ijm-1})$$

A matrix operation $\Gamma(\cdot)$ cyclically shifts right one subpixel in every m subpixels (for a secret pixel) in the shadow image.

When the difference of whiteness "$h - l$" is odd, we can design an ideal contrast VC scheme even when the underlying VC scheme with reversing is nonperfect black. At this time the decoding needs an XOR operation.

Yang et al.'s scheme [25] is described in a pseudo-code style below in terms of its construction procedure and the revealing procedure.

Yang et al.'s algorithm for a *(k,n)*-VC scheme

Distribution phase.

Step 1: Given a secret image, the dealer performs a (k, n) nonperfect black VC scheme($NPBVCS$) to generate n shadows, $s_1^1, \cdots, s_n^1$ for the first run.

Step 2: The dealer generates the shadows $s_j^r = \Gamma(s_j^{r-1})$ for the rth run, $r = 2, \cdots, m$. Note that the shadow should be labeled as to which run it is, for easy management by the participant.

Step 3: The dealer distributes m shadows $s_j^1, \cdots, s_j^m$ to Participant j, $j = 1, \cdots, n$.

Step 4: Finally, every participant holds m shadows.

Reconstruction phase.

Step 1: To recover the secret within m runs, at least k participants, Participants $j_1, \cdots, j_k$, offer their $(k \times R)$ shadows $s_{j1}^i, \cdots, s_{jk}^i, i = 1, \cdots R$, for reconstruction.

Step 2: Stack the shadows $s_{j1}^r, \cdots, s_{jk}^r$ to reconstruct the image T_r in the rth run.

$$T_r = s_{j1}^r + s_{j2}^r + \cdots + s_{jk}^r, r = 1, \cdots, m$$

Step 3: Finish m runs by using an XOR operation to reconstruct $U' = T_1 \oplus \cdots \oplus T_m$.

Step 4: If "$m - h$" is even (i.e., "$m - l$" is odd) then the reconstructed image is U'; otherwise, the reconstructed image is $\overline{U'}$.

Theorem 9 *[25] The whiteness percentages of the white and black secret pixels for Yang et al.'s algorithm are $P_W = 100\%$ and $P_W = 0\%$, respectively, after finishing m runs.*

Proof There are "$m - h$" B "h" W (respectively "$m - l$" B "l" W) subpixels for the white (respectively black) secret pixel. When shifting right one bit m times, there are $m - h$ (respectively $m - l$) black subpixels for the white (respectively black) secret pixels in U'. Suppose "$m - h$" is even (respectively odd); an XOR operation will result in all white subpixels for the white pixels in U' (respectively $\overline{U'}$). Thus, the P_W of the white secret pixel is 100%. On the other hand, even (respectively odd) "$m - h$" means odd (respectively even) "$m - l$" since '$h - l$' is odd. It is evident that an XOR operation will result in all black subpixels for the black pixels in U' (respectively $\overline{U'}$), i.e., the P_W of the black secret pixel is 0%. ∎

An example $(2, 4) - NPBVCS$ is given below as a demonstration for Yang et al.'s algorithm.

Example 4 *Suppose in $(2,4)$-VCS with basis matrices B_0 and B_1.*

$$B_0 = \begin{bmatrix} 1000 \\ 1000 \\ 1000 \\ 1000 \end{bmatrix}, B_1 = \begin{bmatrix} 1000 \\ 0100 \\ 0010 \\ 0001 \end{bmatrix}.$$

The distribution phase is shown in Table 6.1 and the reconstruction phase is shown in Table 6.2 and Table 6.3.

Distribution phase:

TABLE 6.1

Distribution phase of $(2,4)$-VCS using Yang et al.'s [25] algorithm.

Pixel	First run	Second run	Third run	Fourth Run
White	$s_1^1 = (1000)$	$s_1^2 = (0100)$	$s_1^3 = (0010)$	$s_1^4 = (0001)$
White	$s_2^1 = (1000)$	$s_2^2 = (0100)$	$s_2^3 = (0010)$	$s_2^4 = (0001)$
White	$s_3^1 = (1000)$	$s_3^2 = (0100)$	$s_3^3 = (0010)$	$s_3^4 = (0001)$
White	$s_4^1 = (1000)$	$s_4^2 = (0100)$	$s_4^3 = (0010)$	$s_4^4 = (0001)$
Black	$s_1^1 = (1000)$	$s_1^2 = (0100)$	$s_1^3 = (0010)$	$s_1^4 = (0001)$
Black	$s_2^1 = (0100)$	$s_2^2 = (0010)$	$s_2^3 = (0001)$	$s_2^4 = (1000)$
Black	$s_3^1 = (0010)$	$s_3^2 = (0001)$	$s_3^3 = (1000)$	$s_3^4 = (0100)$
Black	$s_4^1 = (0001)$	$s_4^2 = (1000)$	$s_4^3 = (0100)$	$s_4^4 = (0010)$

Reconstruction phase

TABLE 6.2

Reconstruction of participant 1 and 2 using Yang et al.'s [25] algorithm.

Pixel	First run $T_1 = s_1^1 + s_2^1$	Second run $T_2 = s_1^2 + s_2^2$	Third run $T_3 = s_1^3 + s_2^3$	Fourth run $T_4 = s_1^4 + s_2^4$	$\overline{U'}$
White	(1000)	(0100)	(0010)	(0001)	(0000)
Black	(1100)	(0110)	(0011)	(1001)	(1111)

TABLE 6.3

Reconstruction of participant 3 and 4 using Yang et al.'s [25] algorithm.

Pixel	First run $T_1 = s_3^1 + s_4^1$	Second run $T_2 = s_3^2 + s_4^2$	Third run $T_3 = s_3^3 + s_4^3$	Fourth run $T_4 = s_3^4 + s_4^4$	$\overline{U'}$
White	(1000)	(0100)	(0010)	(0001)	(0000)
Black	(0011)	(1001)	(1100)	(0110)	(1111)

We can see that the white pixel is reconstructed as four white subpixels and the black pixel is reconstructed as four black subpixels.

6.4.2   A Scheme for General Access Structure

Let $P = \{1, 2, \cdots, n\}$ be a set of participants, 2^P represents the set of all subsets of P. Let $\Gamma_{qual} \subseteq 2^P$ and $\Gamma_{forb} \subseteq 2^P$, where $\Gamma_{qual} \cap \Gamma_{forb} = \phi$. The members of Γ_{qual}(resp. Γ_{forb}) is called qualified sets (resp. forbidden sets). The $\Gamma = (\Gamma_{qual}, \Gamma_{forb})$ is called the access structure. Define $\Gamma_0 = \{A \in \Gamma_{qual} : A' \notin \Gamma_{qual} \ for \ all \ A' \subset A\}$ be all the minimal qualified sets [1].

Suppose $\Gamma_0 = \{\Gamma_{Q_1}, \cdots, \Gamma_{Q_t}\}$, by employing the optimal (k, k)-scheme [16], the basis matrices L_0 and L_1 are constructed as follows:

Let $k_p = |\Gamma_{Q_p}|$, and $\Gamma_{Q_p} = \{p_1, \cdots, p_{k_p}\}$, for $1 \le p \le t$. We will construct a $n \times 2^{k_p - 1}$ matrix E_p^i, $i \in \{0, 1\}$ according to the following steps: the p_i row of E_p^0 is the i-th row of the basis matrix B_0 of the (k_p, k_p)-scheme. The elements of other rows of E_p^0 are all 1's.

Then $L_0 = E_1^0$. The construction of E_p^1 is similar to E_p^0 except we replace the p_i row of E_p^1 from the basis matrix B_1 of the (k_p, k_p)-scheme instead of B_0. Then $L_1 = E_1^1 \| \cdots \| E_t^1$.

Lemma 10 *[10] The L_0 and L_1 are a pair of basis matrices of a perfect black VC scheme for Γ_0 such that the expansion rate is $m = 2^{|Q_1 - 1|} + \cdots + 2^{|Q_t - 1|}$ and $GRAY(white) = 1 - 1/m$.*

We now construct an $n \times 2^{k_p - 1}$ matrix F_p, which has the property that the elements in p_i row of F_p are all 0's and the other rows of F_p are all 1's, here $1 \le p \le t$. Then an auxiliary basis matrix $A_0 = F_1 | \cdots | F_i$.

Hu and Tzeng [10] algorithm for minimal access structure

Input:

1. Γ_0 on a set ρ of n participants.

2. Let C_p^0 and C_p^1 be the collection of basis Boolean matrices E_p^0 and E_p^1, where $1 \le p \le |\Gamma_0|$.

3. Let C_p^A be the collection of matrix F_p.

Output:
The reconstructed secret image U.

Distribution phase:
The dealer encodes each transparency t_i as $|\Gamma_0|$ subtransparecies $t_{i,p}$ and each subblock consists of one secret image. For $1 \le p \le |\Gamma_0|$, each white (resp. black) pixel on subblock $t_{i,p}$ is encoded using $n \times 2^{k_p-1}$ matrices E_p^0 (resp. E_p^1). To share a white (resp. black) pixel, the dealer performs the following steps:

Step 1: Randomly choose a matrix S_p^0 in C_p^0 (resp. S_p^1 in C_p^1), and a matrix $A_p^0 = [a_{i,j}]$ in C_p^A.

Step 2: For each participant i, put a white (resp. black) pixel on the subblock $t_{i,p}$ if $s_{i,j} = 0$ (resp. $s_{i,j} = 1$).

Step 3: For each participant i, put a white (resp. black) pixel on the subblock $A_{i,p}$ if $a_{i,j} = 0$ (resp. $a_{i,j} = 1$).

Reconstruction phase:
Let $Q_p = \{i_1, \cdots, i_{k_p}\}$ be the minimal qualified set in Γ_0, participants in Q_p reconstruct the secret image by:

Step 1: XORing all the shares t_j and stacking all the shares A_j for $j = 1, \cdots, k_p$ and obtain T and A: $T = t_1 \oplus \cdots \oplus t_{k_p}$, $A = A_1 + A_2 + \cdots + A_{k_p}$ respectively.

Step 2: Computing $U = (T + A) \oplus A$.

Lemma 11 *[10] The (k,k)-VC scheme [16] is an ideal contrast (k,k)-VC scheme with reversing.*

Proof k participants perform XOR operations on the k transparencies by computing $t_1 \oplus t_2 \oplus \cdots \oplus t_k$. It is easy to see that the white pixels are all white since S_0 has an even number of 1's; whereas the black pixels are all black since S_1 has an odd number of 1's. ∎

Theorem 10 *[10] Let $\Gamma = (P, Q, F)$ be an access structure on a set ρ of n participants. Then the basis matrices B_0, B_1, and A_0 constitute a compatible ideal contrast VC scheme with reversing in two runs.*

Proof It is obvious that the VC scheme is security. The basis matrix A_0 also reveals absolutely no information about the secret image since no secret is encoded into the shares A_j for $j = 1, \cdots, k_p$.

Let $L_0 = E_1^0 \| \cdots \| E_t^0$, $L_1 = E_1^1 \| \cdots \| E_t^1$ and $A_0 = F_1 \| \cdots \| F_t$ be the basis matrices for a VC scheme with reversing, constructed using the previously described technique. Without loss of generality, let $\Gamma_0 = \{Q_0, \cdots, Q_t\}$ and $X = Q_1$, X be a subset of qualified participants. Since the secret image is reconstructed by computing $(T + A) \oplus A$, we will prove that the general access structure has an ideal contrast, i.e., $H((E_1^0 + F_1) \oplus F_1) = 0$, $H((E_1^1 + F_1) \oplus F_1) = 2^{|Q_1|-1}$ and $H((E_i^b + F_i) \oplus F_i)$ for $i = 2, \cdots, |\Gamma|$ and $b = 0, 1$. It results that $H((E_1^0 + F_1) \oplus F_1) = H((E_1^0 + 0) \oplus 0) = H(E_1^0 \oplus 0) = 0$ by Lemma 11, and $H((E_1^1 + F_1) \oplus F_1) = H((E_1^1 + 0) \oplus 0) = H(E_1^1 \oplus 0) = H(E_1^1) = 2^{|Q_1|-1}$ by Lemma 11, whereas, $H((E_i^b + F_i) \oplus F_i) = H((E_i^b + 1) \oplus 1) \oplus 1 = w(1 \oplus 1) = 0$ for $i = 2, \cdots, |\Gamma_0|$ and $b = 0, 1$. ∎

We give the following example to illustrate the construction method above.

Example 5 *Let* $p = \{1, 2, 3, 4\}$ *and* $\Gamma_0 = \{\{1, 2\}, \{1, 3\}, \{2, 3\}\}$. *Then the basis matrices* $L_0, L_1,$ *and* A *are constructed as follows according to the method above.* B_0 *and* B_1 *are basis matrices of a* $(2, 2) - VC$ *scheme.*

$$B_0 = \begin{bmatrix} 10 \\ 10 \end{bmatrix}, B_1 = \begin{bmatrix} 10 \\ 01 \end{bmatrix}.$$

$$E_1^0 = \begin{bmatrix} 10 \\ 10 \\ 10 \end{bmatrix}, E_2^0 = \begin{bmatrix} 11 \\ 10 \\ 10 \end{bmatrix}, E_3^0 = \begin{bmatrix} 10 \\ 11 \\ 10 \end{bmatrix}, E_1^1 = \begin{bmatrix} 10 \\ 01 \\ 11 \end{bmatrix}, E_2^1 = \begin{bmatrix} 11 \\ 10 \\ 01 \end{bmatrix}, E_3^1 = \begin{bmatrix} 10 \\ 11 \\ 01 \end{bmatrix}.$$

$$L_0 = \begin{bmatrix} 101110 \\ 101011 \\ 111010 \end{bmatrix}, L_1 = \begin{bmatrix} 101110 \\ 011011 \\ 110101 \end{bmatrix}.$$

$$F_1 = \begin{bmatrix} 00 \\ 00 \\ 11 \end{bmatrix}, F_2 = \begin{bmatrix} 11 \\ 00 \\ 00 \end{bmatrix}, F_3 = \begin{bmatrix} 00 \\ 11 \\ 00 \end{bmatrix}, A_0 = \begin{bmatrix} 001100 \\ 000011 \\ 110000 \end{bmatrix}.$$

The distribution phase is shown in Table 6.4 and the reconstruction phase is shown in Tables 6.5, 6.6, and 6.7.

Distribution phase

TABLE 6.4

Distribution phase of the visual cryptography with minimal access structure $\Gamma_0 = \{\{1, 2\}, \{1, 3\}, \{2, 3\}\}$ using Hu and Tzeng [10] algorithm.

Pixel	First run	Second run
Black	$t_1 = (101110)$	$A_1 = (001100)$
Black	$t_2 = (011011)$	$A_2 = (000011)$
Black	$t_3 = (110101)$	$A_3 = (110000)$
White	$t_1 = (101110)$	$A_1 = (001100)$
White	$t_2 = (101011)$	$A_2 = (000011)$
White	$t_3 = (111010)$	$A_3 = (110000)$

Reconstruction phase

TABLE 6.5
Reconstruction of participant 1 and 2 using Hu and Tzeng [10] algorithm.

Pixel	First run $T = t_1 \oplus t_2$	Second run $A = A_1 + A_2$	$U = (T + A) \oplus A$
Black	(110101)	(001111)	(110000)
White	(000101)	(001111)	(000000)

TABLE 6.6
Reconstruction of participant 1 and 3 using Hu and Tzeng [10] algorithm.

Pixel	First run $T = t_1 \oplus t_3$	Second run $A = A_1 + A_3$	$U = (T + A) \oplus A$
Black	(011011)	(111100)	(000011)
White	(010100)	(111100)	(000000)

TABLE 6.7
Reconstruction of participant 2 and 3 using Hu and Tzeng [10] algorithm.

Pixel	First run $T = t_2 \oplus t_3$	Second run $A = A_2 + A_3$	$U = (T + A) \oplus A$
Black	(101110)	(110011)	(001100)
White	(010001)	(110011)	(000000)

6.5 Secret Sharing Scheme Using Boolean Operation

This section gives an introduction to some methods using an XOR Boolean operation directly without using basis matrices to construct secret sharing schemes. The algorithms of three schemes: the probabilistic $(2, n)$ secret sharing scheme, (n, n) secret sharing scheme, and (k, n) secret sharing scheme will be described.

Consider a secret image A with size $N_R \times N_C$. Each pixel of A can take any one of c different colors or gray levels. Image A is represented by an integer matrix A. $A = [a_{ij}]_{N_R \times N_C}$, where $i = 1, 2, \cdots, N_R$, $j = 1, 2, \cdots, N_C$, and $a_{ij} \in \{0, 1, \cdots, c - 1\}$. We have $c = 2$ for a binary image and $c = 256$ for grayscale image with one byte per pixel. In a color image with one byte per pixel, the pixel value can be an index to a color table, thus, $c = 256$. In a color image using an RGB model, each pixel has three integers: R(red),

G(green), and B(blue). If each R, G, or B takes a value between 0 and 255, we have $c = 256^3$. This integer matrix is also called A and will be treated as equivalent to the secret image A itself. The symbol "&" denotes an AND operation in the following construction.

6.5.1 (2, *n*) Scheme

To solve the pixel expansion problem with VSS schemes, Ito et al. [11], Yang [24] and Cimato et al. [6] proposed probabilistic VSS models. The frequency of white pixels in a white (or black) area is used to display the contrast of the recovered image. In reconstructing the secret image, the "OR"-ed operation of pixels of the shadows is the same as the stacking operation of subpixels in the nonprobabilistic VSS schemes. They defined p_0 (resp. p_1) as the appearance probability of a white pixel in a white (resp. black) area of the recovered image. For a fixed threshold probability $0 \le p_{TH} \le 1$ and relative contrast $\alpha \ge 0$, if $p_0 \ge p_{TH}$ and $p_1 \le p_{TH} - \alpha$, the frequency of white pixels in a white area of the recovered image should be higher than that in a black area.

Wang et al. [23] proposed a probabilistic $(2, n)$ secret sharing scheme for binary images. Boolean XOR and AND operations are employed, and $n + 1$ distinct random matrices are generated as intermediate results. The scheme is described in a pseudo-code style below in terms of its input, output, the construction procedure, and the revealing procedure.

Wang et al.'s [23] algorithm for (2,*n*) scheme

Input: an integer n with $n \ge 2$, and the secret image A.
Output: n distinct matrices $A_1 \cdots, A_n$, called shadow images.
Construction:
   Step 1: Generate $n + 1$ random matrices $B_1, \cdots, B_{n+1}$.
   Step 2: Compute n intermediate matrices $C_1, \cdots, C_n$ with $C_i = B_i \& A$ for $i = 1, \cdots, n$.
   Step 3: Compute n shadow images $A_1, \cdots, A_n$ with $A_i = B_{n+1} \oplus C_i$ for $i = 1, \cdots, n$.
Revealing: $A' = A_i \oplus A_j$ where $i, j \in \{1, 2, \cdots, n\}$ and $i \ne j$.

For integer scalar inputs between 0 and $c-1$, each operand is represented in binary and the operation is carried out bit-by-bit. For example, when $a = 125$ and $b = 18$, the XOR between these two integers is

$$a \oplus b = (125)_{10} \oplus (18)_{10} = (01111101)_2 \oplus (00010010)_2 = (01101111)_2 = (111)_{10}$$

For matrix inputs, the XOR operation of two $N_R \times N_C$ matrices is defined pixel-wise. That is,

$$A \oplus B = [a_{ij} \oplus b_{ij}], \text{ where } i = 1, 2, \cdots, N_R, j = 1, 2, \cdots, N_c.$$

The AND operation for integer scalar operands and matrix operands can be defined similarly.

In all computations, every pixel is handled individually, separated from other pixels. Therefore, to make the context clear, we denote pixel $A_i(s,t)$ simply as A_i. With the above construction procedure, for a "0" pixel in A and any i, we have $C_i = B_i \& 0 = 0$ and $A_i = B_{n+1} \oplus C_i = B_{n+1}$, thus

$$A' = A_i \oplus A_j = B_{n+1} \oplus B_{n+1} = 0.$$

For a "1" pixel in A, $C_i = B_i \& 1 = B_i$ and $A_i = B_{n+1} \oplus B_i$, thus

$$A' = A_i \oplus A_j = B_{n+1} \oplus B_{n+1} \oplus B_i \oplus B_j = B_i \oplus B_j$$

which could be 0 or 1. In other words, between the original image A and a reconstructed image A', the "0" bits are kept the same and the "1" bits may or may not changed. With any single shadow image, no information of A is revealed because of the random nature of the matrices B's. It is easy to verify that the n matrices $A_1, A_2, \cdots, A_n$ are n distinct random matrices from construction method above, each A_i $(i = 1, \cdots, n)$ does not contain any information of the original matrix A.

Associated Shamir's secret sharing scheme and a gradual search algorithm for a single bitmap block truncation coding with the above $(2, n)$ scheme, the secret sharing scheme for color images was proposed in Reference [3]. Combined the above $(2, n)$ scheme with voting strategy, the probabilistic visual secret sharing scheme for grayscale images was provided in Reference [4]. Based on the above proposed $(2, n)$ scheme, the matrices $B_1, B_2, \cdots, B_n$ are chosen according to Yang's into "probabilistic VC scheme and the $(2, n)$ probabilistic scheme with improved contrast was given in Reference [19].

6.5.2 (n, n) Scheme

The construction steps are given in the context of grayscale images. It is trivially applicable to binary images and can be easily extended to color images.

Wang et al.'s [23] algorithm for (n,n) scheme

Input: an integer n with $n \geq 2$, and the secret image A.
Output: n distinct matrices $A_1, \cdots, A_n$, called shadow images.
Construction:
 Step 1: Generate $n - 1$ random matrices $B_1, \cdots, B_{n-1}$.
 Step 2: Compute the shadow images as below:
 $A_1 = B_1, A_2 = B_1 \oplus B_2, \cdots\cdots, A_{n-1} = B_{n-2} \oplus B_{n-1}$,
 $A_n = B_{n-1} \oplus A$.
Revealing: $A' = A_1 \oplus A_2 \oplus \cdots \oplus A_n$.

Theorem 11 *[23]* $A_1 \oplus A_2 \oplus \cdots \oplus A_n = A$.

Proof Because the "$\oplus$" operation is associative and $B_i \oplus B_i$ is a zero matrix for any i, we have

$$A' = B_1 \oplus (B_1 \oplus B_2) \oplus \cdots \oplus (B_{n-2} \oplus B_{n-1}) \oplus (B_{n-1} \oplus A)$$
$$= (B_1 \oplus B_1) \oplus \cdots \oplus (B_{n-1} \oplus B_{n-1}) \oplus A$$
$$= A.$$

∎

Theorem 12 *[23] $A_{i1} \oplus A_{i2} \oplus \cdots \oplus A_{ik} \neq A$ for any set of integers $\{i_1, \cdots, i_k\}$ when $k < n$.*

Proof We consider two cases. Case 1 is for $n \in \{i_1, \cdots, i_k\}$ and Case 2 is for $n \notin \{i_1, \cdots, i_k\}$.

Case 1: $n \in \{i_1, \cdots, i_k\}$. In this case, $A_n \oplus (\oplus_{j=s}^t A_j) = A \oplus B_{n-1} \oplus (\oplus_{j=s}^t A_j)$ where $\oplus_{j=s}^t$ means $A_s \oplus \cdots \oplus A_t$ with $s, \cdots, t$ being the indices in $\{i_1, \cdots, i_k\}$ besides n. Since there are an odd number of $n-2$ term random matrices involved, at least one of them cannot be absorbed into a zero matrix, thus $A_{i_1} \oplus A_{i_2} \oplus \cdots \oplus A_{i_k}$ must be random, thus not equal to A.

Case 2: $n \notin \{i_1, \cdots, i_k\}$. Since no matrix A involved in $A_{i_1} \oplus A_{i_2} \oplus \cdots \oplus A_{i_k}$ to begin with, $A_{i_1} \oplus A_{i_2} \oplus \cdots \oplus A_{i_k}$ is constructed from the $n-1$ term random matrices only and it must be random. ∎

Next, we analyze the steps of Wang et al.'s [23] (n, n) algorithm in the context of the grayscale image below. It is trivially applicable to a binary image and can be easily extended to a color image. Now we give an example to demonstrate the computation steps in the construction/revealing process using the above algorithm.

Example 6 *Let $n = 3$ and the secret image A be a single pixel.*

Input: $A = (240)_{10} = (1111\ 0000)_2$
Construction:
  Step 1:
    $B_1 = (125)_{10} = (0111\ 1101)_2$,
    $B_2 = (10)_{10} = (0001\ 0010)_2$
  Step 2:
    $A_1 = B_1 = (125)_{10} = (0111\ 1101)_2$,
    $A_2 = B_1 \oplus B_2 = (0110\ 1111)_2$,
    $A_3 = B_2 \oplus A = (1110\ 0010)_2$
Revealing: $A' = A_1 \oplus A_2 \oplus A_3 = (1111\ 0000)_2 = (240)_{10}$.

From the above algorithm, it is known that the proposed (n, n) scheme reconstructs the secret image exactly and when fewer than n shadows are used, the original secret image A will not be revealed. Moreover, only simple Boolean XOR operations are used and the size of each shadow image is the same as the original image, thus no pixel expansion. The above (n, n) secret sharing scheme was applied to deal with gray scale and color secret images respectively in Reference [22].

6.5.3   (k, n) Scheme

Based on above Wang et al.'s [23] (n,n) secret sharing scheme [23], Chao and Lin [5] have attempted to construct a (k,n) secret sharing scheme. To design a threshold (k,n) scheme, one may first directly utilize Wang et al.'s [23] (m,m) scheme for some carefully chosen parameter $m = \binom{n}{k-1}$. The (k,n,m) shadows-assignment matrix has n rows and m columns. Its n rows represent n persons; and its m columns represent the m (distinct) temporary shadows produced by the above (m,m) scheme. The element of H is either 0 or 1. The i-th person (row) has a copy of the j-th shadow image (column) if and only if H_{ij}. Each column of H have exactly $k-1$ zeros and $n-k+1$ ones.

Now we will introduce Chao et al.'s (k,n)-VC scheme based on the (k,n,m) shadows assignment matrix H.

Chao and Lin [5] algorithm for (k,n) scheme

Input: the secret image A

Output: the final shadow S_i

Construction:

　　Step 1: Generate a random image B_1 with the same size as A.

　　Step 2: Generate another image B_2 using $B_2 = B_1 \oplus A$.

　　Step 3: Construct a $n \times m$ shadows-assignment matrix H, each column of H have exactly $k-1$ zeros and $n-k+1$ ones, so $m = \binom{n}{k-1}$.

　　Step 4: Partition B_1 and B_2 into m nonoverlapping blocks $C_{11}, C_{21}, \cdots, C_{m1}$ and $C_{12}, C_{22}, \cdots, C_{m2}$.

　　Step 5: Let $C^* = C_{11} \oplus C_{21} \oplus \cdots \oplus C_{m1}$. Compute $C_{i3} = C_{i2} \oplus C^*$.

　　Step 6: Construct m temporary shadows $C_1, C_2, \cdots, C_m$, where the upper half of each C_i is the block C_{i1} and the lower half of each C_i is the block C_{i3}.

　　Step 7: Assign the duplicated copies of the m temporary shadows $C_1, C_2, \cdots, C_m$ to the n persons according to the shadows-assignment matrix H. For each participant i, the final shadow S_i is exactly the union of those copies assigned to him.

Revealing:

　　Step 1: k participants using their shares $S_{i_1}, \cdots, S_{i_k}$ to reconstruct the secret image. Referring to H, all m temporary shadows $C_1, C_2, \cdots, C_m$ can be extracted from these k final shadows.

　　Step 2: The upper half of each C_i is the block C_{i1} and the lower half of each C_i is the block C_{i3}. So we get all $C_{11}, C_{21}, \cdots, C_{m1}$ and $C_{13}, C_{23}, \cdots, C_{m3}$.

　　Step 3: Compute $C^* = C_{11} \oplus C_{21} \oplus \cdots \oplus C_{m1}$, then $C_{i2} = C_{i3} \oplus C^*$.

　　Step 4: Recombination $C_{11}, C_{21}, \cdots, C_{m1}$ and $C_{12}, C_{22}, \cdots, C_{m2}$ into B_1 and B_2.

　　Step 5: Reveal the secret image A' by $A' = B_1 \oplus B_2$.

　　Let the size of secret image A be $N_R \times N_C$. Since the size of every temporary shadow $C_i (1 \le i \le m)$ is $2 \times (N_R \times N_C)/m$, the size of every final shadow $S_i (1 \le i \le n)$ is $\left\lceil \frac{2 \times (N_R \times N_C)}{m} \right\rceil \times \binom{n-1}{k-1} = \frac{2(N_R \times N_C)(n-k+1)}{n}$. Notice that each final shadow $S_i (1 \le i \le n)$ contains $\binom{n-1}{k-1}$ temporary shadows. After we divide by the size of secret image A, we obtain the pixel expansion rate $per = 2 \times \frac{n-k+1}{n} < 2$. Each shadow will be at most two times larger than secret image A.

　　Next, we will give an example to analyze the steps of Chao and Lin [5] (k,n) algorithm scheme.

Example 7 *Let* $k = 3, n = 4$, *and a grayscale secret image* $A = \begin{bmatrix} 235 & 45 & 239 \\ 188 & 103 & 234 \end{bmatrix}$.

Input: the secret image $A = \begin{bmatrix} 235 & 45 & 239 \\ 188 & 103 & 234 \end{bmatrix}$

Output: the final shadows S_1, S_2, S_3, S_4

Construction:

Step 1: Generate random image $B_1 = \begin{bmatrix} 105 & 14 & 208 \\ 228 & 90 & 2 \end{bmatrix}$.

Step 2: Generate $B_2 = B_1 \oplus A = \begin{bmatrix} 105 & 14 & 208 \\ 228 & 90 & 2 \end{bmatrix} \oplus \begin{bmatrix} 235 & 45 & 239 \\ 188 & 103 & 234 \end{bmatrix} = \begin{bmatrix} 130 & 35 & 63 \\ 88 & 61 & 232 \end{bmatrix}$.

Step 3: $m = \binom{n}{k-1} = \binom{4}{2}$. Construct a 4×6 shadows-assignment matrix H, each column of H have exactly 2 zeros and 2 ones, so

$$H = \begin{pmatrix} 0 & 0 & 0 & 1 & 1 & 1 \\ 0 & 1 & 1 & 0 & 0 & 1 \\ 1 & 0 & 1 & 0 & 1 & 0 \\ 1 & 1 & 0 & 1 & 0 & 0 \end{pmatrix}$$

Step 4: Partition B_1 into 6 nonoverlapping blocks $C_{11} = 105, C_{21} = 14, C_{31} = 208, C_{41} = 228, C_{51} = 90, C_{61} = 2$, Partition into 6 nonoverlapping blocks $C_{12} = 130, C_{22} = 35, C_{32} = 63, C_{42} = 88, C_{52} = 61, C_{62} = 232$.

Step 5: $C^* = C_{11} \oplus C_{21} \oplus \cdots \oplus C_{61} = 105 \oplus 14 \oplus 208 \oplus 228 \oplus 90 \oplus 2 = 11$. Compute $C_{13} = C_{12} \oplus C^* = 130 \oplus 11 = 137$, similarly $C_{23} = 40, C_{33} = 52, C_{43} = 83, C_{53} = 54, C_{63} = 227$.

Step 6: Construct 6 temporary shadows $C_1 = \begin{bmatrix} 105 \\ 137 \end{bmatrix}, C_2 = \begin{bmatrix} 14 \\ 40 \end{bmatrix}, C_3 = \begin{bmatrix} 208 \\ 52 \end{bmatrix}, C_4 = \begin{bmatrix} 228 \\ 83 \end{bmatrix}, C_5 = \begin{bmatrix} 90 \\ 54 \end{bmatrix}, C_6 = \begin{bmatrix} 2 \\ 227 \end{bmatrix}$,, where the upper half of each C_i is the block C_{i1} and the lower half is the block C_{i3}.

Step 7: According to the shadows-assignment matrix H, assign $C_4, C_5, C_6 \rightarrow S_1$, $C_2, C_3, C_6 \rightarrow S_2$, $C_1, C_3, C_5 \rightarrow S_3$, $C_1, C_2, C_4 \rightarrow S_4$.

Revealing:

Step 1: Suppose 3 participants using their shares S_1, S_2, S_3 to reconstruct the secret image. Referring to H, all 6 temporary shadows $C_1 = \begin{bmatrix} 105 \\ 137 \end{bmatrix}, C_2 = \begin{bmatrix} 14 \\ 40 \end{bmatrix}, C_3 = \begin{bmatrix} 208 \\ 52 \end{bmatrix}, C_4 = \begin{bmatrix} 228 \\ 83 \end{bmatrix}, C_5 = \begin{bmatrix} 90 \\ 54 \end{bmatrix}, C_6 = \begin{bmatrix} 2 \\ 227 \end{bmatrix}$, can be extracted from S_1, S_2, S_3.

Step 2: The upper half of each C_i is the block C_{i1} and the lower half of each C_i is the block C_{i3}. So we get all $C_{11} = 105, C_{21} = 14, C_{21} = 14, C_{31} = 208, C_{41} = 228, C_{51} = 90, C_{61} = 2$ and $C_{13} = 137, C_{23} = 40, C_{33} = 52, C_{43} = 83, C_{53} = 54, C_{63} = 227$.

Step 3: Compute $C^* = C_{11} \oplus C_{21} \oplus \cdots \oplus C_{61} = 105 \oplus 14 \oplus 208 \oplus 228 \oplus 90 \oplus 2 = 11$, then $C_{12} = C_{13} \oplus C^* = 137 \oplus 11 = 130$. Similarly, $C_{22} = 35, C_{32} = 63, C_{42} = 88, C_{52} = 61, C_{62} = 232$.

Step 4: $B_1 = \begin{bmatrix} C_{11} & C_{21} & C_{31} \\ C_{41} & C_{51} & C_{61} \end{bmatrix} = \begin{bmatrix} 105 & 14 & 208 \\ 228 & 90 & 2 \end{bmatrix}$ and $B_2 = \begin{bmatrix} C_{12} & C_{22} & C_{32} \\ C_{42} & C_{52} & C_{62} \end{bmatrix} = \begin{bmatrix} 130 & 35 & 63 \\ 88 & 61 & 232 \end{bmatrix}$.

Step 5: Reveal the secret image

$$A' = B_1 \oplus B_2 = \begin{bmatrix} 105 & 14 & 208 \\ 228 & 90 & 2 \end{bmatrix} \oplus \begin{bmatrix} 130 & 35 & 63 \\ 88 & 61 & 232 \end{bmatrix} = \begin{bmatrix} 235 & 45 & 239 \\ 188 & 103 & 234 \end{bmatrix}.$$

6.6 Conclusion

XOR-based visual cryptography schemes use XOR operation to decrypt the secret. These schemes have higher contrast compared with OR-based visual cryptography. This chapter introduced three types of different XOR-based visual cryptography schemes. The three type schemes have their virtues respectively. We also notice there is XOR-based audio cryptography, which uses music to embed message and relies on the human auditory system to decrypt secret messages. It is a very interesting topic, however we do not cover audio cryptography schemes because the topic in this chapter is limited to image, and readers can see more details in Reference [8].

Acknowledgments

The research work related to this chapter was supported by the National Natural Science Foundation of China under Grant No. 60873249, 863 Project of China under Grant 2008AA01Z419.

Bibliography

[1] G. Ateniese, C. Blundo, and A. D. Santis. Visual cryptography for general access structures. In *Information and Computation*, volume 129, pages 86–106, 1996.

[2] E. Biham and A. Itzkovitz. Visual cryptography with polarization. In *The Dagstuhl Seminar on Cryptography, September 1997, and in the RUMP session of CRYPTO'98*, 1997.

[3] C. C. Chang, C. C. Lin, T. H. N. Le, and H. B. Le. A new probabilistic visual secret sharing scheme for color images. In *4th International Conference on Intelligent Information Hiding and Multimedia Signal Processing*, pages 1305–1308, 2008.

[4] C. C. Chang, C. C. Lin, T. H. N. Le, and H. B. Le. A probabilistic visual secret sharing scheme for grayscale images with voting strategy. In *Proceedings of the International Symposium on Electronic Commerce and Security*, pages 184–188, 2008.

[5] K. Y. Chao and J. C. Lin. Secret image sharing: A Boolean-operations-based approach combining benefits of polynomial-based and fast approaches. *International Journal of Pattern Recognition and Artificial Intelligence*, 23(2):263–285, 2009.

[6] S. Cimato, R. De Prisco, and A. De Santis. Probabilistic visual cryptography schemes. *Computer Journal*, 49(1):97–107, 2006.

[7] S. Cimato, A. De Santis, A. L. Ferrara, and B. Masucci. Ideal contrast visual cryptography schemes with reversing. *Information Processing Letters*, 93(4):199–206, 2005.

[8] Y. Desmedt, S. Hou, and J. J. Quisquater. Audio and optical cryptography. In *ASIACRYPT '98, Lecture Notes in Computer Science*, volume 1514, pages 392–404, 1998.

[9] S. Droste. New results on visual cryptography. In *Advances in Cryptology-CRYPTO '96, Lecture Notes in Computer Science*, volume 1109, pages 401–415, 1996.

[10] C. M. Hu and W. G. Tzeng. Compatible ideal contrast visual cryptography schemes with reversing. *Information Security*, 3650:300–313, 2005.

[11] R. Ito, H. Kuwakado, and H. Tanaka. Image size invariant visual cryptography. *IEICE Trans. Fundamentals*, E82-A(10):2172–2177, 1999.

[12] S. S. Lee, J. C. Na, S. W. Sohn, C. Park, D. H. Seo, and S. J. Kim. Visual cryptography based on an interferometric encryption technique. *ETRI Journal*, 24(5):373–380, 2002.

[13] F. Liu, C. K. Wu, and X. J. Lin. Colour visual cryptography schemes. *IET Information Security*, 2(4):151–165, 2008.

[14] F. Liu, C. K. Wu, and X. J. Lin. Some extensions on threshold visual cryptography schemes. *Computer Journal*, 53:107–119, 2010.

[15] F. Liu, C. K. Wu, and X. J. Lin. Step construction of visual cryptography schemes. In *IEEE Transactions on Information Forensics and Security*, volume 5, pages 27–28, 2010.

[16] M. Naor and A. Shamir. Visual cryptography. In *Visual Cryptography, Proceedings of Eurocrypto' 94, Lecture Note in Computer Science*, volume 950, pages 1–12, 1995.

[17] P. Tuyls, H. D. L. Hollmann, J. H. v. Lint, and L. Tolhuizen. Xor-based visual cryptography schemes. *Designs Codes and Cryptography*, 37:169–186, 2005.

[18] P. Tuyls, H.D.L. Hollmann, H.H.v. Lint, and L. Tolhuizen. A polarisation based visual crypto system and its secret sharing schemes. http://eprint.iacr.org, 2002.

[19] M. Ulutas, V. V. Nabiyev, and G. Ulutas. A PVSS scheme based on Boolean operations with improved contrast. In *2009 International Conference on Network and Service Security*, pages 1–5, 2009.

[20] E. R. Verheul and H. C. A. van Tilborg. Constructions and properties of k-out-of-n visual secret sharing schemes. *Designs, Codes and Cryptography*, 11:179–196, 1997.

[21] D. Q. Viet and K. Kurosawa. Almost ideal contrast visual cryptography with reversing. In *Proceeding of Topics in Cryptology-CT-RSA2004*, *Lecture Notes in Computer Science*, volume 2964, pages 353–365, 2004.

[22] D. S. Wang, X. Li, and F. Yi. Probabilistic (n,n) visual sharing scheme for grayscale images. In *SKLOIS Conference on Information Security and Cryptology, Inscrypt 2007, Lecture Notes in Computer Science*, volume 4990, pages 192–200, 2008.

[23] D. S. Wang, L. Zhang, N. Ma, and X. Li. Two secret sharing schemes based on Boolean operations. *Pattern Recognition*, 40:2776–2785, 2007.

[24] C. N. Yang. New visual secret sharing schemes using probabilistic method. *Pattern Recognition Letters*, 25:481–494, 2004.

[25] C. N. Yang, C. C. Wang, and T. S. Chen. Real perfect contrast visual secret schemes with reversing. In *Lecture Notes in Computer Science*, volume 3989, pages 433–447, 2006.

[26] C. N. Yang, C. C. Wang, and T. S. Chen. Visual cryptography schemes with reversing. *Computer Journal*, 51(6):710–722, 2008.

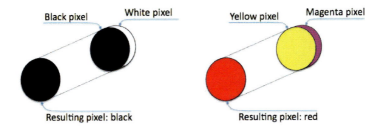

FIGURE 2.1
Pixels superposition: black and white (left) and colored (right).

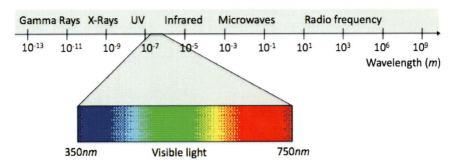

FIGURE 2.2
Electromagnetic spectrum.

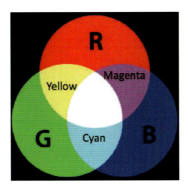

FIGURE 2.3
Additive color model with primaries red, green, and blue.

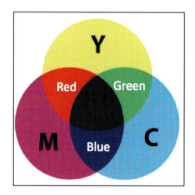

FIGURE 2.4
Subtractive color model with primaries cyan, magenta, and yellow.

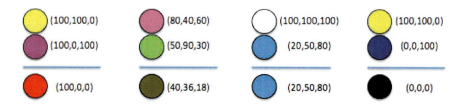

FIGURE 2.5
Examples of pixels superposition.

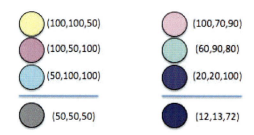

FIGURE 2.6
More examples of pixels superposition.

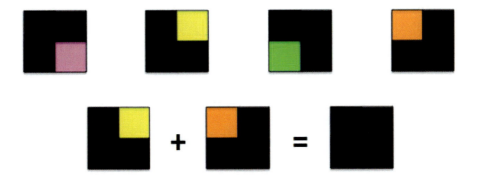

FIGURE 2.9

The vv trick for the case of 4 colors. Subpixels with different colors are never superposed.

FIGURE 7.9

Results of Steps 1 and 2 of Algorithm 8 for VCRG-3 with respect to color image P in Experiment 3: (a) P; (b) $P^{\mathbf{c}}$, (c) $P^{\mathbf{m}}$, (d) $P^{\mathbf{y}}$; (e) $P^{\mathbf{c}}$, (f) $P^{\mathbf{m}}$, (g) $P^{\mathbf{y}}$.

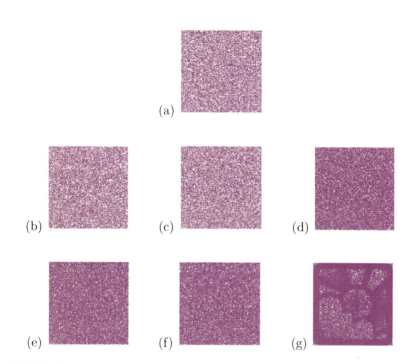

FIGURE 7.10
Results of Step 3 of Algorithm 8 with respect to $P^{\mathbf{m}}$ where *Ey-crypt_cVCRG*$(P^{\mathbf{m}}, \mathbf{m}, 3)$ was based upon Algorithm 4: (a) $\boldsymbol{R}_1^{\mathbf{m}}$, (b) $\boldsymbol{R}_2^{\mathbf{m}}$, (c) $\boldsymbol{R}_3^{\mathbf{m}}$; (d) $\boldsymbol{R}_1^{\mathbf{m}} \otimes \boldsymbol{R}_2^{\mathbf{m}}$, (e) $\boldsymbol{R}_1^{\mathbf{m}} \otimes \boldsymbol{R}_3^{\mathbf{m}}$, (f) $\boldsymbol{R}_2^{\mathbf{m}} \otimes \boldsymbol{R}_3^{\mathbf{m}}$; (g) $\boldsymbol{R}_1^{\mathbf{m}} \otimes \boldsymbol{R}_2^{\mathbf{m}} \otimes \boldsymbol{R}_3^{\mathbf{m}}$.

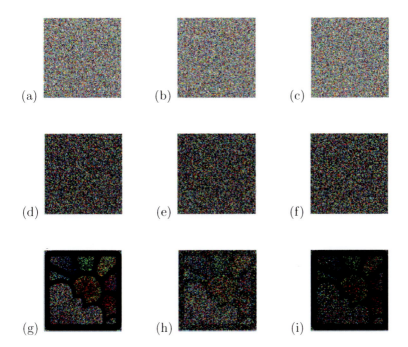

FIGURE 7.11
Results of Algorithm 8 for VCRG-3 with respect to P: (a) $\boldsymbol{R}_1$, (b) $\boldsymbol{R}_2$, (c) $\boldsymbol{R}_3$; (d) $\boldsymbol{R}_1 \otimes \boldsymbol{R}_2$, (e) $\boldsymbol{R}_1 \otimes \boldsymbol{R}_3$, (f) $\boldsymbol{R}_2 \otimes \boldsymbol{R}_3$; (g) $\boldsymbol{R}_1 \otimes \boldsymbol{R}_2 \otimes \boldsymbol{R}_3$ (based upon Algorithm 4); (h) $\boldsymbol{R}_1' \otimes \boldsymbol{R}_2' \otimes \boldsymbol{R}_3'$ (based upon Algorithm 5); (i) $\boldsymbol{R}_1'' \otimes \boldsymbol{R}_2'' \otimes \boldsymbol{R}_3''$ (based upon Algorithm 6).

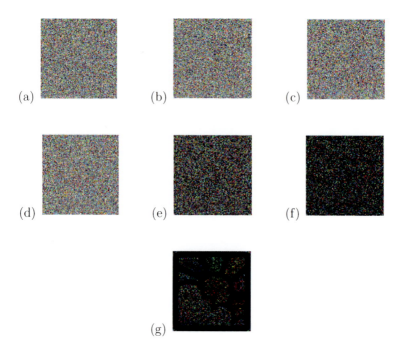

FIGURE 7.13
Results of Algorithms 8 where $Encrypt\_cVCRG(P^x, \boldsymbol{x}, 4)$ was based upon Algorithm 4 for VCRG-4 with respect to P (Figure 7.9(a)): (a) $\boldsymbol{R}_1$, (b) $\boldsymbol{R}_2$, (c) $\boldsymbol{R}_3$, (d) $\boldsymbol{R}_4$; (e) $\boldsymbol{R}_1 \otimes \boldsymbol{R}_2$; (f) $\boldsymbol{R}_1 \otimes \boldsymbol{R}_2 \otimes \boldsymbol{R}_3$; (g) $\boldsymbol{R}_1 \otimes \boldsymbol{R}_2 \otimes \boldsymbol{R}_3 \otimes \boldsymbol{R}_4$.

(a) Original image and shares 1 and 2

(b) Original image and shares 1, 2, and 3

FIGURE 14.3
Experimental results of image sharing based on the Lagrange interpolation in (a) and (b).

(a) Original image   (b) Shares 1   (c) Shares 2

FIGURE 14.4
The image sharing by using a high degree polynomial interpolation in (a)-(c).

FIGURE 14.8
The experimental results of image sharing by moving lines.

FIGURE 14.9
The experimental results of image sharing by moving lines.

FIGURE 14.10
The experimental results of image sharing by moving lines.

FIGURE 14.11
Breaking the correlation of neighboring blocks in an image.

7

Visual Cryptography and Random Grids

Shyong Jian Shyu

Ming Chuan University, Taiwan

CONTENTS

7.1 Introduction

A *random grid* was defined by Kafri and Keren in 1987 [7] as a transparency comprising a two-dimensional array of pixels. Each pixel is either fully *transparent* or simply *opaque* and the choice between the alternatives is made by a coin-flip procedure. Thus, there is no correlation between the values of different pixels in the array.

We could encrypt binary pictures or shapes into two random grids such that only the areas containing information in the two grids are intercorrelated, while the others are purely random. When the two grids are superimposed together, the correlated areas will be resolved from the random background due to the difference in *light transmission* so that the secret picture or shape can be seen visually. Just like conventional schemes in visual cryptography, the decoding process is done by the human visual system where no computation is needed; however, no extra pixel expansion is required using random grids.

Observing the interesting features of random grids in image encryption, which were discussed right before Naor and Shamir's visual cryptographic scheme (VCS) [9], Shyu [12, 13] generalized the random grids-based approaches into *visual cryptograms of random grids* (VCRG) for achieving visual secret sharing recently. The most appealing benifits using random grids lie in that the pixel expansion needed is merely one and no basis matrix is needed. With the same contrast in the reconstructed results, the optimal pixel expnasion in (n, n)-VCS is 2^{n-1}; while that in (n, n)-VCRG is still 1.

We study the random grids-based schemes, analyze the performances, and demonstrate their feasibilities in this chapter. The rest of the chapter is organized as follows. The fundamental characteristics of random grids are discussed in Section 7.2. Section 7.3 discusses how to apply visual cryptograms of random grids in visual cryptography where the formal definition of VCRG is given, and the designs, analyses, and implementations of $(2, 2)$-VCRGs and (n, n)-VCRGs for binary, gray-level and color images are examined. Section 7.4 exhibits some concluding remarks.

7.2   Random Grids

7.2.1   Random Pixel, Random Grid, and Average Light Transmission

We refer to a binary pixel r as a *random pixel* if the choice for r to be transparent or opaque in R is totally random; or equivalently, the probability for r to be transparent is equal to that for r to be opaque,

$$\mathscr{P}\mathit{rob}(r = 0) = \mathscr{P}\mathit{rob}(r = 1) = \frac{1}{2}, \tag{7.1}$$

where 0 (1) denotes a transparent (opaque) pixel. Since a transparent pixel lets through the light while an opaque one stops it, the *average light transmission* of random pixel r is $\frac{1}{2}$, denoted as

$$\ell(r) = \frac{1}{2}. \tag{7.2}$$

Definition 1 *R is a binary random grid if each pixel r in R is a binary random pixel.*

Regarding a random grid R, the number of transparent pixels is probabilistically the same as that of opaque ones so that the average light transmission of R is also $\frac{1}{2}$, denoted as

$$\mathscr{T}(R) = \frac{1}{2}. \tag{7.3}$$

7.2.2 Superimposition of Random Grids

Let $\otimes$ denote the generalized "or" operation, which describes the relation of the superimposition of two random pixels or grids pixel by pixel. It is obvious that $r \otimes r$ $(R \otimes R)$ is entirely the same as r (R), that is

$$\mathscr{t}(r \otimes r) = \frac{1}{2} \text{ or } \mathscr{T}(R \otimes R) = \frac{1}{2} \tag{7.4}$$

for each pixel r in R.

Let R_1 and R_2 be two independent random grids with the same size. When R_1 and R_2 are superimposed pixel by pixel, each pixel (either transparent or opaque) in R_1 has an equal possibility to be stacked by a transparent pixel or an opaque pixel in R_2. We call $r_1 = R_1[i,j]$ the corresponding pixel of $r_2 = R_2[i',j']$ if and only if $i = i'$ and $j = j'$ (the positions of r_1 at R_1 and r_2 at R_2 are the same). It is easy to see that the order of the two random grids does not affect the superimposed result, i.e.,

$$R_1 \otimes R_2 = R_2 \otimes R_1. \tag{7.5}$$

Indeed, $\otimes$ is a *commutative* operation.

TABLE 7.1
Results of the superimposition of two random pixels.

r_1	r_2	$r_1 \otimes r_2$
0	0	0
0	1	1
1	0	1
1	1	1

Table 7.1 shows the superimposed results of the corresponding random pixels r_1 and r_2. There is only one outcome among the four possible combinations of $r_1 \otimes r_2$ showing transparency. Since the four possible combinations occur with an equal probability, the probability for $r_1 \otimes r_2$ to be transparent is $\frac{1}{4}$. That is, the average light transmission of the superimposition of R_1 and R_2 $(r_1$ and $r_2)$ is $\frac{1}{4}$.

We summarize the aforementioned properties in Lemma 1.

Lemma 1 *If R_1 and R_2 are two independent random grids with $\mathscr{T}(R_1) = \mathscr{T}(R_2) = 1/2$,*

(1) $\mathscr{T}(R_1 \otimes R_2) = \mathscr{T}(R_2 \otimes R_1) = 1/4;$

(2) $\mathscr{T}(R_1 \otimes R_1) = 1/2.$

We define $\bar{R}$ to be the *inverse random grid* of R if and only if

$$r' = \bar{r}$$

for each r' in $\bar{R}$ where r is the corresponding pixel of r' in R and $\bar{r}$ denotes the inverse of r. It is easy to see that $r \otimes \bar{r} = 1$ and $R \otimes \bar{R} = \mathbf{1}$ ($\mathbf{1}$ denotes a grid in which all pixels are opaque), that is

$$\ell(r \otimes \bar{r}) = 0 \text{ and } \mathscr{T}(R \otimes \bar{R}) = 0 \tag{7.6}$$

For each pixel r' in $\bar{R}$, since $\mathscr{P}\!\mathit{rob}(r' = 0) = \mathscr{P}\!\mathit{rob}(\bar{r} = 0) = \mathscr{P}\!\mathit{rob}(r = 1) = \frac{1}{2}$, we obtain

$$\ell(\bar{r}) = \frac{1}{2} \text{ and } \mathscr{T}(\bar{R}) = \frac{1}{2} \tag{7.7}$$

In general, the relationship of the average light transmissions of $R \in \bar{R}$ is given in Lemma 2.

Lemma 2 *If R is a random grid with $\mathscr{T}(R) = \lambda$, $\mathscr{T}(\bar{R}) = 1 - \lambda$.*

Proof From $\mathscr{T}(R) = \lambda$, we know that for any pixel $r \in R \, \mathscr{P}\!\mathit{rob}(r = 0) = \lambda$ and $\mathscr{P}\!\mathit{rob}(r = 1) = 1 - \lambda$. Let $r' \in \bar{R}$ be the corresponding pixel of r. Thus, $\mathscr{P}\!\mathit{rob}(r' = 0) = \mathscr{P}\!\mathit{rob}(r = 1) = 1 - \lambda$ and $\mathscr{P}\!\mathit{rob}(r' = 1) = \mathscr{P}\!\mathit{rob}(r = 0) = \lambda$. We have $\mathscr{T}(\bar{R}) = 1 - \lambda$. ∎

Consider two independent random grids X and Y. There is another important property called the *principle of combination* : if we cut a section, say A, from X and replace it with section B (with the same size as A) from Y, the result, denoted as $Z = (X \setminus A) \cap B$, is another random grid, i.e.,

$$\mathscr{T}(Z) = \mathscr{T}(X) = \mathscr{T}(Y) = \frac{1}{2} \tag{7.8}$$

We shall expose how to use the aforemnetioned characteristics of random grids to accomplish visual secret sharing in the next section.

7.3 Visual Cryptograms of Random Grids

First of all, we give a formal definition to VCRG for binary images in Section 7.3.1. Section 7.3.2 presents three algorithms for producing (2, 2)-VCRGs for binary images. The implementation results of these algorithms are reported in Section 7.3.3. Section 7.3.4 evaluates the performances of these algorithms in terms of light contrast. Then, algorithms for producing (n, n)-VCRGs for binary images are discussed in Section 7.3.5. Further, those for gray-level and color images are examined in Section 7.3.6. Section 7.3.7 reports the experimental results of (n, n)-VCRGs. The correctness of these algorithms would be proved and the corresponding light contrasts would be analyzed accordingly.

7.3.1 Definition of VCRG

Consider a set of n random grids $\mathscr{E} = \{R_1, R_2, \ldots, R_n\}$, which are encrypted to share a a secret image B such that $R_1 \otimes R_2 \otimes \ldots \otimes R_n$ reveals B to our eyes while any group of less than n grids obtains no information about B. We call $\mathscr{E}$ a set of *visual cryptograms of random grids* with respect to B.

Let $B(0)$ $(B(1))$ denote the area of white (black) pixels in B and $S^{\mathscr{E}}[B(0)]$ $(S[B(1)])$ denotes the area of pixels in $S^{\mathscr{E}}$ corresponding to $B(0)$ $(B(1))$ where $S^{\mathscr{E}} = R_1 \otimes R_2 \otimes \cdots \otimes R_n$. Formally, a set of (n, n) visual cryptograms of random grids for a binary image is now defined as follows.

Definition 2 *Given a binary image B, n random grids $\{R_1, R_2, \ldots, R_n\}$ comprise a set of visual cryptograms, referred to as (n,n)-VCRG of B if the following conditions hold:*

(1) *R_i is a random grid with $\mathscr{T}(R_i) = 1/2$ for $1 \leqslant i \leqslant n$;*

(2) *$\mathscr{T}(S^{\mathscr{D}}[B(0)]) = \mathscr{T}(S^{\mathscr{D}}[B(1)])$ where $S^{\mathscr{D}} = R_{i_1} \otimes R_{i_2} \otimes \cdots \otimes R_{i_d}$ which is the superimposed result of some set $\mathscr{D} = \{R_{i_1}, R_{i_2}, \ldots, R_{i_d}\}$ of d distinct random grids in $\mathscr{E}$, i.e., $\mathscr{D} \subset \mathscr{E}$, $2 \leqslant d \leqslant n-1$ and $1 \leqslant i_1 < i_2 < \cdots < i_d \leqslant n$; and*

(3) *$\mathscr{T}(S^{\mathscr{E}}[B(0)]) > \mathscr{T}(S^{\mathscr{E}}[B(1)])$ where $S^{\mathscr{E}} = R_1 \otimes R_2 \otimes \cdots \otimes R_n$ and $\mathscr{E} = \{R_1, R_2, \ldots, R_n\}$.*

Both Conditions 1 and 2 are "security" conditions which ensure that each individual grid and the superimposed result of any group of less than n random grids are merely random grids so that no information of B can be obtained. Condition 3 is the "contrast" condition, which claims that once the light transmission of $S^{\mathscr{E}}[B(0)]$ is larger than that of $S^{\mathscr{E}}[B(1)]$, our visual perception is able to distinguish $B(0)$ from $B(1)$ by observing $S^{\mathscr{E}}$ due to the difference of their light transmissions in $S^{\mathscr{E}}$.

Note that this definition is different from that in Naor and Shamir's model. Suppose that C_1 and C_2 are the two encoded shares produced by some (2, 2)-VCS using basis matrices M^0 and M^1 with respect to binary image B. Then the security condition in (2, 2)-VCS is $w(M^b[1]) = (M^b[2])$ for $b \in \{0, 1\}$ where $M^b[i]$ is the ith row of M^b for $i \in \{1, 2\}$, while the contrast condition becomes $w(M^1[1]$ or $M^1[2]) - (M^0[1]$ or $M^0[2]) > 0$. Let c_1 in C_1 and c_2 in C_2 be the corresponding pixels of b in B. Let $d = c_1 \otimes c_2$ where d is in $D = C_1 \otimes C_2$. Let $b(0)$ $(b(1))$ denote the pixel of $b = 0(1)$ and $d[b(0)]$ $(d[b(1)])$ denote such d corresponding to $b(0)$ $(b(1))$. Since each $b = 0$ (1) is encoded according to M^0 (M^1), the contrast condition in (2, 2)-VCS assures $w(d[b(1)]) - w(d[b(0)]) > 1$ so that $d(1)$ can be identified from $d(0)$ for all $d(0)$'s and $d(1)$'s in D. However, VCRG only demands $\mathscr{T}(S[B(0)]) - \mathscr{T}(S[B(1)]) > 0$, or equivalently, $\mathit{r}(s[b(0)]) - \mathit{r}(s[b(1)]) > 0$.

7.3.2   (2, 2)-VCRG Algorithms for Binary Images

Let *random_pixel*(0, 1) be a function that returns a binary value 0 or 1 to represent a transparent or opaque pixel, respectively, by a coin-flip procedure and $\overline{R_1[i,j]}$ denote the complement of $R_1[i,j]$. Each of the following three algorithms successfully encodes a secret binary image B into two random grids R_1 and R_2 which constitute a set of (2, 2)-VCRG.

Algorithms 1–3. Sharing a binary image by two random grids

Input: A $w \times h$ binary image B where $B[i, j] \in \{0, 1\}$ (white or black), $1 \leqslant i \leqslant w$ and $1 \leqslant j \leqslant h$

Output: Two shares of random grids R_1 and R_2 which reveal B when superimposed where $R_k[i, j] \in \{0, 1\}$ (transparent or opaque), $1 \leqslant i \leqslant w$, $1 \leqslant j \leqslant h$ and $k \in \{1, 2\}$

Encryption(B)

Algorithm 1.

1. Generate R_1 as a random grid, $\mathscr{T}(R_1) = 1/2$

    // **for** (each pixel $R_1[i,j]$, $1 \leqslant i \leqslant w$, and $1 \leqslant j \leqslant h$) **do**

    // $R_1[i,j] = \textit{random\_pixel}(0, 1)$

2. **for** (each pixel $B[i, j]$, $1 \leqslant i \leqslant w$ and $1 \leqslant j \leqslant h$) **do**

2.1 {   **if** $(B[i, j] = 0)$ **then** $R_2[i,j] = R_1[i,j]$

        else $R_2[i,j] = \overline{R_1[i,j]}$

     }

3. **output**(R_1, R_2)

Algorithm 2.

1. Generate R_1 as a random grid, $\mathcal{T}(R_1) = 1/2$

2. **for** (each pixel $B[i, j]$, $1 \leqslant i \leqslant w$ and $1 \leqslant j \leqslant h$) **do**

2.1 $\{$ **if** $(B[i, j] = 0)$ **then** $R_2[i, j] = R_1[i, j]$

         else $R_2[i, j] = random\_pixel(0, 1)$

    $\}$

3. **output**(R_1, R_2)

Algorithm 3.

1. Generate R_1 as a random grid, $\mathcal{T}(R_1) = 1/2$

2. **for** (each pixel $B[i, j]$, $1 \leqslant i \leqslant w$ and $1 \leqslant j \leqslant h$) **do**

2.1 $\{$ **if** $(B[i, j] = 0)$ **then** $R_2[i, j] = random\_pixel(0, 1)$

         else $R_2[i, j] = \overline{R_1[i, j]}$

    $\}$

3. **output**(R_1, R_2)

The three algorithms are capsulated into a generic procedure named *Encryption* so that when *Encryption*(B) is called, each of Algorithms 1, 2, or 3 can be applied onto B. Also note that in this paper, 0 (1) denotes a white (black) pixel in the secret binary image or a transparent (opaque) pixel in the encrypted share interchangeably.

Table 7.2 summarizes the encoding process of pixel b in secret image B into r_1 and r_2 by Algorithms 1, 2, and 3 respectively, the result of $r_1 \otimes r_2$ and its average light transmission.

Let $B(0)$ $(B(1))$ denote the area of all of the transparent (opaque) pixels in B, that is, pixel b is in $B(0)$ $(B(1))$ if and only if $b = 0$ $(b = 1)$ where $B = B(0) \cup B(1)$ and $B(0) \cap B(1) = \varnothing$. We denote the area of pixels in random grid R corresponding to $B(0)$ $(B(1))$ by $R[B(0)]$ $(R[B(1)])$, that is, pixel r is in $R[B(0)]$ $(R[B(1)])$ if and only if r's corresponding pixel b is in $B(0)$ $(B(1))$. Surely, $R = R[B(0)] \cup R[B(1)]$ and $R[B(0)] \cap R[B(1)] = \varnothing$.

Based upon the above notations, we have the following theorem.

Theorem 1 *Given a binary image B, $\{R_1, R_2\}$ produced by Algorithms 1–3, respectively, is a set of $(2, 2)$-VCRG of B.*

TABLE 7.2

Encoding b into r_1 and r_2 and results of $s = r_1 \otimes r_2$ by Algorithms 1, 2, and 3.

	b	Probability	r_1	r_2	$s = r_1 \otimes r_2$	$\mathscr{P}\mathit{rob}(s = 0)\ (\mathscr{T}(s))$
Algorithm 1	□	$\frac{1}{2}$	□	□	□	$\frac{1}{2}$
		$\frac{1}{2}$	■	■	■	
	■	$\frac{1}{2}$	□	■	■	0
		$\frac{1}{2}$	■	□	■	
Algorithm 2	□	$\frac{1}{2}$	□	□	□	$\frac{1}{2}$
		$\frac{1}{2}$	■	■	■	
	■	$\frac{1}{4}$	□	□	□	$\frac{1}{4}$
		$\frac{1}{4}$	□	■	■	
		$\frac{1}{4}$	■	□	■	
		$\frac{1}{4}$	■	■	■	
Algorithm 3	□	$\frac{1}{4}$	□	□	□	$\frac{1}{4}$
		$\frac{1}{4}$	□	■	■	
		$\frac{1}{4}$	■	□	■	
		$\frac{1}{4}$	■	■	■	
	■	$\frac{1}{2}$	□	■	■	0
		$\frac{1}{2}$	■	□	■	

Proof To prove that R_1 and R_2 constitute a set of (2, 2)-VCRG, we should validate whether the following two conditions (setting $n = 2$ in Definition 2) hold:

(1) $\mathscr{T}(R_1) = \mathscr{T}(R_2) = \frac{1}{2}$; and

(2) $\mathscr{T}(S[B(0)]) > \mathscr{T}(S[B(1)])$ where $S = R_1 \otimes R_2$.

Let us examine Algorithm 1 first. Let R_1 [B(b)] denote the area of pixels in R_1 that corresponds to B(b) for b = 0 or 1. Note that $R_1 = R_1[B(0)] \cup R_1[B(1)]$ where $R_1[B(0)] \cap R_1[B(1)] = \varnothing$. Since Step 1 in Algorithm 1 composes R_1 as a random grid with $\mathscr{T}(R_1) = 1/2$, we have $\mathscr{T}(R_1) = \mathscr{T}(R_1[B(0)]) = \mathscr{T}(R_1[B(1)]) = 1/2$. When $B[i, j] = 0$, we have $R_2[i, j] = R_1[i, j]$. That means $\mathscr{T}(R_2[B(0)]) = \mathscr{T}(R_1[B(0)]) = 1/2$. Moreover, when $B[i, j] = 1, R_2[i, j] = \overline{R_1[i, j]}$. By Lemma 2, we have $\mathscr{T}(R_2[B(1)]) = \mathscr{T}(\overline{R_1}[B(1)]) = 1 - \mathscr{T}(R_1[B(1)]) = 1 - 1/2 = 1/2$. Due to the facts that $R_2 = R_2[B(0)] \cup R_2[B(1)]$ and $\mathscr{T}(R_2[B(0)]) = \mathscr{T}(R_2[B(1)]) = 1/2$, we have $\mathscr{T}(R_2) = 1/2$ by the principle of combination. Therefore both of R_1 and R_2

are merely random grids and none of them individually leaks any information about B. The security condition of Definition 2 is met.

Consider $B(0)$ (the ares of transparent pixels in B). Since $R_2[i,j] = R_1[i,j]$ for each $B[i,j] = 0$ (or $B[i,j] \in B(0)$), thus $S[B(0)] = R_1[B(0)] \otimes R_2[B(0)] = R_1[B(0)]$. Thus $\mathscr{T}(S[B(0)]) = 1/2$. Regarding $B(1)$, $R_2[i,j] = \overline{R_1[i,j]}$ for each $B[i,j] = 1$ (or $B[i,j] \in B(1)$). We have $S[B(1)] = R_1[B(1)] \otimes R_2[B(1)] = \mathbf{1}$ (an area with all opaque pixels), i.e., $\mathscr{T}(S[B(1)]) = 0$. Thus $\mathscr{T}(S[B(0)]) > \mathscr{T}(S[B(1)])$. The light contrast condition of Definition 2 is satisfied. Therefore, $\{R_1, R_2\}$ produced by Algorithm 1 is a set of (2, 2)-VCRG of B and $(\mathscr{T}(S[B(0)]), \mathscr{T}(S[B(1)])) = (1/2, 0)$.

For Algorithm 2, we have $\mathscr{T}(R_2[B(0)]) = \mathscr{T}(R_1[B(0)]) = 1/2$ and $R_2[B(1)]$ is purely a random grid with $\mathscr{T}(R_2[B(1)]) = 1/2$. Thus, R_2 is a random grid with $\mathscr{T}(R_2) = 1/2$. Since $R_2[B(0)] = R_1[B(0)]$, $\mathscr{T}(S[B(0)]) = \mathscr{T}(R_1[B(0)] \otimes R_2[B(0)]) = \mathscr{T}((R_1[B(0)]) = 1/2$; while $R_1[B(1)]$ and $R_2[B(1)]$ are two independent random grids so that $\mathscr{T}(S[B(1)]) = \mathscr{T}(R_1[B(1)] \otimes R_2[B(1)]) = \mathscr{T}(R_1[B(1)]) \times \mathscr{T}(R_2[B(1)]) = 1/4$ (by Lemma 1(1)). We obtain $\mathscr{T}(S[B(0)]) > \mathscr{T}(S[B(1)])$.

For Algorithm 3, $\mathscr{T}(R_2[B(0)]) = 1/2$ because $R_2[B(0)]$ is a purely random grid and $\mathscr{T}(R_2[B(1)]) = \mathscr{T}(\overline{R_1}[B(1)]) = 1 - \mathscr{T}((R_1[B(1)]) = 1/2$. We have $\mathscr{T}(R_2) = 1/2$. Further, due to the fact that $R_1[B(0)]$ and $R_2[B(0)]$ are independent, $\mathscr{T}(S[B(0)]) = \mathscr{T}(R_1[B(0)] \otimes \mathscr{T}R_2[B(0)]) = \mathscr{T}(R_1[B(0)]) \times \mathscr{T}(R_2[B(0)]) = 1/4$; on the other hand, since $R_2[B(1)] = \overline{R_1}[B(1)]$, $S[B(1)] = R_1[B(1)] \otimes (R_2[B(1)] = \mathbf{1}$, i.e. $\mathscr{T}(S[B(1)] = 0$. We have $\mathscr{T}(S[B(0)]) > \mathscr{T}(S[B(1)])$.

Based upon the above statements, we realize that both security and contrast conditions in Definition 2 hold for all of the three algorithms. We conclude that no information of B can be obtained from random grids R_1 or R_2 individually, while S reveals B in our visual system for all of the three algorithms. Theorem 1 is proved.

The following corollary is an immediate consequence from the statements in the proof of Theorem 1.

Corollary 1 $(\mathscr{T}(S[B(0)]), \mathscr{T}(S[B(1)])) = (1/2, 0), (1/2, 1/4)$ or $(1/4, 0)$ for Algorithms 1–3, respectively where $S = R_1 \otimes R_2$ and $\{R_1, R_2\}$ is a set of (2, 2)-VCRG produced by Algorithms 1–3 with respect to secret image B.

7.3.3 Experiments for (2, 2)-VCRG

Figure 7.1 illustrates the results of the implementation of the above three algorithms. Figure 7.1(a) is secret binary image B, Figures 7.1(b) and (c) present the two random grids produced by Algorithm 1 and Figure 7.1(d) is the superimposed result of these two shares ((b) and (c)). Figures 7.1(e)–g) illustrate the corresponding results by Algorithm 2, while Figures 7.1(h)–(j) are the corresponding results by Algorithm 3. It can be easily seen from Figure

7.1 that the encrypted shares (see Figure 7.1(b), (c), (e), (f), (h), and (i)) are merely random pictures and no information about B can be obtained. Only when the two shares are superimposed (see Figures (d), (g), and (j)), can we see B by our visual system. It is worthy of notifying that there is no extra pixel expansion in Figures 7.1(b)–(j).

Figure 7.2 shows the reconstructed results by using Naor and Shamir's approach for encrypting B in Figure 7.1(a). The pixel expansion of Figure 7.2(a) is 2, while that of Figure 7.2(b) is 4 (by applying $S^0 = \begin{bmatrix} 0 & 1 & 0 & 1 \\ 0 & 1 & 0 & 1 \end{bmatrix}$ and $S^1 = \begin{bmatrix} 0 & 1 & 0 & 1 \\ 1 & 0 & 1 & 0 \end{bmatrix}$). The former does not retain the aspect ratio with respect to B, while the latter does. Both sizes of the results in Figure 7.2 are larger than that of B.

7.3.4 Definition of Light Contrast and Performance Evaluation

To evaluate the relative difference of the light transmissions between the transparent and opaque pixels in reconstructed image S by these random grid-based algorithms, we define the *light contrast* of S with respect to B as follows.

Definition 3 *The light contrast of a set $\mathscr{E}$ of VCRG produced by an encryption algorithm for a binary image B is defined as*

$$c(\mathscr{E}) = \frac{\mathscr{T}(S[B(0)]) - \mathscr{T}(S[B(1)])}{1 + \mathscr{T}(S[B(1)])}$$

where S is the superimposed result of all visual cryptograms in $\mathscr{E}$.

Let $\mathscr{E}_1$, $\mathscr{E}_2$, and $\mathscr{E}_3$ denote the three sets of (2, 2)-VCRG produced by Algorithms 1–3, respectively. By Definition 3, the light contrasts of $\mathscr{E}_1$, $\mathscr{E}_2$, and $\mathscr{E}_3$ are $c(\mathscr{E}_1) = 1/2 \ (= (1/2 - 0)/(1 + 0))$, $c(\mathscr{E}_2) = 1/5 \ (= (1/2 - 1/4)/(1 + 1/4))$, and $c(\mathscr{E}_3) = 1/4 \ (= (1/4 - 0)/(1 + 0))$. That is, Algorithm 1 achieves the highest light contrast among the three. This outcome can also be observed by comparing the reconstructed images in Figures 7.1(d), (g), and (j) in which (d) is more recognizable than (g) and (j). Note that $1 + \mathscr{T}(S[B(1)])$ is introduced as the denominator of $c(\mathscr{E})$ in favor of a less $\mathscr{T}(S[B(1)])$ (than a larger one) when two schemes have a same numerator (i.e., $\mathscr{T}(S[B(0)]) - \mathscr{T}(S[B(1)])$). For instance, both of $\mathscr{E}_2$ and $\mathscr{E}_3$ have the same result of $\mathscr{T}(S[B(0)]) - \mathscr{T}(S[B(1)])$, yet their $\mathscr{T}(S[B(1)])$'s are 1/4 and 0, respectively. Thus, $c(\mathscr{E}_3) > c(\mathscr{E}_2)$ according to Definition 3. As we can see, the reconstructed image by $\mathscr{E}_3$ (Figure 7.1(j)) is indeed more recognizable than that by $\mathscr{E}_2$ (Figure 7.1(g)) by our visual system. In general, we prefer a set of VCRG with a larger light contrast that helps our visual perception to recognize the result. Thus, Algorithm 1 is more preferable than the other two.

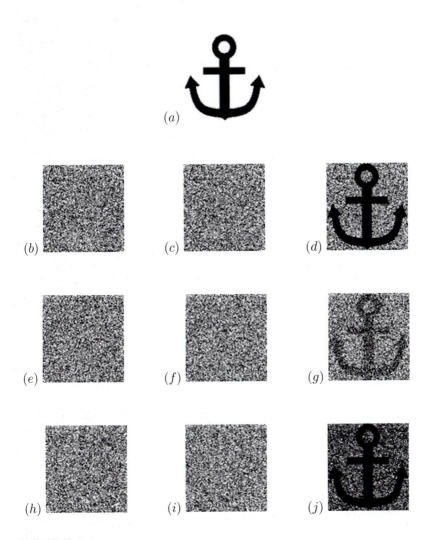

FIGURE 7.1

Implementation results of Algorithms 1, 2, and 3 for encrypting binary image *B*: (a) *B*; (b), (c), and (d) two encrypted shares and reconstructed image by Algorithm 1; (e), (f), and (g) two encrypted shares and reconstructed image by Algorithm 2; (h), (i), and (j) two encrypted shares and reconstructed image by Algorithm 3.

(a)   (b)

FIGURE 7.2
Reconstructed results by Naor and Shamir's approach for binary image B in Figure 7.1(a): (a) $m = 2$, (b) $m = 4$.

We summarize the reconstructed light contrasts by these three algorithms in Table 7.3.

TABLE 7.3
Light contrasts by Algorithms 1–3 in producing VCRG-2.

$\mathscr{E}$	$\mathscr{T}(S[B(0)])$	$\mathscr{T}(S[B(1)])$	$c(\mathscr{E})$
Algorithm 1	$\frac{1}{2}$	0	$\frac{1}{2}$
Algorithm 2	$\frac{1}{2}$	$\frac{1}{4}$	$\frac{1}{5}$
Algorithm 3	$\frac{1}{4}$	0	$\frac{1}{4}$

It is noticed that various definitions of contrast in [9, 14, 1] all depend on pixel expansion so that they are not suitable any more to measure the effectiveness of our schemes. On the contrary, since we consider the light transmissions through different areas of the transparency, the measurement of light contrast is more generalized and can be applied to all kinds of schemes.

We discuss how to generate (n, n)-VCRG in the next subsection.

7.3.5   Algorithms of (n, n)-VCRG for Binary Images

Consider an $h \times w$ binary image B and n random grids $R_1, R_2, \ldots, R_n$ with the same dimension. We call $r_1, r_2, \ldots, r_n$ the *corresponding pixels* of $b \in B$ if all r_k's in R_k's for $1 \leqslant k \leqslant n$ have the same coordinates as b (if $b = B[i, j]$, then $r_k = R_k[i, j]$ where $1 \leqslant i \leqslant h$ and $1 \leqslant j \leqslant w$).

By extending the idea of Algorithm 2 directly, we may obtain a set of two out of n visual cryptograms of random grids where any pair of two out of the n shares can reveal the secret when superimposed. Given a binary image B, we first find a random grid R_1 with $\mathcal{T}(R_1) = 1/2$. Then, we define R_k for $2 \leqslant k \leqslant n$ as follows:

$$r^k = \{ \begin{array}{ll} r_1 & \text{if } b = 0; \\ random\_pixel() & \text{otherwise,} \end{array} \tag{7.9}$$

for each pixel $r_k \in R_k$ where $r_1, r_2, \ldots, r_n$ are corresponding pixels of $b \in B$. It is easy to see that all R_k's are random grids and when $b = 0$, $t(r_i \otimes r_j) = \mathcal{\ell}(r_1 \otimes r_1) = \mathcal{\ell}(r_1) = 1/2$; while $b = 1$, $\mathcal{\ell}(r_i \otimes r_j) = 1/4$ by Lemma 1 for each pair of i and j where $1 \leqslant i \neq j \leqslant n$. That is, $\mathcal{T}(R_i[B(0)] \otimes R_j[B(0)]) = 1/2 > 1/4 = \mathcal{T}(R_i[B(1)] \otimes R_j[B(1)])$. As a result, $\{R_1, R_2, \ldots, R_n\}$ produced in this way is a set of two out of n visual cryptograms of random grids. Yet, it is not a set of (n, n)-VCRG, since Condition 2 in Definition 2 fails.

Let us extract some informative features from the idea in Algorithm 1. Let $B = \{0, 1\}$ be a binary set. We introduce a function $f \colon \mathscr{B} \times \mathscr{B} \to \mathscr{B}$ defined as follows to transcribe the basic idea in Algorithm 1:

$$f(x, s) = \{ \begin{array}{ll} s & \text{if } x = 0; \\ \overline{s} & \text{otherwise,} \end{array} \tag{7.10}$$

for $x, s, \overline{s} \in \mathscr{B}$ where $\overline{s}$ is the inverse value of s. We may say that function $f(x, s)$ *preserves* the value of s if $x = 0$, and *reverses* it otherwise ($x = 1$). In the subject of sequential circuits, the behavior of $f(x, s)$ is equivalent to that of a T flip-flop where s, x and $f(x, s)$ are the current state, toggle input, and next state, respectively. In the area of logical operations, $f(x, s)$ can be implemented by using the Exclusive-OR operation ($\oplus$), that is, $f(x, s) = x \oplus s$.

Let B be a binary image and R_1 be a random grid with $\mathcal{T}(R_1) = 1/2$. By introducing function f, the essential idea of Algorithm 1 for generating each pixel $r_2 \in R_2$ corresponding to $r_1 \in R_1$ and $b \in B$ can be formulated as

$$r_2 = f(b, r_1). \tag{7.11}$$

Note that when $b = 0$, $r_2 = r_1$; while $b = 1$, $r_2 = \overline{r_1}$. This is exactly the same as the manipulation of Step 2 in Algorithm 1.

This critical observation is emphasized as a corollary as follows.

Corollary 2 *If R_1 is a random grid with $\mathcal{T}(R_1) = 1/2$ and the corresponding pixel $r_2 \in R_2$ of $r_1 \in R_1$ is obtained by $r_2 = f(b, r_1)$ for each $r_1 \in R_1$ with any $b \in \mathscr{B}$, then R_2 is a random grid with $\mathcal{T}(R_2) = 1/2$.*

By using function f, our first construction of a set of (n, n)-VCRG ($\mathscr{E} = \{R_1, R_2, \ldots, R_n\}$) with respect to B is now introduced as follows. We first generate $n - 1$ random grids $R_1, R_2, \ldots, R_{n-1}$ independently with $\mathcal{T}(R_k) = 1/2$ for $1 \leqslant k \leqslant n - 1$. That is, for every pixel $b \in B$ its $n - 1$ corresponding pixels $r_1, r_2, \ldots, r_{n-1}$ are totally random where $r_k \in R_k$ for $1 \leqslant k \leqslant n - 1$. Based upon $r_1, r_2, \ldots, r_k$, we compute a_k for $1 \leqslant k \leqslant n - 1$ by using the recursive formula:

$$a_k = \{ \begin{array}{ll} r_1 & \text{if } k = 1; \\ f(r_k, a_{k-1}) & \text{otherwise.} \end{array} \tag{7.12}$$

Then, we find r_n according to b and a_{n-1} by

$$r_n = f(b, a_{n-1}). \tag{7.13}$$

It is noticed that formula (7.11) is a special case of formula (7.13) by setting $n = 2$. After all r_n's $\in R_n$ corresponding to all b's $\in B$ are computed, we obtain R_n. Then, $\mathscr{E} = \{R_1, R_2, \ldots, R_n\}$ is reported as a set of (n, n)-VCRG of B. The whole idea is formally illustrated in Algorithm 4.

Algorithm 4. Encrypting a secret image into a set of (n, n)-VCRG
Input: an $h \times w$ binary image B and an integer n
Output: $\mathscr{E} = \{R_1, R_2, \ldots, R_n\}$ constituting (n, n)-VCRG of B

1.   **for** $(1 \leqslant k \leqslant n - 1)$ **do**
            $\{$ generate R_k as a random grid, $\mathscr{T}(R_k) = 1/2$
            $\}$

2.   **for** (each pixel $B[i, j]$, $1 \leqslant i \leqslant h$ and $1 \leqslant j \leqslant w$) **do**

2.1 $\{$ $a_1 = R_1[i, j]$

2.2   **for** $(2 \leqslant k \leqslant n - 1)$ **do**
               $\{$ $a_k = f(R_k[i, j], a_{k-1})$ $//f(x, s)$ is defined in formula (7.10)
               $\}$

2.3 $R_n[i, j] = f(B[i, j], a_{n-1})$
         $\}$

3.   **output**$(R_1, R_2, \ldots, R_n)$

Before we prove that $\mathscr{E} = \{R_1, R_2, \ldots, R_n\}$ generated by Algorithm 4 with respect to B is indeed a set of (n, n)-VCRG of B, we explore more useful features among multiple random grids in the following.

It is easy to see that two independent random grids produce another random grid as they are superimposed even when their light transmission is not $1/2$. We formalize this in Lemma 3, which is a generalized version of Lemma 1.

Lemma 3 *Given two independent random grids R_1 and R_2 with $\mathscr{T}(R_1) = \lambda_1$ and $\mathscr{T}(R_2) = \lambda_2$, $R_1 \otimes R_2$ is a random grid with $\mathscr{T}(R_1 \otimes R_2) = \lambda_1 \lambda_2$.*

Proof *For any pixel $r_1 \in R_1$ and its corresponding pixel $r_2 \in R_2$, we have $\mathscr{P}rob(r_1 = 0) = \lambda_1$ and $\mathscr{P}rob(r_2 = 0) = \lambda_2$. $r_1 \otimes r_2$ is transparent if and only if both of r_1 and r_2 are transparent, that is, $\mathscr{P}rob(r_1 \otimes r_2 = 0) = \mathscr{P}rob(r_1 = 0) \times \mathscr{P}rob(r_2 = 0) = \lambda_1 \lambda_2$. Therefore, $\mathscr{T}(R_1 \otimes R_2) = \mathscr{T}(R_1) \times \mathscr{T}(R_2) = \lambda_1 \lambda_2$.*

The theme of Lemma 3 can be extended for more than two random grids.

Lemma 4 *If $\mathscr{U} = \{R_1, R_2, \ldots, R_u\}$ is a set of $u \geqslant 2$ independent random grids with $\mathscr{T}(R_i) = \lambda$ for $1 \leqslant i \leqslant u$, then $S^{\mathscr{U}}$ is a random grid with $\mathscr{T}(S^{\mathscr{U}}) = \lambda^u$ where $S^{\mathscr{U}} = R_1 \otimes R_2 \otimes \cdots \otimes R_u$.*

Proof We prove by mathematical induction. Consider the case of $\mathscr{V} = \{R_1, R_2\}$. The statement (i.e., $\mathscr{T}(S^{\mathscr{V}}) = \lambda^2$) is true by Lemma 3. Assume that the statement holds for the case of $\mathscr{V} = \{R_1, R_2, \ldots, R_{u-1}\}$, that is, $S^{\mathscr{V}}$ is a random grid with $\mathscr{T}(S^{\mathscr{V}}) = \lambda^{u-1}$ where $S^{\mathscr{V}} = R_1 \otimes R_2 \otimes \cdots \otimes R_{u-1}$. Then, we further consider $\mathscr{U} = \{R_1, R_2, \ldots, R_u\} = \mathscr{V} \cup \{R_u\}$. $S^{\mathscr{U}}$ is simply the superimposed result of $S^{\mathscr{V}}$ and R_u, i.e., $S^{\mathscr{U}} = S^{\mathscr{V}} \otimes R_u$. By Lemma 3, we have $\mathscr{T}(S^{\mathscr{U}}) = \mathscr{T}(S^{\mathscr{V}} \otimes R_u) = \mathscr{T}(S^{\mathscr{V}}) \times \mathscr{T}(R_u) = (\lambda^{u-1}) \times \lambda = \lambda^u$.

Since the superimposition operation $\otimes$ is commutative, it is not hard to obtain that the results of different superimposing orders of $R_1, R_2, \ldots, R_u$ are all the same (i.e., $S^{\mathscr{U}}$). As a matter of fact, the superimposed result of any subset of $\mathscr{U}$ is also a random grid as indicated in Lemma 5.

Lemma 5 *Let $\mathscr{U} = \{R_1, R_2, \ldots, R_u\}$ be a set up $u \geqslant 2$ independent random grids and $\mathscr{V} = \{R_{i_1}, R_{i_2}, \ldots, R_{i_v}\} \subset \mathscr{U}$ where $1 \leqslant i_1 < i_2 < \cdots < i_v \leqslant u$ and $\mathscr{T}(R_i) = \lambda$ for $1 \leqslant i \leqslant u$. $S^{\mathscr{V}}$ is a random grid with $\mathscr{T}(S^{\mathscr{V}}) = \lambda^v$ where $S^{\mathscr{V}} = R_{i_1} \otimes R_{i_2} \otimes \ldots \otimes R_{i_v}$.*

The proof of Lemma 4 can be easily applied to prove Lemma 5. Consider a set of u independent random grids $\mathscr{U} = \{R_1, R_2, \ldots, R_u\}$ where $\mathscr{T}(R_k) = 1/2$ for $1 \leqslant k \leqslant u$. We may generate a set of u binary images $\mathscr{A} = \{A_1, A_2, \ldots, A_u\}$ with respect to $\mathscr{U}$ in such a way that each pixel $a_k \in A_k$ is defined by formula (7.12), that is,

$$a_k = \begin{cases} r_1 & \text{if } k = 1; \\ f(r_k, a_{k-1}) & \text{otherwise,} \end{cases}$$

where $a_k \in A_k$ and $r_k \in R_k$ are corresponding pixels for $1 \leqslant k \leqslant u$. Lemma 6 claims that all of $A_1, A_2, \ldots, A_u$ are random grids with $\mathscr{T}(A_k) = 1/2$ for $1 \leqslant k \leqslant u$.

Lemma 6 *For $\mathscr{A} = \{A_1, A_2, \ldots, A_u\}$ generated by formula (7.12) with respect to $\mathscr{U} = \{R_1, R_2, \ldots, R_u\}$ where $\mathscr{T}(R_k) = 1/2$, A_k is a random grid with $\mathscr{T}(A_k) = 1/2$ for $1 \leqslant k \leqslant u$.*

Proof We prove by mathematical induction. For the case of $|\mathscr{A}| = 1$ (i.e., $u = 1$), since $a_1 \in A_1$ is equal to $r_1 \in R_1$ by formula (7.12), A_1 is exactly the same as R_1. Thus, A_1 is a random grid with $\mathscr{T}(A_1) = \mathscr{T}(R_1) = 1/2$. Assume that the statement holds for the case of $|\mathscr{A}| = u - 1$, that is, $A_1, A_2, \ldots, A_{u-1}$ are random grids with $\mathscr{T}(A_k) = 1/2$ for $1 \leqslant k \leqslant u - 1$. Due to the reasons that $r_u(\in R_u) = 0$ or 1, $\mathscr{T}(A_{u-1}) = 1/2$, and for each pixel $a_u \in A_u$, $a_u = f(r_u, a_{u-1})$, we obtain that A_u is also a random grid with $\mathscr{T}(A_k) = 1/2$ by Corollary 2.

According to formula (7.12), for each pixel $a_k \in A_k$,

$$
\begin{aligned}
a_k &= f(r_k, a_{k-1}) = f(r_k, f(r_{k-1}, a_{k-2})) \\
&= \cdots = f(r_k, f(r_{k-1}, f(\ldots, f(r_2, a_1) \ldots))) \\
&= f(r_k, f(r_{k-1}, f(\ldots, f(r_2, r_1) \ldots))).
\end{aligned}
$$

That is, the value of a_k is determined by its k corresponding pixels $r_1, r_2, \ldots, r_k$ for $1 \leqslant k \leqslant u$. From Lemma 6, we know that $a_k \in A_k$ for $1 \leqslant k \leqslant u$ is a random pixel with $\mathscr{r}(a_k) = \frac{1}{2}$.

Let $\mathscr{V} = \{R_{i_1}, R_{i_2}, \ldots, R_{i_v}\}$ be a set of v random grids randomly selected from $\mathscr{U}$, (i.e. $\mathscr{V} \subset \mathscr{U}$) where $1 \leqslant v \leqslant u$. Let $S^{\mathscr{U}}(0)(S(0))$ denote the area of transparent pixels in $S^{\mathscr{U}}(S^{\mathscr{V}})$ where $S^{\mathscr{U}} = R_1 \otimes R_2 \otimes \cdots \otimes R_u$ ($S^{\mathscr{V}} = R_{i_1} \otimes R_{i_2} \otimes \cdots \otimes R_{i_v}$). Now we would like to examine the light transmission of the area of pixels in A_u corresponding to $S^{\mathscr{U}}(0)(S(0))$, referred to as $A_u[S^{\mathscr{U}}(0)]$ ($A_u[S^{\mathscr{V}}(0)]$).

Lemma 7

(1) $\mathscr{T}(A_u[S^{\mathscr{U}}(0)]) = 1$;

(2) $\mathscr{T}(A_u[S^{\mathscr{V}}(0)]) = 1/2$.

Proof

(1) From Lemma 4, we have $\mathscr{T}(S^{\mathscr{U}}) = 1/2^{\mathscr{U}}$. Consider pixel $s = 0$(or $s \in S^{\mathscr{U}}(0)$). The only chance for $s = 0$ is that its corresponding pixels $r_1, r_2, \ldots, r_u$ should all be transparent, i.e., $r_1 = r_2 = \cdots = r_u = 0$. Under this circumstance, by formula (7.14) s's corresponding pixel a_u in A_u becomes

$$
\begin{aligned}
a_u &= f(r_u, f(r_{u-1}, f(\ldots, f(r_2, r_1) \ldots))) \\
&= f(0, f(0, f(\ldots, f(0, 0) \ldots))) \\
&= 0.
\end{aligned}
$$

That is, if $s = 0$ (or $s \in S^{\mathscr{U}}(0)$). then $a_u = 0$. As a result, $\mathscr{T}(A_u[S^{\mathscr{U}}(0)]) = 1$.

(2) Let $\mathscr{U} - \mathscr{V} = \mathscr{W} = \{R_{j_1}, R_{j_2}, \ldots, R_{j_w}\}$ where $w = u - v$ (i.e., $\{j_1, j_2, \ldots, j_w\} = \{1, 2, \ldots, u\} - \{i_1, i_2, \ldots, i_v\}$). From Lemma 5, we know that both $S^{\mathscr{V}}$ and $S^{\mathscr{W}}$ are random grids with $\mathscr{T}(S^{\mathscr{V}}) = 1/2^v$ and $\mathscr{T}(S^{\mathscr{W}}) = 1/2^{u-v}$. Since the order of random grids does not affect the result of superimposition, without losing generality, let $(j_1, j_2, \ldots, j_w) = (1, 2, \ldots, w)$ and $(i_1, i_2, \ldots, i_v) = (w + 1, w + 2, \ldots, u)$. That is, $\mathscr{W} = \{R_1, R_2, \ldots, R_w\}$ and $\mathscr{V} = \{R_{w+1}, R_{w+2}, \ldots, R_u\}$. (This can easily be done by renaming all of the

random grids.) We focus on pixel $t = 0$ in $S^{\mathscr{V}}$ (or $t \in S^{\mathscr{V}}(0)$) only. Its corresponding pixels $r_{w+1} \in R_{w+1}, r_{w+2} \in R_{w+2}, \ldots, r_u \in R_u$ should all be transparent, i.e., $r_{w+1} = r_{w+2} = \cdots = r_u = 0$. Consequently, its corresponding pixel $a_u \in A_u$ becomes

$$
\begin{aligned}
a_u &= f(r_u, f(r_{u-1}, \ldots, f(r_{w+1}, f(r_w, f(r_{w-1}, \ldots, f(r_2, r_1)) \ldots)))) \\
&= f(0, f(0, \ldots, f(0, f(r_w, f(r_{w-1}, \ldots, f(r_2, r_1)) \ldots)))) \\
&= f(r_w, f(r_{w-1}, \ldots, f(r_2, r_1) \ldots)).
\end{aligned}
$$

From Lemma 6, we have $\ell(a_u) = 1/2$. In summary, if $t = 0$ in $S^{\mathscr{V}}$ (or $t \in S^{\mathscr{V}}(0)$), then $\ell(a_u) = 1/2$; therefore, $\mathscr{T}(A_u[S^{\mathscr{V}}(0)]) = 1/2$.

Lemmas 5–7 investigate the properties of u random grids in $\mathscr{U} = \{R_1, R_2, \ldots, R_u\}$ and another u random grids $A_1, A_2, \ldots, A_u$ produced by formula (7.12) with respect to $\mathscr{U}$ for $u \geqslant 2$. Now we pay attention to Algorithm 4. Give a secret image B and n participants, Algorithm 4 first produces a set of $n - 1$ independent random grids, namely $\mathscr{R} = \{R_1, R_2, \ldots, R_{n-1}\}$, based upon which $A_1, A_2, \ldots, A_{n-1}$ are generated, and then R_n is produced according to A_{n-1} and B. We claim that R_n is a random grid and the superimposed result of any group of less than n shares is also a random grid.

Lemma 8 *Give a secret image B and a set of independent random grids* $\mathscr{R} = \{R_1, R_2, \ldots, R_{n-1}\}$,

(1) R_n *generated by formula (7.13) is a random grid with* $\mathscr{T}(R_n) = 1/2$*; and*

(2) $S^{\mathscr{D}}$ *is a random grid with* $\mathscr{T}(S^{\mathscr{D}}) = 1/2^d$ *where* $\mathscr{D} = \{R_{i_1}, R_{i_2}, \ldots, R_{i_d}\} \subset \mathscr{R} \cup \{R_n\}$ *and* $S^{\mathscr{D}} = R_{i_1} \otimes R_{i_2} \otimes \cdots \otimes R_{i_d}$.

Proof

(1) By setting $u = n - 1$ in Lemma 6, we know that A_{n-1} is a random grid with $\mathscr{T}(A_{n-1}) = 1/2$ (or $\ell(a_{n-1}) = 1/2$) for $a_{n-1} \in A_{n-1}$. For $r_n \in R_n[B(0)]$, $r_n = f(0, a_{n-1}) = a_{n-1}$; while $r_n \in R_n[B(1)]$, $r_n = f(1, a_{n-1}) = \overline{a}_{n-1}$. It implies that both $R_n[B(0)]$ and $R_n[B(1)]$ are areas of random grids with $R_n[B(0)] = 1/2 = R_n[B(1)]$. By the principle of combination, R_n is a random grid with $\mathscr{T}(R_n) = 1/2$.

(2) Since the generation of R_n is different for $R_1, R_2, \ldots, R_{n-1}$ (in fact, R_n is generated from B and A_{n-1} which depends upon $R_1, R_2, \ldots, R_{n-1}$ (see formulae (7.12) and (7.13))), there are two cases with regard to the constituents of $\mathscr{D}$: Case 1: $R_n \notin \mathscr{D}$; and Case 2: $R_n \in \mathscr{D}$.

For Case 1, since all random grids in $\mathscr{D}$ are independently generated, $S^{\mathscr{D}}$ is a random grid according to Lemma 5 with $\mathscr{T}(S^{\mathscr{D}}) = \mathscr{T}(S^{\mathscr{D}}[B(0)]) = \mathscr{T}(S^{\mathscr{D}}[B(1)]) = 1/2^d$.

Regarding Case 2, assume that $\mathscr{D} = \{R_{i_1}, R_{i_2}, \ldots, R_{i_{d-1}}, R_n\}$. Let $\mathscr{L} = \{R_{i_1}, R_{i_2}, \ldots, R_{i_{d-1}}\}$ (i.e., $\mathscr{D} = \mathscr{L} \cup \{R_n\}$) and $S^{\mathscr{L}} = R_{i_1} \otimes R_{i_2} \otimes \cdots \otimes R_{i_{d-1}}$. From Lemma 5, we realize that $S^{\mathscr{L}}$ is a random grid with $\mathscr{T}(S^{\mathscr{L}}) = 1/2^{d-1}$. Besides, by setting $u = n - 1$ and $v = d - 1$ in Lemma 7(2), we obtain

$\mathcal{T}(A_{n-1}[S^{\mathcal{L}}(0)]) = 1/2$. Consider any pixel $a_{n-1} \in A_{n-1}[S^{\mathcal{L}}(0)]$ and its corresponding pixels $s^l \in S^{\mathcal{L}}(0)$ (i.e., $S^l = r_{i_1} \otimes r_{i_2} \otimes \cdots \otimes r_{i_{d-1}} = 0$ where $r_{i_k} \in R_{i_k}$ for $1 \leqslant k \leqslant d-1$), $b \in B$ and $r_n \in R_n$. If $b = 0$, $r_n = f(b, a_{n-1}) = f(0, a_{n-1}) = a_{n-1}$; otherwise ($b = 1$), $r_n = f(1, a_{n-1}) = \bar{a}_{n-1}$. We obtain that $\mathcal{t}(r_n) = 1/2$ on condition that $s^l \in S^{\mathcal{L}}(0)$. It means that $\mathcal{T}(R_n[S^{\mathcal{L}}(0)]) = 1/2$.

To explore the light transmission of $S^{\mathcal{D}}(= S^{\mathcal{L}} \otimes R_n = (S^{\mathcal{L}}(0) \otimes R_n[S^{\mathcal{L}}(0)]) \cup (S^{\mathcal{L}}(1) \otimes R_n[S^{\mathcal{L}}(1)]))$, we only need to consider the light transmission of $S^{\mathcal{L}}(0) \otimes R_n[S^{\mathcal{L}}(0)]$, since no light can pass through the area of $S^{\mathcal{L}}(1) \otimes R_n[S^{\mathcal{L}}(1)]$ (note that $S^{\mathcal{L}}(1)$ is the area of opaque pixels in $S^{\mathcal{L}}$). Due to the facts that $\mathcal{T}(S^{\mathcal{L}}) = 1/2^{d-1} = \mathcal{T}(S^{\mathcal{L}}(0))$ (by Lemma 5) and $\mathcal{T}(R_n[S^{\mathcal{L}(0)}]) = 1/2$ (proved previously). We have $\mathcal{T}(S^{\mathcal{D}}) = \mathcal{T}(S^{\mathcal{L}}(0) \otimes R_n[S^{\mathcal{L}}(0)]) = \mathcal{T}(S^{\mathcal{L}}(0)) \times \mathcal{T}(R_n[S^{\mathcal{L}}(0)]) = (1/2^{d-1}) \times (1/2) = 1/2^d$ by Lemma 3.

Based upon the above discussions, we prove the correctness of Algorithm 4 in Theorem 2.

Theorem 2 *Given a binary image* B, $\mathcal{E} = \{R_1, R_2, \ldots, R_n\}$ *produced by Algorithm 4 with respect to* B *is a set of* (n, n)-*VCRG of* B.

Proof We prove that $\mathcal{E} = \{R_1, R_2, \ldots, R_n\}$ meets the three conditions in Definition 2 respectively in the following.

(1) From Step 1 of the algorithm, we realize that $\mathcal{R} = \{R_1, R_2, \ldots, R_{n-1}\}$ are truly random grids with $\mathcal{T}(R_i) = 1/2$ for $1 \leqslant i \leqslant n-1$. From Lemma 8(1), R_n is also a random grid with $\mathcal{T}(R_n) = 1/2$. Thus, all of $R_1, R_2, \ldots, R_n$ are random grids with a light transmission of $1/2$. Condition 1 of Definition 2 holds.

(2) Regarding any subset of $\mathcal{E}$ with $d(< n)$ random grids, namely $\mathcal{D} = \{R_{i_1}, R_{i_2}, \ldots, R_{i_d}\} \subset \mathcal{E}$, $1 \leqslant i_1 < i_2 < \cdots < i_d \leqslant n$, we have that $S^{\mathcal{D}}$ is a random grid with $\mathcal{T}(S^{\mathcal{D}}) = 1/2^d$ from Lemma 8(2). Condition 2 holds.

(3) Let $\mathcal{R} = \{R_1, R_2, \ldots, R_{n-1}\}$, $S^{\mathcal{R}} = R_1 \otimes R_2 \otimes \cdots \otimes R_{n-1}$ and $s^r \in S^{\mathcal{R}}$, $r_n \in R_n$, $b \in B$ as well as $r_i \in R_i$ for $1 \leqslant i \leqslant n-1$ be corresponding pixels. By Lemma 4, we know $\mathcal{T}(S^{\mathcal{R}}) = \mathcal{T}(S^{\mathcal{R}}[B(1)]) = \mathcal{T}(S^{\mathcal{R}}[B(0)]) = 1/2^{n-1}$. We focus on pixel $s^r \in S^{\mathcal{R}}$ that is transparent. The only case for $s^r = 0$ is $r_1 = r_2 = \cdots = r_{n-1} = 0$. In this case, their corresponding pixel $a_{n-1} \in A_{n-1}$ is also 0 (or $\mathcal{T}(A_{n-1}[S^{\mathcal{R}}(0)]) = 1$, see also Lemma 7(1) by setting $u = n-1$ and $\mathcal{U} = \mathcal{R}$). Thus, when $b = 0$, $r_n = f(p, a_{n-1}) = f(0, 0) = 0$, while $b = 1$, $r_n = f(p, a_{n-1}) = f(1, 0) = 1$. We obtain that $r_1 \otimes r_2 \otimes \cdots \otimes r_{n-1} \otimes r_n = s^r \otimes r_n = s^r \otimes 0 = s^r$ for $b = 0$; while $r_1 \otimes r_2 \otimes \cdots \otimes r_{n-1} \otimes r_n = s^r \otimes r_n = s^r \otimes 1 = 1$ for $b = 1$. That is, $\mathcal{T}(S^{\mathcal{E}}[B(0)]) = \mathcal{T}(S^{\mathcal{R}}[B(0)]) = 1/2^{n-1} > \mathcal{T}(S^{\mathcal{E}}[B(1)]) = 0$. Condition 3 holds.

Since all of the three conditions in Definition 2 hold, we conclude that $\mathcal{E} = \{R_1, R_2, \ldots, R_n\}$ generated by Algorithm 4 is a set of (n, n)-VCRG of B.

Corollary 3 is an immediate result from Theorem 2.

Corollary 3 $(\mathscr{T}(S^{\mathscr{E}}[B(0)]), \mathscr{T}(S^{\mathscr{E}}[B(1)])) = (1/2^{n-1}, 0)$ *where* $\mathscr{E} = \{R_1, R_2, \ldots, R_n\}$ *is produced by Algorithm 4 with respect to B and $S^{\mathscr{E}} = R_1 \otimes R_2 \otimes \cdots R_n$.*

We give a simple example for $n = 3$ to explain the relationship among corresponding pixels $b \in B$, $r_1 \in R_1$, $r_2 \in R_2$, $r_3 \in R_3$, $a_1 \in A_1$, and $a_2 \in A_2$ in Algorithm 4.

Example 1. Consider $n = 3$ and a secret image B. Algorithm 4 first generates R_1 and R_2 independently; then produces A_1 and A_2 by formula (7.12); at last finds R_3 depending on B and A_2 according to formula (7.13). Let $\mathscr{E} = \{R_1, R_2, R_3\}$. Table 7.4 summarizes all of the possible combinations of corresponding pixels $b \in B$, $r_1 \in R_1$, and $r_2 \in R_2$; and their corresponding results of $a_1(= r_1) \in A_1$, $a_2(= f(r_2, a_1)) \in A_2$, and $r_3(= f(b, a_2)) \in R_3$; as well as the superimposed results of $r_1 \otimes r_2$, $r_1 \otimes r_3$, $r_2 \otimes r_3$, and $r_1 \otimes r_2 \otimes r_3$. Table 7.5 lists the light transmissions of the corresponding results in Table 7.4.

It is seen from Tables 7.4 and 7.5 that all of R_1, R_2, R_3, A_1, A_2 are random grids with a light transmission of $1/2$. Besides, $\mathscr{T}(S^{\mathscr{D}}[B(0)]) = \mathscr{T}(S^{\mathscr{D}}[B(1)]) = 1/4$ and $\mathscr{T}(S^{\mathscr{D}}[B(0)]) = 1/4 > 0 = \mathscr{T}(S^{\mathscr{D}}[B(1)])$ where $\mathscr{D}(\subset \mathscr{E}) = \{R_1, R_2\}, \{R_1, R_3\}$ or $\{R_2, R_3\}$. It means that each group of two random grids obtains no information about $B(0)$ or $B(1)$ when superimposed (thus, no information about the colors of pixels in B can be found), while $B(0)$ and $B(1)$ can be identified from $S^{\mathscr{E}}$ due to the difference of their light transmissions, that is, we see B out of $S^{\mathscr{E}}$ because of $\mathscr{T}(S^{\mathscr{E}}[B(0)]) > \mathscr{T}(S^{\mathscr{E}}[B(1)])$.

TABLE 7.4

All possible combinations of b, r_1, and r_2, the corresponding results of a_1 $(= r_1)$, $a_2 = f(r_2, a_1)$, $r_3 = f(b, a_2)$, and $r_1 \otimes r_2$, $r_1 \otimes r_3$, $r_2 \otimes r_3$ as well as $r_1 \otimes r_2 \otimes r_3$ by Algorithm 4 for (3, 3)-VCRG.

b	r_1	a_1	r_2	a_2	r_3	$r_1 \otimes r_2$	$r_1 \otimes r_3$	$r_2 \otimes r_3$	$r_1 \otimes r_2 \otimes r_3$
0	0	0	0	0	0	0	0	0	0
			1	1	1	1	1	1	1
	1	1	0	1	1	1	1	1	1
			1	0	0	1	1	1	1
1	0	0	0	0	1	0	1	1	1
			1	1	0	1	0	1	1
	1	1	0	1	0	1	1	0	1
			1	0	1	1	1	1	1

Algorithm 4 can be viewed as the generalization from the idea in Algorithm 1. Such knowledge can be easily adapted to generalize the ideas in Algorithms 2 and 3. Algorithms 5 and 6 are the results accordingly.

TABLE 7.5

Light transmissions of the corresponding pixels in Table 7.3.

b	$\ell(r_1)$	$\ell(a_1)$	$\ell(r_2)$	$\ell(a_2)$	$\ell(r_3)$
0	1/2	1/2	1/2	1/2	1/2
1	1/2	1/2	1/2	1/2	1/2

$\ell(r_1 \otimes r_2)$	$\ell(r_1 \otimes r_3)$	$\ell(r_2 \otimes r_3)$	$\ell(r_1 \otimes r_2 \otimes r_3)$
1/4	1/4	1/4	1/4
1/4	1/4	1/4	0

Algorithm 5–6. Encrypting a secret image into n random grid as a set of (n,n)-VCRG (based upon Algorithms 2 and 3, respectively)

Input: an $h \times w$ binary image B and an integer n
Output: a set of n random grids $\mathscr{E} = \{R_1, R_2, \ldots, R_n\}$ constituting (n,n)-VCRG of B

Algorithm 5.

1.   **for** $(1 \leqslant k \leqslant n-1)$ **do**
     {   generate R_k as a random grid, $\mathscr{T}(R_k) = 1/2$
     }

2.   **for** (each pixel $B[i,j]$, $1 \leqslant i \leqslant h$ and $1 \leqslant j \leqslant w$) **do**

2.1 { $a_1 = R_1[i,j]$

2.2   **for** $(2 \leqslant k \leqslant n-1)$ **do**
               { $a_k = f(R_k[i,j], a_{k-1})$
               }

2.3   **if** $(B[i,j] = 0)$ **then** $R_n[i,j] = f(B[i,j], a_{n-1})$
               $(= f(0, a_{n-1}) = a_{n-1})$
               else $R_n[i,j]$ =random_pixel()

     }

3.   output$(R_1, R_2, \ldots, R_n)$

Algorithm 6.

1.   **for** $(1 \leqslant k \leqslant n-1)$ **do**
     {   generate R_k as a random grid, $\mathscr{T}(R_k) = 1/2$
     }

2. **for** (each pixel $B[i, j]$, $1 \leqslant i \leqslant h$ and $1 \leqslant j \leqslant w$) **do**

2.1 { $\quad a_1 = R_1[i, j]$

2.2 $\qquad$ **for** $(2 \leqslant k \leqslant n - 1)$ **do**

$\qquad\qquad$ { $\quad a_k = f(R_k[i, j], a_{k-1})$

$\qquad\qquad$ }

2.3 $\qquad$ **if** $(B[i, j] = 0)$ **then** $R_n[i, j]$ =random_pixel()

$\qquad\qquad$ **else** $R_n[i, j] = f(B[i, j], a_{n-1})$ $(= f(1, a_{n-1}) = \overline{a}_{n-1}$

$\qquad$ }

3. **output**$(R_1, R_2, \ldots, R_n)$

Theorem 3 declares the validity of Algorithm 5 and 6 in producing (n, n)-VCRG.

Theorem 3 *Given a secret binary image B, $\mathscr{E} = \{R_1, R_2, \ldots, R_n\}$ produced by Algorithms 5 or 6 with respect to B is a set of (n, n)-VCRG of B.*

The statements in the proof of Theorem 2 can be easily applied to prove Theorem 3. We omit the details here.

In Algorithm 5, if $b = 0$, $r_n = f(b, a_{n-1}) = f(0, a_{n-1}) = a_{n-1}$, otherwise $(b = 1)$ r_n =random_pixel( ). Consequently, we obtain $\mathscr{T}(S^{\mathscr{E}}[B(0)]) = 1/2^{n-1}$ by Corollary 3 and $\mathscr{T}(S^{\mathscr{E}}[B(1)]) = 1/2^n$ because all pixels in $R_k[B(1)]$ are random pixels for $1 \leqslant k \leqslant n$. Regarding Algorithm 6, if $b = 0$, r_n =random_pixel(); otherwise $(b = 1)$, $r_n = f(p, a_{n-1}) = f(1, a_{n-1}) = \overline{a}_{n-1}$. Thus, we have $\mathscr{T}(S^{\mathscr{E}}[B(0)]) = 1/2^n$ since $\mathscr{T}(S_k[B(0)]) = \mathscr{T}(R_k) = 1/2$ for $1 \leqslant k \leqslant n$, and $\mathscr{T}(S^{\mathscr{E}}[B(1)]) = 0$ since the only condition for $r_1 \otimes r_2 \otimes \cdots \otimes r_n$ to let through the light (i.e., $r_1 = r_2 = \cdots = r_{n-1} = r_n = 0$) would never occur (once $r_1 = r_2 = \cdots = r_{n-1} = 0$, we have $a_{n-1} = 0$; but $b = 1$ causes $r_n = \overline{a}_{n-1} = 1$). We obtain the following corollary.

Corollary 4 $(\mathscr{T}(S^{\mathscr{E}}[B(0)]), \mathscr{T}(S^{\mathscr{E}}[B(1)])) = (1/2^{n-1}, 1/2^n)$ *or* $(1/2^n, 0)$ *where $\mathscr{E} = \{R_1, R_2, \ldots, R_n\}$ is produced by Algorithms 5 or 6, respectively, with respect to B and $S^{\mathscr{E}} = R_1 \otimes R_2 \otimes \cdots \otimes R_n$.*

The light contrasts of Algorithms 4–6 in point of Definition 3 are summarized in Table 7.6. It is seen from Table 7.6 that among the three approaches, the light contrast obtained by Algorithm 4 is the best.

TABLE 7.6

Light contrasts by Algorithms 4–6 for (n, n)-VCRG.

$\mathscr{E}$	$\mathscr{T}(S[B(0)])$	$\mathscr{T}(S[B(1)])$	$c(\mathscr{E})$
Algorithm 4	$1/2^{n-1}$	0	$1/2^{n-1}$
Algorithm 5	$1/2^{n-1}$	$1/2^n$	$1/(2^n + 1)$
Algorithm 6	$1/2^n$	0	$1/2^n$

7.3.6   Algorithms of (n, n)-VCRG for Gray-Level and Color Images

In Shyu [12], the encryption algorithms of $(2, 2)$-VCRG for gray-level and color images were generalized from those for a binary image. We apply the same reasoning and skills adopted by Shyu in [12] to extend the binary (n, n)-VCRG algorithms to cope with gray-level and color images.

According to Ref. [12], we simply transform a gray-level image G into its binary version H by some halftone technology and encrypt the binary equivalent by using the aforementioned algorithms directly. Let $Encryption\_VCRG(B, n)$ denote the procedure of applying Algorithms 4, 5, or 6 to obtain (n, n)-VCRG with respect to binary image B. Algorithm 7 describes the idea formally.

Algorithm 7. Encrypting a gray-level image into a set of (n, n)-VCRG
Input: an $h \times w$ gray-level image G and an integer n
Output: a set of n random grids $\mathscr{E} = \{R_1, R_2, \ldots, R_n\}$ constituting a VCRG-n of G

1. $H = \mathscr{H}(G)$ // $\mathscr{H}(G)$ is a halftone function with respect to G

2. $(R_1, R_2, \ldots, R_n) = Encryption\_VCRG(H, n)$

    // Encrypt H by Algorithms 4, 5, or 6 directly

3. $\text{output}(R_1, R_2, \ldots, R_n)$

Regarding the (n, n)-VCRG for a color image, we follow the experience from Ref. [12] where the binary $(2, 2)$-VCRG encryption algorithms were extended to their color versions by utilizing skills including color decomposition, halftoning, and color composition. Specifically, we decompose a color image P into the $\mathbf{c}$, $\mathbf{m}$ and $\mathbf{y}$ components (which are the *primitive colors* in the *subtractive model*), namely $P_C, P_m,$ and P_y; halftone each of them to be P^c, P^m, and P^y such that each pixel $\boldsymbol{p}^x \in P^x$ is either $\boldsymbol{x}$ or 0 where $\boldsymbol{x} \in \{\mathbf{c}, \mathbf{m}, \mathbf{y}\}$ and encrypt the three halftone images into $(\boldsymbol{R}_1^{\mathbf{c}}, \boldsymbol{R}_2^{\mathbf{c}}, \ldots, \boldsymbol{R}_n^{\mathbf{c}}), (\boldsymbol{R}_1^{\mathbf{m}}, \boldsymbol{R}_2^{\mathbf{m}}, \ldots, \boldsymbol{R}_n^{\mathbf{m}})$, and $(\boldsymbol{R}_1^{\mathbf{y}}, \boldsymbol{R}_2^{\mathbf{y}}, \ldots, \boldsymbol{R}_n^{\mathbf{y}})$ by using any of Algorithms 4, 5, or 6 where the corresponding sets of binary colors are $\{\mathbf{c}, 0\}, \{\mathbf{m}, 0\},$ and $\{\mathbf{y}, 0\}$, respectively (instead of $\{1, 0\}$). Let $\boldsymbol{R}_i$ denote the image composed by $\boldsymbol{R}_i^{\mathbf{c}}, \boldsymbol{R}_i^{\mathbf{m}}$ and $\boldsymbol{R}_i^{\mathbf{y}}$, i.e., $\boldsymbol{R}_i = (\boldsymbol{R}_i^{\mathbf{c}}, \boldsymbol{R}_i^{\mathbf{m}}, \boldsymbol{R}_i^{\mathbf{y}})$ for $1 \leqslant i \leqslant n$. Then, $\mathscr{E} = \{\boldsymbol{R}_1, \boldsymbol{R}_2, \ldots, \boldsymbol{R}_n\}$ can be reported as a set of *color* (n, n)-VCRG of P. It means that only when all $\boldsymbol{R}_i$'s are superimposed can we see P, while any group of less than n shares obtains nothing but a color random grid.

Let $Encrypt\_cVCRG(P^x,\ \boldsymbol{x},\ n)$ denote the procedure of encrypting P^x into n shares $\boldsymbol{R}_1^x, \boldsymbol{R}_2^x, \ldots, \boldsymbol{R}_n^x$ in terms of binary color set $\{\boldsymbol{x}, 0\}$ where $\boldsymbol{x} \in \{\mathbf{c}, \mathbf{m}, \mathbf{y}\}$. It is easy to implement $Encrypt\_cVCRG(P^x$ by using Algorithms 4, 5, or 6 as long as we take 0 and $\boldsymbol{x}$ as the inverse to each other and modify random_pixel() to return 0 or $\boldsymbol{x}$ randomly. Algorithm 8 summarizes the whole idea of producing color (n, n)-VCRG for a color image.

Algorithm 8. Encrypting a color image into a set of (n, n)-VCRG
Input: an $h \times w$ color image P and an integer n
Output: a set of n color random grids $\mathscr{E} = \{\boldsymbol{R}_1, \boldsymbol{R}_2, \ldots, \boldsymbol{R}_n\}$ constituting a (n, n)-VCRG of P

1. Decompose P into $(P^{\mathbf{c}}, P^{\mathbf{m}}, P^{\mathbf{y}})$

2. **for** $(\boldsymbol{x} \in \{\mathbf{c}, \mathbf{m}, \mathbf{y}\})$ **do** $P^x = \mathscr{H}(P^x)$

   // $\mathscr{H}(P^x)$ is an x-colored halftone function so that for each $\boldsymbol{p}^x \in P^x$, $\boldsymbol{p}^x = \boldsymbol{x}$ or 0

3. **for** $(\boldsymbol{x} \in \{\mathbf{c}, \mathbf{m}, \mathbf{y}\})$ **do** $(\boldsymbol{R}_1^x, \boldsymbol{R}_2^x, \ldots, \boldsymbol{R}_n^x) = Encrypt\_cVCRG(P^x,\ \boldsymbol{x},\ n)$

4. **for** $(1 \leqslant i \leqslant n)$ **do** $\boldsymbol{R}_i = (\boldsymbol{R}_i^{\mathbf{c}}, \boldsymbol{R}_i^{\mathbf{m}}, \boldsymbol{R}_i^{\mathbf{y}})$

   // color composition of $\boldsymbol{R}_i^{\mathbf{c}}$, $\boldsymbol{R}_i^{\mathbf{m}}$, and $\boldsymbol{R}_i^{\mathbf{y}}$

5. **output**$(\boldsymbol{R}_1, \boldsymbol{R}_2, \ldots, \boldsymbol{R}_n)$

Based upon the statements in the proofs of Theorems 2 and 3 as well as those for color $(2, 2)$-VCRG in Ref. [12], we have the following consequence.

Theorem 4 *Given a color image P, $\mathscr{E} = \{\boldsymbol{R}_1, \boldsymbol{R}_2, \ldots, \boldsymbol{R}_n\}$ produced by Algorithm 8 with respect to P is a set of color (n, n)-VCRG of P.*

7.3.7 Experiments for (n, n)-VCRG

We designed four experiments by computer simulations to verify the feasibility and applicability of the VCRG algorithms. Experiment 1 was designed to test Algorithms 4, 5, and 6 to obtain sets of VCRG-3 for a binary image. Experiments 2 and 3 focused on producing sets of VCRG-3 for gray-level and color images, respectively. The sets of VCRG-4 were tested in Experiment 4. All computer programs in these experiments were coded in Borland C++ Builder and run in a PC with Windows.

Experiment 1: Encrypting a binary image to obtain VCRG-3.
In this experiment, we adopted Algorithms 4, 5, and 6 to produce three sets

of VCRG-3, respectively, for binary image B as in Figure 7.3. Figure 7.4 illustrates the implementation results of Algorithm 4 where (a), (b), and (c) are the three random grids produced, namely R_1^4, R_2^4, and R_3^4, respectively; (d), (e) and (f) are the superimposed results of $R_1^4 \otimes R_2^4$, $R_1^4 \otimes R_3^4$, and $R_2^4 \otimes R_3^4$, respectively; and (g) is the result of $R_1^4 \otimes R_2^4 \otimes R_3^4$.

FIGURE 7.3
Binary image B in Experiment 1.

We see from Figure 7.4 that (a)–(f), including R_1^4, R_2^4, R_3^4 and the superimposed results of all groups of two out of the three shares, are merely random grids from which no secret can be obtained, whereas (g) $R_1^4 \otimes R_2^4 \otimes R_3^4$, the superimpositiion of the three random grids, reveals B to our visual system. As a result, $\{R_1^4, R_2^4, R_3^4\}$ produced by Algorithm 4 is indeed a set of VCRG-3 of B. It is lucid that Figure 7.4 provides some visualized evidence to Theorem 2.

Figures 7.5 and 7.6 show the corresponding results of Algorithms 5 and 6 for VCRG-3 with respect to B where $\{R_1^5, R_2^5, R_3^5\}$ and $\{R_1^6, R_2^6, R_3^6\}$ are the two sets of random grids produced by Algorithms 5 and 6, respectively. It is seen from Figure 7.5 and 7.6, $\{R_1^5, R_2^5, R_3^5\}$ and $\{R_1^6, R_2^6, R_3^6\}$ are surely two sets of VCRG-3 of B. The correctness of Theorem 3 is visually shown here.

In summary, based upon Figures 7.4–6, R_i^a and $R_i^a \otimes R_i^a$ are random grids with $\mathscr{T}(R_i^a) = 1/2$ (see Figures 7.4(a)–(c), 7.5(a)–(c), and 7.6(a)–(c)) and $\mathscr{T}(R_i^a \otimes (R_j^a) = 1/4$ (see Figures 7.4(d)–(f), 7.5(d)–(f), and 7.6(d)–(f)) for $1 \leqslant i \neq j \leqslant 3$ and $a \in \{4, 5, 6\}$. By comparing the reconstructed images, i.e., $R_1^4 \otimes R_2^4 \otimes R_3^4$, $R_1^5 \otimes R_2^5 \otimes R_3^5$ and $R_1^6 \otimes R_2^6 \otimes R_3^6$ (see Figures 7.3(g), 7.4(g), and 7.5(g), respectively), we realize that $R_1^4 \otimes R_2^4 \otimes R_3^4$ (by Algorithm 4) attains the highest light contrast, while $R_1^5 \otimes R_2^5 \otimes R_3^5$ (by Algorithm 5) the lowest. In fact, these results are foretold by Table 7.6: $c(\mathscr{E}_4) = 1/4$, $c(\mathscr{E}_5) = 1/9$, $c(\mathscr{E}_6) = 1/8$ for $n = 3$ where $\mathscr{E}_a = \{R_1^a, R_2^a, R_3^a\}$ for $a \in \{4, 5, 6\}$.

Experiment 2: Encrypting a gray-level image to obtain VCRG-3.
Figure 7.7 illustrates the experimental results of applying Algorithm 7 to produce a set of VCRG-3 for a gray-level image. Figure 7.7(a) is the gray-level image G to be encrypted; (b) shows the halftone version H of G by using the *error diffusion* technology; (c), (d), and (e) present the outcomes of Algorithm 7 where *Encryption_VCRG*$(H, 3)$ was implemented by Algorithm 4, namely R_1^4, R_2^4 and R_3^4, respectively; (f), (g), and (h) give the superimposed results of $R_1^4 \otimes R_2^4$, $R_1^4 \otimes R_3^4$ and $R_2^4 \otimes R_3^4$, respectively; and (i) depicts that of $R_1^4 \otimes R_2^4 \otimes R_3^4$. As seen from Figure 7.7, R_1^4, R_2^4, and R_3^4 and the superimposed

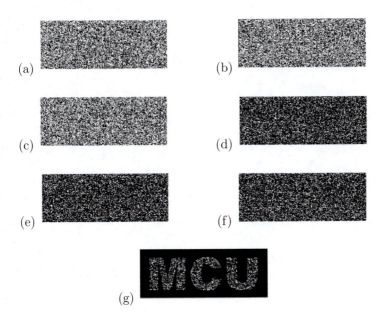

FIGURE 7.4
Implementation results of Algorithm 4 for VCRG-3 with respect to B: (a) R_1^4,
(b) R_2^4, (c) R_3^4; (d) $R_1^4 \otimes R_2^4$, (e) $R_1^4 \otimes R_3^4$, (f) $R_2^4 \otimes R_3^4$; (g) $R_1^4 \otimes R_2^4 \otimes R_3^4$.

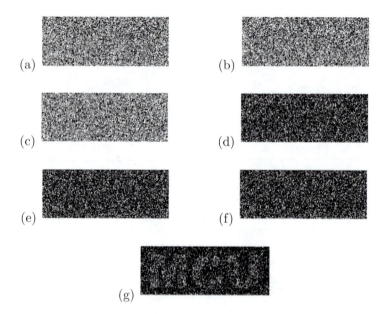

FIGURE 7.5
Implementation results of Algorithm 5 for VCRG-3 with respect to B: (a) R_1^5, (b) R_2^5, (c) R_3^5; (d) $R_1^5 \otimes R_2^5$, (e) $R_1^5 \otimes R_3^5$, (f) $R_2^5 \otimes R_3^5$; (g) $R_1^5 \otimes R_2^5 \otimes R_3^5$.

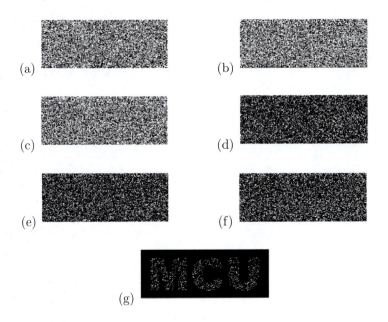

FIGURE 7.6

Implementation results of Algorithm 6 for VCRG-3 with respect to B: (a) R_1^6, (b) R_2^6, (c) R_3^6; (d) $R_1^6 \otimes R_2^6$, (e) $R_1^6 \otimes R_3^6$, (f) $R_2^6 \otimes R_3^6$; (g) $R_1^6 \otimes R_2^6 \otimes R_3^6$.

results from any two out of them are nothing but random pictures, where $R_1^4 \otimes R_2^4 \otimes R_3^4$ reveals H, and consequently G. We realize that $\{R_1^4, R_2^4, R_3^4\}$ is a set of VCRG-3 with respect to G.

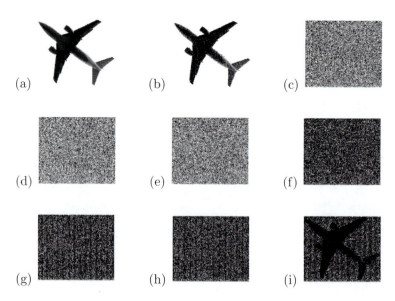

FIGURE 7.7
Results of Algorithm 7 where $Encryption\_VCRG(H, 3)$ was implemented by Algorithm 4 with respect to gray-level image G in Experiment 2: (a) G; (b) halftone version H of G; (c) R_1^4, (d) R_2^4, (e) R_3^4; (f) $R_1^4 \otimes R_2^4$, (g) $R_1^4 \otimes R_3^4$, (h) $R_2^4 \otimes R_3^4$; (i) $R_1^4 \otimes R_2^4 \otimes R_3^4$.

Let $\{R_1^5, R_2^5, R_3^5\}$ and $\{R_1^6, R_2^6, R_3^6\}$ denote the outcomes of Algorithm 7 when $Encryption\_VCRG(H, 3)$ was implemented by Algorithms 5 and 6, respectively, with respect to G. Figures 7.8(a) and (b) are the superimposed results of $R_1^5 \otimes R_2^5 \otimes R_3^5$ and $R_1^6 \otimes R_2^6 \otimes R_3^6$, respectively. It is noted that the three encrypted shares and the superimposed result of any group of two out of the three shares are indeed random grids that are omitted here.

The feasibility and applicability for Algorithm 7 to encrypt a gray-level image into VCRG-3 are demonstrated in a visual sense from Figures 7.7 and 7.8. Obviously, implementing $Encryption\_VCRG(H, 3)$ based upon Algorithm 4 makes Algorithm 7 achieve the highest contrast (while that based upon Algorithm 5 is the worst).

Experiment 3: Encrypting a color image to obtain color VCRG-3.
We tested Algorithm 8 with respect to a color image for obtaining color VCRG-3 in this experiment. Figure 7.9(a) is the color image P to be encrypted; (b), (c), and (d) are P^c, P^m, and P^y, which are the **c**, **m**, and **y**

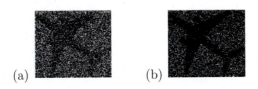

(a)   (b)

FIGURE 7.8
Reconstructed results of VCRG-3 with respect to G: (a) $R_1^5 \otimes R_2^5 \otimes R_3^5$; (b) $R_1^6 \otimes R_2^6 \otimes R_3^6$.

components of P, respectively; (e), (f), and (g) are $P^\mathbf{c}$, P^m, and P^y which are the **c**-, **m**-, and **y**-colored halftone images (by error diffusion) of $P^\mathbf{c}$, $P^\mathbf{m}$ and $P^\mathbf{y}$, respectively.

By Step 3 of Algorithm 8, P^x would be encrypted into $\{\boldsymbol{R}_1^x, \boldsymbol{R}_2^x, \boldsymbol{R}_3^x\}$ by $Eycrypt\_cVCRG(P^x, \boldsymbol{x}, 3)$ for $\boldsymbol{x} \in \{\mathbf{c}, \mathbf{m}, \mathbf{y}\}$. When $Eycrypt\_cVCRG(P^\mathbf{m}, \mathbf{m}, 3)$ was based upon Algorithm 4, Figure 7.10 evidences the effectiveness of $\{\boldsymbol{R}_1^m, \boldsymbol{R}_2^m, \boldsymbol{R}_3^m\}$ for being a set of **m**-colored VCRG-3 with respect to $P^{\mathbf{m}}$ in which (a)–(f) are the results of $\boldsymbol{R}_1^\mathbf{m}$, $\boldsymbol{R}_2^\mathbf{m}$, $\boldsymbol{R}_3^\mathbf{m}$, $\boldsymbol{R}_1^\mathbf{m} \otimes \boldsymbol{R}_2^\mathbf{m}$, $\boldsymbol{R}_1^\mathbf{m} \otimes \boldsymbol{R}_3^\mathbf{m}$, and $\boldsymbol{R}_2^\mathbf{m} \otimes \boldsymbol{R}_3^\mathbf{m}$, respectively; and (g) is that of $\boldsymbol{R}_1^\mathbf{m} \otimes \boldsymbol{R}_2^\mathbf{m} \otimes \boldsymbol{R}_3^\mathbf{m}$. Regarding $P^\mathbf{c}$ and $P^\mathbf{y}$, the corresponding results are similar to Figure 7.10 so that we simply omit them.

The encrypted monochromatic-colored shares $P_i^\mathbf{c}$, $P_i^\mathbf{m}$ and $P_i^\mathbf{y}$ were further composed to obtain color random grid $\boldsymbol{R}_i = \{P_i^\mathbf{c}, P_i^\mathbf{m}, P_i^\mathbf{y}\}$ for $1 \leqslant i \leqslant 3$ in Step 4 of Algorithm 8. We show the corresponding results in Figure 7.11 where (a)–(c) are color random grids $\boldsymbol{R}_1$, $\boldsymbol{R}_2$, $\boldsymbol{R}_3$; (d)–(f) give results of $\boldsymbol{R}_1 \otimes \boldsymbol{R}_2$, $\boldsymbol{R}_1 \otimes \boldsymbol{R}_3$, $\boldsymbol{R}_2 \otimes \boldsymbol{R}_3$, respectively; and (g) shows that of $\boldsymbol{R}_1 \otimes \boldsymbol{R}_2 \otimes \boldsymbol{R}_3$. When we implemented $Eycrypt\_cVCRG(P^x, \boldsymbol{x}, 3)$ is based upon Algorithm 4, 5, or 6. The correctness of Theorem 4 holds in a visual sense from the results of Figures 7.8–10.

Experiment 4: Obtaining VCRG-4.
From the above analytic and experimental results, we know that Algorithm 4 achieves the higher light contrast as compared to Algorithms 5 and 6 for binary and gray-level images, so does Algorithm 8 based upon Algorithm 4 as compared to those based upon Algorithms 5 and 6 for a color image. We only tested Algorithm 4 for binary VCRG-4 and Algorithm 8 based upon Algorithm 4 for color VCRG-4 here. Figure 7.12 summarizes the results of binary VCRG-4 where (a) is the binary image B encrypted; (b)–(e) are R_1, R_2, R_3, R_4 produced by Algorithm 4; (f) presents the result of $R_1 \otimes R_2$; (g) shows that of $R_1 \otimes R_2 \otimes R_3$; and (h) gives that of $R_1 \otimes R_2 \otimes R_3 \otimes R_4$. Even though the superimposed results from other groups of less than four shares are not shown here, they are really random grids as expected. It is seen from Figure 7.12 that only when all four shares are superimposed can we see B, while no group

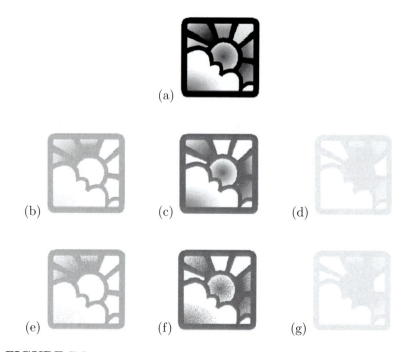

FIGURE 7.9
(See color insert.) Results of Steps 1 and 2 of Algorithm 8 for VCRG-3 with respect to color image P in Experiment 3: (a) P; (b) P^c, (c) P^m, (d) P^y; (e) P^c, (f) P^m, (g) P^y.

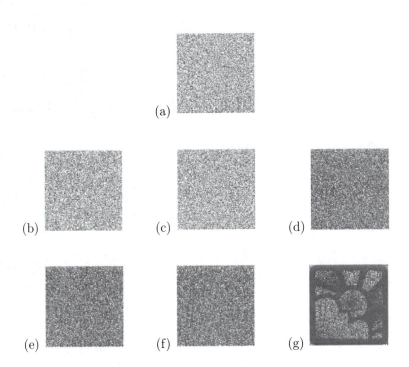

FIGURE 7.10
(See color insert.) Results of Step 3 of Algorithm 8 with respect to $P^{\mathbf{m}}$ where $Eycrypt\_cVCRG(P^{\mathbf{m}}, \mathbf{m}, 3)$ was based upon Algorithm 4: (a) $\boldsymbol{R}_1^{\mathbf{m}}$, (b) $\boldsymbol{R}_2^{\mathbf{m}}$, (c) $\boldsymbol{R}_3^{\mathbf{m}}$; (d) $\boldsymbol{R}_1^{\mathbf{m}} \otimes \boldsymbol{R}_2^{\mathbf{m}}$, (e) $\boldsymbol{R}_1^{\mathbf{m}} \otimes \boldsymbol{R}_3^{\mathbf{m}}$, (f) $\boldsymbol{R}_2^{\mathbf{m}} \otimes \boldsymbol{R}_3^{\mathbf{m}}$; (g) $\boldsymbol{R}_1^{\mathbf{m}} \otimes \boldsymbol{R}_2^{\mathbf{m}} \otimes \boldsymbol{R}_3^{\mathbf{m}}$.

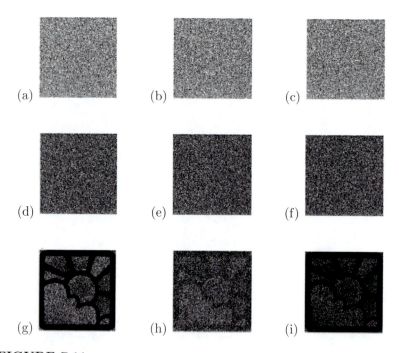

FIGURE 7.11
(See color insert.) Results of Algorithm 8 for VCRG-3 with respect to P: (a)
$\boldsymbol{R}_1$, (b) $\boldsymbol{R}_2$, (c) $\boldsymbol{R}_3$; (d) $\boldsymbol{R}_1 \otimes \boldsymbol{R}_2$, (e) $\boldsymbol{R}_1 \otimes \boldsymbol{R}_3$, (f) $\boldsymbol{R}_2 \otimes \boldsymbol{R}_3$; (g) $\boldsymbol{R}_1 \otimes \boldsymbol{R}_2 \otimes \boldsymbol{R}_3$
(based upon Algorithm 4); (h) $\boldsymbol{R}_1^{'} \otimes \boldsymbol{R}_2^{'} \otimes \boldsymbol{R}_3^{'}$ (based upon Algorithm 5); (i)
$\boldsymbol{R}_1^{''} \otimes \boldsymbol{R}_2^{''} \otimes \boldsymbol{R}_3^{''}$ (based upon Algorithm 6).

of less than four shares tells anything about B. $\{R_1, R_2, R_3, R_4\}$ is indeed a set of VCRG-4.

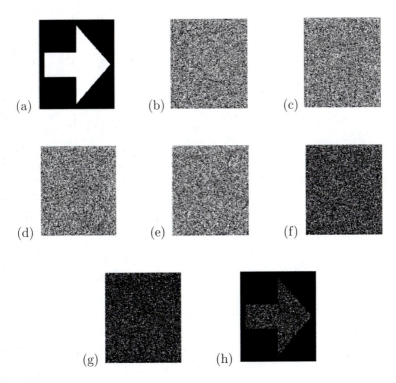

FIGURE 7.12
Results of Algorithms 4 for VCRG-4 with respect to binary image B: (a) B; (b) R_1, (c) R_2, (d) R_3, (e) R_4; (f) $R_1 \otimes R_2$; (g) $R_1 \otimes R_2 \otimes R_3$; (h) $R_1 \otimes R_2$; (g) $R_1 \otimes R_2 \otimes R_3 \otimes R_4$.

Since $n = 4$ in this experiment, we have $\mathscr{T}(R_i) = 1/2$ (see Figures 7.12(b)–(e)), $\mathscr{T}(R_i \otimes R_j) = 1/4$ (see (f)), and $\mathscr{T}(R_i \otimes R_j \otimes R_k) = 1/8$ (see (g)) for $1 \leqslant i \neq j \neq k \leqslant 4$. Further, from Table 7.6 we know that the light contrast of Algorithm 4 is $c(\mathscr{E}_4) = 1/8$ for $n = 4$. Figure 7.13 gives some experimental results of Algorithm 8 for color VCRG-4 where $Eycrypt\_cVCRG(P^x, \boldsymbol{x}, 4)$ was based upon Algorithm 4 with respect to P (see Figure 7.9(a)). Figures 7.13(a)–(d) are the color random grids $\boldsymbol{R}_1, \boldsymbol{R}_2, \boldsymbol{R}_3, \boldsymbol{R}_4$ produced; (e) presents the result of $\boldsymbol{R}_1 \otimes \boldsymbol{R}_2$; (f) shows that of $\boldsymbol{R}_1 \otimes \boldsymbol{R}_2 \otimes \boldsymbol{R}_3$; and (g) gives that of $\boldsymbol{R}_1 \otimes \boldsymbol{R}_2 \otimes \boldsymbol{R}_3 \otimes \boldsymbol{R}_4$. As seen from Figure 7.13, we realize that $\{\boldsymbol{R}_1 \otimes \boldsymbol{R}_2 \otimes \boldsymbol{R}_3 \otimes \boldsymbol{R}_4\}$ is a set of color VCRG-4 of P.

The results of the above four experiments are visualized evidences to the correctness of Theorems 2–4. The proposed schemes (Algorithms 4-8) in producing VCRG-n for a secret image are feasible and applicable. Among the

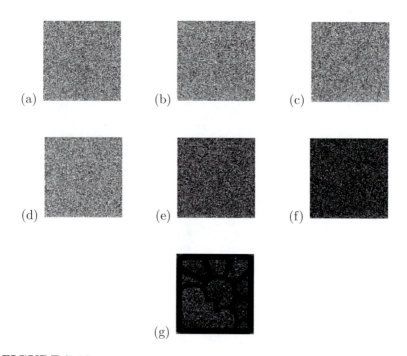

FIGURE 7.13

(See color insert.) Results of Algorithms 8 where $Encrypt\_cVCRG(P^x, \boldsymbol{x}, 4)$ was based upon Algorithm 4 for VCRG-4 with respect to P (Figure 7.9(a)): (a) $\boldsymbol{R}_1$, (b) $\boldsymbol{R}_2$, (c) $\boldsymbol{R}_3$, (d) $\boldsymbol{R}_4$; (e) $\boldsymbol{R}_1 \otimes \boldsymbol{R}_2$; (f) $\boldsymbol{R}_1 \otimes \boldsymbol{R}_2 \otimes \boldsymbol{R}_3$; (g) $\boldsymbol{R}_1 \otimes \boldsymbol{R}_2 \otimes \boldsymbol{R}_3 \otimes \boldsymbol{R}_4$.

approaches, Algorithm 4 (Algorithm 8 based upon Algorithm 4) is the most effective scheme in terms of light contrast for binary or gray-level (color) images.

It is noticed that when we adopt the light contrast to measure the reconstructed result for a conventional n out of n visual secret sharing scheme [9], the best result is $1/2^{n-1}$. That means that Algorithm 4 achieves the same best light contrast. Moreover, it is so appealing that these schemes neither induce any extra pixel expansion nor require encoding basis matrix. Those approaches in conventional visual cryptography suffer from the disadvantage of the inevitable pixel expansion, which increases exponentially as n increases (and additionally increases as c or $\lceil \log_2 c \rceil$ for the color cases where c is the number of colors in the secret image), in the basis matrices. Table 7.7 summarizes the pixel expansions needed by some efficient n out of n visual cryptographic schemes in the literature and our (n, n)-VCRG schemes. Owing to the reason that the size of the encrypted shares and reconstructed image would not be expanded, our schemes based upon random grids are more attractive than those in conventional visual cryptography for both theoretical concerns and practical applications.

TABLE 7.7

Comparison of pixel expansions for conventional (n, n) visual cryptographic and (n, n)-VCRG schemes.

binary		color		
Naor and Shamir [9]	Shyu [13]	Blundo et al. [3]	Shyu [11]	Shyu [13]
2^{n-1}	1	$\begin{cases} (c-1)2^{n-1} - c + 2 & \text{if } n \text{ is odd;} \\ c(c-1)2^{n-2} - c & \text{otherwise} \end{cases}$	$\lceil \log_2 c \rceil \times 2^{n-1}$	1

7.4 Concluding Remarks

In this chapter, we propose novel schemes for visual secret sharing using random grids. At first, we give a new definition for the visual cryptograms of $n(\geqslant 2)$ random grids with respect to a binary image. Based upon the definition, we design effective algorithms, prove their correctness formally, analyze the light contrast in the reconstructed image, and demonstrate their feasibility by computer simulations. By exploiting the skills of halftone and color decomposition in Ref. [12], the enhanced algorithms are also developed and verified to deal with gray-level and color images.

The proposed schemes incorporate human visual intelligence with information security in such a way that no computation but only human vision is needed in the decryption process. They provide cost effective, handy, and portable solutions to image encryption or sharing even for inexperienced users, especially for circumstances where no computer can be accessed. Since a secret image can be encrypted/shared among $n(\geq 2)$ participants (instead of only two), our approaches extend and generalize the studies in Refs. [12, 7] such that the applicability of image encryption or sharing can be broadened to a greater extent.

The n shares of random grids generated by our VCRG-n algorithms work well just like those shares by the conventional n out of n visual secret sharing schemes based upon the definition of Naor and Shamir [9]. However, the pixel expansion in our schemes is 1 for both of the binary and color images, while that in Ref. [9] is 2^{n-1} for the binary case and that in Ref. [11] is $\lceil \log_2 c \rceil \times 2^{n-1}$ for the color image containing c colors. Therefore, the size of the encoded shares by our schemes would be much smaller (the same as the secret image). Further, the sophisticated encoding basis matrices in Refs. [9, 5, 4, 8, 14, 2, 1, 3, 10, 6, 11] are no more needed in our schemes. Regarding the visual perception in the reconstructed images, our Algorithm 4 achieves the same light contrast as good as the best n out of n visual secret sharing scheme devised in Ref. [9]. These clarify the superiority of our schemes.

Our encryption algorithms can be easily hardwired by incorporating a 0/1 random number generator with T flip-flops or Exclusive-OR gates. It would be an interesting challenge to design a special VCRG hardware for image encryption. In fact, many research topics in conventional visual cryptography could be reexamined in view of random grids.

Bibliography

[1] C.-N. Yang and C.-S. Laih. New colored visual secret sharing schemes. *Des. Codes Cryptogr.*, 20:325–335, 2000.

[2] C.Blundo, A. De Santis, and D.R.Stinson. On the contrast in visual cryptography schemes. *J. Cryptogr*, 12:261–289, 1999.

[3] C. Blundo, A. De Santis, and D.R. Stinson. Improved schemes for visual cryptography. *Des. Codes Cryptogr.*, 24:255–278, 2001.

[4] S. Droste. New results on visual cryptography. *in: N. Koblitz (Ed.), Advances in Cryptology: CRYPTO96, Lecture Notes in Computer Science*, 1109:401–415, 1996.

[5] G. Ateniese, C. Blundo, A. De Santis, and D. R. Stinson. Constructions and bounds for visual cryptography. *in: F.M. auf der Heide, B.*

Monien (Eds.), Automata, Languages and Programming: ICALP'96, Lecture Notes in Computer Science, 1099, Springer, Berlin:416–428, 1996.

[6] Y.-C. Hou. Visual cryptography for color images. *Pattern Recognition*, 36:1619–1629, 2003.

[7] O. Kafri and E. Keren. Encryption of pictures and shapes by random grids. *Opt. Lett*, 12:377–379, 1987.

[8] M. Naor and B. Pinkas. Visual authentication and identification. *in: B.S. Kaliski Jr. (Ed.), Advances in Cryptology: CRYPTO97, Lecture Notes in Computer Science*, 1294:322–336, 1997.

[9] M. Naor and A. Shamir. Visual cryptography. *in: A. De Santis (Ed.), Advances in Cryptology: Eurpocrypt'94, Lecture Notes in Computer Science*, 950:1–12, 1995.

[10] P.A. Eisen and D.R. Stinson. Threshold visual cryptography schemes with specified whiteness levels of reconstructed pixels. *Des. Codes Cryptogr.*, 25:15–61, 2002.

[11] S.J. Shyu. Efficient visual secret sharing scheme for color images. *Pattern Recognition*, 39:866–880, 2006.

[12] S.J. Shyu. Image encryption by random grids. *Pattern Recognition*, 40:1014–1031, 2007.

[13] S.J. Shyu. Image encryption by multiple random grids. *Pattern Recognition*, 42:1582–1596, 2009.

[14] E.R. Verheul and H.C.A. Van Tilborg. Constructions and properties of k out of n visual secret sharing schemes. *Designs Codes Cryptography*, 11:179–196, 1997.

8

Visual Cryptography and Contrast Bounds

Andreas Klein

Ghent University, Belgium

CONTENTS

8.1 Introduction

To describe a visual cryptography scheme we encode the encryption of a pixel by Boolean matrices. The rows represent the slides and the columns represent the different subpixels.

For example, the encoding illustrated in Figure 8.1 is described by

$$M = \begin{pmatrix} 0 & 1 & 0 & 1 \\ 0 & 0 & 1 & 1 \\ 0 & 1 & 1 & 0 \end{pmatrix} .$$

To describe the random decision, which must be made when encoding a pixel, we use multisets that contain all possible choices with the right frequency.

This leads to the formal definition of a visual cryptography scheme.

Definition 1 *Let Γ be any access structure for n persons. A visual cryptography scheme is a pair of multisets $\mathfrak{M}_b$ and $\mathfrak{M}_w$ of Boolean $n \times m$ matrices. Which satisfy:*

FIGURE 8.1
Example of encoding a visual cryptography scheme.

1. $|\mathfrak{M}_b| = |\mathfrak{M}_w| = r$

2. *For each $G \notin \Gamma$ the restriction of all matrices in $\mathfrak{M}_b$ to the rows indexed by $i \in G$ gives the same multiset of $|G| \times n$ matrices as the restriction of all matrices in $\mathfrak{M}_w$ to the rows indexed by $i \in G$. (This formalizes the requirement that an unqualified set of persons gain no information about the secret.)*

3. *$G \in \Gamma$ is a minimal qualified set if every proper subset $G' \subset G$ is not an element of Γ.*

For each minimal qualified set of persons G the restriction of a matrix in $\mathfrak{M}_b$ to the rows indexed by $i \in G$ give a $|G| \times m$ matrix with at least $h(G)$ nonzero columns and the restriction of a matrix in $\mathfrak{M}_w$ to the rows indexed by $i \in G$ give a $|G| \times m$ matrix with at least $l(G)$ nonzero columns. The inequality $h(G) > l(G)$ must be satisfied. (This formalizes the idea that we see a black pixel as a darker gray than a white pixel.)

Note: The condition (a) can be relaxed a bit, we choose $|\mathfrak{M}_b| = |\mathfrak{M}_w|$ for simplicity. Some authors require that condition (c) holds for all qualified sets $G \in \Gamma$ not only for the minimal qualified sets.

The performance of visual cryptography scheme is described by 3 parameters.

• The randomness r, that measures the amount of entropy needed to encode an image.

• The number of subpixels m need to encode an pixel (pixel expansion).

• The contrast $\alpha = \min\{(h(G) - l(G))/m | G \in \Gamma\}$ that measures the quality of the reconstructed image.

In applications the contrast is the most important factor. (Anything below $1/10$ can be hardly used if you really stack transparencies.) Thus, finding contrast optimal schemes is an important task.

The goal of this chapter is to prove upper bounds on the possible contrast and to construct schemes near that bounds.

8.2 Preliminaries

For the moment we will ignore the pixel expansion and the randomness. Then contrast optimal visual cryptography schemes can be simplified.

Lemma 1 *Let Γ be any access structure then there exists a contrast optimal visual cryptography scheme $(\mathfrak{M}_W, \mathfrak{M}_B)$ for Γ with the property that $\mathfrak{M}_W$ is the multiset of all column permutations of a matrix M_W and $\mathfrak{M}_B$ is the multiset of all column permutations of a matrix M_B.*

Proof
Let $(\mathfrak{M}'_W, \mathfrak{M}'_B)$ be a contrast optimal visual cryptography scheme for Γ with pixel expansion m and randomness r.

Let $M_W = (M'_W{}^{(1)}, \ldots, M'_W{}^{(r)})$ be the $n \times (mr)$ matrix which consist of all matrices $M'_W{}^{(i)} \in \mathfrak{M}'_W$ $(i = 1, \ldots, r)$ written in sequence, one after the other. Similarly let M_B be the $n \times (mr)$ matrix that consists of the matrices in $\mathfrak{M}'_B$ written after each other.

Let $G \notin \Gamma$, by Definition 1 (b) the restriction of M_W and M_B to the rows $i \in \Gamma$ must give return two matrices that different only for a permutation of their columns, i.e., the scheme $(\mathfrak{M}_W, \mathfrak{M}_B)$ satisfies the security requirement (Definition 1 (b)).

Now let $G \in \Gamma$ by Definition 1 (c) the restriction of every matrix in $\mathfrak{M}_B$ to the rows $i \in \Gamma$ contains at least αm more nonzero columns than the restriction of a matrix in $\mathfrak{M}_W$ to the same rows. Thus, the restriction of M_B to the rows $i \in \Gamma$ contains at least $\alpha m r$ more nonzero columns than the restriction of M_W to the rows $i \in \Gamma$. That proves that the scheme $(\mathfrak{M}_W, \mathfrak{M}_B)$ reconstructs the secret image. $\square$

The advantage of the scheme constructed by Lemma 1 is that the visual cryptography scheme is now described by just two Boolean matrices M'_W and M'_B instead of two multisets of Boolean matrices. Moreover the order of the columns in M'_W and M'_B does not matter. For each $S \subseteq \{1, \ldots, n\}$ let $x_S^{(W)}$ denote the number of columns in M'_W with 1 in the rows corresponding to S and 0 in the other rows. Similarly define the variables $x_S^{(B)}$. The scheme is described completely by the values $x_S^{(B)}, x_S^{(W)}$.

The security requirements of the visual cryptography scheme translates to the linear equations

$$\sum_{\substack{S \subseteq \{1, \ldots n\} \\ S \cap G \neq 0}} x_s^{(W)} = \sum_{\substack{S \subseteq \{1, \ldots n\} \\ S \cap G \neq 0}} x_s^{(B)} \tag{8.1}$$

for all $G \notin \Gamma$ and the requirement that a qualified subset can see the image translates to

$$\sum_{\substack{S \subseteq \{1, \ldots n\} \\ S \cap G \neq 0}} x_s^{(W)} < \sum_{\substack{S \subseteq \{1, \ldots n\} \\ S \cap G \neq 0}} x_s^{(B)} \tag{8.2}$$

for all minimal qualified subsets G.

The equations (8.1) and (8.2) lead to the linear program. Replace (8.1) by

$$\sum_{\substack{S \subseteq \{1,...n\} \\ S \cap T \neq 0}} x_s^{(W)} + \alpha \leq \sum_{\substack{S \subseteq \{1,...n\} \\ S \cap T \neq 0}} x_s^{(B)}$$

and require

$$\sum_{S \subseteq \{1,...n\}} x_s^{(B)} = \sum_{S \subseteq \{1,...n\}} x_s^{(W)} = 1$$

and maximize α. For small values of n it is no problem to feed that program into a computer and get the optimal visual cryptography scheme. (I prepared such a program for my book [11]. You can download it from my homepage http://cage.ugent.be/~klein/vis-crypt/buch/).

For the k-out-of-n access structure it is possible to simplify the linear program and solve it directly.

Lemma 2 *Let Γ be the k-out-of-n access structure. Then in addition to Lemma 1 we can require that any row permutation of M_W is also a column permutation of M_W and, similarly, every row permutation of M_B is also a column permutation of M_B.*

Proof

Let M_W and M_B be the two Boolean matrices that describe the visual cryptography scheme constructed in Lemma 1.

For every row permutation σ the matrices M_W^σ and M_B^σ also describe a k-out-of-n visual cryptography scheme.

Let $\hat{M}_W$ be the $n \times (mn!)$ matrix that consists of all row permutations M_W^σ of M_W written after each other. Similarly define $\hat{M}_B$. As we have seen already in the proof of Lemma 1 this also gives a solution of the k-out-of-n visual cryptography scheme. In addition $\hat{M}_W$ and $\hat{M}_B$ satisfy the permutation invariance stated in the Lemma.   □

Lemma 2 allows us to simplify the linear program for a contrast optimal k-out-of-n visual cryptography scheme. With the notations from above we get $x_S^{(B)} = x_{S'}^{(B)}$ and $x_S^{(W)} = x_{S'}^{(W)}$ for $|S| = |S'|$. Let for $i \in \{0, \dots, n\}$ be $x_i^{(B)} = x_{\{1,...,i\}}^{(B)}$ and $x_i^{(W)} = x_{\{1,...,i\}}^{(W)}$.

The linear program simplifies to:

$$\sum_{i=0}^{n} \binom{n}{i} x_i^{(W)} = \sum_{i=0}^{n} \binom{n}{i} x_i^{(B)} = 1 \qquad (8.3)$$

This equation express that the variables denotes the fractions of black subpixels and that all fractions add up to 1.

$$\sum_{i=0}^{n-j} \binom{n-j}{i} x_i^{(W)} = \sum_{i=0}^{n-j} \binom{n-j}{i} x_i^{(B)} \qquad (8.4)$$

for $j = 0, \ldots, k-1$. This equation expresses that the stack of j transparencies always has the same number of white subpixels and replaces (8.1).

Maximize

$$\alpha = \sum_{i=0}^{n-k} \binom{n-k}{i} x_i^{(W)} - \sum_{i=0}^{n-k} \binom{n-k}{i} x_i^{(B)} \qquad (8.5)$$

In this form the linear program is simple enough to solve it directly. We will track this problem in Sections 8.5 and 8.6.

The linear program has been independently deduced by several researchers. (Variable transformations like $\hat{x}_i^{(W)} = x_{n-i}^{(W)}$ or $\hat{x}_i^{(W)} = \binom{n}{i} x_i^{(W)}$ are common. This makes the results look quite different in various articles. Always check the meaning of the variables.)

8.3 Approximate Inclusion Exclusion

The well-known inclusion-exclusion-formula states.

$$|A_1 \cup A_2 \cup \ldots \cup A_n| = \sum_i |A_i| - \sum_{i<j} |A_i \cap A_j| +$$
$$\sum_{i<j<k} |A_i \cap A_j \cap A_k| - \ldots - (-1)^n |A_1 \cap \ldots \cap A_n| \, .$$

Obviously, every term on the right-hand side is needed to determine the size of the union. At this point we can ask whether it is possible to give an approximate inclusion-exclusion formula. More formally we ask:

Given integers m, n with $m \leq n$ and sets $A_1, \ldots, A_n$ and $B_1, \ldots, B_n$ where not all B_i are empty and where

$$\left| \bigcap_{i \in S} A_i \right| = \left| \bigcap_{i \in S} B_i \right|$$

for every subset $S \subseteq \{1, \ldots, n\}$ such that $|S| < m$, what is the smallest (or largest) possible value for the fraction

$$\frac{|A_1 \cup \ldots \cup A_n|}{|B_1 \cup \ldots \cup B_n|}?$$

The problem of approximate inclusion-exclusion is closely related to contrast bounds in visual cryptography schemes. For n-out-of-n and $(n-1)$-out-of-n scheme the solution of the approximate inclusion-exclusion problem translates directly to a contrast optimal visual cryptography scheme (see Theorem 4 for the details of the translation). Other applications of the approximate inclusion-exclusion are constant depth circuits and Boolean functions [15].

It is convenient to replace the size of the sets in the approximated inclusion exclusion problem by arbitrary measures to get a continuous problem. This leads to the following problem. Let $(\Omega, \mathcal{A}, \mu)$ be a measurable space and let $A_1, \ldots, A_n$ and $B_1, \ldots, B_n$ be measurable sets with

$$\mu\left(\bigcap_{i \in S} A_i\right) = \mu\left(\bigcap_{i \in S} B_i\right)$$

for every subset $S \subseteq \{1, \ldots, n\}$ such that $|S| < m$, what is the smallest (or largest) possible value for the fraction

$$\frac{\mu(A_1 \cup \ldots \cup A_n)}{\mu(B_1 \cup \ldots \cup B_n)}?$$

Similarly to Lemma 1 we can restrict the approximate inclusion-exclusion problem to symmetric collections. With

$$x_i = \binom{n}{j}\left[\mu\left(\bigcap_{j \in S} A_j \cap \bigcap_{j \notin S} \overline{A_j}\right) - \mu\left(\bigcap_{j \in S} B_j \cap \bigcap_{j \notin S} \overline{B_j}\right)\right]$$

for $|S| = i$ the approximate inclusion-exclusion problem leads to the linear program

$$\text{Maximize:} \quad \sum_{i=1}^{n} x_i \tag{8.6}$$

subject to:

$$\sum_{i=j}^{n} \binom{i}{j} x_i = 0 \quad \text{for } 1 \leq j < m \tag{8.7}$$

$$-1 \leq \sum_{i \in S} x_i \leq 1 \quad \text{for all} S \subseteq \{1, \ldots, n\} . \tag{8.8}$$

(See [15] Lemma 3 for a proof).

Dualizing the linear program leads to the following problem of approximation theory (see [15] Lemma 5 and Section 8.6 where we use that technique to determine the asymptotic behavior of k-out-of-n schemes). Determine

$$\inf_{q} \max_{x=1,\ldots,n} 1 - q(x) \tag{8.9}$$

where the infimum ranges over all polynomials q of less than m that have zero constant terms and satisfies $q(x) \leq 1$ for all $x \in \{1, \ldots, n\}$.

The special case $m = n$ leads to Krawtchuck polynomials (see [15] Theorem 3). But for this special case a elementary combinatorial proof exists (see [12]).

Theorem 3 *Let $A_1, \ldots, A_n$ and $B_1, \ldots, B_n$ be two collections of sets satisfying*

$$\left| \bigcap_{i \in S} A_i \right| = \left| \bigcap_{i \in S} B_i \right|$$

for all proper subsets S of $\{1, \ldots, n\}$. Then

$$\frac{\left| \bigcup_{i=1}^{n} B_i \right| - \left| \bigcup_{i=1}^{n} A_i \right|}{\left| \bigcup_{i=1}^{n} B_i \right|} \leq \frac{1}{2^{n-1}} .$$

The bound is sharp.

Proof

We prove by induction on n that the conditions

$$\left| \bigcap_{i \in S} A_i \right| = \left| \bigcap_{i \in S} B_i \right|$$

for all $S \subsetneq \{1, \ldots, n\}$ and the condition

$$\left| \bigcup_{i=1}^{n} A_i \right| + k = \left| \bigcup_{i=1}^{n} B_i \right|$$

with $k > 0$ imply that

$$\left| \bigcup_{i=1}^{n} B_i \right| \geq k 2^{n-1} .$$

For $n = 1$ this is trivial. Now suppose that the theorem holds for n and let the sets $A_1, \ldots, A_{n+1}$ and $B_1, \ldots, B_{n+1}$ satisfy

$$\left| \bigcap_{i \in S} A_i \right| = \left| \bigcap_{i \in S} B_i \right|$$

for all $S \subsetneq \{1, \ldots, n+1\}$ and

$$\left| \bigcup_{i=1}^{n+1} A_i \right| + k = \left| \bigcup_{i=1}^{n+1} B_i \right| .$$

The collections $A_i' = A_i \backslash A_{n+1}$ and $B_i' = B_i \backslash B_{n+1}$ satisfy $|\bigcup_{i=1}^{n} A_i'| + k = |\bigcup_{i=1}^{n} B_i'|$ and for every proper subset $S \subsetneq \{1, \ldots, n\}$ we have

$$\left| \bigcap_{i \in S} A_i' \right| = \left| \bigcap_{i \in S} A_i \right| - \left| \bigcap_{i \in S} A_i \cap A_{n+1} \right|$$

$$= \left| \bigcap_{i \in S} B_i \right| - \left| \bigcap_{i \in S} B_i \cap B_{n+1} \right| = \left| \bigcap_{i \in S} B_i' \right| .$$

Thus the collections A'_i, B'_i satisfy the induction hypothesis, i.e., we have $|\bigcup_{i=1}^{n} B'_i| \geq k2^{n-1}$.

On the other hand, we have the collections $A''_i = A_i \cap A_{n+1}$ and $B''_i = B_i \cap B_{n+1}$. Since

$$\left| \bigcup_{i=1}^{n} A_i \right| = \left| \bigcup_{i=1}^{n} B_i \right|$$

$$\Longleftrightarrow \left| \bigcup_{i=1}^{n} A'_i \right| + \left| \bigcup_{i=1}^{n} A''_i \right| = \left| \bigcup_{i=1}^{n} B'_i \right| + \left| \bigcup_{i=1}^{n} B''_i \right|$$

and $|\bigcup_{i=1}^{n} A'_i| + k = |\bigcup_{i=1}^{n} B'_i|$ we find that the collections A''_i and B''_i satisfy the induction hypothesis with $|\bigcup_{i=1}^{n} B''_i| + k = |\bigcup_{i=1}^{n} A''_i|$. Thus, $|B_{n+1}| = |A_{n+1}| \geq |\bigcup_{i=1}^{n} A''_i| \geq k2^{n-1}$. This proves

$$\left| \bigcup_{i=1}^{n+1} B_i \right| = \left| \bigcup_{i=1}^{n} B'_i \right| + |B_{n+1}| \geq k2^{n-1} + k2^{n-1} = k2^n$$

as desired.

To see that the bound is sharp consider the sets

$$A_i = \{ S \subset \{1, \ldots, n\} \mid |S| \text{ even}, i \in S \}$$

and

$$B_i = \{ S \subset \{1, \ldots, n\} \mid |S| \text{ odd}, i \in S \} .$$

Since for each nonempty set the number of subsets of even cardinality is the same as the number subsets of odd cardinality $(0 = (1-1)^n = \sum_{j=0}^{n}(-1)^j \binom{n}{j})$ we get

$$\left| \bigcap_{i \in S} A_i \right| = \left| \bigcap_{i \in S} B_i \right| = 2^{n-|S|-1}$$

for each proper subset S of $\{1, \ldots, n\}$. Hence, the sets A_i and B_i satisfies the requirement conditions of the approximate inclusion-exclusion problem and

$$\frac{|\bigcup_{i=1}^{n} B_i| - |\bigcup_{i=1}^{n} A_i|}{|\bigcup_{i=1}^{n} B_i|} = \frac{2^{n-1} - (2^{n-1} - 1)}{2^{n-1}} = \frac{1}{2^{n-1}} ,$$

i.e., the bound given by the Theorem is sharp.   □

Theorem 3 translates directly to an contrast optimal n-out-of-n visual cryptography scheme.

Theorem 4 (see [16]) *The optimal contrast of a n-out-of-n visual cryptography scheme is $\alpha = 2^{1-n}$ and the minimal pixel-expansion is $m = 2^{n-1}$.*

Proof

By Lemma 1 a n-out-of-n visual cryptography scheme can be described by two Boolean matrices M_W and M_B.

Interpret the row vectors of M_W and M_B as incidence vectors of the sets A_i and B_i.

The security requirement ((b) in Defintion 1) says $|\bigcup_{i\in S} A_i| = |\bigcup_{i\in S} B_i|$ for all proper subsets S of $\{1,\ldots,n\}$. By the inclusion-exclusion formula this is equivalent to $|\bigcap_{i\in S} A_i| = |\bigcap_{i\in S} B_i|$ for all proper subsets S of $\{1,\ldots,n\}$. By Theorem 3 we have

$$\alpha = \frac{\left|\bigcup_{i=1}^n B_i\right| - \left|\bigcup_{i=1}^n A_i\right|}{\left|\bigcup_{i=1}^n B_i\right|} \leq 2^{1-n} \ .$$

Vice versa the incidence vectors of sets solving the approximated inclusion-exclusion problem define a contrast optimal n-out-of-n visual cryptography scheme. Note that $1/\alpha$ is also a lower bound for the pixel expansion m and the example shows that this bound is sharp. □

Using the proof technique of Theorem 3 one could also solve the approximate inclusion-exclusion problem for the case that only the size intersections of up to $n-2$ sets are known.

Result 5 (see [12] Theorem 3.6) *Let* $A_1,\ldots,A_n$ *and* $B_1,\ldots,B_n$ *be two collections of sets satisfying*

$$\left|\bigcap_{i\in S} A_i\right| = \left|\bigcap_{i\in S} B_i\right|$$

for all proper subsets S *of* $\{1,\ldots,n\}$ *with* $|S| \leq n-2$. *Then*

$$\frac{\left|\bigcup_{i=1}^n A_i\right|}{\left|\bigcup_{i=1}^n B_i\right|} \geq 1 - \binom{n-1}{\lfloor\frac{n-1}{2}\rfloor}^{-1} \ .$$

The bound is sharp.

For the proof see [12]. At this point we show only the construction that proves the sharpness.

We give for every subset $S \subseteq \{1,\ldots,n\}$ the size

$$\left|\bigcap_{i\in S} A_i \setminus \bigcup_{i\notin S} A_i\right| \quad \text{and} \quad \left|\bigcap_{i\in S} B_i \setminus \bigcup_{i\notin S} B_i\right| \ .$$

The construction is best understood if we look at the example $n = 9$ first.

$|S|$	1	2	3	4	5	6	7	8	9
$\left|\bigcap_{i\in S} A_i \setminus \bigcup_{i\notin S} A_i\right|$	0	3	0	1	0	0	2	0	4
$\left|\bigcap_{i\in S} B_i \setminus \bigcup_{i\notin S} B_i\right|$	4	0	2	0	0	1	0	3	0

In general we will have a zigzag line of numbers starting on the left side in the B-row with the value $\lfloor \frac{n-1}{2} \rfloor$, going down to 1, then has one gap and restart with 1. The general rule is as follows:

$$\left| \bigcap_{i \in S} A_i \backslash \bigcup_{i \notin S} A_i \right| = \begin{cases} |S| - \lceil \frac{n}{2} \rceil & \text{if } |S| > n/2, \text{ and } |S| \text{ is odd} \\ \lceil \frac{n}{2} \rceil - |S| & \text{if } |S| < n/2, \text{ and } |S| \text{ is even} \\ 0 & \text{in all other cases} \end{cases} \qquad (8.10)$$

$$\left| \bigcap_{i \in S} B_i \backslash \bigcup_{i \notin S} B_i \right| = \begin{cases} |S| - \lceil \frac{n}{2} \rceil & \text{if } |S| > n/2, \text{ and } |S| \text{ is even} \\ \lceil \frac{n}{2} \rceil - |S| & \text{if } |S| < n/2, \text{ and } |S| \text{ is odd} \\ 0 & \text{in all other cases} \end{cases} \qquad (8.11)$$

Some elementary combinatorial calculations ([12] Theorem 3.8) show that this example satisfies Result 5 with equality.

Similar to Theorem 4 we can use Result 5 to prove that the contrast α of the optimal $(n-1)$-out-of-n visual cryptography scheme satisfies $\alpha \leq \frac{2}{n\binom{n-1}{\lfloor \frac{n-1}{2} \rfloor}}$.

The example given above can be used to construct a scheme that satisfies the bound with equality (see [12] Theorem 3.10).

8.4   Designs and Codes

2-out-of-n visual cryptography schemes are a very special case. In a symmetric, contrast optimal scheme we encode a white pixel by giving all participants the same share. For a black pixel we want to minimize the overlap of black subpixels on the transparencies. This is a typical coding theory problem.

The link to coding theory becomes clear by comparing the following two theorems and their proofs.

Theorem 6 (Plotkin bound) *A binary code of length n with minimal distance $d > \frac{1}{2}n$ has at most $2 \left\lfloor \frac{d}{2d-n} \right\rfloor$ codewords.*

Proof

Let m be the number of codewords. Let $A = \sum_{c \neq c'} d(c, c')$ where the sum ranges over all pairs of codewords. Since the code has minimal distance d the sum is bounded below by $\binom{m}{2}d$.

Let m_i be the number of codewords that have 1 in the ith column. The contribution of those columns to the total distance A is $m_i(m - m_i) \leq \lfloor m/2 \rfloor \lceil m/2 \rceil$.

Hence, $\lfloor m/2 \rfloor \lceil m/2 \rceil \geq A \geq \frac{1}{2}m(m-1)d$. For $d > \frac{1}{2}n$ this gives the bound stated by the Theorem. $\square$

Theorem 7 ([3] Theorem 4.2) *The contrast α of an 2-out-of-n visual cryptography scheme is bounded by*

$$\alpha \leq \frac{\lceil n/2 \rceil \lfloor n/2 \rfloor}{n(n-1)} = \begin{cases} \frac{n}{4n-4} & \text{for } n \text{ even} \\ \frac{n+1}{4n} & \text{for } n \text{ odd} \end{cases}$$

Proof
Let m be the number of subpixels and α be the contrast. When stacking transparencies the fraction of black subpixels cannot become smaller. Thus, the number of black subpixels on each transparencies must be increased at least by αm, i.e. for every transparency t_1 and every transparency t_2 there must be at least αm subpixels that are write on t_1 and black on t_2. All pairs of transparencies give, therefore, at least $n(n-1)\alpha m$ black-white subpixel combinations.

Now look at the number of black-white combinations that come from a single subpixel. If the subpixel is white on i transparencies and black on $n-i$ transparencies it contributes $i(n-i)$ black-white combinations. The term $i(n-1)$ is bounded by $\lfloor n/2 \rfloor \lceil n/2 \rceil$.

This lead to the inequality

$$n(n-1)\alpha m \leq \lfloor n/2 \rfloor \lceil n/2 \rceil m \ .$$

Solving the inequality gives the theorem. $\square$

The bound of Theorem 7 is sharp and the contrast optimal 2-out-of-n schemes are related to designs. Let's recall the definition of a $2-(v,k,\lambda)$ design. (I recommend [1] as a reference for design theory.)

Definition 2 *A $2-(v,k,\lambda)$ design is a an incidence structure $(\mathcal{V},\mathcal{B},I)$ with*

- $|\mathcal{V}| = v$ *points*

- *Every block $B \in \mathcal{B}$ is incident with k points.*

- *Every pair of points $p, q \in V$ is joined by exactly λ blocks.*

Theorem 8 (see [3] Theorem 4.4) *Let n be even. A 2-out-of-n visual cryptography scheme with optimal contrast $\alpha = \frac{n+1}{4n}$ and pixel expansion m exists if and only if there exists a $2-(n, \frac{n}{2}, \frac{m(n-2)}{4n-4})$ design.*

Proof
If we have equality in Theorem 7 we must have equality in every step.

Thus, for each two transparencies there are exactly αm subpixels that are white on the first and black on the second transparency. Furthermore, each subpixel is black on exactly $n/2$ transparencies.

Let M_B be the Boolean matrix that describes the encoding of a black pixel. Interpret M_B as an incidence matrix of incidence structure $(\mathcal{V},\mathcal{B},I)$ (Columns of M_B corresponds to block and rows of M_B to points.)

As we have seen above every block consists of $\frac{n}{2}$ points.

Since for every two transparencies the number of subpixels that is white on the first and black on the second is constant. Each transparency must have the same number of black subpixels. Since each subpixel is black on exactly half of the transparencies the number of black subpixels per transparency must be $\frac{m}{2}$.

Thus, two transparencies have exactly $\frac{m}{2} - \alpha m = \frac{m(n-1)}{4n-4}$ common black subpixels. In the interpretation as incidence structure: Each two points are joined by $\frac{m(n-1)}{4n-4}$ blocks.

In other words M_B is the incidence matrix of a $2 - (n, \frac{n}{2}, \frac{m(n-2)}{4n-4})$ design. Conversely the incidence matrix of a $2 - (n, \frac{n}{2}, \frac{m(n-2)}{4n-4})$ design satisfies all requirements of a 2-out-of-n visual cryptography scheme.   □

Similarly, we can deal with the case of n odd. The Proof of Theorem 7 shows that every column in M_B must contain either $\lfloor n/2 \rfloor$ or $\lceil n/2 \rceil$ zero entries. So we do not get a design (every block is of the same size), but a structure called *pairwise balanced design* (PDB) where the size of block may vary between some values. We skip the details and just state the final result.

Result 9 (see [3] Theorem 4.5) *Let n be odd. A 2-out-of-n visual cryptography scheme with optimal contrast $\alpha = \frac{n}{4n-4}$ and pixel expansion m exists if and only if there exists a $2 - (n, \{\frac{n-1}{2}, \frac{n+1}{2}\}, w - \frac{m(n+1)}{4n})$ PBD such that every point lies in exactly w blocks, where w is an integer in the range*

$$\frac{(n-1)m}{2n} \leq w \leq \frac{(n+1)m}{2n} \quad .$$

A well-known result from coding theory states that Hadamard codes archive equality in the Plotkin bound. So we are not surprised that Hadamard matrices can also be used to construct optimal 2-out-of-n visual cryptography schemes.

Recall the Definition of Hadamard matrices.

Definition 3 *A* Hadamard matrix *is a $n \times n$ matrix H with entries ± 1 and $HH^t = nI$.*

It is well known that except for $n = 1, 2$ the order of a Hadamard matrix must be divisible by 4. The famous Hadamard conjecture states that for every n a $(4n) \times (4n)$ Hadamard matrix exists. See [8] for an overview of constructions of Hadamard matrices.

The connection to visual cryptography is established by the next theorem.

Theorem 10 (see [3] Theorem 4.7) *The pixel expansion of a contrast optimal 2-out-of-$(4n - 1)$ visual cryptography scheme is at least $m \geq 4n - 1$. A scheme with $m = 4n - 1$ is equivalent to the existence of a $2 - (4n - 1, 2n - 1, n-1)$ design (called a* Hadamard design*). This is equivalent to the existence of a Hadamard matrix of order $4n$.*

Proof

Let M_B be the matrix describing the encoding of a black pixel. As in the proof of Theorem 8 we find that each transparency must have the same number w of black subpixels and each two transparencies must share the same number $\lambda < w$ of black subpixels. Interpret M_B as an integer matrix, then

$$M_B M_B^t = (w - \lambda)I + \lambda J \ .$$

This matrix has a full rank $(\det(w - \lambda)I + \lambda J = (w - \lambda)^{n-1}(n\lambda - \lambda + w))$ and hence,

$$4n - 1 = \operatorname{rank} M_B M_B^t \leq \operatorname{rank} M_B \leq m \ .$$

(This is the Fisher's inequality.)

Now assume $m = 4n - 3$ then w must be either $2n - 1$ or $2n$. In the first case every column of M_W must contain $2n - 1$ entries 1 and M_W is the incidence matrix of a $2 - (4n - 1, 2n - 1, n - 1)$ design. In the second case every column of M_W must contain $2n$ entries 1 and M_W is the incidence matrix of a $2 - (4n - 1, 2n, n)$ design. But then the complement of M_W is the incidence matrix of a $2 - (4n - 1, 2n - 1, n - 1)$ design. This shows that a contrast optimal 2-out-of-$(4n - 1)$ visual cryptography scheme with pixel expansion $m = 4n - 3$ implies the existence of a $2 - (4n - 1, 2n - 1, n - 1)$ design. But conversely you get a 2-out-of-$(4n - 1)$ visual cryptography scheme from a $2 - (4n - 1, 2n - 1, n - 1)$ design by defining M_B to be the incidence matrix of the design and encode it with a pixel by choosing on all transparencies the same subpixels.

Now we prove that the existence of $2 - (4n - 1, 2n - 1, n - 1)$ design is equivalent to the existence of a Hadamard matrix of order $4n$.

Multiplying a row or a column of a Hadamard matrix with -1 again gives a Hadamard matrix. So without loss of generality we may assume that the first row and the first column of a Hadamard matrix has only 1 as entries. Let H be a Hadamard matrix of order $4n$ in that normal from. All rows except the first row of H must contain $2n$ entries -1 and for two rows different rows must differ in exactly $2n$ columns. Hence, after deleting the first row and first column H forms the incidence matrix of a $2 - (4n - 1, 2n - 1, n - 1)$ design (with -1 instead of 0 to mark that a point is not incident with a block). Conversely the incidence matrix of a $2 - (4n - 1, 2n - 1, n - 1)$ becomes a Hadamard matrix, if one replaces all 0 by -1 and adds an extra column and an extra row that contains only 1. $\qquad\square$

Other 2-out-of-n visual cryptography schemes are also connected to Hadamard matrices.

Result 11 (see [3] Theorem 4.9) *The minimal pixel expansion m of contrast optimal 2-out-of-n visual cryptography schemes satisfies:*

$$m \geq \begin{cases} 2n - 2 & \text{if } n \text{ even} \\ n & \text{if } n \equiv 3 \mod 4 \\ 2n & \text{if } n \equiv 1 \mod 4 \end{cases}$$

If the Hadamard conjecture is true the bounds are sharp.

We remark that there are 2-out-of-n visual cryptography schemes with smaller pixel-expansion that are not contrast optimal. In [3] (Theorem 4.12) 2-out-of-n visual cryptography schemes with pixel expansion $m \approx \sqrt{n}$ and contrast $\alpha \approx \frac{1}{4}$ are constructed. For $n = \binom{m}{\lfloor m/2 \rfloor}$ there exists a 2-out-of-n visual cryptography scheme with pixel expansion m and contrast $1/m$ (just color on each transparency is a different set of $\lfloor m/2 \rfloor$ subpixels). This is the minimal possible pixel expansion.

8.5   Optimal 3-out-of-n Schemes

We construct a scheme that has the normal form of Lemma 2. Remember that we denoted by $x_i^{(W)}$ the number of subpixels when encoding a white pixel that are black of the slides $1, \ldots, i$ and white on the other slides. Similarly $x_i^{(B)}$ describes the encoding rule for a black pixel. As we have seen in the introduction, a k-out-of-n scheme must satisfy

$$\alpha m = \sum_{i=0}^{n-k} \binom{n-k}{i} x_i^{(W)} - \sum_{i=0}^{n-k} \binom{n-k}{i} x_i^{(B)} \tag{8.12}$$

and

$$\sum_{i=0}^{n-j} \binom{n-j}{i} x_i^{(W)} = \sum_{i=0}^{n-j} \binom{n-j}{i} x_i^{(B)} \tag{8.13}$$

for $j = 0, \ldots, k - 1$.

Let $S(3, g, n)$ be the visual cryptography scheme that is described by the values $x_0^{(W)} = x_n^{(B)} = \binom{n-1}{g} - \binom{n-1}{g-1}$, $x_{n-g}^{(W)} = x_g^{(B)} = 1$ and all other variables are 0.

Theorem 12 ([2] Theorem 4.6) $S(3, g, n)$ *is a 3-out-of-n visual cryptography scheme with pixel expansion* $m = 2\binom{n-1}{g}$ *and contrast*

$$\alpha = \frac{g(n - 2g)}{2(n - 1)(n - 2)} .$$

Proof
For $j = 0$ we have in (8.13):

$$\binom{n}{0} \left[\binom{n-1}{g} - \binom{n-1}{g-1} \right] + \binom{n}{n-g} \cdot 1 = \binom{n}{g} \cdot 1 + \binom{n}{n} \left[\binom{n-1}{g} - \binom{n-1}{g-1} \right] = r$$

For $j = 1$ we get

$$\binom{n-1}{0} \left[\binom{n-1}{g} - \binom{n-1}{g-1} \right] + \binom{n-1}{n-g} \cdot 1 = \binom{n-1}{g} \cdot 1$$

which is also true, since $\binom{n-1}{n-g} = \binom{n-1}{g-1}$. For $j = 2$ we get

$$\binom{n-2}{0}\left[\binom{n-1}{g} - \binom{n-1}{g-1}\right] + \binom{n-2}{n-g} \cdot 1 = \binom{n-2}{g} \cdot 1$$

which is true since $\binom{n-1}{g} - \binom{n-1}{g-1} = \binom{n-2}{g}$ and $\binom{n-2}{g} + \binom{n-2}{n-g} = \binom{n-2}{g} + \binom{n-2}{q-1} = \binom{n-2}{g}$.

Thus, $S(3, g, n)$ satisfies the security requirements of an 3-out-of-n visual cryptography scheme.

The contrast of $S(3, g, n)$ is

$$\begin{aligned}
\alpha m &= \binom{n-3}{0}\left[\binom{n-1}{g} - \binom{n-1}{g-1}\right] + \binom{n-3}{n-g} \cdot 1 - \binom{n-3}{g} \cdot 1 \\
&= \binom{n-2}{g} + \binom{n-2}{g-1} - \binom{n-1}{g-1} + \binom{n-3}{g-3} - \binom{n-3}{3} \\
&= \binom{n-3}{g-1} - \binom{n-2}{g-3} + \binom{n-3}{g-3} \\
&= \binom{n-3}{g-1} - \binom{n-3}{g-2}
\end{aligned}$$

So the image is indeed recovered. $\qquad\square$

The scheme $S(3, g, n)$ archives optimal contrast if we choose $g = \lfloor \frac{n+1}{4} \rfloor$. This is the best possible contrast for a 3-out-of-n scheme. To prove this result we introduce the canonical form of a k-out-of-n scheme.

Remember that a k-out-of-n scheme is described by the following linear program.

Maximize:

$$\alpha = \sum_{i=0}^{n-k} \binom{n-k}{i} x_i^{(W)} - \sum_{i=0}^{n-k} \binom{n-k}{i} x_i^{(B)} \qquad (8.14)$$

Subject to

$$\sum_{i=0}^{n} \binom{n}{i} x_i^{(W)} = \sum_{i=0}^{n} \binom{n}{i} x_i^{(B)} = 1 \qquad (8.15)$$

and

$$\sum_{i=0}^{n-j} \binom{n-j}{i} x_i^{(W)} = \sum_{i=0}^{n-j} \binom{n-j}{i} x_i^{(B)} \qquad (8.16)$$

for $j = 0, \ldots, k - 1$.

Lemma 13 *There is an optimal solution of the linear program defined by* (8.14), (8.15), *and* (8.16) *that satisfies:*

1.If k is even, then $x_i^{(W)} = x_{n-i}^{(W)}$ and $x_i^{(B)} = x_{n-i}^{(B)}$ for $i = 0, \ldots, n$.

   2.*If k is odd, then $x_i^{(W)} = x_{n-i}^{(B)}$ for $i = 0, \ldots, n$.*

   3.$x_i^{(W)} x_i^{(B)} = 0$ *for all $i \in \{0, \ldots, n\}$.*

We will say the solution of the linear program is in a canonical form.

Proof

We first show that replacing each transparency of a k-out-of-n visual cryptography scheme by its complement again gives a solution of a k-out-of-n visual cryptography scheme. In the language of linear programming:

   Let $x_i^{(W)}$, $x_i^{(B)}$ be a solution of the linear program (8.14)–(8.16). We claim that

$$\hat{x}_i^{(W)} = \begin{cases} x_{n-i}^{(W)} & \text{if } k \text{ is even} \\ x_{n-i}^{(B)} & \text{if } k \text{ is odd} \end{cases}$$

and

$$\hat{x}_i^{(B)} = \begin{cases} x_{n-i}^{(B)} & \text{if } k \text{ is even} \\ x_{n-i}^{(W)} & \text{if } k \text{ is odd} \end{cases}$$

is also a solution of the linear program.

   Since $\binom{n}{i} = \binom{n}{n-i}$ the variables $\hat{x}_i^{(W)}$m $\hat{x}_i^{(B)}$ satisfy (8.16).

We claim that equation 8.16 implies

$$\sum_{i=0}^{n-j} \binom{n-j}{i-h} x_i^{(W)} = \sum_{i=0}^{n-j} \binom{n-j}{i-h} x_i^{(B)}$$

for all $h \le j$. For $j = 0$ this is trivial and for $j \ge 1$, $h \ge 1$ it follows from $\binom{n-j}{i-h} = \binom{n-j+1}{i-h+1} - \binom{n-j}{i-h+1}$ by induction

$$\sum_{i=0}^{n-j} \binom{n-j+1}{i-h+1} x_i^{(W)} = \sum_{i=0}^{n-j} \binom{n-j+1}{i-h+1} x_i^{(B)}$$

and

$$\sum_{i=0}^{n-j} \binom{n-j}{i-h+1} x_i^{(W)} = \sum_{i=0}^{n-j} \binom{n-j}{i-h+1} x_i^{(B)} .$$

To simplify the notation let us assume in the following that k is even.

When k is odd, it is necessary to exchange $x_i^{(W)}$ by $x_i^{(B)}$ and vice versa.

$$\sum_{i=0}^{n-j} \binom{n-j}{i} \hat{x}_i^{(W)} = \sum_{i=0}^{n-j} \binom{n-j}{i} x_{n-i}^{(W)}$$

$$= \sum_{i=j}^{n} \binom{n-j}{n-i} x_i^{(W)}$$

$$= \sum_{i=j}^{n} \binom{n-j}{i-j} x_i^{(W)}$$

$$= \sum_{i=j}^{n} \binom{n-j}{i-j} x_i^{(B)}$$

$$= \sum_{i=0}^{n-j} \binom{n-j}{i} x_i^{(B)} \ ,$$

i.e., the variables $\hat{x}_i^{(W)}$ m $\hat{x}_i^{(B)}$ satisfy (8.16).

By the same argument we get

$$\sum_{i=0}^{n-k} \binom{n-k}{i-h} x_i^{(B)} + (-1)^h \alpha = \sum_{i=0}^{n-k} \binom{n-k}{i-h} x_i^{(W)} \qquad (8.17)$$

for $h \leq k$.

For $h = k$ this shows that $\hat{x}_i^{(W)}$ m $\hat{x}_i^{(B)}$ is an optimal solution of the linear program.

Let $x_i^{(W)}, x_i^{(B)}, \hat{x}_i^{(W)}$ and $\hat{x}_i^{(B)}$ be two solutions of the linear program, then $\tilde{x}_i^{(W)} = \frac{1}{2}(x_i^{(W)} + \hat{x}_i^{(W)})$, $\tilde{x}_i^{(B)} = \frac{1}{2}(x_i^{(B)} + \hat{x}_i^{(B)})$ is also a solution of the linear program.

After this transformations we have a solution that satisfies (a) and (b).

Assume that we have a scheme that does not satisfy (c), i.e., the matrices M_W and M_B have columns in common. Since the corresponding subpixels occur independent of the encoded color they can be omitted. Deleting all columns that occur in M_W and M_B we get a scheme with smaller pixel expansion and higher contrast. That proves that every optimal k-out-of-n scheme must satisfy (c). □

Now we are ready to determine the optimal contrast of a 3-out-of-n visual cryptography scheme.

Theorem 14 ([2] Theorem 4.7) *The contrast α of a 3-out-of-n visual cryptography scheme satisfies*

$$\alpha \leq \frac{(n - 2\lfloor \frac{n+1}{4} \rfloor)\lfloor \frac{n+1}{4} \rfloor}{2(n-1)(n-2)}$$

Proof

Assume that the scheme is in canonical form (see Lemma 13) and let M_W and M_B be the $n \times m$ Boolean matrices that describe the encoding.

Since the scheme is in canonical form, m is even and exactly $m' = \frac{m}{2}$ subpixels on each slide are black. In terms of the linear program this means

$$\sum_{j=0}^{n-1} x_j^{(W)} \binom{n-1}{j} = \sum_{j=0}^{n-1} x_j^{(B)} \binom{n-1}{j} = \frac{1}{2}. \tag{8.18}$$

By equation 8.17 (with $h = 1$) the contrast α is

$$\alpha = \sum_{i=0}^{n-k} \binom{n-3}{i-1} x_i^{(B)} - \sum_{i=0}^{n-k} \binom{n-3}{i-1} x_i^{(W)}$$

Since the scheme is in canonical form we have $x_i^{(W)} = x_{n-i}^{(B)}$ and hence,

$$\alpha = \sum_{i=0}^{n-k} \left(\binom{n-3}{i-1} x_i^{(B)} - \binom{n-3}{i-1} x_{n-i}^{(B)} \right)$$

$$= \sum_{i=0}^{n} x_i^{(B)} \left(\binom{n-3}{i-1} - \binom{n-3}{n-i-1} \right)$$

$$= \sum_{i=0}^{n} x_i^{(B)} \left(\binom{n-3}{i-1} - \binom{n-3}{i-2} \right)$$

Now use equation (8.18) to multiply with 1 and we get

$$\alpha = \frac{\sum_{i=0}^{n} x_i^{(B)} \left(\binom{n-3}{i-1} - \binom{n-3}{i-2} \right)}{2 \sum_{i=0}^{n} x_i^{(B)} \binom{n-1}{j}}$$

The inequality $\frac{\sum_x f(x)}{\sum_x g(x)} \leq \max_x \frac{f(x)}{g(x)}$ holds for any functions f and g. Hence,

$$\alpha \leq \max_{0 \leq i \leq n} \frac{\binom{n-3}{i-1} - \binom{n-3}{i-2}}{2\binom{n-1}{j}}$$

The maximum is reached for $i = \lfloor \frac{n+1}{4} \rfloor$, which proves the theorem. $\qquad \square$

In a recent article M. Bose and R. Mukerjee [5] show how to use group divisible designs and balanced incomplete block designs to construct 3-out-of-n visual cryptography schemes with optimal contrast and small pixel expansion.

8.6 Asymptotic Optimal k-out-of-n Schemes

In the previous sections we solved the problem of contrast optimal visual cryptography schemes exactly. The result for general k-out-of-n is a little bit weaker; we have only a tight bound for the optimal contrast.

We start with the linear program
Maximize:

$$\alpha = \sum_{i=0}^{n-k} \binom{n-k}{i} x_i^{(W)} - \sum_{i=0}^{n-k} \binom{n-k}{i} x_i^{(B)}$$

Subject to:

$$\sum_{i=0}^{n} \binom{n}{i} x_i^{(W)} = \sum_{i=0}^{n} \binom{n}{i} x_i^{(B)} = 1$$

and

$$\sum_{i=0}^{n-j} \binom{n-j}{i} x_i^{(W)} = \sum_{i=0}^{n-j} \binom{n-j}{i} x_i^{(B)}$$

for $j = 0, \ldots, k-1$. The sign conditions are $x_i^{(W)} \geq 0$, $x_i^{(B)} \geq 0$.

For the following it is convenient to use the variable transform $\hat{x}_i^{(W)} = \binom{n}{i} x_i^{(W)}$ and $\hat{x}_i^{(B)} = \binom{n}{i} x_i^{(B)}$. The linear program takes the form
Maximize $\alpha = c'(\hat{x}^{(W)} - \hat{x}^{(B)})$ subject to

$$\sum_{i=0}^{n} \hat{x}_i^{(W)} = \sum_{i=0}^{n} \hat{x}_i^{(B)} = 1$$

and

$$A(\hat{x}^{(W)} - \hat{x}^{(B)}) = 0$$

where A is the $k \times (n+1)$ matrix with entries $A = (a_{j,i})$

$$a_{j,i} = \binom{n-j}{i} \binom{n}{i}^{-1} = q_j(i)$$

Note that the j-th row of A is the evaluation of a polynomial q_j of degree j at the positions $i = 0, \ldots, n$. Similarly the vector c is the evaluation of k-th degree polynomial $p(i) = \binom{n-k}{i} \binom{n}{i}^{-1}$ at the places $i = 0, \ldots, n$.

Now dualize the linear program. For each constraint we get a variable in the dual program and since all constraints of the primal program are equalities the variables in the dual program have no sign restriction. Each variable of the primal program gives a constraint in the dual program and since all variables have a sign condition we get inequalities as constraints in the dual program. The dual program is

Minimize $s + t$ subject to

$$A'u + (s, \ldots, s) \geq c \tag{8.19}$$
$$-A'u + (t, \ldots, t) \geq c \quad \Longleftrightarrow \quad A' - (t, \ldots, t) \leq c \tag{8.20}$$

Equations (8.19) and (8.20) just state that

$$s \geq \max_{i=0,\ldots,n} (c_i - A'_i u)$$

$$t \geq \max_{i=0,\ldots,n} (A'_i u - c_i).$$

In the optimal solution these inequalities must be equalities, i.e.,

$$s = \max_{i=0,\ldots,n} (c_i - A'_i u) \tag{8.21}$$

$$t = \max_{i=0,\ldots,n} (A'_i u - c_i). \tag{8.22}$$

Now we translate these equations into the language of polynomials. Remember that $c_i = p(i)$ for a k-th degree polynomial and that the j-th row of A is the evaluation of a j-th degree polynomial q_j. The polynomials q_j $(j = 0, \ldots, k-1)$ form a basis of the vector space P_{k-1} of all polynomials of a degree up to $k-1$.

So we may write the dual problem as:
Determine

$$\alpha = \min_{q \in P_{k-1}} \left(\max_{i=0,\ldots,n} (p(i) - q(i)) + \max_{i=0,\ldots,n} (q(i) - p(i)) \right). \tag{8.23}$$

By adding the right constant to q, we can choose $\max_{i=0,\ldots,n}(p(i) - q(i)) = \max_{i=0,\ldots,n}(q(i) - p(i))$, which simplifies (8.23) to

$$\alpha = \min_{q \in P_{k-1}} \max_{i=0,\ldots,n} |p(i) - q(i)|. \tag{8.24}$$

The low terms of p lie in P_{k-1}, so the problem is just to approximate the term of degree k. Since

$$p(i) = \binom{n-k}{i}\binom{n}{i}^{-1}$$

$$= \frac{(n-i)!}{(n-i-k)!}\frac{(n-k)!}{n!}$$

$$= \frac{(n-i)(n-1-i)\cdots(n-(k-1)-i)}{n(n-1)\cdots(n-(k-1))}$$

$$= \left(1 - \frac{i}{n}\right)\left(1 - \frac{i}{n-1}\right)\cdots\left(1 - \frac{i}{n-(k-1)}\right)$$

the highest degree term of $p(i)$ is $(-1)^k / n^{\underline{k}} i^k$ where $n^{\underline{k}} = n(n-1)\cdots(n-k+1)$.

This simplifies (8.24) to

$$\alpha = \min_{q \in P_{k-1}} \max_{i=0,\ldots,n} \left| \frac{x^k}{n^{\underline{k}}} - q(i) \right| . \qquad (8.25)$$

Equation 8.25 is almost identical to the following classical problem in approximation theory:

$$\min_{q \in P_{k-1}} \max_{x \in [-1,1]} |x^k - q(x)| \qquad (8.26)$$

The only difference is that the approximation points in our problem are discrete.

A well-known result from approximation theory states:

Result 15 (see for example Theorem 5.7 of [18]) *Of all monic polynomials of degree k the best approximation to 0 in $[-1,1]$ with the $\|.\|_\infty$-norm is $2^{1-k}T_k$, where $T_k(x) = \cos(k \arccos x)$ denotes the Chebychev polynomial of the first kind.*

Hence, the solution of (8.26) is

$$q(x) = x^k - 2^{1-k}T_k(x) .$$

The minimum in (8.26) is 2^{k-1}. Transforming it back to (8.25) we get an upper bound for α (only an upper bound since we switch from continuous evaluation points to discrete points.)

Thus, we have the contrast $\alpha_{n,k}$ of a k-out-of-n visual cryptography scheme is bounded by:

$$\alpha_{n,k} \leq 4^{1-k} \frac{n^k}{n^{\underline{k}}}$$

Asymptotically continuous evaluation points and discrete evaluation points are the same, thus we have also

$$\lim_{n\to\infty} \alpha_{n,k} \leq \lim_{n\to\infty} 4^{1-k} \frac{n^k}{n^{\underline{k}}} = 4^{1-k} .$$

Obtaining a lower bound for $\alpha_{n,k}$ need a bit extra work. We just cite without proof.

Result 16 (see [14]) *The contrast $\alpha_{n,k}$ of a contrast optimal k-out-of-n visual secret sharing scheme satisfies*

$$4^{k-1} \leq \alpha_{n,k} \leq 4^{k-1} \frac{n^k}{n^{\underline{k}}} .$$

8.7 Contrast Trade-Offs for Extended Visual Cryptography Schemes

In this section we look at a visual cryptography scheme with extended capacity. In [16] Naor and Shamir show that it is possible to create two transparencies such that each transparency shows an image and the stack of the two transparencies reveals another image. The three images must not satisfy any relation.

In general, we have n slides and a subset $\mathfrak{S}$ of $\mathcal{P}(\{1,\dots,n\})\backslash\{\emptyset\}$. For an $\mathfrak{S}$-extended visual cryptography scheme we require that for every $S \in \mathfrak{S}$ the stack of the transparencies $i \in S$ reveals an image I_S. Furthermore, we require that there is no other way to get information about the image I_S.

As in the case of the simple visual cryptography schemes we formalize these ideas by describing the encoding algorithm by a multiset of Boolean matrices. This leads to Definition 4.

Definition 4 *Let $\mathfrak{S} \subseteq \mathcal{P}(\{1,\dots,n\})\backslash\{\emptyset\}$.*

An $\mathfrak{S}$-extended visual cryptography scheme is described by multisets $\mathfrak{M}^{\mathfrak{T}}$ of $n \times m$ Boolean matrices for $\mathfrak{T} \subseteq \mathfrak{S}$. (For given $\mathfrak{T}$ each Boolean matrix in $\mathfrak{M}^{\mathfrak{T}}$ describes the colors of the subpixels on each transparency, where the corresponding pixel in image I_T is black if and only if $T \in \mathfrak{T}$. For encoding, each matrix in $\mathfrak{M}^{\mathfrak{T}}$ is chosen with the same probability.)

The multisets $\mathfrak{M}^{\mathfrak{T}}$ must satisfy the following conditions:

> 1. *Let $B \in \mathfrak{M}^{\mathfrak{T}}$. For $\{i_1,\dots,i_q\} \in \mathfrak{S}$ the Hamming weight of the OR of the rows $i_1,\dots,i_q$ of B is $h_{\{i_1,\dots,i_q\}}$ if $\{i_1,\dots,i_q\} \in \mathfrak{T}$ and $l_{\{i_1,\dots,i_q\}}$ otherwise, i.e.,*

$$w_{Ham}((b_{i_1,1},\dots,b_{i_1,m})\ OR\ \dots\ OR\ (b_{i_q,1},\dots,b_{i_q,m}))$$
$$= \begin{cases} h_{\{i_1,\dots,i_q\}} & \text{if } \{i_1,\dots,i_q\} \in \mathfrak{T} \\ l_{\{i_1,\dots,i_q\}} & \text{if } \{i_1,\dots,i_q\} \notin \mathfrak{T} \end{cases} \qquad .2$$

> *(This means stacking the transparencies $i_1,\dots,i_q$ together we recover the image $I_{\{i_1,\dots,i_q\}}$.)*

> 2. *For $\{i_1,\dots,i_q\} \subseteq \{1,\dots,n\}$ and $\mathfrak{T},\mathfrak{T}' \subseteq \mathfrak{S}$ with $\mathfrak{T} \cap \mathcal{P}(\{i_1,\dots,i_q\}) = \mathfrak{T}' \cap \mathcal{P}(\{i_1,\dots,i_q\})$ we obtain the same multisets if we restrict the matrices in $\mathfrak{M}^{\mathfrak{T}}$ and $\mathfrak{M}^{\mathfrak{T}'}$, respectively, to the rows $i_1,\dots,i_q$.*

> *(This condition guarantees the security of the different images.)*

If $\mathfrak{S} = \mathcal{P}(\{1,\dots,n\})\backslash\{\emptyset\}$ we simply call this an extended visual cryptography scheme.

$\mathfrak{S}$-extended schemes exists as the following theorem shows.

Theorem 17 (see [9]) *An $\mathfrak{S}$-extended schemes with pixel expansion*

$$m = \sum_{S \in \mathfrak{S}} 2^{|S|-1}$$

and contrast $\alpha = 1/m$ exists.

Proof
The $\mathfrak{S}$-extended scheme is built from several k-out-of-k schemes. For each $S \in \mathfrak{S}$ we reserve $2^{|S|-1}$ subpixels. On slides with number $i \notin S$ these subpixels are colored black. For the slides $i \in S$ these $2^{|S|-1}$ subpixels are used to construct an $|S|$-out-of-$|S|$ visual cryptography scheme to encode the image I_S (see Theorem 4 for the construction).

To illustrate the construction let $\mathfrak{S} = \{\{1,2\},\{1,3\}\}$ then $m = 4$. The first two pixels are used to encode the image $I_{\{1,2\}}$ and the next two subpixels are for the image $I_{\{2,3\}}$.

A typical matrix of $\mathfrak{M}^{\emptyset}$ is

$$M^{\emptyset} = \begin{pmatrix} 1 & 0 & 1 & 1 \\ 1 & 0 & 1 & 0 \\ 1 & 1 & 1 & 0 \end{pmatrix}$$

and a typical matrix of $\mathfrak{M}^{\{\{1,2\}\}}$ is

$$M^{\emptyset} = \begin{pmatrix} 1 & 0 & 1 & 1 \\ 0 & 1 & 1 & 0 \\ 1 & 1 & 1 & 0 \end{pmatrix}.$$

In the upper right corner you see the 2-out-of-2 scheme for the image $I_{\{1,2\}}$.

We claim that this construction yields an $\mathfrak{S}$-extended scheme.

Let $T \subseteq \{1,\dots,n\}$, what happens if we stack the transparencies $i \in T$?

In the $2^{|T|-1}$ subpixels associated with the image I_T we have $|T|$-out-of-$|T|$ scheme and hence either all $2^{|T|-1}$ subpixels in the stack are black or only $2^{|T|-1} - 1$ subpixels are black depending on the color in the image I_T.

For $S \neq T$ and $T \not\subseteq S$ we find an $i \in T$ with $i \notin S$. On transparency i all $2^{|S|-1}$ subpixels associated with I_S are black. Hence, in the stack of the transparencies $i \in T$ these $2^{|S|-1}$ subpixels are always black and do not contribute to the encoded color.

For $S \neq T$ and $T \subseteq S$ we have a $|S|$-out-of-$|S|$ scheme for the image I_S. By the security requirement stacking only the transparencies $i \in T$ the number of black subpixels must be independent of the color of image I_S. (There will be exactly $2^{|S|-1} - 2^{|S|-|T|-1}$ black subpixels associated with the image I_S.)

As we can see, only the $2^{|T|-1}$ subpixels associated with the image I_T contribute to the image that we see when stacking the transparencies $i \in T$. Hence, the stack of the transparencies $i \in T$ reveals image I_T.

Since the scheme is built from several standard k-out-of-k schemes, it inherits the security from the standard visual cryptography scheme. $\square$

Similar to Lemma 1 we can restrict ourselves to the case that $\mathfrak{M}^{\mathfrak{T}}$ is the multiset that consists of all column permutations of a matrix $M^{\mathfrak{T}}$.

For $S \subseteq \{1, \ldots, n\}$ we denote by $x_S^{\mathfrak{T}}$ the number of columns of $M^{\mathfrak{T}}$ that have a 1 in the rows $i \in S$ and a zero in the rows $i \notin S$. The number of black subpixels in the image I_S is either h_S (when a black pixel is encoded) or l_s when a white pixel is encoded. Analog to the linear program for ordinary visual cryptography we get

$$M x^{\mathfrak{T}} = r^{\mathfrak{T}} \tag{8.27}$$

where $r^{\mathfrak{T}} = (r_S^{\mathfrak{T}})_{\emptyset \neq S \subseteq \{1,\ldots,n\}}$ with $r_S^{\mathfrak{T}} = h_S$ if $S \in \mathfrak{T}$ and $r_S^{\mathfrak{T}} = l_S$ if $S \notin \mathfrak{T}$.

The $(2^n - 1) \times (2^n - 1)$ matrix $M = (m_{S,T})_{\emptyset \neq S, T \subseteq \{1,\ldots,n\}}$ is defined by $m_{S,T} = 1$ if $S \cap T \neq \emptyset$ and $m_{S,T} = 0$ otherwise.

It possible to solve the program explicitly and the starting point is the observation that M has an simple inverse.

Lemma 18 *The matrix $M = (m_{S,T})_{\emptyset \neq S, T \subseteq \{1,\ldots,n\}}$ with $m_{S,T} = 1$ if $S \cap T \neq \emptyset$ and $m_{S,T} = 0$ otherwise is invertible and M^{-1} has only the entries ± 1 and 0.*

Proof

We prove this by induction on the number n of transparencies.

Sort the variables x_T by the following order: First, we enumerate all subsets that do not contain n. The next subset is the set $\{n\}$. Then, the other subsets containing n follow.

Writing $M_1 = (1)$, we obtain the following recursion formula that follows directly from the definition:

$$M_{n+1} = \begin{pmatrix} M_n & \mathbf{0}_{n,1} & M_n \\ \mathbf{0}_{1,n} & 1 & \mathbf{1}_{1,n} \\ M_n & \mathbf{1}_{n,1} & \mathbf{1}_{n,n} \end{pmatrix}.$$

Here the index denotes the number of transparencies. $\mathbf{0}_{i,j}$ or $\mathbf{1}_{i,j}$, respectively, denoting a $(2^i - 1) \times (2^j - 1)$ matrix with all entries 0 or 1, respectively.

With $M_1^{-1} = (1)$ we obtain the following recursion formula for M_n^{-1}:

$$M_{n+1}^{-1} = \begin{pmatrix} \mathbf{0}_{n,n} & -M_n^{-1}\mathbf{1}_{n,1} & M_n^{-1} \\ -\mathbf{1}_{1,n}M_n^{-1} & 0 & \mathbf{1}_{1,n}M_n^{-1} \\ M_n^{-1} & M_n^{-1}\mathbf{1}_{n,1} & -M_n^{-1} \end{pmatrix}$$

Thus, M is invertible, i.e., Equation (8.27) has a unique solution.

We notice that the components of $M_n^{-1}\mathbf{1}_{n,1}$ are only $-1, 0$, and 1 and that $\mathbf{1}_{1,n}M_n^{-1}\mathbf{1}_{n,1} = 1$. Then the formula can be proved by induction.

Thus, M_n^{-1} contains only the entries $-1, 0$, and 1 and therefore the solution of equation (8.27) is integral. $\square$

By Lemma 18 the solution $x^{\mathfrak{T}}$ of equation (8.27) is an integer vector if the right-hand side $r^{\mathfrak{T}}$ is an integer vector.

We can explicitly solve equation (8.27) by multiplying with M^{-1} and get:

$$x_S^{\mathfrak{T}} = \sum_{\{1,\dots,n\}\setminus S \subseteq T \subseteq \{1,\dots,n\}} (-1)^{|T|+|S|+n+1} r_T^{\mathfrak{T}} \qquad (8.28)$$

To verify equation (8.28) we can plug in the values of $x_S^{\mathfrak{T}}$ in equation (8.27). The calculation is elementary and uses only the identity

$$\sum_{i=0}^{n} (-1)^i \binom{n}{i} = \begin{cases} 1 & \text{for } n = 0 \\ 0 & \text{for } n > 0 \end{cases}.$$

In a solution of an extended visual cryptography scheme every $x_S^{\mathfrak{T}}$ must be nonnegative. This leads to

$$\sum_{S \subseteq T \subseteq \{1,\dots,n\}} (-1)^{|S|+|T|} r_T^{\mathfrak{T}} \leq 0 \qquad (8.29)$$

for all possible subsets $\mathfrak{T}$ of $\mathcal{P}(\{1,\dots,n\})\setminus\{\emptyset\}$.

In left-hand side of inequality (8.29) we have the sum over r_T for $|S|+|T|$ even and over $-r_T$ for $|S|+|T|$ odd. This becomes maximal if all r_T with $|S|+|T|$ even are as large as possible (i.e., equal h_T) and all r_t with $|S|+|T|$ odd are as small as possible (i.e., equal l_T).

Thus, an extended visual cryptography scheme exists if and only if

$$\sum_{\substack{S \subseteq T \subseteq \{1,\dots,n\} \\ |S| \equiv |T| \mod 2}} h_T \leq \sum_{\substack{S \subseteq T \subseteq \{1,\dots,n\} \\ |S| \not\equiv |T| \mod 2}} l_T \qquad (8.30)$$

holds for each $S \subsetneq \{1,\dots,n\}$.

The linear program given by equation (8.30) is very simple and can be solved directly.

Theorem 19 ([13] Theorem 3.4) *An extended visual cryptography scheme with n transparencies needs at least $\frac{1}{2}(3^n - 1)$ subpixels. Hence, the construction of Theorem 17 is optimal with respect to the pixel expansion, i.e.,*

$$M(\mathcal{P}(\{1,\dots,n\})\setminus\{\emptyset\}) = \frac{1}{2}(3^n - 1) .$$

Proof
For each $\emptyset \neq T \subseteq \{1,\dots,n\}$, let $\delta_T = h_T - l_T$. Our goal is to prove that equation (8.30) implies

$$m \geq h_{\{1,\dots,n\}} \geq \sum_{\emptyset \neq T \subseteq \{1,\dots,n\}} \delta_T 2^{|T|-1} . \qquad (8.31)$$

As the first step in this proof we show that, for given values δ_T (for $\emptyset \neq$

$T \subseteq \{1, \ldots, n\}$), the number $h_{\{1,\ldots,n\}}$ is minimal if for all $S \subsetneq \{1, \ldots, n\}$ inequality (8.30) is satisfied with equality.

To this end, suppose

$$\sum_{\substack{S \subseteq T \subseteq \{1,\ldots,n\} \\ |S| \equiv |T| \mod 2}} h_T \;<\; \sum_{\substack{S \subseteq T \subseteq \{1,\ldots,n\} \\ |S| \not\equiv |T| \mod 2}} l_T$$

for some $S \subsetneq \{1, \ldots, n\}$. But the contrast levels

$$\bar{h}_T = \begin{cases} h_T & \text{for } T \subseteq S \\ h_T - 1 & \text{otherwise} \end{cases}$$

and

$$\bar{l}_T = \begin{cases} l_T & \text{for } T \subseteq S \\ l_T - 1 & \text{otherwise} \end{cases}$$

satisfy (8.30), since

$$|\{T \mid S \subseteq T \subseteq \{1, \ldots, n\}; |T| \equiv |S| \mod 2\}| =$$
$$|\{T \mid S \subseteq T \subseteq \{1, \ldots, n\}; |T| \not\equiv |S| \mod 2\}| \quad .$$

This proves that if inequality (8.30) is not satisfied with equality we can find smaller values for the parameters l_T and h_T, which also satisfy (8.30). Thus, in an optimal scheme (i.e., a scheme with the smallest possible values for l_T and h_T) inequality (8.30) is satisfied with equality for each $T \subsetneq \{1, \ldots, n\}$.

Next we claim that

$$h_T = \sum_{\emptyset \neq T' \subseteq \{1,\ldots,n\}} \delta_{T'} 2^{|T'|-1} - \sum_{T \subsetneq T' \subseteq \{1,\ldots,n\}} \delta_{T'} 2^{|T'|-1-|T|} \qquad (8.32)$$

for $\emptyset \neq S \subseteq \{1, \ldots, n\}$ satisfy (8.30) with equality.

To prove this we have to show that

$$\sum_{\substack{S \subseteq T \subseteq \{1,\ldots,n\} \\ |S| \equiv |T| \mod 2}} \left[\sum_{\emptyset \neq T' \subseteq \{1,\ldots,n\}} \delta_{T'} 2^{|T'|-1} - \sum_{T \subsetneq T' \subseteq \{1,\ldots,n\}} \delta_{T'} 2^{|T'|-1-|T|} \right] =$$

$$\sum_{\substack{S \subseteq T \subseteq \{1,\ldots,n\} \\ |S| \not\equiv |T| \mod 2}} \left(\left[\sum_{\emptyset \neq T' \subseteq \{1,\ldots,n\}} \delta_{T'} 2^{|T'|-1} - \sum_{T \subsetneq T' \subseteq \{1,\ldots,n\}} \delta_{T'} 2^{|T'|-1-|T|} \right] - \delta_T \right)$$

or equivalently

$$\sum_{\emptyset \neq T' \subseteq \{1,\ldots,n\}} \delta_{T'} \left[\sum_{\substack{S \subseteq T \subseteq \{1,\ldots,n\} \\ |S| \equiv |T| \mod 2}} 2^{|T'|-1} - \sum_{\substack{S \subseteq T \subsetneq T' \\ |S| \equiv |T| \mod 2}} 2^{|T'|-1-|T|} \right] =$$

$$\sum_{\emptyset \neq T' \subseteq \{1,\ldots,n\}} \delta_{T'} \left[\sum_{\substack{S \subseteq T \subseteq \{1,\ldots,n\} \\ |S| \not\equiv |T| \mod 2}} 2^{|T'|-1} - \sum_{\substack{S \subseteq T \subsetneq T' \\ |S| \not\equiv |T| \mod 2}} 2^{|T'|-1-|T|} \right] + \delta_{T'} \frac{(-1)^{|T'|+|S|}-1}{2}.$$

(Note that the last summand is equal to $-\delta_{T'}$ for $|T'| \not\equiv |S| \mod 2$ and equal to 0 otherwise.)

Comparing coefficients for each $\delta_{T'}$ we obtain

$$2^{n-|S|-1} \cdot 2^{|T'|-1} - \sum_{\substack{S \subseteq T \subsetneq T' \\ |S| \equiv |T| \mod 2}} 2^{|T'|-1-|T|} =$$

$$2^{n-|S|-1} \cdot 2^{|T'|-1} - \left(\sum_{\substack{S \subseteq T \subsetneq T' \\ |S| \not\equiv |T| \mod 2}} 2^{|T'|-1-|T|} \right) + \frac{(-1)^{|T'|+|S|}-1}{2}, \quad (8.33)$$

but this is true since

$$(-1)^{|T'|-|S|} = (1-2)^{|T'|-|S|}$$

$$= \sum_{i=0}^{|T'|-|S|} \binom{|T'|-|S|}{i} (-2)^i$$

$$= \sum_{\substack{i=0 \\ i \text{ even}}}^{|T'|-|S|} \binom{|T'|-|S|}{i} 2^i - \sum_{\substack{i=0 \\ i \text{ odd}}}^{|T'|-|S|} \binom{|T'|-|S|}{i} 2^i$$

$$= \sum_{\substack{S \subseteq T \subseteq T' \\ |T'| \equiv |T| \mod 2}} 2^{|T'|-|T|} - \sum_{\substack{S \subseteq T \subseteq T' \\ |T'| \not\equiv |T| \mod 2}} 2^{|T'|-|T|}$$

$$\text{, since } \binom{|T'|-|S|}{i} = \sum_{\substack{S \subseteq T \subseteq T' \\ |T|-|S|=i}} 1.$$

Thus,

$$\sum_{\substack{S \subseteq T \subseteq T' \\ |T'| \equiv |T| \mod 2}} 2^{|T'|-|T|} = \left(\sum_{\substack{S \subseteq T \subseteq T' \\ |T'| \not\equiv |T| \mod 2}} 2^{|T'|-|T|} \right) + (-1)^{|T'|-|S|}$$

and therefore,

$$\left(\sum_{\substack{S \subseteq T \subsetneq T' \\ |T'| \equiv |T| \mod 2}} 2^{|T'|-|T|} \right) + 1 = \left(\sum_{\substack{S \subseteq t3 T \subsetneq T' \\ |T'| \not\equiv |T| \mod 2}} 2^{|T'|-|T|} \right) + (-1)^{|T'|+|S|} .$$

(Note that $(-1)^{|T'|-|S|} = (-1)^{|T'|+|S|}$.) Division by 2 gives

$$\left(\sum_{\substack{S \subseteq T \subsetneq T' \\ |T'| \equiv |T| \mod 2}} 2^{|T'|-1-|T|} \right) = \left(\sum_{\substack{S \subseteq T \subsetneq T' \\ |T'| \not\equiv |T| \mod 2}} 2^{|T'|-1-|T|} \right) + \frac{(-1)^{|T'|+|S|} - 1}{2}$$

as required for equation (8.33).

Suppose that $\bar{h}_T$ (for $\emptyset \neq T \subseteq \{1, \ldots, n\}$) satisfy (8.30) with equality, too.

If inequality (8.30) is satisfied with equality for all subsets S, we can solve these equations recursively and get

$$h_S = h_{\{1,\ldots,n\}} + F_S(\delta_T \mid T \subseteq \{1, \ldots, n\}) ,$$

for some function F_S. Since inequality (8.30) is satisfied with equality for the contrast values h_T and h'_T this yields

$$\bar{h}_S = h_S + \bar{h}_{\{1,\ldots,n\}} - h_{\{1,\ldots,n\}}.$$

But for $S = \emptyset$ inequality (8.30) yields $\bar{h}_{\{1,\ldots,n\}} = h_{\{1,\ldots,n\}}$ and therefore $\bar{h}_T = h_T$ for all $\emptyset \neq T \subseteq \{1, \ldots, n\}$.

This proves that (8.32) is the only solution of (8.30) that satisfies all inequalities with equality.

Thus, we find

$$m \geq h_{\{1,\ldots,n\}} \geq \sum_{\emptyset \neq T' \subseteq \{1,\ldots,n\}} \delta_{T'} 2^{|T'|-1} \geq \sum_{\emptyset \neq T' \subseteq \{1,\ldots,n\}} 2^{|T'|-1} = \frac{1}{2}(3^n - 1)$$

what proves the theorem. $\square$

Without proof we mention the following result about the contrast of the different images.

Result 20 ([13] Theorem 3.6) *For $\emptyset \neq T \subseteq \{1, \ldots, n\}$ let $\alpha_T = \frac{h_T - l_T}{m}$ be the contrast of the image I_T. The contrast levels of the images satisfy*

$$\sum_{\emptyset \neq T \subseteq \{1,\ldots,n\}} 2^{|T|-1} \alpha_T \leq 1 \quad . \tag{8.34}$$

Further, let $\alpha'_T \geq 0$ (for $\emptyset \neq T \subseteq \{1, \ldots, n\}$) satisfy (8.34). Then for every $\varepsilon > 0$ there exists a generalized visual cryptography scheme with contrast levels α_T (for $\emptyset \neq T \subseteq \{1, \ldots, n\}$) where $|\alpha_T - \alpha'_T| < \varepsilon$ for all nonempty subsets T of $\{1, \ldots, n\}$.

The situation becomes more complicated if we no longer require that all combinations of transparencies reveal an image. The only case that is completely solved is Result 21.

Result 21 ([13] Theorem 4.4) *Let* $\mathfrak{S} = \mathcal{P}(\{1, \ldots, n\}) \backslash \{\emptyset, \{1, \ldots, n\}\}$ *then the minimal pixel expansion* m *is given by*

$$
m = \begin{cases} \frac{1}{2}(3^n - 1) - 2^{n-1} - 1 & \text{for } n \ \text{even} \\ \frac{1}{2}(3^n - 1) - 2^{n-1} & \text{for } n \ \text{odd} \end{cases} .
$$

8.8 Enhancing the Contrast by Nonstandard Models

All results in this chapter worked with the original model of visual cryptography as defined in [16]. In several papers extended models were discussed. Without going into details we mention:

- In [19] the authors run an edge-detecting algorithm on the original image and use a higher number of subpixels on the edges for improved contrast.

- Several authors studied colored visual cryptography (see [6, 17, 20, 7] and Chapter 7 of [11]). One can use additive and subtractive color mixing in combination to obtain good results. A possible variation of the classical 2-out-of-2 scheme replace the white and black subpixels by any two opposing colors.

- It was also suggested to use polarization filters of different orientations to encrypt an image.

- A interesting variation is the segment based visual cryptography [4]. For a 7-segment display an image may be encoded as:

- In [10] the authors note that the contrast definition $\alpha = \frac{h-l}{m}$ of Naor and Shamir do not always fit the human intuition. They suggest variations like $\frac{h-l}{h+l}$ and show how to optimize visual cryptography schemes with respect to the new goal.

8.9 Conclusion

We constructed contrast optimal visual cryptography schemes for various access structures. The constructions exhibit links to different fields like linear programming, approximation theory, the theory of error correcting codes, design theory, and elementary combinatorics.

━━━━━━━━━━

Bibliography

[1] T. Beth, D. Jungnickel, and H. Lenz. *Design Theory*. Cambridge University Press, 1985.

[2] C. Blundo, P. D'Arco, A. De Santis, and D. R. Stinson. Contrast optimal threshold visual cryptography schemes. *SIAM J. Discrete Math.*, 16(2):224–261 (electronic), 2003.

[3] C. Blundo, A. De Santis, and D. R. Stinson. On the contrast in visual cryptography schemes. *J. Cryptology*, 12(4):261–289, 1999.

[4] B. Borchert. Segment-based visual cryptography. Technical report, 17 Wilhelm-Schickard-Institut für Informatik, Univeristät Tübingen, 2007. http://w210.ub.uni-tuebingen.de/volltexte/2007/3048/.

[5] M. Bose and R. Mukerjee. Optimal (n, k) visual cryptography schemes for general k. *Des. Codes Cryptogr.*, 55:19–35, 2010.

[6] S. Cimato, R. De Prisco, and A. De Santis. Optimal colored threshold visual cryptography schemes. *Des. Codes Cryptogr.*, 35(3):311–335, 2005.

[7] S. Cimato, R. De Prisco, and A. De Santis. Colored visual cryptography without color darkening. *Theoret. Comput. Sci.*, 374(1-3):261–276, 2007.

[8] R. Craigen. Hadamard matrices and designs. In C. J. Colbourn and J. H. Dinitz, editors, *The CRC Handbook of Combinatorial Designs*, pages 370–380. CRC Press, Boco Raton, New York, London, Tokyo, 1996.

[9] S. Droste. New results in visual cryptography. In *Advances in cryptology – CRYPTO '96*, volume 1109 of *Lect. Notes Comput. Sci.*, pages 401–415. Springer, Berlin, 1996.

[10] P. A. Eisen and D. R. Stinson. Threshold visual cryptography schemes with specified whiteness levels of reconstructed pixels. *Des. Codes Cryptogr.*, 25(1):15–61, 2002.

[11] A. Klein. *Visuelle Kryptographie*. Springer-Verlag, Berlin, 2007.

[12] A. Klein and K. Metsch. On approximate inclusion-exclusion. *Innov. Incidence Geom.*, 6/7:249–270, 2007/08.

[13] A. Klein and M. Wessler. Extended visual cryptography schemes. *Information and Computation*, 205(5):716–732, 2007.

[14] M. Krause and H. U. Simon. Determining the optimal contrast for secret sharing schemes in visual cryptography. *Combin. Probab. Comput.*, 12(3):285–299, 2003. Combinatorics, probability and computing (Oberwolfach, 2001).

[15] N. Linial and N. Nisan. Approximate inclusion-exclusion. *Combinatorica*, 10(4):349–365, 1990.

[16] M. Naor and A. Shamir. Visual cryptography. In Alfredo De Santis, editor, *Advances in cryptology - EUROCRYPT '94*, volume 950 of *Lect. Notes Comput. Sci.*, pages 1–12. Springer-Verlag, 1995.

[17] E. R. Verheul and H. C. A. van Tilborg. Constructions and properties of k out of n visual secret sharing schemes. *Des. Codes Cryptogr.*, 11(2):179–196, 1997.

[18] G. A. Watson. *Approximation Theory and Numerical Methods*. John Wiley & Sons, 1980.

[19] C.-N. Yang and T.-S. Chen. Visual secret sharing scheme: Improving the contrast of a recovered image via different pixel expansions. In A. Campilho and M. Kamel, editors, *ICIAR 2006*, volume 4141 of *LNCS*, pages 468–479, Berlin Heidelberg, 2006. Springer-Verlag.

[20] C.-N. Yang and C.-S. Laih. New colored visual secret sharing schemes. *Des. Codes Cryptogr.*, 20(3):325–336, 2000.

9

Visual Cryptography Schemes with Reversing

Alfredo De Santis

Università di Salerno, Italy

Anna Lisa Ferrara

Università di Salerno, Italy

Barbara Masucci

Università di Salerno, Italy

CONTENTS

9.1 Introduction

A visual cryptography scheme for a set of n participants is a method to encode a secret image, consisting of black and white pixels, into n shadow images called *shares*, one for each participant. Each share is a collection of black and white subpixels, which are printed in close proximity to each other, so that the human visual system averages their individual black/white contributions. The encoding is done in such a way that certain subsets of participants, called *qualified sets*, can "visually" recover the secret image, but other subsets of participants, called *forbidden sets*, cannot gain any information (in an information-theoretic sense) about the secret image by inspecting their shares. A "visual" recover for a set of qualified participants consists of xeroxing each share onto a separate transparency, of stacking together the transparencies and projecting the result with an overhead projector. If the transparencies are aligned carefully, then the participants will be able to see the secret image (without any knowledge of cryptography and without performing any cryptographic computation).

This cryptographic paradigm was introduced by Naor and Shamir [11]. They analyzed the case of (k, n)-threshold visual cryptography schemes, in which the qualified subsets of participants have cardinality k, whereas, the forbidden subsets of participants have cardinality less than k. Some results on (k, n)-threshold visual cryptography schemes $((k, n)$-threshold visual cryptography scheme (VCS), for short) can be found in [1, 2, 5, 8, 10, 12]. The model by Naor and Shamir has been extended in [1] to general access structures (an access structure is a specification of all qualified and forbidden subsets of participants), where general techniques to construct visual cryptography schemes for any access structure have been proposed.

Visual cryptography schemes are characterized by two parameters: The *pixel expansion*, corresponding to the number of subpixels contained in each share (transparency) and the *contrast*, which measures the "difference" between a black and a white pixel in the reconstructed image. Visual cryptography schemes such that, in the reconstructed image, all the subpixels associated to a black pixel are black, are referred to as *visual cryptography schemes with perfect reconstruction of black pixels*. Such schemes have been considered in [12, 4, 3]. Unfortunately, it is not possible to obtain visual cryptography schemes with perfect reconstruction of both black and white pixels. Such schemes are said to have *ideal contrast*.

In order to obtain perfect black visual cryptography schemes whose reconstruction of white pixels is *almost* perfect, Viet and Kurosawa [13] proposed a different kind of VCS, called *VCS with reversing*. Such VCSs require the introduction of an extra noncryptographic operation, that participants can use to reconstruct the image. Such an operation, which can be easily performed by many copy machines, is applied to a transparency and creates another transparency in which black pixels are reversed to white pixels and vice versa.

Specifically, Viet and Kurosawa [13] proposed to run c times (with c arbitrary constant) the distribution phase of a VCS with perfect reconstruction of black pixels and pixel expansion m, hence requiring each participant to store $c \cdot m$ subpixels for each pixel of the original image. The larger the number of runs c, the better the contrast of the resulting VCS with reversing. The drawback of such a solution is that the number of pixels in the reconstructed image, which depends on both the number of runs and on the pixel expansion of the underlying VCS, is greater than that in the original secret image, i.e., there is a loss of resolution.

Subsequently, Cimato, De Santis, Ferrara, and Masucci [6] showed how to construct VCSs with reversing where reconstruction of *both* black and white pixels is perfect. In particular, Cimato et al. [6] proposed two different constructions. The first solution uses the fact that the introduction of the reversing operation, in addition to the stacking operation, allows participants to compute any Boolean function of their transparencies, since these two operations, corresponding to NOT and OR, respectively, represent a complete basis for Boolean functions. In particular, the construction uses a binary secret sharing scheme and requires each participant to hold r transparencies, where r denotes the size of the largest share in the underlying secret sharing scheme. The second solution uses as a building block a VCS with perfect reconstruction of black pixels, having a certain pixel expansion m and requires each participant to store m transparencies, each having the same number of pixels as the original image. Compared to the scheme of Viet and Kurosawa, such a scheme requires each participant to store m pixels instead than $c \cdot m$, for each pixel of the original image, where c is the number of runs of the underlying VCS needed in [13] to obtain an asymptotically ideal contrast. By using a sequence of stacking and reversing operations on their transparencies, in both schemes proposed by Cimato et al. [6], each qualified set of participants recover the original secret image with *no loss of resolution*.

Later, Hu and Tzeng [9] considered the problem of reducing the number of the transparencies held by each participant in VCSs with reversing. They proposed a construction for ideal contrast VCSs with reversing requiring each participant to store only two transparencies. In particular, the first one contains an encoding of the secret image, whereas, the second one is an auxiliary transparency needed for the reconstruction phase. However, the size of each transparency is $|\Gamma_0|$ times larger than the size of the secret image, where $|\Gamma_0|$ denotes the minimum number of subsets qualified to reconstruct the secret image. Indeed, each transparency contains $|\Gamma_0|$ blocks and each qualified subset of participants reconstructs, without loss of resolution, the secret image in a single block, whereas the other reconstructed blocks contain only white pixels.

Then, Yang et al. [14] proposed two different constructions for ideal contrast VCS with reversing. In particular, one of their schemes removes the need of using as a building block a VCS with perfect reconstruction of black pixels. However, there is a loss of resolution in both their schemes.

Organization. In Section 9.2 we recall the definition and security requirements of VCSs. In Section 9.3 we describe the almost ideal contrast VCS with reversing by Viet and Kurosawa. In Section 9.4 we show and compare some ideal contrast VCSs with reversing that use as a building block any perfect black VCS. Finally, a VCS with reversing constructed upon any VCS is described in Section 9.4.3.

9.2 Visual Cryptography Schemes

Let $\mathcal{P} = \{1, \ldots, n\}$ be a set of elements called *participants*, and let $2^{\mathcal{P}}$ denote the set of all subsets of $\mathcal{P}$. Let $\Gamma_{\mathsf{Qual}} \subseteq 2^{\mathcal{P}}$ and $\Gamma_{\mathsf{Forb}} \subseteq 2^{\mathcal{P}}$, where $\Gamma_{\mathsf{Qual}} \cap \Gamma_{\mathsf{Forb}} = \emptyset$. We refer to members of Γ_{Qual} as *qualified sets* and we call members of Γ_{Forb} *forbidden sets*. The pair $(\Gamma_{\mathsf{Qual}}, \Gamma_{\mathsf{Forb}})$ is called the *access structure* of the scheme. Define Γ_0 to consist of all the minimal qualified sets: $\Gamma_0 = \{X \in \Gamma_{\mathsf{Qual}} : X' \notin \Gamma_{\mathsf{Qual}} \text{ for all } X' \subset X\}$. In the case where Γ_{Qual} is monotone increasing, Γ_{Forb} is monotone decreasing, and $\Gamma_{\mathsf{Qual}} \cup \Gamma_{\mathsf{Forb}} = 2^{\mathcal{P}}$, the access structure is said to be *strong*. In a strong access structure, $\Gamma_{\mathsf{Qual}} = \{C \subseteq \mathcal{P} : B \subseteq C \text{ for some } B \in \Gamma_0\}$, and we say that Γ_{Qual} is the *closure* of Γ_0. A participant $i \in \mathcal{P}$ is an *essential* participant for $(\Gamma_{\mathsf{Qual}}, \Gamma_{\mathsf{Forb}})$ if there exists a set $X \subseteq \mathcal{P}$ such that $X \cup \{i\} \in \Gamma_{\mathsf{Qual}}$ but $X \notin \Gamma_{\mathsf{Qual}}$. If a participant is not essential then we can construct a visual cryptography scheme giving him nothing as his share. In fact, a nonessential participant does not need to participate actively in the reconstruction of the image, since the information he has is not needed by any set in $\mathcal{P}$ in order to recover the secret image. Therefore, unless otherwise specified, we assume throughout this chapter that all participants are essential.

The secret image consists of black and white[1] pixels. In order to share each pixel of the secret image the owner of the secret, usually called the *dealer*, provides each participant with a *share* (transparency), which is an enlarged version of the secret pixel consisting of a certain number m of subpixels, which are printed in close proximity to each other, so that the human visual system averages their individual black/white contributions. Notice that the term "subpixel" is misleading since a pixel is the smallest unit we can control on an image and thus we cannot further divide the pixel into subpixels. So the shared version of the original secret pixel will consist of m pixels, which are called *subpixels* because all together they represent the original secret pixel. The value m is called *pixel expansion*. The shares can be conveniently represented with an $n \times m$ matrix S where each row represents one share, i.e., m subpixels, and each element is either 0, for a white subpixel, or 1 for a black subpixel.

To reconstruct the secret image a group of participants stacks together their shares. The grey level of the combined share, obtained by stacking the transparencies $i_1, \ldots, i_s$, is proportional to the Hamming weight $w(V)$ of the m-vector $V = OR(r_{i_1}, \ldots, r_{i_s})$, where $r_{i_1}, \ldots, r_{i_s}$ are the rows of S associated with the transparencies we stack. This grey level is interpreted by the visual system of the users as black or as white in accordance with some rule of contrast. Since each secret pixel is represented by m pixels in the shares, the reconstructed image will be bigger than the original (depending on m

---

[1] Where white should really be interpreted as transparent. So we use white as a synonym for transparent.

and on the actual positions of the pixels, the image can also be distorted; a perfect square is a good choice for m because it avoids distortion). The characteristics of the model are encapsulated in the following definition, which is a generalization of the definition of (k, n)-threshold VCSs due to [12].

Definition 1 *Let* $(\Gamma_{\mathsf{Qual}}, \Gamma_{\mathsf{Forb}})$ *be an access structure on a set of n participants. Two collections (multisets) of $n \times m$ Boolean matrices $\mathcal{C}_0$ and $\mathcal{C}_1$ constitute a* visual cryptography scheme $(\Gamma_{\mathsf{Qual}}, \Gamma_{\mathsf{Forb}})$-VCS *with pixel expansion m if there exist two integers ℓ and h such that $h > \ell$ satisfying:*

> *1. Any (qualified) set $X = \{i_1, i_2, \ldots, i_p\} \in \Gamma_{\mathsf{Qual}}$ can recover the shared image by stacking their transparencies.*
> *Formally, for any $M \in \mathcal{C}_0$, the "or" V of rows $i_1, i_2, \ldots, i_p$ satisfies $w_H(V) \leq m - h$; whereas, for any $M \in \mathcal{C}_1$ it results that $w_H(V) \geq m - \ell$.*

> *2. Any (forbidden) set $X = \{i_1, i_2, \ldots, i_p\} \in \Gamma_{\mathsf{Forb}}$ has no information on the shared image.*
> *Formally, the two collections of $p \times m$ matrices $\mathcal{D}_t$, with $t \in \{0, 1\}$, obtained by restricting each $n \times m$ matrix in $\mathcal{C}_t$ to rows $i_1, i_2, \ldots, i_p$ are indistinguishable in the sense that they contain the same matrices with the same frequencies.*

To share a white (black, resp.) pixel, the dealer randomly chooses one of the matrices in $\mathcal{C}_0$ ($\mathcal{C}_1$, resp.), and distributes row i to participant i. The chosen matrix defines the m subpixels in each of the n transparencies.

The first property of Definition 1 is related to the *contrast* of the image. It states that when a qualified set of participants stack their transparencies, they can correctly recover the image shared by the dealer. A pixel will be seen as a white pixel if sufficiently many subpixels (at least h) in the reconstructed image are white; whereas, it will be seen as a black one if not too many (at most ℓ) are white. The value $(h - \ell)/(h + \ell)$ is referred to as the *contrast* of the reconstructed image. Contrast gives a measurement of how clear the reconstructed image is in relation to the original one. The second property of Definition 1 is related to the security of the scheme, since it implies that, even by inspecting all their shares, any forbidden set of participants cannot gain any information in deciding whether the shared pixel was white or black.

Several visual cryptography schemes have been realized by using two $n \times m$ matrices, S^0 and S^1, called *basis matrices*. The collections $\mathcal{C}_0$ and $\mathcal{C}_1$ are obtained by permuting the columns of the corresponding basis matrix (S^0 for $\mathcal{C}_0$, and S^1 for $\mathcal{C}_1$) in all possible ways. This technique has been introduced in [11].

9.2.1 (k, k)-Threshold Visual Cryptography Schemes

In a (k, k)-threshold visual cryptography scheme the secret image is visible if and only if all k transparencies are stacked together, but totally invisible if fewer than k transparencies are stacked together or analyzed by any other method.

Naor and Shamir [11] proposed a construction for (k, k)-threshold VCSs using two basis matrices S^0 and S^1 defined as follows: S^0 is the matrix whose columns are all the Boolean k-vectors having an even number of 1's, and S^1 is the matrix whose columns are all the Boolean k-vectors having an odd number of 1's. The pixel expansion m of such a scheme is equal to 2^{k-1} and the relative difference $(h - \ell)/m$ between reconstructed white and reconstructed black is equal to $1/2^{k-1}$. The above construction is optimal with respect to the pixel expansion and the relative difference between reconstructed white and reconstructed black.

Example 1 *The basis matrices S^0 and S^1 in a $(4, 4)$-threshold VCS are*

$$S^0 = \begin{bmatrix} 0 & 0 & 0 & 0 & 1 & 1 & 1 & 1 \\ 0 & 0 & 1 & 1 & 0 & 0 & 1 & 1 \\ 0 & 1 & 0 & 1 & 0 & 1 & 0 & 1 \\ 0 & 1 & 1 & 0 & 1 & 0 & 0 & 1 \end{bmatrix} \quad S^1 = \begin{bmatrix} 0 & 0 & 0 & 0 & 1 & 1 & 1 & 1 \\ 0 & 0 & 1 & 1 & 0 & 0 & 1 & 1 \\ 0 & 1 & 0 & 1 & 0 & 1 & 0 & 1 \\ 1 & 0 & 0 & 1 & 0 & 1 & 1 & 0 \end{bmatrix}.$$

The collections $\mathcal{C}_0$ and $\mathcal{C}_1$ are obtained by permuting the columns of the corresponding basis matrix (S^0 for $\mathcal{C}_0$, and S^1 for $\mathcal{C}_1$) in all possible ways. In this scheme each pixel of the secret image is encoded into $m = 8$ subpixels.

It is easy to see that the integers $h = 1$ and $\ell = 0$ satisfy Property 1 of Definition 1. Let S^0 be the matrix chosen by the dealer to share a white pixel; by stacking the transparencies held by all four participants we get the vector $(0, 1, 1, 1, 1, 1, 1, 1)$. On the other hand, let S^1 be the matrix chosen by the dealer to share a black pixel; by stacking the transparencies held by all four participants we get the vector $(1, 1, 1, 1, 1, 1, 1, 1)$. Property 2 of Definition 1 can also be easily verified. Indeed, consider what happens when less than four participants stack their together transparencies. For example, consider the first three participants and notice that, by stacking their shares, we get the vector $(0, 1, 1, 1, 1, 1, 1, 1)$ in both cases when the shared pixel is either white or black. Thus, the participants are not able to distinguish the color of the shared pixel by inspecting their shares.

9.2.2 Perfect Black Visual Cryptography Schemes

Visual cryptography schemes such that, in the reconstructed image, all the subpixels associated with a black pixel are black, i.e., $\ell = 0$, are referred either to as *visual cryptography schemes with perfect reconstruction of black pixels* or to as *maximal contrast schemes*. For example, the (k, k)-threshold VCS described in Section 9.2.1 has perfect reconstruction of black pixels. A perfect reconstruction of both black and white pixels, which would correspond to visual cryptography schemes having *ideal contrast*, is impossible. Indeed, in [11] it has been shown that any (k, k)-threshold VCS has a pixel expansion of at least 2^{k-1} and the relative difference $(h - \ell)/m$ between reconstructed white and reconstructed black is at most $1/2^{k-1}$. This means that, in the reconstructed white pixel, for any 2^{k-1} subpixels, there is at most a white subpixel.

Visual cryptography schemes with perfect reconstruction of black pixels have been analyzed in [12, 4, 3]. In particular, in [3] it was shown how to construct $(\Gamma_{\mathsf{Qual}}, \Gamma_{\mathsf{Forb}})$-VCSs with perfect reconstruction of black pixels having a pixel expansion $m = \sum_{X \in \Gamma_{\mathsf{Qual}}} 2^{|X|-1}$, by using a technique proposed in [1]. More precisely, such a technique to constructs visual cryptography schemes with a perfect reconstruction of black pixels using small schemes as building blocks in the construction of larger schemes, as explained in the following. For $i = 1, \ldots, q$, let $(\Gamma_{\mathsf{Qual}}^i, \Gamma_{\mathsf{Forb}}^i)$ be an access structure on a set $\mathcal{P}$ of n participants. If a participant $j \in \mathcal{P}$ is not essential for the i-th access structure, we assume that $j \notin \Gamma_{\mathsf{Forb}}^i$ and that j does not receive any share. Suppose there exists a $(\Gamma_{\mathsf{Qual}}^i, \Gamma_{\mathsf{Forb}}^i)$-VCS with a pixel expansion m_i and basis matrices T_i^0 and T_i^1, for each $i = 1, \ldots, q$. The basis matrix S^0 (S^1, resp.) of a VCS for the access structure $(\Gamma_{\mathsf{Qual}}, \Gamma_{\mathsf{Forb}})$ where $\Gamma_{\mathsf{Qual}} = \Gamma_{\mathsf{Qual}}^1 \cup \ldots \cup \Gamma_{\mathsf{Qual}}^q$ and $\Gamma_{\mathsf{Forb}} = \Gamma_{\mathsf{Forb}}^1 \cap \ldots \cap \Gamma_{\mathsf{Forb}}^q$ is constructed as the concatenation of some auxiliary matrices $\hat{T}_i^0$ ($\hat{T}_i^1$, resp.), for each $i = 1, \ldots, q$. Such matrices are obtained as follows: for each $j = 1, \ldots, n$, the j-th row of $\hat{T}_i^0$ ($\hat{T}_i^1$, resp.) has all ones as entries if the participant j is not essential for $(\Gamma_{\mathsf{Qual}}^i, \Gamma_{\mathsf{Forb}}^i)$, otherwise it is the row of T_i^0 (T_i^1, resp.) corresponding to participant j. Hence, $S^0 = \hat{T}_1^0 \circ \hat{T}_2^0 \circ \ldots \circ \hat{T}_q^0$ and $S^1 = \hat{T}_1^1 \circ \hat{T}_2^1 \circ \ldots \circ \hat{T}_q^1$, where $\circ$ denotes the *concatenation* of matrices. The resulting VCS has a pixel expansion $m = \sum_{i=1}^q m_i$. For a special class of access structures, such as threshold access structures and graph-based access structures, it is possible to design VCSs with the perfect reconstruction of black pixels achieving a smaller value of m, as shown in [1, 12, 3].

Example 2 *Let* $\mathcal{P} = \{1, 2, 3, 4\}$ *and* $\Gamma_0 = \left\{ \{1, 2\}, \{2, 3\}, \{3, 4\} \right\}$. *We can construct a VCS with a perfect reconstruction of black pixels for the access structure having basis* Γ_0 *by using three* $(2, 2)$*-threshold VCSs on the sets of participants* $\{1, 2\}, \{2, 3\}$, *and* $\{3, 4\}$. *The basis matrices* T^0 *and* T^1 *for the* $(2, 2)$*-threshold VCS described in Section 9.2.1 are*

$$T^0 = \begin{bmatrix} 1 & 0 \\ 1 & 0 \end{bmatrix} \quad T^1 = \begin{bmatrix} 1 & 0 \\ 0 & 1 \end{bmatrix}.$$

From T^0 *and* T^1 *we construct the pair of matrices* $(\hat{T}_i^0, \hat{T}_i^1)$, *for* $i = 1, \ldots, 3$ *having four rows each, as follows:*

$$\hat{T}_1^0 = \begin{bmatrix} 1 & 0 \\ 1 & 0 \\ 1 & 1 \\ 1 & 1 \end{bmatrix} \quad \hat{T}_1^1 = \begin{bmatrix} 1 & 0 \\ 0 & 1 \\ 1 & 1 \\ 1 & 1 \end{bmatrix},$$

$$\hat{T}_2^0 = \begin{bmatrix} 1 & 1 \\ 1 & 0 \\ 1 & 0 \\ 1 & 1 \end{bmatrix} \qquad \hat{T}_2^1 = \begin{bmatrix} 1 & 1 \\ 1 & 0 \\ 0 & 1 \\ 1 & 1 \end{bmatrix},$$

$$\hat{T}_3^0 = \begin{bmatrix} 1 & 1 \\ 1 & 1 \\ 1 & 0 \\ 1 & 0 \end{bmatrix} \qquad \hat{T}_3^1 = \begin{bmatrix} 1 & 1 \\ 1 & 1 \\ 1 & 0 \\ 0 & 1 \end{bmatrix}.$$

Concatenating the matrices $\hat{T}_1^0, \hat{T}_2^0,$ and $\hat{T}_3^0$ we obtain the basis matrix S^0

$$S^0 = \begin{bmatrix} 1 & 0 & 1 & 1 & 1 & 1 \\ 1 & 0 & 1 & 0 & 1 & 1 \\ 1 & 1 & 1 & 0 & 1 & 0 \\ 1 & 1 & 1 & 1 & 1 & 0 \end{bmatrix},$$

whereas, concatenating the matrices $\hat{T}_1^1, \hat{T}_2^1,$ and $\hat{T}_3^1$ we obtain the basis matrix S^1

$$S^1 = \begin{bmatrix} 1 & 0 & 1 & 1 & 1 & 1 \\ 0 & 1 & 1 & 0 & 1 & 1 \\ 1 & 1 & 0 & 1 & 1 & 0 \\ 1 & 1 & 1 & 1 & 0 & 1 \end{bmatrix}.$$

The pixel expansion of the above construction is $m = \sum_{X \in \Gamma_0} 2^{|X|-1} = 6$.

9.3   Almost Ideal Contrast VCS with Reversing

In order to improve the contrast in VCSs, Viet and Kurosawa [13] introduced another noncryptographic operation, called *reversing*, which can be used by participants in the reconstruction phase. Such an operation, which can be easily performed by many copy machines, is applied to a transparency and creates another transparency in which black pixels are reversed to white pixels and viceversa. In the following, we will denote by $\bar{t}$, the transparency obtained after applying the reversing operation to the transparency t.

Viet and Kurosawa [13] showed how to construct a $(\Gamma_{\text{Qual}}, \Gamma_{\text{Forb}})$-VCS with reversing, where black pixels are perfectly reconstructed, whereas, white pixels are almost perfectly reconstructed. The idea behind their construction is to run several times the distribution phase of a $(\Gamma_{\text{Qual}}, \Gamma_{\text{Forb}})$-VCS with a perfect reconstruction of black pixels. The larger the number c of runs, the better the contrast of the resulting VCS with reversing. In particular, Viet and Kurosawa showed that if the contrast of the underlying VCS is $q < 1$, then the contrast in their VCS with reversing is on the average q^c. The price to pay to get an *almost ideal contrast* is the number of transparencies stored by each participant, which corresponds to the number c of runs. Moreover, there is a loss of resolution in their scheme, since each pixel in the original image corresponds to m subpixels in the reconstructed image, where m denotes the pixel expansion of the underlying VCS.

In the distribution phase of the VCS with reversing in [13], the encoding of the secret image is handled pixel by pixel, where each pixel is considered independently of the others. For each pixel of the original image, the dealer runs c times independently a $(\Gamma_{\mathsf{Qual}}, \Gamma_{\mathsf{Forb}})$-VCS with a perfect reconstruction of black pixels and pixel expansion m; we denote by s_i^ℓ the share for participant i in run ℓ, for $i = 1, \ldots, n$ and $\ell = 1, \ldots, c$ (notice that the transparency corresponding to such a share contains m subpixels for each pixel of the original image). In the reconstruction phase, any qualified set of participants can recover the original secret image by performing a sequence of stacking and reversing operations on their transparencies. The construction is described in Figure 9.1.

Assume there exists a $(\Gamma_{\mathsf{Qual}}, \Gamma_{\mathsf{Forb}})$-VCS with a perfect reconstruction of black pixels.

Distribution phase. For each pixel of the secret image, the dealer:

- runs c times independently of the distribution phase of the underlying $(\Gamma_{\mathsf{Qual}}, \Gamma_{\mathsf{Forb}})$-VCS; let s_i^ℓ be the share distributed to participant i in the ℓ-th run, for $i = 1, \ldots, n$ and $\ell = 1, \ldots, c$;

- distributes the c-tuple $(s_i^1, \ldots, s_i^c)$ to participant i.

Reconstruction phase. A qualified set $\{i_1, \ldots, i_p\}$ of participants reconstructs the secret pixel as follows:

- superimpose their shares to get $\alpha^\ell = OR(s_{i_1}^\ell, \ldots, s_{i_p}^\ell)$, for $\ell = 1, \ldots, c$;

- reverse the results of the previous step to obtain $\overline{\alpha}^\ell$, for $\ell = 1, \ldots, c$;

- superimpose the results of the previous step to get $\beta = OR(\overline{\alpha}^1, \ldots, \overline{\alpha}^c)$;

- finally, reverse the result of the previous step to obtain $\overline{\beta}$, which is the reconstructed pixel.

FIGURE 9.1

A construction for almost ideal contrast $(\Gamma_{\mathsf{Qual}}, \Gamma_{\mathsf{Forb}})$-VCS with reversing.

Notice that in the VCS with reversing described in Figure 9.1, the computation of the reconstructed pixel $\overline{\beta}$ corresponds to performing $c - 1$ AND operations, since $\overline{\beta} = \overline{OR(\overline{\alpha}^1, \ldots, \overline{\alpha}^c)} = AND(\alpha^1, \ldots, \alpha^c)$. From the truth table of the AND operator, it follows that the reconstructed pixel $\overline{\beta}$ will be black only when $\alpha^1, \ldots, \alpha^c$ are all black. Since each α^ℓ corresponds to a reconstructed pixel in the ℓ-th run of the underlying VCS with a perfect reconstruction of black pixels, it follows that in the case a black pixel has been shared by the dealer, the reconstructed pixel $\overline{\beta}$ will be black no matter how many AND operations have been performed. On the other hand, in the case a white pixel has been shared, the reconstructed pixel $\overline{\beta}$ will increasily become whiter with the execution of the runs. Thus, if the contrast of the underlying VCS with a perfect reconstruction of black pixels is $q < 1$, then the contrast in the VCS with reversing described in Figure 9.1 is on the average q^c.

The security of the scheme directly follows from the security of the underlying VCS.

It is easy to see that the reconstruction phase requires a qualified set of p participants to perform exactly $c + 1$ reversing operations and $cp - 1$ stacking operations. Finally, notice that in the construction of Figure 9.1 there is a loss of resolution, since each pixel in the original image corresponds to m subpixels in the reconstructed image, where m denotes the pixel expansion of the underlying VCS. For example, in the (k, k)-threshold VCS with reversing resulting from the above construction, each pixel of the original image corresponds to 2^{k-1} subpixels in the reconstructed image.

Example 3 *Let* $\mathcal{P} = \{1, 2, 3, 4\}$ *and* $\Gamma_0 = \left\{ \{1, 2\}, \{2, 3\}, \{3, 4\} \right\}$. *The basis matrices* S^0 *and* S^1 *in a VCS realizing the access structure* Γ_{Qual} *whose basis is* Γ_0 *are:*

$$S^0 = \begin{bmatrix} 0 & 1 & 1 & 0 \\ 0 & 1 & 1 & 1 \\ 0 & 1 & 1 & 1 \\ 0 & 1 & 0 & 1 \end{bmatrix} \qquad S^1 = \begin{bmatrix} 1 & 0 & 0 & 1 \\ 1 & 1 & 1 & 0 \\ 1 & 1 & 0 & 1 \\ 1 & 0 & 1 & 0 \end{bmatrix}.$$

The collections $\mathcal{C}_0$ *and* $\mathcal{C}_1$ *are obtained by permuting the columns of the corresponding basis matrix (S^0 for $\mathcal{C}_0$, and S^1 for $\mathcal{C}_1$) in all possible ways. First, consider two rounds where the shares generated for participant 1 are* $s_1^1 = 0110$, $s_1^2 = 1100$ *whereas, the shares generated for the participant 2 are* $s_2^1 = 0111$, $s_2^2 = 1101$. *During the reconstruction phase, the two participants stack their shares and retrieve* $\alpha^1 = OR(s_1^1, s_2^1) = 0111$, $\alpha^2 = OR(s_1^2, s_2^2) = 1101$. *By applying the reversing operation to each* α^j *and stacking the results, they obtain* $\beta = OR(\overline{0111}, \overline{1101}) = 1010$. *By reversing* β, *they obtain* $\overline{\beta} = 0101$, *and reconstruct a white pixel that consists of two out of four white subpixels. Now, let's add a third round. Assume that the extra shares for participants 1 and 2 are* $s_1^3 = 1001$ *and* $s_2^3 = 1011$, *respectively. During the reconstruction phase, the two participants stack their shares and retrieve* $\alpha^1 = OR(s_1^1, s_2^1) = 0111$, $\alpha^2 = OR(s_1^2, s_2^2) = 1101$, $\alpha^3 = OR(s_1^3, s_2^3) = 1011$. *By applying the reversing operation to each* α^j *and stacking the results, they obtain* $\beta = OR(\overline{0111}, \overline{1101}, \overline{1011}) = 1110$. *By reversing* β, *they obtain* $\overline{\beta} = 0001$, *and reconstruct a white pixel that consists of three out of four white* *subpixels. Then, with an extra round the whiteness of the reconstructed pixel increases.*

Let S^1 be the matrix chosen by the dealer to share a black pixel. Consider two rounds. The shares generated for participant 1 are $s_1^1 = 1001$, $s_1^2 = 1100$, whereas, the shares generated for the participant 2 are $s_2^1 = 1110$, $s_2^2 = 0111$. During the reconstruction phase, the two participants stack their shares and retrieve $\alpha^1 = OR(s_1^1, s_2^1) = 1111$, $\alpha^2 = OR(s_1^2, s_2^2) = 1111$. By applying the reversing operation to each α^j and stacking the results, they obtain $\beta = OR(\overline{1111}, \overline{1111}) = 0000$. By reversing β, they obtain $\overline{\beta} = 1111$, and perfectly reconstruct a black pixel. It is easy to see that, by increasing the amount of rounds, the reconstruction of a black pixel continues to be perfect.

9.4 Ideal Contrast VCS with Reversing

The scheme proposed by Viet and Kurosawa [13] is said to have an *almost ideal contrast* because the black pixels are perfectly reconstructed, whereas, the white ones are *almost* perfectly reconstructed. The larger the number of runs, the more the whiteness of the reconstructed white pixels. However, a perfect reconstruction of both black and white pixels, corresponding to an *ideal contrast*, cannot be achieved by their scheme, even for a large number of runs. Moreover, the scheme proposed by Viet and Kurosawa [13] also has the following drawbacks:

- Each participant is required to store c transparencies, where c denotes the number of runs for the underlying VCS;

- The size of each transparency is m times the size of the original image, where m denotes the pixel expansion of the underlying VCS;

- There is a loss of resolution in the reconstructed image;

- A large number of runs is required to obtain an almost ideal contrast;

- The underlying VCS has to be perfect black.

In this section we describe different solutions for VCSs with reversing all achieving ideal contrast. Each solution further improves on the proposal by Viet and Kurosawa [13] by overcoming some of the above drawbacks. The first solution, described in Section 9.4.1, uses the fact that the introduction of the reversing operation allows participants to compute any Boolean function of their transparencies and uses as a building block a binary secret sharing scheme. The schemes described in Section 9.4.2 all use as a building block a perfect black VCS, whereas, the scheme in Section 9.4.3 is based on a non-perfect black VCS.

9.4.1   A Construction Using a Binary Secret Sharing Scheme

In this section we describe an ideal contrast VCS with reversing for any access structure $(\Gamma_{\mathsf{Qual}}, \Gamma_{\mathsf{Forb}})$, proposed in [6]. The scheme uses as a building block a *binary secret sharing scheme* (BSS for short).

A BSS for an access structure $(\Gamma_{\mathsf{Qual}}, \Gamma_{\mathsf{Forb}})$ on a set of n participants is a method to share a secret $s \in \{0, 1\}$ among the n participants in such a way that only subsets of participants in Γ_{Qual} can recover the secret, whereas, subsets of participants in Γ_{Forb} have no information about the secret. A BSS consists of two collections $\mathcal{B}_0$ and $\mathcal{B}_1$ of distribution functions. A distribution function $f \in \mathcal{B}_0 \cup \mathcal{B}_1$ is a function associating each participant i to the share $f(i)$. In the reconstruction phase, any qualified set of participants $X = \{i_1, \ldots, i_p\} \in \Gamma_{\mathsf{Qual}}$ run a reconstruction algorithm $Rec(f(i_1), \ldots, f(i_p))$, which on inputs the shares they hold, outputs the secret s. See [7] for a formal definition of BSSs. Let r be the number of bits in the binary representation of the largest share $f(i)$. Without loss of generality (W.l.o.g.), we consider the r-bits binary representation of each share $f(j)$, where $j \neq i$, obtained by prefixing the r'-bits binary representation of $f(j)$, where $r' < r$, with $r - r'$ zeroes. The reconstruction algorithm $Rec(f(i_1), \ldots, f(i_p))$ computes a Boolean function on its inputs. Such a function can be computed by using a Boolean circuit. It is well known that OR and NOT represent a complete basis for Boolean functions, i.e., any Boolean function can be computed by a binary circuit composed only of OR and NOT gates. Therefore, $Rec(f(i_1), \ldots, f(i_p))$ can be computed by a Boolean circuit composed only of OR and NOT gates, corresponding to stacking and reversing operations, respectively.

In the distribution phase of the VCS with reversing, the encoding of the secret image is handled pixel by pixel, where each pixel is considered independently of the others. For each pixel of the original image and for each participant i, the dealer generates the corresponding pixel in each transparency $t_{i,1}, \ldots, t_{i,r}$, where r is the size of the shares distributed by the underlying BSS. In the reconstruction phase, any qualified set of participants recover the original secret image with *no loss of resolution* by performing a sequence of stacking and reversing operations on their transparencies. Such a sequence of operations corresponds to simulating parallel runs of the reconstruction algorithm of the BSS, that is, one run for each pixel of the original image. Such a parallel execution enables the reconstruction of the original image, because each transparency contains the same number of pixels of the original image and each pixel can be reconstructed by using same Boolean circuit. The construction is described in Figure 9.2.

It is easy to see that the construction of Figure 9.2 gives an ideal contrast VCS with reversing with no loss of resolution. Indeed, let us consider the encoding pixel by pixel. Let $X \in \Gamma_{\mathsf{Qual}}$ be a qualified set of participants. From the reconstruction property of the underlying BSS, the participants in X can reconstruct the secret pixel, by performing the sequence of stacking and reversing operations on their transparencies corresponding to the binary

Let $(\Gamma_{\text{Qual}}, \Gamma_{\text{Forb}})$ be an access structure on a set of n participants. Let $\mathcal{B}_0$ and $\mathcal{B}_1$ be the collections of distribution functions realizing a BSS for $(\Gamma_{\text{Qual}}, \Gamma_{\text{Forb}})$. Let r be the size of the shares distributed by the BSS.

Distribution phase. To share a white (black, resp.) pixel of the original image, the dealer has to:

- randomly choose a distribution function $f \in \mathcal{B}_0$ (resp. $f \in \mathcal{B}_1$),

- for each participant i, consider the binary representation $s_{i,1}, \ldots, s_{i,r}$ of the share $f(i)$ and, for each $j = 1, \ldots, r$, put a white (black, resp.) pixel on the transparency $t_{i,j}$ if $s_{i,j} = 0$ ($s_{i,j} = 1$, resp.).

Reconstruction phase. Let $X = \{i_1, \ldots, i_p\} \in \Gamma_{\text{Qual}}$. Participants in X reconstruct the secret image by performing the sequence of reversing and stacking operations on their transparencies, corresponding to the NOT and OR gates of the Boolean circuit computing in parallel $Rec(f(i_1), \ldots, f(i_p))$, for each pixel of the original image, where f is the distribution function chosen to share that pixel.

FIGURE 9.2
A construction for ideal contrast VCS with reversing using a BSS.

circuit simulating the reconstruction algorithm of the underlying BSS. The security of the scheme directly follows from the security of the underlying BSS.

As shown in the following, we can construct an ideal contrast (k, k)-threshold VCS with reversing, with $k \geq 2$, where each participant has to store only one transparency, if we use as a building block a (k, k)-threshold BSS distributing shares of one bit. This improves on the construction for a (k, k)-threshold VCS with reversing of Figure 9.1, where each participant has to store c transparencies, each containing 2^{k-1} subpixels for each pixel of the original secret image, c being the number of runs of the underlying VCS needed to obtain an asymptotically ideal contrast. Indeed, consider the following (k, k)-threshold BSS: to share a secret $s \in \{0, 1\}$, the dealer randomly chooses $k - 1$ random bits $s_1, \ldots, s_{k-1}$ and computes $s_k = s \oplus s_1 \oplus \cdots \oplus s_{k-1}$, where $\oplus$ denotes the XOR operation. For $i = 1, \ldots, k$, the share for participant i is the bit s_i. In order to reconstruct the secret s in the BSS the k participants are required to compute the XOR of their shares $s_1 \oplus \cdots \oplus s_k$. By arranging the $k > 2$ bits $s_1, \ldots, s_k$ as the leaves of a binary tree of height $\lceil \log k \rceil$, where internal nodes have two children and the leaves are distributed on at most two levels, the problem of computing the XOR of k bits can be reduced to the problem of computing $k - 1$ pairwise XORs, corresponding to the internal nodes of the tree. Since each pairwise XOR operation can be expressed in terms of 3 ORs and 4 NOTs as follows:

$$s_i \oplus s_j = OR(\overline{OR(\overline{s_i}, s_j)}, \overline{OR(s_i, \overline{s_j})}),$$

then, the computation of each pairwise XOR can be performed by using a Boolean circuit, having a depth 4, constituted only by OR and NOT gates. Therefore, the computation of $s_1 \oplus \cdots \oplus s_k$ can be performed by a Boolean circuit having a depth $4\lceil \log k \rceil$.

In the corresponding VCS with reversing, for each pixel of the original image, the dealer runs the distribution phase of the BSS, and, for each participant i, generates on the transparency t_i the pixel corresponding to the share s_i distributed by the BSS. In order to reconstruct the secret image, the k participants are involved in a computation consisting of stacking and reversing their transparencies, simulating the circuit computing the XOR, which is composed only by OR and NOT gates. It is easy to see that the number of reversing and stacking operations needed to reconstruct the secret image is equal to $4(k-1)$ and $3(k-1)$, respectively.

Example 4 *Let $k = 4$. If the dealer wants to share a white pixel, he runs the distribution phase of the BSS for the secret $s = 0$. Hence, he randomly chooses three bits s_1, s_2, and s_3, and computes $s_4 = s \oplus s_1 \oplus s_2 \oplus s_3$. For example, let $s_1 = 1$, $s_2 = 0$, $s_3 = 0$, then $s_4 = 1$. The corresponding pixels in the transparencies t_1, t_2, t_3, t_4 are equal to $1, 0, 0, 1$, respectively. In order to reconstruct the secret pixel, the participants have to simulate the computation of $s_1 \oplus s_2 \oplus s_3 \oplus s_4$ by performing a sequence of stacking and reversing operations on their transparencies. Notice that $s = s_1 \oplus s_2 \oplus s_3 \oplus s_4 = (s_1 \oplus s_2) \oplus (s_3 \oplus s_4)$. Since $a = s_1 \oplus s_2 = OR(\overline{OR(\overline{s_1}, s_2)}, \overline{OR(s_1, \overline{s_2})}) = 1$ and $b = s_3 \oplus s_4 = OR(\overline{OR(\overline{s_3}, s_4)}, \overline{OR(s_3, \overline{s_4})}) = 1$, it follows that the four participants can reconstruct the secret pixel corresponding to the secret $s = a \oplus b = OR(\overline{OR(\overline{a}, b)}, \overline{OR(a, \overline{b})}) = 0$ by performing 12 reversing operations and 9 stacking operations.*

9.4.2   Constructions Using Perfect Black VCSs

In this section we describe different constructions using as a building block a perfect black VCS. In particular, the solution described in Section 9.4.2.1 requires each participant to store m transparencies, where m denotes the pixel expansion of the underlying perfect black VCS, and offers no loss of resolution. The scheme described in Section 9.4.2.2 still offers no loss of resolution, but requires each participant to hold two transparencies having a large size, i.e., $|\Gamma_0|$ times the size of the secret image. Finally, the scheme in Section 9.4.2.3 reduces the number of transparencies stored by each participant to $m - h + 1$, but there is a loss of resolution on the reconstructed image.

9.4.2.1 The Scheme by Cimato, De Santis, Ferrara, and Masucci

In this section we describe a general technique, due to Cimato, De Santis, Ferrara, and Masucci [6], to construct an ideal contrast VCS with reversing for any access structure $(\Gamma_{\mathsf{Qual}}, \Gamma_{\mathsf{Forb}})$. The scheme uses as a building block a VCS with perfect reconstruction of black pixels, having a certain pixel expansion m, and requires each participant to store m transparencies, each having the same number of pixels as the original image.

In the distribution phase of the scheme, the encoding of the secret image is handled pixel by pixel, where each pixel is considered independently of the others. For each pixel of the original image and for each participant i, the dealer generates the corresponding pixel in each transparency $t_{i,1}, \ldots, t_{i,m}$. In the reconstruction phase, any qualified set of participants recover the original secret image with *no loss of resolution* by performing a sequence of stacking and reversing operations on their transparencies. The construction is described in Figure 9.3.

Let $(\Gamma_{\mathsf{Qual}}, \Gamma_{\mathsf{Forb}})$ be an access structure on a set of n participants. Let $\mathcal{C}_0$ and $\mathcal{C}_1$ be the collections of Boolean matrices constituting a $(\Gamma_{\mathsf{Qual}}, \Gamma_{\mathsf{Forb}})$-VCS with a perfect reconstruction of black pixels and pixel expansion m.

Distribution phase. To share a white (black, resp.) pixel, the dealer has to:

- randomly choose a matrix $S = [s_{i,j}]$ in $\mathcal{C}_0$ (S in $\mathcal{C}_1$, resp.);

- for each participant i, consider the m bits $s_{i,1}, \ldots, s_{i,m}$ composing the i-th row of S and, for each $j = 1, \ldots, m$, put a white (black, resp.) pixel on the transparency $t_{i,j}$ if $s_{i,j} = 0$ ($s_{i,j} = 1$, resp.).

Reconstruction phase. Let $X = \{i_1, \ldots, i_p\} \in \Gamma_{\mathsf{Qual}}$. Participants in X reconstruct the secret pixel by computing:

- $\alpha_j = OR(s_{i_1,j}, \ldots, s_{i_p,j})$, for $j = 1, \ldots, m$;

- $\overline{\alpha_j}$, for $j = 1, \ldots, m$;

- $\beta = OR(\overline{\alpha_1}, \ldots, \overline{\alpha_m})$;

- $\overline{\beta}$, which is the reconstructed pixel.

FIGURE 9.3
Cimato, De Santis, Ferrara, and Masucci's ideal contrast VCS with reversing.

It is easy to see that the construction of Figure 9.3 gives an ideal contrast $(\Gamma_{\mathsf{Qual}}, \Gamma_{\mathsf{Forb}})$-VCS with reversing. Indeed, let us consider the encoding pixel by pixel and analyze separately the reconstruction phase in case the dealer shared a white pixel or a black pixel. Let $X = \{i_1, \ldots, i_p\} \in \Gamma_{\mathsf{Qual}}$ be a qualified subset of participants and assume that the secret pixel shared by the dealer was white. From Property 1. of Definition 1, there exists an index $j \in \{1, \ldots, m\}$ for which the encoded pixels in each transparency $t_{i_1,j}, \ldots, t_{i_p,j}$ are all white. It follows that for the same j, $\alpha_j = 0$ and $\beta = OR(\overline{\alpha_1}, \ldots, \overline{\alpha_m})$ is equal to 1. Then, $\overline{\beta}$ is zero, i.e., the reconstructed pixel will be white. Now, assume that the secret pixel shared by the dealer was black. Since the underlying scheme is a $(\Gamma_{\mathsf{Qual}}, \Gamma_{\mathsf{Forb}})$-VCS with a perfect reconstruction of black pixels, it holds that $\alpha_j = 1$, for $j = 1, \ldots, m$. Hence, $\beta = OR(\overline{\alpha_1}, \ldots, \overline{\alpha_m})$ is zero and $\overline{\beta} = 1$, i.e., the reconstructed pixel will be black. The security of the scheme directly follows from the security of the underlying VCS.

Compared to the scheme described in Figure 9.1, the one of Figure 9.3 requires each participant to store m transparencies, each having the same number of pixels as the original image. Furthermore, the scheme requires a qualified set of p participants to execute exactly $m + 1$ reversing operations and $mp - 1$ stacking operations. Finally, the reconstructed image has no loss of resolution.

Example 5 *Let $\mathcal{P} = \{1, 2, 3, 4\}$ and $\Gamma_0 = \left\{\{1, 2\}, \{2, 3\}, \{3, 4\}\right\}$. The basis matrices S^0 and S^1 in a VCS realizing the access structure Γ_{Qual} whose basis is Γ_0 are:*

$$
S^0 = \begin{bmatrix} 0 & 1 & 1 & 0 \\ 0 & 1 & 1 & 1 \\ 0 & 1 & 1 & 1 \\ 0 & 1 & 0 & 1 \end{bmatrix} \quad S^1 = \begin{bmatrix} 1 & 0 & 0 & 1 \\ 1 & 1 & 1 & 0 \\ 1 & 1 & 0 & 1 \\ 1 & 0 & 1 & 0 \end{bmatrix}.
$$

The collections $\mathcal{C}_0$ and $\mathcal{C}_1$ are obtained by permuting the columns of the corresponding basis matrix (S^0 for $\mathcal{C}_0$, and S^1 for $\mathcal{C}_1$) in all possible ways. Let S^0 be the matrix chosen by the dealer to share a white pixel. The corresponding pixels generated for the transparencies $t_{1,1}, \ldots, t_{1,4}$ for participant 1 are 0, 1, 1, and 0, whereas, the pixels generated for the transparencies $t_{2,1}, \ldots, t_{2,4}$ for participant 2 are 0, 1, 1, and 1. During the reconstruction phase, the two participants stack their shares and retrieve $\alpha_1 = 0, \alpha_2 = 1, \alpha_3 = 1, \alpha_4 = 1$. By applying the reversing operation to each α_j and stacking the results, they obtain $\beta = OR(\overline{0}, \overline{1}, \overline{1}, \overline{1}) = 1$. By reversing β, they obtain $\overline{\beta} = 0$, and reconstruct a white pixel.

Let S^1 be the matrix chosen by the dealer to share a black pixel. The corresponding pixels generated for the transparencies $t_{1,1}, \ldots, t_{1,4}$ for participant 1 are 1, 0, 0, and 1, whereas, the corresponding pixels generated for the transparencies $t_{2,1}, \ldots, t_{2,4}$ for participant 2 are 1 1 1, and 0. During the reconstruction phase, the two participants will perform the same operation sequence as before, retrieving $\alpha_1 = 1, \alpha_2 = 1, \alpha_3 = 1, \alpha_4 = 1$, $\beta = OR(\overline{1}, \overline{1}, \overline{1}, \overline{1}) = 0$, and $\overline{\beta} = 1$, reconstructing a black pixel.

Notice that if the qualified set $X \in \Gamma_{\mathsf{Qual}}$ is a superset of an $X' \in \Gamma_0$, then, in the reconstruction phase, participants in X can recover the secret image by performing the given sequence of operations on the subset of transparencies held by participants in X' only, thus reducing the total number of operations. A further improvement can be obtained by considering as a building block a $(\Gamma_{\mathsf{Qual}}, \Gamma_{\mathsf{Forb}})$-VCS with a perfect reconstruction of black pixels having a smaller pixel expansion than $m = \sum_{X \in \Gamma_{\mathsf{Qual}}} 2^{|X|-1}$. In particular, by considering the set Γ_0 of all the minimal qualified subsets, and by using the same technique proposed in [1], it is possible to construct a $(\Gamma_{\mathsf{Qual}}, \Gamma_{\mathsf{Forb}})$-VCS with a perfect reconstruction of black pixels, having pixel expansion $m' = \sum_{X \in \Gamma_0} 2^{|X|-1}$, thus reducing the number of transparencies held by each participant.

9.4.2.2 The Scheme by Hu and Tzeng

In this section we describe an ideal contrast VCS with reversing for any access structure $(\Gamma_{\mathsf{Qual}}, \Gamma_{\mathsf{Forb}})$, due to Hu and Tzeng [9]. The construction uses the basis matrices of the Naor-Shamir's (k, k)-VCS described in Section 9.2.1 along with the properties of the XOR operator.

Let S^0 and S^1 be the basis matrices of the Naor-Shamir's (k, k)-VCS. The key idea behind Hu and Tzeng's construction relies upon the fact that the XOR of the bits belonging to a column of S^0 (S^1, respectively) is zero (one, respectively). With such an observation in mind, a (k, k)-VCS with reversing may be easily constructed by issuing to the participant i a transparency of the same size as the original image constructed as follows: for each white (black, respectively) pixel the dealer chooses a column in S^0 (S^1, respectively) and puts on the transparency the i-th item of the chosen column. Since each column of S^0 (S^1, respectively) has an even (odd, respectively) number of ones, it is easy to see that by XORring k transparencies the original image is reconstructed without any loss of resolution. In the following we refer to this construction as the XOR-(k, k)-VCS. As seen in Section 9.4.1, the problem of computing the XOR of k transparencies can be reduced to a sequence of stacking and reversing operations.

By using $|\Gamma_0|$ executions of the XOR-(k, k)-VCS, a naive VCS with reversing for any access structure $(\Gamma_{\mathsf{Qual}}, \Gamma_{\mathsf{Forb}})$ can be obtained. Indeed, we simply execute the above XOR-$(|X|, |X|)$-VCS for each qualified set $X \in \Gamma_0$. Such a solution requires each participant to store as many transparencies as the number of qualified sets he belongs to. Each transparency has the same size as the original image, on the other hand, each participant needs to keep track of the correspondence between a transparency and the related qualified set. In order to overcome such a drawback, Hu and Tzeng proposed a way to combine the transparencies distributed to a participant in a single transparency. Such a transparency is then used by a participant in a qualified set along with an additional transparency to reconstruct the original image. Both transparencies held by each participant are $|\Gamma_0|$ times larger than the size of the secret image. Indeed, each transparency contains $|\Gamma_0|$ blocks and each qualified subset of participants reconstructs, without loss of resolution, the secret image in a single block, whereas the other reconstructed blocks contain only white pixels.

Compared to the naive solution, the scheme introduces some computational overhead by requiring additional stacking and reversing operations.

In the distribution phase, for each qualified set X in Γ_0 and each participant $i \in X$ the dealer generates the subtransparency $t_{X,i}$ resulting by applying the XOR-$(|X|, |X|)$-VCS on the set of participants X. For each set X in Γ_0 such that the participant i does not belong to X, the dealer generates the subtransparency $t_{X,i}$ consisting of all ones. Each participant i receives a transparency t_i corresponding to the concatenation of its subtransparency according to an ordering over the qualified sets in Γ_0. Hence, let $X_1, \ldots, X_{|\Gamma_0|}$ be the qualified sets in Γ_0, each participant i receives the transparency $t_i = t_{X_1,i} \circ \ldots \circ t_{X_{|\Gamma_0|},i}$, where $\circ$ denotes the concatenation between transparencies. Each participant i also receives an additional transparency t_i' of the same size as t_i such that each subtransparency $t_{X,i}'$ has all ones if the participant i does not belong to the qualified set X whereas it (to repeat the subject that is each sub transparency) has all zeros, otherwise. The construction is described in Figure 9.4.

Let $(\Gamma_{\mathsf{Qual}}, \Gamma_{\mathsf{Forb}})$ be an access structure on a set of n participants.

Distribution phase. For each qualified set $X \in \Gamma_0$ the dealer has to:

- execute the XOR-$(|X|, |X|)$-VCS on X to generate the subtransparency $t_{X,i}$ for each participant $i \in X$;

- for each $i \notin X$, generate the subtransparency $t_{X,i}$ consisting of all ones;

- for each $i \in X$, generate a subtransparency $t_{X,i}'$ of the same size of the original image, having all zeros;

- for each $i \notin X$, generate a subtransparency $t_{X,i}'$ of the same size of the original image, having all ones;

- distribute to participant i the transparencies $t_i = t_{X_1,i} \circ \ldots \circ t_{X_{|\Gamma_0|},i}$ and $t_i' = t_{X_1,i}' \circ \ldots \circ t_{X_{|\Gamma_0|},i}'$.

Reconstruction phase. Let $X = \{i_1, \ldots, i_p\}$ be a qualified set in Γ_{Qual}. Participants in X reconstruct the original image by computing:

- $T = XOR(t_{i_1}, \ldots, t_{i_p})$;
- $T' = OR(t_{i_1}', \ldots, t_{i_p}')$;
- $U = OR(T, T')$;
- $U' = XOR(U, T')$, which corresponds to the original image.

FIGURE 9.4
Hu and Tzeng's ideal contrast VCS with reversing.

It is easy to see that the i-th qualified set obtains the original image in place of the i-th subtransparency of U' while all other subtransparencies contains all zeros. Indeed, the matrix corresponding to the i-th subtransparency of U' is the same as the i-th subtransparency of T, which from the property of the XOR operator and the composition of the Naor-Shamir basis matrices corresponds to the original image. Moreover, the matrices corresponding to the subtransparency of U but the i-th one contains all ones. Hence, from the property of the XOR operator such subtransparencies contains all zeros in U'.

The reconstruction phase requires a qualified set of p participants to perform exactly $4p$ reversing operations and $4p - 1$ stacking operations, since, as seen in Section 9.4.1, the XOR operation can be implemented by means of 3 ORs and 4 NOTs operations.

Example 6 *Let* $\mathcal{P} = \{1, 2, 3, 4\}$, $\Gamma_0 = \left\{X_1, X_2\right\} = \left\{\{1, 4\}, \{2, 3, 4\}\right\}$. *Let* S_1^0 *and* S_1^1 *be the basis matrices associated to the XOR-(2,2)-VCS for* X_1 *and let* S_2^0 *and* S_2^1 *be the basis matrices associated with the XOR-(3,3)-VCS for* X_2 *defined as follows:*

$$
S_1^0 = \begin{bmatrix} 1 & 0 \\ 1 & 1 \\ 1 & 1 \\ 1 & 0 \end{bmatrix} \quad S_1^1 = \begin{bmatrix} 1 & 0 \\ 1 & 1 \\ 1 & 1 \\ 0 & 1 \end{bmatrix} ;
$$

$$
S_2^0 = \begin{bmatrix} 1 & 1 & 1 & 1 \\ 0 & 0 & 1 & 1 \\ 0 & 1 & 0 & 1 \\ 0 & 1 & 1 & 0 \end{bmatrix} \quad S_2^1 = \begin{bmatrix} 1 & 1 & 1 & 1 \\ 1 & 1 & 0 & 0 \\ 1 & 0 & 1 & 0 \\ 1 & 0 & 0 & 1 \end{bmatrix} .
$$

Let $I = \begin{bmatrix} 1 & 0 & 1 \\ 0 & 1 & 0 \end{bmatrix}$ *be the original image. The transparencies are computed as follows:*

$$
t_1 = \begin{bmatrix} 1 & 1 & 0 & 1 & 1 & 1 \\ 0 & 1 & 1 & 1 & 1 & 1 \end{bmatrix} \quad t_1' = \begin{bmatrix} 0 & 0 & 0 & 1 & 1 & 1 \\ 0 & 0 & 0 & 1 & 1 & 1 \end{bmatrix}
$$

$$
t_2 = \begin{bmatrix} 1 & 1 & 1 & 1 & 0 & 1 \\ 1 & 1 & 1 & 0 & 0 & 1 \end{bmatrix} \quad t_2' = \begin{bmatrix} 1 & 1 & 1 & 0 & 0 & 0 \\ 1 & 1 & 1 & 0 & 0 & 0 \end{bmatrix}
$$

$$
t_3 = \begin{bmatrix} 1 & 1 & 1 & 1 & 0 & 0 \\ 1 & 1 & 1 & 1 & 1 & 0 \end{bmatrix} \quad t_3' = \begin{bmatrix} 1 & 1 & 1 & 0 & 0 & 0 \\ 1 & 1 & 1 & 0 & 0 & 0 \end{bmatrix}
$$

$$
t_4 = \begin{bmatrix} 0 & 1 & 1 & 1 & 0 & 0 \\ 0 & 0 & 1 & 1 & 0 & 1 \end{bmatrix} \quad t_4' = \begin{bmatrix} 0 & 0 & 0 & 0 & 0 & 0 \\ 0 & 0 & 0 & 0 & 0 & 0 \end{bmatrix}
$$

Participants in $X_2 = \{2, 3, 4\}$ *reconstruct the original image by computing*

$$
T = \begin{bmatrix} 1 & 0 & 1 & 0 & 1 & 1 \\ 0 & 1 & 0 & 0 & 1 & 0 \end{bmatrix} \quad T' = \begin{bmatrix} 0 & 0 & 0 & 1 & 1 & 1 \\ 0 & 0 & 0 & 1 & 1 & 1 \end{bmatrix}
$$

$$
U = \begin{bmatrix} 1 & 0 & 1 & 1 & 1 & 1 \\ 0 & 1 & 0 & 1 & 1 & 1 \end{bmatrix} \quad U' = \begin{bmatrix} 1 & 0 & 1 & 0 & 0 & 0 \\ 0 & 1 & 0 & 0 & 0 & 0 \end{bmatrix} .
$$

9.4.2.3   The Scheme by Yang, Wang, and Chen

Yang, Wang, and Chen [14] proposed a different method to construct a VCS with reversing starting from any perfect black VCS with pixel expansion m. In their scheme each participant receives $m - h + 1$ shares, where the first one corresponds to the one obtained by the underlying perfect black VCS, whereas the i-th share is obtained by cyclically shifting the $(i - 1)$-th share one bit to the right, for each $i = 2, \ldots, m - h + 1$. Notice that the right shift operation can be implemented by means of OR and NOT operations, since OR and NOT represent a complete basis for Boolean functions.

The scheme is shown in Figure 9.5.

Assume there exists a $(\Gamma_{\mathsf{Qual}}, \Gamma_{\mathsf{Forb}})$-VCS with a perfect reconstruction of black pixels. Let $\Phi(\cdot)$ be a one-bit cyclical right-shift function.

Distribution phase. For each pixel of the secret image, the dealer:

- runs the distribution phase of the underlying $(\Gamma_{\mathsf{Qual}}, \Gamma_{\mathsf{Forb}})$-VCS; let s_i^1 be the share distributed to participant i, for $i = 1, \ldots, n$;

- for each $\ell = 2, \ldots, m - h + 1$, compute $s_i^\ell = \Phi(s_i^{\ell-1})$;

- distributes the $(m - h + 1)$-tuple $(s_i^1, \ldots, s_i^{m-h+1})$ to participant i.

Reconstruction phase. A qualified set $\{i_1, \ldots, i_p\}$ of participants reconstructs the secret pixel as follows:

- superimpose their shares to get $\alpha^\ell = OR(s_{i_1}^\ell, \ldots, s_{i_p}^\ell)$, for $\ell = 1, \ldots, m - h + 1$;

- reverse the results of the previous step to obtain $\overline{\alpha}^\ell$, for $\ell = 1, \ldots, m - h + 1$;

- superimpose the results of the previous step to get $\beta = OR(\overline{\alpha}^1, \ldots, \overline{\alpha}^{m-h+1})$;

- finally, reverse the result of the previous step to obtain $\overline{\beta}$, which is the reconstructed pixel.

FIGURE 9.5
Yang, Wang, and Chen's ideal contrast VCS with reversing.

The reconstruction phase is the same as that of the scheme shown in Figure 9.1. Recall that the computation of the reconstructed pixel $\overline{\beta}$ corresponds to performing $m - h + 1$ AND operations, since $\overline{\beta} = \overline{OR(\overline{\alpha}^1, \ldots, \overline{\alpha}^{m-h+1})} = AND(\alpha^1, \ldots, \alpha^{m-h+1})$. It is easy to see that the construction of Figure 9.5 achieves a perfect reconstruction of both white and black pixels. Indeed, in the case a white pixel has been shared, α^1 contains h white subpixels and $m - h$ black ones. The maximum interval between two 0s in α^1 is $m - h$, thus, by shifting right one bit $(m - h)$ times, there is at least an α^i having a subpixel equal to 0 at position j, for each $i = 1, \ldots, m - h + 1$ and $j = 1, \ldots, m$.

Therefore, the reconstructed pixel $\overline{\beta} = AND(\alpha^1, \ldots, \alpha^{m-h+1})$ results in all white subpixels. On the other hand, in case a black pixel has been shared, α^i contains all black subpixels, for each $i = 1, \ldots, m - h + 1$, thus also $\overline{\beta}$ perfectly reconstructs the pixel.

The security of the scheme directly follows from the security of the underlying VCS.

The reconstruction phase requires a qualified set of p participants to perform exactly $m - h + 2$ reversing operations and $(m - h + 1)p - 1$ stacking operations. Finally, notice that in the construction of Figure 9.5 there is a loss of resolution, since each pixel in the original image corresponds to m subpixels in the reconstructed image, where m denotes the pixel expansion of the underlying VCS.

Example 7 *Let $\mathcal{P} = \{1, 2, 3, 4\}$ and $\Gamma_0 = \Big\{ \{1, 2\}, \{2, 3\}, \{3, 4\} \Big\}$. The basis matrices S^0 and S^1 in a VCS realizing the access structure Γ_{Qual} whose basis is Γ_0 are:*

$$S^0 = \begin{bmatrix} 0 & 1 & 1 & 0 \\ 0 & 1 & 1 & 1 \\ 0 & 1 & 1 & 1 \\ 0 & 1 & 0 & 1 \end{bmatrix} \qquad S^1 = \begin{bmatrix} 1 & 0 & 0 & 1 \\ 1 & 1 & 1 & 0 \\ 1 & 1 & 0 & 1 \\ 1 & 0 & 1 & 0 \end{bmatrix}.$$

The collections $\mathcal{C}_0$ and $\mathcal{C}_1$ are obtained by permuting the columns of the corresponding basis matrix (S^0 for $\mathcal{C}_0$, and S^1 for $\mathcal{C}_1$) in all possible ways. Let S^0 be the matrix chosen by the dealer to share a white pixel. Notice that $m = 4$, $h = 1$, and $m - h + 1 = 4$. The shares generated for participant 1 are $s_1^1 = 0110$, $s_1^2 = 0011$, $s_1^3 = 1001$, and $s_1^4 = 1100$, whereas, the shares generated for the participant 2 are $s_2^1 = 0111$, $s_2^2 = 1011$, $s_2^3 = 1101$, and $s_2^4 = 1110$. During the reconstruction phase, the two participants stack their shares and retrieve $\alpha^1 = OR(s_1^1, s_2^1) = 0111$, $\alpha^2 = OR(s_1^2, s_2^2) = 1011$, $\alpha^3 = OR(s_1^3, s_2^3) = 1101$, and $\alpha^4 = OR(s_1^4, s_2^4) = 1110$. By applying the reversing operation to each α^j and stacking the results, they obtain $\beta = OR(\overline{0111}, \overline{1011}, \overline{1101}, \overline{1110}) = 1111$. By reversing β, they obtain $\overline{\beta} = 0000$, and reconstruct a white pixel.

Let S^1 be the matrix chosen by the dealer to share a black pixel. The shares generated for participant 1 are $s_1^1 = 1001$, $s_1^2 = 1100$, $s_1^3 = 0110$, and $s_1^4 = 0011$, whereas, the shares generated for the participant 2 are $s_2^1 = 1110$, $s_2^2 = 0111$, $s_2^3 = 1011$, and $s_2^4 = 1101$. During the reconstruction phase, the two participants stack their shares and retrieve $\alpha^1 = OR(s_1^1, s_2^1) = 1111$, $\alpha^2 = OR(s_1^2, s_2^2) = 1111$, $\alpha^3 = OR(s_1^3, s_2^3) = 1111$, and $\alpha^4 = OR(s_1^4, s_2^4) = 1111$. By applying the reversing operation to each α^j and stacking the results, they obtain $\beta = OR(\overline{1111}, \overline{1111}, \overline{1111}, \overline{1111}) = 0000$. By reversing β, they obtain $\overline{\beta} = 1111$, and reconstruct a black pixel.

9.4.2.4   Comparisons

The efficiency of a VCS with reversing is evaluated according to the following parameters: the contrast, the size (expansion) and the number of the transparencies held by each participant, the number of stacking and reversing operations, and the size (expansion) of the reconstructed image. In Table 9.1 we summarize and compare the parameters of the constructions described in Sections 9.3 and 9.4.2., which are all based on perfect black VCSs for a general access structure $(\Gamma_{\mathsf{Qual}}, \Gamma_{\mathsf{Forb}})$.

TABLE 9.1

Comparison between VCSs with reversing based on perfect black VCSs.

Scheme	Contrast	Share exp.	Number of shares	Number OR oper.	Number NOT oper.	Secret exp.
Fig. 9.1	Almost ideal	m	c	$cp-1$	$c+1$	m
Fig. 9.3	Ideal	NO	m	$mp-1$	$m+1$	NO
Fig. 9.4	Ideal	$|\Gamma_0|$	2	$4p-1$	$4p$	NO
Fig. 9.5	Ideal	m	$m-h+1$	$(m-h+1)p-1$	$m-h+2$	m

9.4.3   A Construction Using a Nonperfect Black VCS

Yang, Wang and Chen [14] proposed a way to construct an ideal VCS with reversing when the difference $h - \ell$ is odd, starting from any VCS (i.e., not necessarily a perfect black) with pixel expansion m. Each participant receives m shares, where the first one corresponds to that obtained by the underlying VCS, whereas, the i-th share is obtained by cyclically shifting the $(i-1)$-th share one bit to right, for each $i = 2, \ldots, m$. Notice that the right shift operation can be implemented by means of OR and NOT operations, since OR and NOT represent a complete basis for boolean functions.

The scheme is shown in Figure 9.5.

The construction of Figure 9.5 achieves perfect reconstruction of both white and black pixels. Indeed, in case a white pixel has been shared, α^1 contains h white subpixels and $m - h$ black ones. By shifting right one bit m times, $\alpha^1, \ldots, \alpha^m$ are constructed in such a way that exactly $(m - h)$ out of m have a black subpixel at position j, for each $j = 1, \ldots, m$. Therefore, $\beta = XOR(\alpha^1, \ldots, \alpha^m)$ is the reconstructed white pixel if $(m - h)$ is even, otherwise the reconstructed pixel is $\bar{\beta}$. On the other hand, since $h - \ell$ is odd, i.e. h and ℓ cannot be both odd or both even, if $(m - h)$ is even (odd, resp.) it holds that $m - \ell$ is odd (even, resp.). Hence, in the case a black pixel has been shared, α^1 contains ℓ white subpixels and $m - \ell$ black ones. By shifting right one bit m times, $\alpha^1, \ldots, \alpha^m$ are constructed in such a way that exactly $(m - \ell)$ out of m have a black subpixel at position j, for each $j = 1, \ldots, m$. Therefore, $\beta = XOR(\alpha^1, \ldots, \alpha^m)$ is the reconstructed black pixel if $(m - h)$ is even, i.e., $(m - \ell)$ is odd, otherwise the reconstructed pixel is $\bar{\beta}$.

Assume there exists a $(\Gamma_{\mathsf{Qual}}, \Gamma_{\mathsf{Forb}})$-VCS with pixel expansion m. Let $\Phi(\cdot)$ be a one-bit cyclical right shift function and let $(h - \ell)$ be odd.

Distribution phase. For each pixel of the secret image, the dealer:

- runs the distribution phase of the underlying $(\Gamma_{\mathsf{Qual}}, \Gamma_{\mathsf{Forb}})$-VCS; let s_i^1 be the share distributed to participant i, for $i = 1, \ldots, n$;

- for each $\ell = 2, \ldots, m$, compute $s_i^\ell = \Phi(s_i^{\ell-1})$;

- distributes the m-tuple $(s_i^1, \ldots, s_i^m)$ to participant i.

Reconstruction phase. A qualified set $\{i_1, \ldots, i_p\}$ of participants reconstructs the secret pixel as follows:

- superimpose their shares to get $\alpha^\ell = OR(s_{i_1}^\ell, \ldots, s_{i_p}^\ell)$, for $\ell = 1, \ldots, m$;

- compute $\beta = XOR(\alpha^1, \ldots, \alpha^m)$, which is the reconstructed pixel if $(m - h)$ is even, otherwise the reconstructed pixel is $\overline{\beta}$.

FIGURE 9.6
Ideal contrast VCS with reversing starting from any VCS.

The security of the scheme directly follows from the security of the underlying VCS.

The reconstruction phase requires a qualified set of p participants to perform exactly $4(m - 1)$ reversing operations and $p - 1 + 3(m - 1)$ stacking operations, since as seen in Section 9.4.1 the XOR operation can be implemented by means of 3 ORs and 4 NOTs operations. Finally, notice that in the construction of Figure 9.5 there is a loss of resolution, since each pixel in the original image corresponds to m subpixels in the reconstructed image, where m denotes the pixel expansion of the underlying VCS.

Example 8 *Consider a $(2,3)$-threshold VCS, which is not a perfect black, where $m - h$ is odd. Assume the basis matrices S^0 and S^1 are:*

$$S^0 = \begin{bmatrix} 1 & 0 & 0 \\ 1 & 0 & 0 \\ 1 & 0 & 0 \end{bmatrix} \qquad S^1 = \begin{bmatrix} 1 & 0 & 0 \\ 0 & 1 & 0 \\ 0 & 0 & 1 \end{bmatrix}.$$

The collections $\mathcal{C}_0$ and $\mathcal{C}_1$ are obtained by permuting the columns of the corresponding basis matrix (S^0 for $\mathcal{C}_0$, and S^1 for $\mathcal{C}_1$) in all possible ways. Let S^0 be the matrix chosen by the dealer to share a white pixel. The shares generated for participant 1 are $s_1^1 = 100$, $s_1^2 = 010$ and $s_1^3 = 001$, whereas, the shares generated for the participant 2 are $s_2^1 = 100$, $s_2^2 = 010$, and $s_2^3 = 001$. During the reconstruction phase, the two participants stack their shares and retrieve $\alpha^1 = OR(s_1^1, s_2^1) = 100$, $\alpha^2 = OR(s_1^2, s_2^2) = 010$, and $\alpha^3 = OR(s_1^3, s_2^3) = 001$. By computing $\beta = XOR(\alpha^1, \alpha^2, \alpha^3) = 111$ and its reverse $\overline{\beta} = 000$ the two participants reconstruct a white pixel.

Let S^1 be the matrix chosen by the dealer to share a black pixel. The shares generated for participant 1 are $s_1^1 = 100$, $s_1^2 = 010$ and $s_1^3 = 001$, whereas, the shares generated for the participant 2 are $s_2^1 = 010$, $s_2^2 = 001$ and $s_2^3 = 100$. During the reconstruction phase, the two participants stack their shares and retrieve $\alpha^1 = OR(s_1^1, s_2^1) = 110$, $\alpha^2 = OR(s_1^2, s_2^2) = 011$, and $\alpha^3 = OR(s_1^3, s_2^3) = 101$. By computing $\beta = XOR(\alpha^1, \alpha^2, \alpha^3) = 000$ and its reverse $\overline{\beta} = 111$ the two participants reconstruct a black pixel.

9.5 Conclusions

Visual cryptography schemes are characterized by two parameters: the pixel expansion, i.e., the number of subpixels contained in each share and the contrast, which measures the difference between a black and a white pixel in the reconstructed image. While it is possible to construct schemes with perfect reconstruction of black pixels (or white pixels, respectively), it has been shown that a perfect reconstruction of both black and white pixels is infeasible. In order to improve the contrast in VCSs, Viet and Kurosawa [13] introduced an extra noncryptographic operation: the reversing operation. Specifically, they showed how to construct VCSs with reversing where the reconstruction of black (white, respectively) pixels is perfect, whereas, the reconstruction of white (black, respectively) pixels is almost perfect. Afterwards, Cimato et al. [6] showed how to construct VCSs with reversing where reconstruction of both black and white pixels is perfect. Such schemes are said to have an *ideal contrast*. In particular, Cimato et al. [6] proposed two different constructions. One uses as a building block a VCS with perfect reconstruction of black pixels while the other construction uses as a building block a binary secret sharing scheme. Subsequently, new constructions for visual cryptography with reversing have been described in [9, 14]. In particular, [9] considered the problem of minimizing the number of the shares held by each participant while in [14] the need of using as a building block a VCS with perfect reconstruction of black pixels is removed.

Bibliography

[1] G. Ateniese, C. Blundo, A. De Santis, and D. R. Stinson. Visual cryptography for general access structures. *Information and Computation*, 129(2):86–106, 1996.

[2] C. Blundo, P. D'Arco, A. De Santis, and D. R. Stinson. Contrast optimal threshold visual cryptography schemes. *SIAM Journal on Discrete Mathematics*, 16(2):224–261, 2003.

[3] C. Blundo, A. De Bonis, and A. De Santis. Improved schemes for visual cryptography. *Designs, Codes, and Cryptography*, 24:255–278, 2001.

[4] C. Blundo and A. De Santis. Visual cryptography schemes with perfect reconstruction of black pixels. *Journal for Computers & Graphics*, 22(4):449–455, 1998.

[5] C. Blundo, A. De Santis, and D. R. Stinson. On the contrast in visual cryptography schemes. *Journal of Cryptology*, 12(8):261–289, 1999.

[6] S. Cimato, A. De Santis, A. L. Ferrara, and B. Masucci. Ideal contrast visual cryptography schemes with reversing. *Information Processing Letters*, 93:199–206, 2005.

[7] A. De Bonis and A. De Santis. Randomness in secret sharing and visual cryptography schemes. *Theoretical Computer Science*, 314(3):351–374, 2004.

[8] T. Hofmeister, M. Krause, and H. U. Simon. Contrast-optimal k out of n secret sharing schemes in visual cryptography. *Theoretical Computer Science*, 240(2):471, 2000.

[9] C.-M. Hu and W.-G.. Tzeng. Compatible ideal contrast visual cryptography schemes with reversing. In *Proc. of ISC 2005, Lecture Notes in Computer Science*, volume 3650, pages 300–313, 2005.

[10] M. Krause and H. U. Simon. Determining the optimal contrast for secret sharing schemes in visual cryptography. *Combinatorics, Probability and Computing*, 12(3):285–299, 2003.

[11] M. Naor and A. Shamir. Visual cryptography. In *Proc. of Advances in Cryptology – EUROCRYPT '94, Lecture Notes in Computer Science*, volume 950, pages 1–12, 1995.

[12] E. R. Verheul and H. C. A. van Tilborg. Constructions and properties of k out of n visual secret sharing schemes. *Designs, Codes, and Cryptography*, 11(2):179, 1997.

[13] D. Q. Viet and K. Kurosawa. Almost ideal contrast visual cryptography with reversing. In *Proc. of Topics in Cryptology - CT-RSA 2004, The Cryptographers' Track at the RSA Conference 2004, Lecture Notes in Computer Science*, volume 2964, pages 21–37, 2004.

[14] C.-N. Yang, C.-C. Wang, and T.-S. Chen. Visual cryptography schemes with reversing. *The Computer Journal*, 51(6):710, 2008.

10

Cheating Prevention in Visual Cryptography

Yu-Chi Chen

National Chung Hsing University, Taiwan

Gwoboa Horng

National Chung Hsing University, Taiwan

Du-Shiau Tsai

Hsiuping Institute of Technology, Taiwan

CONTENTS

10.1   Introduction

In 1994, Naor and Shamir proposed a variant of secret sharing called Visual Cryptography (VC) [9], where the shares given to participants are xeroxed onto transparencies. If X is an authorized subset, then the participants in X can visually recover the secret image by stacking their transparencies together without performing any computation. One special property distinguishes VC from conventional secret sharing scheme [16, 17] is that the security of VC is achieved by losing the contrast and the resolution of the secret image. In other words, the quality of the reconstructed secret image is inferior to that of the original secret image. Since the invention of VC, many researchers have devoted themselves to enhancing the contrast and resolution of the reconstructed images [1, 3] and to extending it to general access structures [7]. Moreover, many schemes to visually share nonbinary secret images, such as gray-level secret images [2, 5] and color secret images [12, 13], were also proposed. There are lots of applications based on VC, for example, visual authentication and identification [10], steganography [4, 6, 8], and image encryption [14].

In 2006, Horng et al. showed that cheating is possible in the k-out-of-n VC [8]. The cheating activity could cause unpredictable damage to victims because they will accept a forged image different from the actual secret image as authentic. In this chapter, we survey some recently proposed cheating prevention schemes in VC and provide a comparative evaluation of their advantages and disadvantages.

The rest of this chapter is organized as follows. Section 10.2 provides preliminary background on VC and the cheating activities. Section 10.3 surveys some cheating prevention schemes. Their comparative analyses are in Section 10.4. Finally, conclusions and some research issues are given in Section 10.5.

10.2   Preliminaries

10.2.1   Visual Cryptography

A visual secret sharing scheme is a special variant of a k-out-of-n secret sharing scheme where the shares given to participants are xeroxed onto transparencies. Therefore, a share is also called a transparency. If X is a qualified subset, then the participants in X can visually recover the secret image by stacking their transparencies without performing any cryptographic computation. Usually, the secret is an image. To create the transparencies, each black and white pixel of the secret image is handled separately. It appears as a collection of m black and white subpixels in each of the n transparencies. We will call these m subpixels a *block*. Therefore, a pixel of the secret image corresponds to nm subpixels. We can describe the nm subpixels by an $n \times m$ Boolean matrix, called a *base matrix*, $S = [S_{ij}]$ such that $S_{ij} = 1$ if and only if the jth subpixel of the ith share is black and $S_{ij} = 0$ if and only if the jth subpixel of the ith share is white. The gray level of the stack of k shared blocks is determined by the Hamming weight $H(V)$ of the "OR"ed m-vector V of the corresponding k rows in S. This gray level is interpreted by the visual system of the users as black if $H(V) \geq d$ and as white if $H(V) \leq d - \alpha * m$ for some fixed threshold d and relative difference α. We would like m to be as small as possible and α to be as large as possible.

More formally, a solution to the k-out-of-n VC consists of two collections C^0 and C^1 of $n \times m$ base matrices. To share a white pixel, the dealer randomly

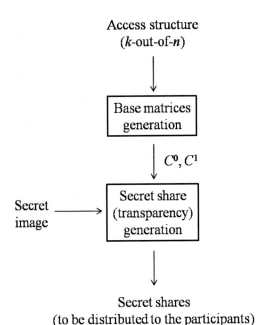

FIGURE 10.1
Visual cryptography.

chooses one of the matrices from C^0, and to share a black pixel, the dealer randomly chooses one of the matrices from C^1. The chosen matrix determines the m subpixels in each one of the n transparencies. Figure 10.1 shows the basic concept of visual cryptography.

Definition 1 A solution to the k-out-of-n VC consists of two collections C^0 and C^1 of $n \times m$ base matrices. The solution is considered valid if the following conditions are met:

Contrast conditions:

1. For any matrix S^0 in C^0, the "or" V of any k of the n rows satisfies $H(V) \leq d - \alpha * m$.

2. For any matrix S^1 in C^1, the "or" V of any k of the n rows satisfies $H(V) \geq d$.

Security condition:

3. For any subset $\{i_1, i_2, \ldots, i_q\}$ of $\{1, 2, \ldots, n\}$ with $q < k$, the two collections D^0, D^1 of $q \times m$ matrices obtained by restricting each $n \times m$ matrix in C^0, C^1

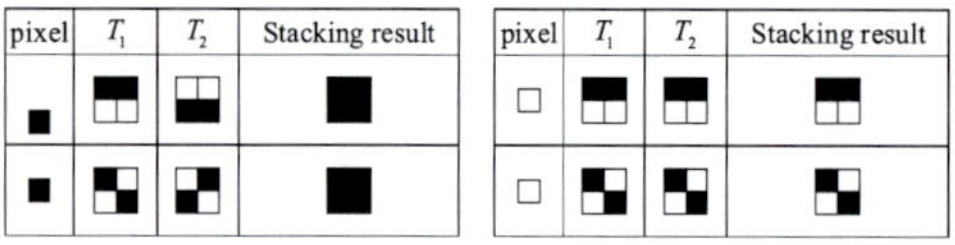

pixel	T_1	T_2	Stacking result
■			■
■			■

pixel	T_1	T_2	Stacking result
□			
□			

FIGURE 10.2
The concept of 2-out-of-2 VC.

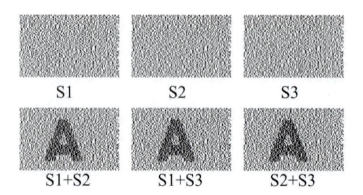

FIGURE 10.3
A 2-out-of-3 visual secret sharing scheme.

to rows i_1, i_2, K, i_q are indistinguishable in the sense that they contain the same matrices with the same frequencies.

Figure 10.2 shows the basic concept of a 2-out-of-2 visual secret sharing scheme. The secret image is shared through two noise-like transparencies: T_1 and T_2. In this case, a block consists of 4 subpixels. Each black(white) pixel of the secret image is turned into two black and two white subpixels of T_1 and T_2. Figure 10.3 shows an example of a 2-out-of-3 visual secret sharing scheme.

10.2.2 Cheating in VC

In [8], Horng et al. showed that cheating is possible in k-out-of-n VC. Let's take the 2-out-of-3 visual secret sharing scheme shown in Figure 10.3 as an example. A secret image is encoded into three distinct transparencies, denoted T_1, T_2, and T_3. Then, the three transparencies are respectively delivered to Alice, Bob, and Carol. Without loss of generality, Bob and Carol are assumed to be collusive cheaters and Alice is the victim. During the cheating activity, Bob and Carol use S_2 and S_3 to create forged transparency S_2' such that superimposing S_2' and S_3 will visually recover the cheating image as in Figure 10.4. Precisely, by observing the following collections of 3×3 matrices that are used to generate transparencies, collusive cheaters can predict the actual structure of the victim's transparency so as to create S_2'.

$$C^0 = \{\text{all the matrices obtained by permuting the columns of } \begin{bmatrix} 1 & 0 & 0 \\ 1 & 0 & 0 \\ 1 & 0 & 0 \end{bmatrix}\}$$

$$C^1 = \{\text{all the matrices obtained by permuting the columns of } \begin{bmatrix} 1 & 0 & 0 \\ 0 & 1 & 0 \\ 0 & 0 & 1 \end{bmatrix}\}$$

By observing the above matrices, two rows of above C^0 or C^1 matrix are determined by collusive cheaters. Therefore, the structure of the corresponding block of S_1 is exact the remaining row. For presenting a white pixel of cheating image, the block of S_2' is set to be the same structure of S_1. For presenting a black pixel of cheating image, the block of S_2' is set to be the different structure of S_1. For example, if the block of S_1 is [010], then S_2' is set to be [010] for a white pixel or it is set to be [001] for a black pixel.

We note that if the forged shares deviate too much from the original shares then the victim will notice the difference and suspect being cheated. Therefore, a necessary condition for a cheating activity to be successful is that the forged shares must be indistinguishable from the original shares. There are different cheating activities defined in [15].

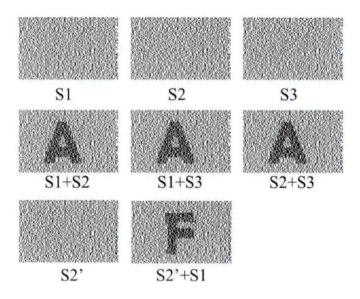

FIGURE 10.4
Cheating in visual cryptography.

10.3 Cheating Prevention Schemes

A visual secret sharing scheme is said to be a cheating prevention scheme if the probability of successful cheating is negligible. We will not consider cheating prevention in computational visual cryptographic schemes [21]. We can divide the cheating prevention schemes into two classes. One is based on share authentication where another share (transparency) is used to authenticate other shares (transparencies) and the other is based on blind authentication where some property of the image is used to authenticate the reconstructed secret image. For example, in [8], it is assumed that a smooth image such that its boundary of black and white regions is clearly perceptible is regarded as authentic. Due to practical reasons, most cheating prevention schemes focus on 2-out-of-n schemes. In the following subsections, we will survey 6 recent proposed cheating prevention schemes: two from [8], proposed in 2006 and will be referred to as HCT1 and HCT2, one from [15], proposed in 2007 and will be referred to as HT, one from [18], proposed in 2007 and will be referred to as TCH, and two from [20], proposed in 2009 and will be referred to as PS1 and PS2.

10.3.1 HCT1 and HCT2

In [8], two cheating prevention schemes are proposed: *the Authentication Based Cheating Prevention scheme*, referred to as HCT1, and the 2-out-of-$(n + l)$ cheating prevention scheme, referred to as HCT2. HCT1, shown in Figure 10.5, is a share-authentication-based cheating prevention scheme and HCT2, shown in Figure 10.7, is a blind-authentication-based cheating prevention scheme. Figure 10.6 shows an experiment of HCT1. In HCT1, each participant uses an extra transparency to verify the integrity of other transparencies by means of the appearance of the verification logo. A participant adopts a verification transparency to verify the integrity of other transparencies through the appearance of his own verification logo. Each verification transparency is generated by a 2-out-of-2 VC. Therefore, each participant receives two transparencies, namely, secret share transparency and verification transparency created by the 2-out-of-n and 2-out-of-2 VC, respectively. Figure 10.6 shows an experiment of HCT1.

HCT2 generates $(n+l)$ transparencies but it only delivers n transparencies to participants. The probability that cheaters can correctly guess the structure of each block generated for a black pixel of the victim's transparency is down to $1/(1 + l)$. The secret image is redesigned to consist of two complementary parts. Two binary images are said to be complementary to each other if and only if they have the same size and, for all corresponding pixels, one is black and the other is white. Therefore, the probability that cheaters can correctly

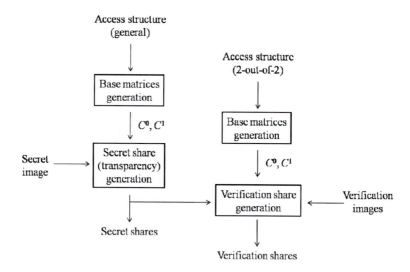

FIGURE 10.5
HCT1.

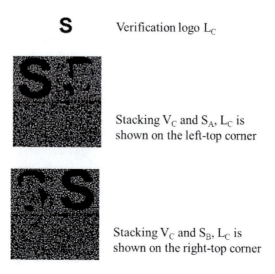

S   Verification logo L_C

Stacking V_C and S_A, L_C is
shown on the left-top corner

Stacking V_C and S_B, L_C is
shown on the right-top corner

FIGURE 10.6
Experiment of HCT1.

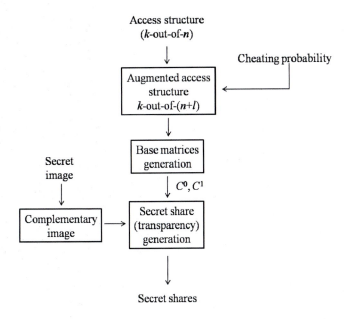

FIGURE 10.7
HCT2.

guess the structure of victim's transparency is negligible. The idea can be extended to design a k-out-of-$(n + l)$ cheating prevention scheme.

10.3.2 HT

The core of HT, shown in Figure 10.8, uses a generic transformation to create new transparencies by adding two subpixels to every block of every original transparency. Then, HT creates a verification transparency for each participant such that the stacking result of the new transparency with the verification transparency will reveal a verification image. Therefore, HT is a share-authentication-based cheating prevention scheme. Basically, HT will perform the following steps with input C^0 and C^1 to create verification transparencies

1. Let $T_0 = \begin{bmatrix} 10 \\ \vdots \\ 10 \end{bmatrix} C^0$ and $T_1 = \begin{bmatrix} 10 \\ \vdots \\ 10 \end{bmatrix} C^1$

2. Use T^0 and T^1 as the base matrices for creating transparencies

3. For each participant P_i, choose a verification image and create a verification transparency v_i as follows:

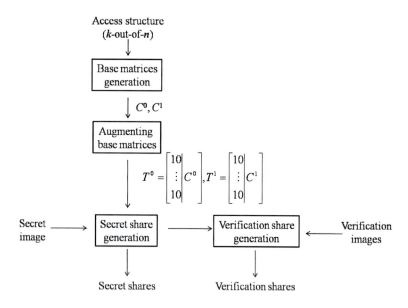

FIGURE 10.8
HT.

   a. For each white pixel in the verification image, the block of v_i created based on $[1\ 0\ 0\ ...\ 0]$ (after corresponding permutation as for generating the secret shares in step 2).

   b. b For each black pixel in the verification image, the block of v_i created based on $[0\ 1\ 0\ ...\ 0]$ (after corresponding permutation as for generating the secret shares in step 2).

In the decoding phase of the secret image, before stacking transparencies, each P_i checks the stacking result of v_i with T_j held by participant P_j revealing his own verification image. Figure 10.9 shows an example of an HT scheme.

10.3.3   TCH

In [18], Tsai et al. proposed a cheating prevention scheme, shown in Figure 10.10, images where shares are generated by Genetic Algorithms. TCH adopts multiple secret images with the same visual meaning. Each qualified subset only reveals the corresponding reconstructed secret image and the others are left unknown to potential cheaters. Any participant accepts the decoding result only if the visually reconstructed secret image is authentic. In TCH, the fitness function of the Generic Algorithm was designed according to a 2-out-of-n visual secret sharing scheme. But, it is not guaranteed that the same quality is obtained because the Generic Algorithm is a kind of heuristic al-

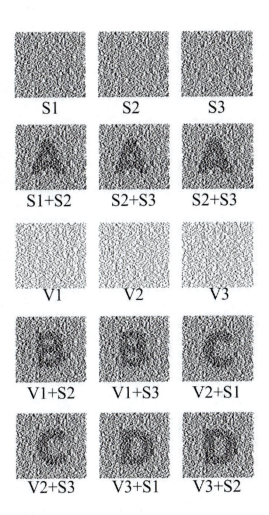

FIGURE 10.9
Experiment of HT.

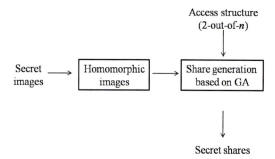

FIGURE 10.10
TCH.

gorithm. Therefore, the dealer should control the quality of all reconstructed secret images before delivering all transparencies. That is, all transparencies should be indistinguishable.

10.3.4   PS1 and PS2

In 2009, De Prisco and De Santis [20] proposed 2-out-of-n and n-out-of-n cheating prevention schemes, as shown in Figure 10.11. They are obtained by simply adding an extra column with all 0s to the base matrices of the schemes of Naor and Shamir.

In the following, we describe another 2-out-of-n cheating prevention scheme, as shown in Figure 10.12 and proposed in the same paper. This scheme does not require the use of a complementary image. The base matrices of the scheme have dimension $n \times (2^n + n + 1)$. The white base matrix C^0 has the following columns: all the possible 2^n binary column vectors of length n, one additional column with all 1s and n additional columns with all 0s. Whereas, the black base matrix C^1 has the following columns: all the possible 2^n binary column-vectors of length n, one additional column with all 0s in the n columns of the identity matrix of dimension $n \times n$.

The base matrices of a 2-out-of-3 PS2 scheme are shown as follows:

$$C^0 = \begin{bmatrix} 0 & 0 & 0 & 0 & 1 & 1 & 1 & 1 & 0 & 1 & 0 & 0 \\ 0 & 0 & 1 & 1 & 0 & 0 & 1 & 1 & 0 & 1 & 0 & 0 \\ 0 & 1 & 0 & 1 & 0 & 1 & 0 & 1 & 0 & 1 & 0 & 0 \end{bmatrix}$$

$$C^1 = \begin{bmatrix} 0 & 0 & 0 & 0 & 1 & 1 & 1 & 1 & 0 & 1 & 0 & 0 \\ 0 & 0 & 1 & 1 & 0 & 0 & 1 & 1 & 0 & 0 & 1 & 0 \\ 0 & 1 & 0 & 1 & 0 & 1 & 0 & 1 & 0 & 0 & 0 & 1 \end{bmatrix}$$

We refer the readers to [20] for more details.

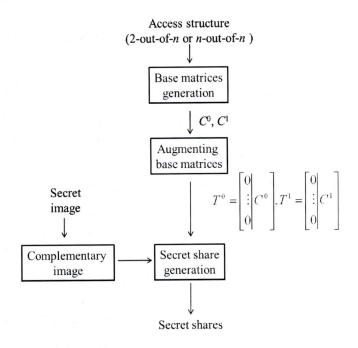

FIGURE 10.11
PS1.

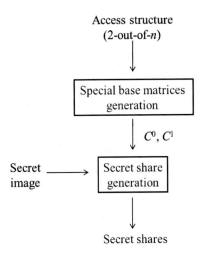

FIGURE 10.12
PS2.

10.4   Analysis of Cheating Prevention Schemes

We begin with a general discussion of the advantages and disadvantages of the share authentication approach and blind authentication approach. The advantages of share-authentication-based cheating prevention approach are twofold. One is that checking the authenticity of shares is optional. It can be done only when someone is suspected of cheating. The other is that the generation of verification shares is done after the generation of secret shares. Therefore, any conventional visual secret sharing schemes (for any access structures) can be turned into a cheating prevention scheme. The quality of the reconstructed secret image is not affected. The disadvantages of this approach are also twofold. One is that additional shares for verification purpose are needed. The other is that these schemes lack a formal proof of security.

The advantages of blind-authentication-based cheating prevention approach are twofold. One is that no additional shares are required. The other is that we can formally prove the security. We can formally argue that the probability of changing a black pixel (or white pixel) by the cheaters is less than 1. The disadvantages of this approach are also twofold. One is that this approach is only suitable for threshold visual secret sharing schemes. The other is that the quality of the reconstructed secret image is degraded.

The comparisons of different cheating prevention schemes are illustrated in Table 10.1. With respect to the total number of subpixels for sharing a pixel, TCH requires the least subpixels for sharing a pixel. HCL1 requires $2n^2$ subpixels, since an extra transparency is used to verify the correctness of other transparencies. HT requires $2n(n + 2)$ subpixels because this scheme also needs verification transparencies and each original transparency is enlarged. Finally, HCT2 needs $2(n + l)^2$ subpixels. With respect to *prevention of shares against cheating*, the following schemes are all designed according to *authentic conditions* (AC). PS and HT both prevent all blocks of each transparency from cheating. HCT1 only prevents these blocks within the corresponding verification logo. And HCT2 only prevents blocks that were created for presenting black pixels. The share-authentication-based cheating prevention approach can be applied to general access structures. Whereas blind-authentication-based cheating prevention approach is more suitable for threshold access structures, with respect to the method for share generation, HT uses a modified VC by the basis matrices T^0 and T^1. HCL1 and HCL2 use the share construction method of traditional VC. Finally, TCH uses GA for share generation. With respect to *security*, all schemes are designed for preventing a known secret cheating attack. However, HCL1, HT, and TCH are not formally proved to be secure.

TABLE 10.1

Comparisons of different cheating prevention visual secret sharing schemes.

Scheme	Approach	Access stru.	Share gen.	subpixels	Security
HCL1	Share auth.	General	Base matrices	$2n^2$	
HCL2	Blind auth.	k-out-of-n	Base matrices	$2(n+1)^2$	Secure
HT	Share auth.	k-out-of-n	Base matrices	$2n(n+2)$	
TCH	Blind auth.	2-out-of-n	GA	n^2	
PS1	Blind auth.	2(n)-out-of-n	Base matrices	$2(n+1)^2$	Secure
PS2	Blind auth.	2-out-of-n	Base matrices	$n \times (2^n + n + 1)$	Secure

10.5 Conclusions

In this chapter, we have surveyed some cheating prevention schemes in visual cryptography and provided a comparative evaluation of their advantages and disadvantages. There are many topics that deserve further instigation, for example, giving formal definition of cheating and cheating attack models and designing new cheating prevention scheme a with less numbers of subpixels for sharing a pixel.

Bibliography

[1] A. Shamir and M. Naor. Visual cryptography II: Improving the contrast via the cover base, security in communication networks, September 16–17, 1996.

[2] C. Blundo, A. De Santis, and M. Naor. Visual cryptography for grey level images. *Information Processing Letters*, Vol. 75, No. 6, (2000) 255–259.

[3] C. Blundo, P. D'Arco, A. De Santis, D. R. Stinson. Contrast optimal threshold visual cryptography schemes, *SIAM J. Discrete Math.* 16(2) (2003) 224–261.

[4] C.C. Chang, and J.C. Chuang. An image intellectual property protection scheme for gray-level image using visual secret sharing strategy, *Pattern Recognition Letters*, Vol. 23 (2002) 931–941.

[5] C.C. Lin, W.H. Tsai. Visual cryptography for gray-level images by dithering techniques, *Pattern Recognition Letters*. Vol. 24 (1–3) (2003) 349–358.

[6] C.C. Wang, S.C. Tai, and C.S. Yu. Repeating image watermarking technique by the visual cryptography, *IEICE Transactions on Fundamentals*, Vol. E83-A (2000) 1589–1598.

[7] G. Ateniese, C. Blundo, A. De Santis, and D. R. Stinson. Visual Cryptography for general access structures, *Information and Computation*, (1996) 86–106.

[8] G. Horng, T. H. Chen, and D. S. Tsai. Cheating in visual cryptography, designs, codes and cryptography, Vol. 38, No. 2 (2006) 219–236.

[9] M. Naor and A. Shamir. Visual cryptography, In Proceedings of Advances in Cryptography-EUROCRYPTO'94, LNCS 950 (1994) 1–12.

[10] M. Naor and B. Pinkas. Visual authentication and identification, advances in cryptology - Proceedings of Crypto 97 (Burton S. Kaliski Jr., ed.), *Lecture Notes in Computer Science*, Springer-Verlag, New York, 1294 (1997) 322–336.

[11] R. Lukac and K.N. Plataniotis, Bit-level based secret sharing for image encryption, *Pattern Recognition* Vol. 38(5) (2005) 767–772.

[12] V. Rijmen and B. Preneel, Efficient colour visual encryption for shared colors of Benetton, EUROCRYPTO'96, Rump Session, Berlin, 1996.

[13] Y. C. Hou. Visual cryptography for color images, *Pattern Recognition*, Vol. 36 (2003) 1619–1629.

[14] T. H. Chen and D. S. Tsai. Owner-customer right protection mechanism using a watermarking scheme and a watermarking protocol, *Pattern Recognition*, Vol. 39, Issue 8, (2006) pp. 1530–1541.

[15] C. Hu and W. Tzeng. Cheating prevention in visual cryptography, *IEEE Transactions on Image Processing*, Vol. 16, No. 1, 2007, pp. 36–45.

[16] A. Shamir. How to share a secret, *Comm. ACM*, Vol. 22 (1979) 612–613.

[17] G. Blakley. Safeguarding cryptographic keys, Proc. AFIPS 1979 Natl. Conf.

[18] D.S. Tsai, T.H. Chen, and G. Horng. A cheating prevention scheme for binary visual cryptography with homogeneous secret images, *Pattern Recognition*, Vol. 40 No. 8, 2007, pp. 2356–2366.

[19] R. C. Gonzalez and R. E. Woods. *Digital Image Processing*, 2nd Ed., Prentice-Hall, 2002.

[20] R. De Prisco and A. De Santis. Cheating immune threshold visual secret sharing, *The Computer Journal*, doi:10.1093/comjnl/bxp068.

[21] C. Chang, T. Chen, and L. Liu. Preventing cheating in computational visual cryptography, *Fundamenta Informaticae* Vol. 92 2009, pp. 27–42.

11

Resolving the Alignment Problem in Visual Cryptography

Feng Liu

Institute of Software, Chinese Academy of Sciences, China

CONTENTS

11.1 Introduction

The basic principle of the Visual Cryptography Scheme (VCS) was first formally introduced by Naor and Shamir [10]. The idea of the VCS proposed in [10] is to split an image into two random shares (printed on transparencies), which separately reveal no information about the original secret image other than the size of the secret image. The image is composed of black and white pixels. The original image can be recovered by stacking the two transparencies together.

One important parameter in VCS is pixel expansion, where the pixel expansion reflects the size of the recovered secret image. Many papers in the literature are dedicated to proposing VCS with smaller pixel expansion. However, all these works are based at the pixel level, i.e., to reduce the number of subpixels that represent a pixel of the original secret image.

We notice that, the final goal of reducing the pixel expansion is to reduce the size of the transparencies that are distributed to the participants, because smaller transparencies are easier to transport. However, the size of the subpixels that are printed on the transparencies affects the final size of the transparencies; in fact, the size of the transparencies is the product of the size of the subpixels and the number of the subpixels in each transparency. Unfortunately, there is a dilemma when one tries to determine the size of the subpixels: When the subpixel size is large, it is easy to align the shares (most publications in the literature require aligning the shares precisely in the decrypting phase), but the large subpixel size will result in large transparencies. On the other hand, when the subpixel size is small, it is relatively hard to align the shares, but smaller transparencies result. From the point of view of the participants of the VCS, the goal is to align the shares easily and have small transparencies as well. Table 11.1 shows the relationship between the size of the subpixels of the transparencies and the ease to align them (more comparisons will be given in Table 11.5 later). Hence, there is a trade-off between the size of the subpixels of the transparencies and the ease to align them.

TABLE 11.1
The advantages and disadvantages of large and small subpixels.

size of the subpixels	advantages	disadvantages
larger	easier to align	larger transparency size
smaller	smaller transparency size	harder to align

The usual way of tackling the alignment problem of the VCS is by adding frames to the shares. To align the shares one just needs to align the frames. Yan et al. [13] employed the Walsh transform to embed marks in both of the shares so as to find the alignment position of these shares. However, both methods need to align the transparencies precisely. Besides, Kobara and Imai [6] considered a different problem. They calculated the visible space when viewing the transparencies. The results are somehow related to the alignment problem, but not exactly, as [6] has no discussion about alignment at all. Kobara and Imai [6] do not consider the stacking of more than two shares. Nakajima and Yamaguchi [9] proposed a $(2,2)$, extended VCS, which the secret image and shares are natural images. Their scheme can simultaneously reduce the alignment difficulty. However, their scheme does not hold the perfect security like a secret sharing scheme.

In fact, the precise alignment of small subpixels is not critical. The secret image can still be recovered visually even if the participants do not align the transparencies precisely. This phenomenon helps to determine the size of the subpixels printed on the transparencies.

This chapter focuses on some recent results about the alignment problem of the visual cryptography scheme [8, 16]. Two kinds of alignment problems of share images are considered. First, the share images are misaligned by integer number of subpixels (less than the pixel expansion). This part mainly comes from [8]. In such a case, the secret image can still be observed as its complementary image. Second, the share images are misaligned by less than one subpixel. This part mainly comes from [16]). Conditions that the secret image can still be recovered are studied, and the different misalignment tolerances of large and small subpixels are compared. The results indicate that the VCS, by itself, has some misalignment tolerance. At last, we provide a misalignment tolerant VCS based on the trade-off between the usage of large and small subpixels.

11.2 Preliminaries

In the VCS, there is a secret image that is encrypted into some shares. The secret image is called the *original secret image* for clarity, and the shares are the encrypted images (and are called the *transparencies* if they are printed). When a qualified set of shares (transparencies) are stacked together, it gives a visual image that is almost the same as the original secret image, we call it the *recovered secret image*. In the case of black and white images, the original secret image is represented as a pattern of black and white pixels. Each of these pixels is divided into subpixels, which themselves are encoded as black and white to produce the shares. The recovered secret image is also a pattern of black and white subpixels that should visually reveal the original secret image if a qualified set of shares is stacked.

In order to simplify the discussion, in this paper, we will only consider the black and white VCS, where black pixels are denoted by 1 and white pixels are denoted by 0.

By a (k, n)-VCS we mean a scheme where the original secret image is divided into n shares, which are distributed to n participants. Any subgroup of k out of these n participants can get a recovered secret image, but any subgroup consisting of less than k participants does not have any information other than the size about the original secret image.

For a vector $v \in GF^m(2)$, we denote by $w(v)$ the Hamming weight of the vector v, i.e., the number of nonzero coordinates in v. A (k, n)-VCS, denoted by (C_0, C_1), consists of two collections of $n \times m$ binary matrices C_0 and C_1. To share a white (*resp.* black) pixel, a dealer (the one who sets up the system) randomly chooses one of the matrices, called a *share matrix*, in C_0 (*resp.* C_1) and distributes its rows (representing a pattern of subpixels in the share) to the n participants of the scheme, giving row i to participant i, $i = 1, \cdots, n$. More precisely, we give a formal definition of the (k, n)-VCS as follows:

Definition 1 *Let k, n, m be nonnegative integers, l and h be positive numbers* *satisfying* $2 \leq k \leq n$ *and* $0 \leq l < h \leq m$. *The two collections of* $n \times m$ *binary matrices* (C_0, C_1) *constitute a visual cryptography scheme* (k, n)-*VCS if the following properties are satisfied:*

1. (*Contrast*) *For any* $s \in C_0$, *the OR of any* k *out of* n *rows of* s *is a vector* v *that satisfies* $w(v) \leq l$.

2. (*Contrast*) *For any* $s \in C_1$, *the OR of any* k *out of* n *rows of* s *is a vector* v *that satisfies* $w(v) \geq h$.

3. (*Security*) *For any* $i_1 < i_2 < \cdots < i_t$ *in* $\{1, 2, \cdots, n\}$ *with* $t < k$, *the two collections of* $t \times m$ *matrices* D_j, $j = 0, 1$, *obtained by restricting each* $n \times m$ *matrix in* C_j, $j = 0, 1$, *to rows* $i_1, i_2, \cdots i_t$, *are indistinguishable in the sense that they contain the same matrices with the same frequencies.*

Note that, in the above definition,

1. m is called the *block length* and determines the pixel expansion of the scheme. A pixel of the original secret image is represented by m subpixels in the recovered secret image. In general, we are interested in schemes with m being as small as possible. h and l are called the *darkness thresholds* of the black and white pixels, respectively.

2. Define the value $\alpha = \frac{h-l}{m}$ to be the contrast of the scheme. Note that, however, there are other definitions of the contrast of VCS. We use this definition to establish our result. The proof is similar for other definitions of contrast.

We consider VCS in which C_0, C_1 are constructed from a pair of $n \times m$ binary matrices M_0, M_1, called *basis matrices*. The set C_i, $i = 0, 1$ consists of the $m!$ matrices obtained by applying all permutations to the columns of M_i. This approach of VCS construction will have small memory requirements (it only keeps the basis matrices) and high efficiency (to choose a matrix in C_0 (*resp.* C_1) as it only needs to generate a permutation of the basis matrix). We will use the basis matrices to simplify the discussions.

The above definition of VCS is also called the Deterministic Visual Cryptography Scheme (DVCS). The original secret image can be deterministically recovered by the qualified shares pixel by pixel in such schemes. In contrast to the DVCS, Yang and Cimato et al. proposed the Probabilistic Visual Cryptography Scheme (PVCS) in [14, 4], where the pixels of the original secret image can only be probabilistically recovered by the qualified shares, however, in the overall view the original secret image can be recovered visually as well.

Definition 2 (Probabilistic VCS [14, 4]) *Let* k, n, *and* m' *be nonnegative integers,* $\bar{l}$ *and* $\bar{h}$ *be positive numbers, satisfying* $2 \leq k \leq n$ *and* $0 \leq \bar{l} < \bar{h} \leq m'$. *The two collections of* $n \times m'$ *binary matrices* (C_0, C_1) *constitute a probabilistic threshold Visual Cryptography Scheme* (k, n)-*PVCS if the following properties are satisfied:*

1. (*Contrast*) *For the collection C_0 and a share matrix $s \in C_0$, by v a vector resulting from the OR of any k out of the n rows of s. If $\bar{w}(v)$ denotes the average of the Hamming weights of v, over all the share matrices in C_0, then $\bar{w}(v) \leq \bar{l}$.*

2. (*Contrast*) *For the collection C_1, the value of $\bar{w}(v)$ satisfies $\bar{w}(v) \geq \bar{h}$.*

3. (*Security*) *For any $i_1 < i_2 < \cdots < i_t$ in $\{1, 2, \cdots, n\}$ with $t < k$, the two collections of $t \times m'$ matrices D_j, $j = 0, 1$, obtained by restricting each $n \times m'$ matrix in C_j, $j = 0, 1$, to rows $i_1, i_2, \cdots i_t$, are indistinguishable in the sense that they contain the same matrices with the same frequencies.*

The definition of PVCS in [14] only considers the case with $n \times 1$ share matrices, we extend their definition to the $n \times m'$ case. And the definition of PVCS in [4] used the factor β to reflect the contrast, we use the values $\bar{l}$ and $\bar{h}$ (darkness grey-levels) to reflect the contrast. The same point of the three definitions of PVCS is that, for a particular pixel in the original secret image, the qualified participants can only correctly represent it in the recovered secret image with a certain probability. Because the human eyes always average the high frequency black and white dots into gray areas, so the average value of the Hamming weight of the black dots in the area reflects the grayness of the area. The PVCS does not require the satisfaction of the difference in grayness for each pixel in the recovered secret image as the DVCS does. It only reflects the difference in grayness in the overall view.

The contrast of the DVCS is fulfilled for each pixel (consisting of m sub-pixels) in the recovered secret image, however, this is quite different in the PVCS. The application of the *average contrast*, denoted by $\bar{\alpha}$, first appeared in [3]. This term is often used in the PVCS, see [4, 14, 11, 7], where the traditional contrast of the PVCS does not exist. Here we define the *average contrast* to be the average value of the overall contrast of the recovered secret image, i.e., the mean value of the contrast of all the pixels in the recovered secret image. According to our definition of the contrast $\alpha = \frac{h-l}{m}$, the average contrast can be calculated by the formula $\bar{\alpha} = \frac{\bar{h}-\bar{l}}{m'}$, where $\bar{h}$ and $\bar{l}$ are the mean values of $w(v)$ for the black and white pixels in the overall recovered secret image respectively, and m' is the pixel expansion of the PVCS. Because the number of pixels is large in the recovered secret image, the values $\bar{h}$ and $\bar{l}$ are equivalent to the mean values of the $w(v)$ in the collections C_1 and C_0, respectively. Note that, the DVCS also has the average contrast, and many proposed DVCS's in the literature have $\bar{\alpha} = \alpha$, see examples in [10, 5, 1], etc. When comparing, the DVCS that has $\bar{\alpha} = \alpha$ then, in the overall view, the clearness of the recovered secret image of the PVCS is the same as the clearness of the recovered secret image of a DVCS. However, because of the probabilistic nature, a PVCS is disadvantaged in displaying the details of the original secret image, for example, a thin line in the original secret image is likely to be displayed as a dotted line in a PVCS.

11.3   Misalignment with Integer Number of Subpixels

According to the traditional view, the subpixels of the transparencies should be aligned precisely. Here, we point out that, to recover the secret image visually, it is not necessary to align the subpixels precisely. In this section, we will only consider the misalignment with integer number of subpixels.

We will show that, by shifting one of the shares by some number (at most $m - 1$) of subpixels to the right (resp. left), one can still recover the secret image visually, for the reason that the average contrast $\bar{\alpha} \neq 0$. This result can naturally be extended to the case when more than one share is shifted. However, we leave the numerical analysis of this case to the interested readers. So, in this section, we will only consider the case with only one share (transparency) being shifted by some number of subpixels. And we call the scheme with a share being shifted the *shifted scheme*, and the basis matrices and share matrices of the shifted scheme are called the *shifted basis matrices* and *shifted share matrices*.

We first give an example to show this phenomenon.

Example 1 *We take the $(2, 2)$-DVCS as an example, where the basis matrices of the scheme are,*

$$M_0 = \begin{bmatrix} 100 \\ 100 \end{bmatrix} \text{ and } M_1 = \begin{bmatrix} 100 \\ 010 \end{bmatrix}$$

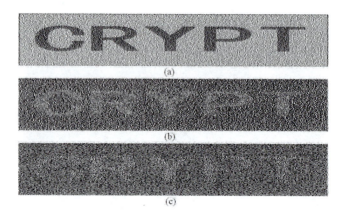

FIGURE 11.1
The stacking results of the (2,2)-DVCS (a) when no share is shifted; (b) when one share is shifted by one subpixels; (c) when one share is shifted by two subpixel. A printed-text "CRYPTO" is tested.

Figure 11.1 are the experimental results of the recovered secret image, where (a) is the recovered secret image without being shifted, and (b) is the recovered secret image with the second share being shifted to the left by one subpixel, and (c) is the recovered secret image with the second share being shifted to the left by 2 subpixels. From the experimental results we notice that, the original secret image can be recovered visually by shifting one or two subpixels for the $(2,2)$-DVCS with pixel expansion $m = 3$. And the clearness of the recovered secret image with two subpixels being shifted is not as clear as the one with one subpixel being shifted. Furthermore, the shifted scheme is not a DVCS anymore; we give an example to show this point:

Example 2 *Take the share matrices* $S^0 = \begin{bmatrix} 010 \\ 010 \end{bmatrix}$ *and* $S^1 = \begin{bmatrix} 100 \\ 010 \end{bmatrix}$ *as an example, which are chosen from the permutations of the basis matrices* M_0 *and* M_1 *in Example 1. By shifting the second share by one subpixel to the left, we get the following four matrices:*

$$S_0^0 = \begin{bmatrix} * & 010 \\ 0 & 100 \end{bmatrix} \quad S_1^0 = \begin{bmatrix} * & 010 \\ 0 & 101 \end{bmatrix} \quad S_0^1 = \begin{bmatrix} * & 100 \\ 0 & 100 \end{bmatrix} \quad S_1^1 = \begin{bmatrix} * & 100 \\ 0 & 101 \end{bmatrix}$$

where S_j^i, $i = 0, 1$, $j = 0, 1$, *is the shifted share matrix that a subpixel* j *is shifted into* S^i. *The left bottom subpixel of* S^i *is shifted out, and the asterisk* $*$ *in the* S_j^i *can be either 1 or 0, which belongs to the pixel on the left of the pixel that we considered. So, here and hereafter, we no longer consider the two subpixels anymore, i.e., we only need to consider the right 3 columns in the shifted share matrix* S_j^i.

For the above four shifted share matrices, the stacking Hamming weights are 2, 3, 1, and 2. In particular, the stacking Hamming weight of S_0^0 *and* S_1^1 *are the same. Here and hereafter, the* stacking Hamming weight *means the Hamming weight of the resulting vector generated by stacking the shares. Hence, the shifted scheme is not a DVCS anymore.*

Generally, we aim at proving the conclusion that, the shifted scheme can visually recover the original secret image based on the (k, n)-DVCS. However, it is noticed that this proof can be reduced to the proof based on the $(2, 2)$-DVCS in the case that only one share is shifted. The reason is as follows:

First, a (k, n)-DVCS consists of $\binom{n}{k}$ (k, k)-DVCS. For a set of k shares, if no share is shifted, then the k shares can recover the secret image obviously. And because we only consider the case when only one of the n shares is shifted, we only need to consider the k shares that contain the shifted share, i.e., we only need to prove our conclusion based on a (k, k)-DVCS.

Second, denote the k shares of a (k, k)-DVCS as $s_1, s_2, \cdots, s_k$, without loss of generality, let s_k be the share that is shifted, and let s_k' be the resulting image of stacking the remaining $k - 1$ shares $s_1, s_2, \cdots, s_{k-1}$ together. Then the scheme becomes a $(2, 2)$-DVCS, where the two shares are s_k' and s_k. Note that, the stacking result of this $(2, 2)$-DVCS is the same as that of the previous (k, k)-DVCS. The previous (k, k)-DVCS can visually recover the secret image if and only if s'_k and s_k can do so. Hence, it is sufficient to prove the conclusion based on a $(2,2)$-DVCS.

We analyze the structure of the basis matrix of the $(2,2)$-DVCS. Denote M_0 and M_1 as the basis matrices of the $(2,2)$-DVCS, then the M_0 and M_1, without loss of generality, are in the following form:

$$
M_0 = \left[\begin{array}{cccc}
1\cdots1\,0\cdots0 & 1\cdots1\,0\cdots0 \\
\underbrace{1\cdots1}_{a}\underbrace{0\cdots0}_{b}\underbrace{0\cdots0}_{c}\underbrace{1\cdots1}_{d}
\end{array} \right],
$$

and

$$
M_1 = \left[\begin{array}{cccc}
1\cdots1\,0\cdots0 & 1\cdots1\,0\cdots0 \\
\underbrace{1\cdots1}_{a'}\underbrace{0\cdots0}_{b'}\underbrace{0\cdots0}_{c'}\underbrace{1\cdots1}_{d'}
\end{array} \right],
$$

where a, b, c, d, a', b', c', and d' are nonnegative integers satisfying $a+c+d = l$ and $a' + c' + d' = h$. According to the contrast and security property of Definition 1, we have,

$$
\begin{cases}
a + b + c + d = a' + b' + c' + d' \\
a + c = a' + c' \\
a + d = a' + d' \\
b > b'
\end{cases}
$$

solving the above equations, we get $a - a' = b - b' = c' - c = d' - d$. Let $e = b - b'$, hence, by deleting identical columns of M_0 and M_1, we get,

$$
M'_0 = \left[\begin{array}{cc}
1\cdots1\,0\cdots0 \\
\underbrace{1\cdots1}_{e}\underbrace{0\cdots0}_{e}
\end{array} \right]
$$

and,

$$
M'_1 = \left[\begin{array}{cc}
1\cdots1\,0\cdots0 \\
\underbrace{0\cdots0}_{e}\underbrace{1\cdots1}_{e}
\end{array} \right]
$$

where the number of columns in M'_0 and M'_1 is $2e$.

Now we know that the basis matrices of an arbitrary $(2,2)$-DVCS M_0 and M_1 contain the same number of identical columns $\begin{bmatrix} 1 \\ 1 \end{bmatrix}, \begin{bmatrix} 1 \\ 0 \end{bmatrix}, \begin{bmatrix} 0 \\ 1 \end{bmatrix}$, and $\begin{bmatrix} 0 \\ 0 \end{bmatrix}$ apart from the submatrices M'_0 and M'_1. Hence, without loss of generality, they can be represented as the following form:

$$
M_0 = \left[\begin{array}{cccccc}
1\cdots1\,0\cdots0 & 1\cdots1\,0\cdots0 & 1\cdots1\,0\cdots0 \\
\underbrace{1\cdots1}_{a'}\underbrace{0\cdots0}_{b'}\underbrace{0\cdots0}_{c}\underbrace{1\cdots1}_{d}\underbrace{1\cdots1}_{e}\underbrace{0\cdots0}_{e}
\end{array} \right]
$$

and

$$M_1 = \left[\begin{array}{c} \underbrace{1 \cdots 1}_{a'} \underbrace{0 \cdots 0}_{b'} \underbrace{1 \cdots 1}_{c} \underbrace{0 \cdots 0}_{d} \underbrace{1 \cdots 1}_{e} \underbrace{0 \cdots 0}_{e} \\ \underbrace{1 \cdots 1}_{a'} \underbrace{0 \cdots 0}_{b'} \underbrace{0 \cdots 0}_{c} \underbrace{1 \cdots 1}_{d} \underbrace{0 \cdots 0}_{e} \underbrace{1 \cdots 1}_{e} \end{array} \right].$$

Let m be the pixel expansion, then it is obvious that $m = a' + b' + c + d + 2e$. The collections C_0 and C_1 contain all the permutations of the basis matrices M_0 and M_1, and hence each has $m!$ share matrices.

The shifted scheme is generated as follows:

Shift the second row of the $m!$ share matrices in C_0 (resp. C_1) to the left (resp. right) by r subpixels, and let $c_1, c_2, \cdots, c_r$ be the r-bit string that is shifted in, where each $c_i \in \{0, 1\}$ represents a subpixel. By the above discussion, we get $m!$ shifted share matrices for C_0 (resp. C_1). Take the share matrix $M_0 \in C_0$ as an example, then the shifted share matrix, denoted by $M_0^{(r)}$, is as follows:

$$M_0^{(r)} = \left[\begin{array}{c} \underbrace{* \cdots * 1 \cdots 1}_{a'} \underbrace{0 \cdots 0}_{b'} \underbrace{1 \cdots 1}_{c} \underbrace{0 \cdots 0}_{d} \underbrace{1 \cdots 1}_{e} \underbrace{0 \cdots 0}_{e} \\ \underbrace{1 \cdots 1}_{a'} \underbrace{0 \cdots 0}_{b'} \underbrace{0 \cdots 0}_{c} \underbrace{1 \cdots 1}_{d} \underbrace{1 \cdots 1}_{e} \underbrace{0 \cdots 0}_{e} \underbrace{c_1 \cdots c_r}_{r} \end{array} \right],$$

where $c_1 \cdots c_r$ of share 2 are the adjacent subpixels of the right pixel that are shifted in.

By going through all $m!$ share matrices of C_0 and C_1 and all the possible string of subpixels $c_1 \cdots c_r \in \{0, 1\}^r$, where $\{0, 1\}^r$ is the set of all the binary strings of length r, the shifted scheme is generated. Hence, we have:

Theorem 1 *The shifted scheme of a DVCS is a PVCS, where the average contrast of the shifted scheme is $\bar{\alpha} = -\frac{(m-r)e}{m^2(m-1)}$, where $1 \leq r \leq m - 1$ is the number of subpixels by which the share 2 (the second share) is shifted.*

Proof: Without loss of generality, we only prove the case when the share 2 is shifted to the left, which is equivalent to the case when the share 1 (the first share) is shifted to the right. Note that, swapping rows 1 and 2 corresponds to swapping the parameters c and d. We observe that, since the share matrices of the DVCS satisfy condition 3 of Definition 1 and the shifting operation is the same for matrices in C_0 and C_1, the share matrices of the shifted scheme satisfy condition 3 of Definition 2.

First, we prove the case that the share 2 is shifted by one subpixel, and then we extend it to the case when it is shifted by r subpixels.

When share 2 is shifted by one subpixel, the adjacent right subpixel of share 2 is shifted in. Let p_1 be the probability that a 1 is shifted in, and p_0 be the probability that a 0 is shifted in. Because the share matrices are all the permutations of the basis matrices, so p_1 and p_0 have fixed values and $p_1 + p_0 = 1$, i.e., the shifted in subpixel is either 1 or 0. More precisely, by the above discussion we have $p_1 = \frac{a'+d+e}{m}$ and $p_0 = \frac{b'+c+e}{m}$. In the general case when r subpixels are shifted in, denote p_c as the probability that a string of subpixels c is shifted in, we also have $\sum_{c \in \{0,1\}^r} p_c = 1$.

Because, under the OR operation, the stacking of two black subpixels results in a black subpixel, i.e., $1\,OR\,1 = 1$, so this means that there is a black subpixel that is ineffective (overlapped). Now we define a black pixel (value 1) to be *ineffective* if it does not contribute to the total number of black pixels in the recovered secret image. There are three cases when a black subpixel 1 is ineffective: (1) when that is in the top right corner of the matrix M_0 (or M_1) and another 1 is shifted in, this results in a overlap; (2) when that is in the bottom left corner of the matrix which is then shifted out (appear below an asterisk '*'); and (3) when an overlap happens after a shift, which is possible on the first $m - 1$ positions (and $m - r$ positions in general). Hence, the total Hamming weight of the stacking of the shifted share matrices can be calculated by the total number of the 1's subtracted by the number of the 1's that are ineffective (when two 1's overlapped, we only count one as ineffective).

In the first case, denote by $s_{0,c}^1$ and $s_{1,c}^1$ the number of 1's that are ineffective for the collections C_0 and C_1, respectively when the subpixel c is shifted in. Since there are $m!$ share matrices in the collection C_0 and C_1, so the total number of 1's that are ineffective in the top right corner of all the share matrices in C_0 and C_1 is $s_{0,1}^1 = s_{1,1}^1 = \frac{a'+c+e}{m}m!$ (when a 1 is shifted in) and $s_{0,0}^1 = s_{1,0}^1 = 0$ (when a 0 is shifted in), where $\frac{a'+c+e}{m}$ is the probability of the 1's in the top right corner of the first row.

In the second case, denote by $s_{0,c}^2$ and $s_{1,c}^2$ the number of 1's that are ineffective for the collections C_0 and C_1, respectively when the subpixel c is shifted in. So the total number of 1's that are ineffective in the bottom left corner of all the share matrices in C_0 and C_1 is $s_{1,1}^2 = s_{0,1}^2 = s_{1,0}^2 = s_{0,0}^2 = \frac{a'+d+e}{m}m!$, where $\frac{a'+d+e}{m}$ is the probability of the 1's in the bottom left corner of the second row.

In the third case, denote by $s_{0,c}^3$ and $s_{1,c}^3$ the number of 1's that are ineffective for the collections C_0 and C_1 respectively when the subpixel c is shifted in. Note that, the pattern $\begin{bmatrix} 1 \\ 1 \end{bmatrix}$ in the shifted share matrices is the shifted result of the following four patterns, $\begin{bmatrix} 11 \\ 01 \end{bmatrix}$, $\begin{bmatrix} 10 \\ 11 \end{bmatrix}$, $\begin{bmatrix} 10 \\ 01 \end{bmatrix}$, and $\begin{bmatrix} 11 \\ 11 \end{bmatrix}$ in the collections C_0 and C_1. We calculate the probability of the first pattern $\begin{bmatrix} 11 \\ 01 \end{bmatrix}$ for example, and the other three patterns can be calculated similarly.

The probability that the pattern $\begin{bmatrix} 1 \\ 0 \end{bmatrix}$ appears at the column i in the matrices of the collection C_1 is $\frac{c+e}{m}$, and, fixing this pattern, the probability that the pattern $\begin{bmatrix} 1 \\ 1 \end{bmatrix}$ appears at the column $i + 1$ of the collection C_1 is $\frac{a'}{m-1}$, where $1 \leq i \leq m - 1$. So the probability that the pattern $\begin{bmatrix} 11 \\ 01 \end{bmatrix}$ appears both at the columns i and $i + 1$ in the collection C_1 is $\frac{c+e}{m}\frac{a'}{m-1}$. Similarly, for the remaining three patterns, the results are shown in Table 11.2.

TABLE 11.2
The probability of the four patterns appearing at the columns i and $i + 1$ in the collections C_0 and C_1.

collections\patterns	$\begin{matrix}10\\11\end{matrix}$		$\begin{matrix}11\\11\end{matrix}$	$\begin{matrix}10\\01\end{matrix}$	$\begin{matrix}11\\01\end{matrix}$
C_1	$\frac{a'}{m}\frac{d+e}{m-1}$		$\frac{a'}{m}\frac{a'-1}{m-1}$	$\frac{c+e}{m}\frac{d+e}{m-1}$	$\frac{c+e}{m}\frac{a'}{m-1}$
C_0	$\frac{a'+e}{m}\frac{d}{m-1}$		$\frac{a'+e}{m}\frac{a'+e-1}{m-1}$	$\frac{c}{m}\frac{d}{m-1}$	$\frac{c}{m}\frac{a'+e}{m-1}$

The collection C_0 and C_1 contain all the column permutations of the basis matrices in all possible ways, and there are only $m - 1$ choices for the value of i, so the total number of 1's that are ineffective of the four patterns of the collection C_1 is $s_{1,1}^3 = s_{1,0}^3 = (\frac{a'}{m}\frac{d+e}{m-1} + \frac{a'}{m}\frac{a'-1}{m-1} + \frac{c+e}{m}\frac{d+e}{m-1} + \frac{c+e}{m}\frac{a'}{m-1})(m-1)m!$, and that of the collection C_0 is $s_{0,1}^3 = s_{0,0}^3 = (\frac{a'+e}{m}\frac{d}{m-1} + \frac{a'+e}{m}\frac{a'+e-1}{m-1} + \frac{c}{m}\frac{d}{m-1} + \frac{c}{m}\frac{a'+e}{m-1})(m-1)m!$.

Denote the total number of 1's in the share matrices in collections C_0 and C_1 plus the number of 1's in the subpixel c that is shifted in as $T_{0,c}$ and $T_{1,c}$, respectively. Then, when a 1 is shifted in, we have $T_{0,1} = T_{1,1} = (2a' + c + d + 2e + 1)m!$, and when a 0 is shifted in, we have $T_{0,0} = T_{1,0} = (2a' + c + d + 2e)m!$.

Denote $T_{c,C}$ as the total stacking Hamming weight of all the matrices of the collection C (C_0 or C_1) when a string of subpixels c are shifted in. The above discussion shows that when a 1 is shifted in, the total stacking Hamming weight of all the matrices of the collection C_1 is

$$
\begin{aligned}
T_{1,C_1} &= T_{1,1} - s_{1,1}^1 - s_{1,1}^2 - s_{1,1}^3 \\
&= [(2a' + c + d + 2e + 1) - \frac{a'+d+e}{m} - \frac{a'+c+e}{m} \\
&\quad -(\frac{a'(d+e)}{m} + \frac{a'(a'-1)}{m} + \frac{(c+e)(d+e)}{m} + \frac{(c+e)a'}{m})]m!
\end{aligned}
$$

and that of the collection C_0 is

$$
\begin{aligned}
T_{1,C_0} &= T_{0,1} - s_{0,1}^1 - s_{0,1}^2 - s_{0,1}^3 \\
&= [(2a' + c + d + 2e + 1) - \frac{a'+d+e}{m} - \frac{a'+c+e}{m} \\
&\quad -(\frac{(a'+e)d}{m} + \frac{(a'+e)(a'+e-1))}{m} + \frac{cd}{m} + \frac{c(a'+e)}{m})]m!
\end{aligned}
$$

When a 0 is shifted in, the total stacking Hamming weight of the collection C_1 is,

$$
\begin{aligned}
T_{0,C_1} &= T_{1,0} - s_{1,0}^1 - s_{1,0}^2 - s_{1,0}^3 \\
&= [(2a' + c + d + 2e) - \frac{a'+d+e}{m} - \\
&\quad (\frac{a'(d+e)}{m} + \frac{a'(a'-1)}{m} + \frac{(c+e)(d+e)}{m} + \frac{(c+e)a'}{m})]m!
\end{aligned}
$$

and that of the collection C_0 is

$$\begin{aligned}
T_{0,C_0} &= T_{0,0} - s_{0,0}^1 - s_{0,0}^2 - s_{0,0}^3 \\
&= [(2a' + c + d + 2e) - \frac{a'+d+e}{m} - \\
&\quad (\frac{(a'+e)d}{m} + \frac{(a'+e)(a'+e-1)}{m} + \frac{cd}{m} + \frac{c(a'+e)}{m})]m!
\end{aligned}$$

We now define *the average stacking Hamming weight* of each share matrix to be the total stacking Hamming weight of all the share matrices being divided by the number of the share matrices, i.e., $\frac{T_{c,C}}{m!}$. Then the difference between the average stacking Hamming weight of each share matrix of the shifted collections C_0 and C_1, denoted by D_A, is

$$D_A = (T_{1,C_1} - T_{1,C_0})p_1 + (T_{0,C_1} - T_{0,C_0})p_0 = -\frac{e(p_1+p_0)}{m} = -\frac{e}{m}.$$

According to the definition of the average contrast in Section 11.2, with $\bar{h} = T_{1,C_1}p_1 + T_{0,C_1}p_0$ and $\bar{l} = T_{1,C_0}p_1 + T_{0,C_0}p_0$, we get the value of the average contrast $\bar{\alpha} = \frac{\bar{h}-\bar{l}}{m} = \frac{D_A}{m} = -\frac{e}{m^2}$.

Now we consider the general case when the share 2 is shifted by r subpixels. For this case there are 2^r possible subpixels that can be shifted in. For example, for $r = 2$, the shifted in strings of subpixels have four cases, 00, 01, 10 and 11. Denote by p_{00}, p_{01} p_{10} and p_{11} the probabilities of these four cases to happen respectively, then we have $p_{00} + p_{01} + p_{10} + p_{11} = 1$.

We consider the string of subpixels c, let the Hamming weight of c be s, i.e., $w(c) = s$. The total number of 1's that are ineffective in the top right corner of all the share matrices in C_0 and C_1 is $s_{0,c}^1 = s_{1,c}^1 = s \cdot \frac{a'+c+e}{m}m!$, and the total number of 1's that are ineffective in the bottom left corner of all the share matrices in C_0 and C_1 is $s_{0,c}^2 = s_{1,c}^2 = r \cdot \frac{a'+d+e}{m}m!$. For the third case, the pattern $\begin{bmatrix} 1 \\ 1 \end{bmatrix}$ in the shifted share matrices is the shifted result of the

following four patterns, $\begin{bmatrix} \overbrace{1 \cdots 1} \\ \underbrace{0 \cdots 1}_{r} \end{bmatrix}$, $\begin{bmatrix} \overbrace{1 \cdots 0} \\ \underbrace{1 \cdots 1}_{r} \end{bmatrix}$, $\begin{bmatrix} \overbrace{1 \cdots 0} \\ \underbrace{0 \cdots 1}_{r} \end{bmatrix}$ and $\begin{bmatrix} \overbrace{1 \cdots 1} \\ \underbrace{1 \cdots 1}_{r} \end{bmatrix}$

in the collections C_0 and C_1, and there are only $m - r$ choices for the value of the position i, so the total number of the 1's that are ineffective of the four patterns of the collection C_1 is

$$(\frac{a'}{m}\frac{d+e}{m-1} + \frac{a'}{m}\frac{a'-1}{m-1} + \frac{c+e}{m}\frac{d+e}{m-1} + \frac{c+e}{m}\frac{a'}{m-1})(m-r)m!$$

and of the collection C_0 is

$$(\frac{a'+e}{m}\frac{d}{m-1} + \frac{a'+e}{m}\frac{a'+e-1}{m-1} + \frac{c}{m}\frac{d}{m-1} + \frac{c}{m}\frac{a'+e}{m-1})(m-r)m!$$

Hence, when a string of subpixels of c is shifted in, the total stacking Hamming weight of all the matrices of the collection C_1 is

$$\begin{aligned}
T_{c,C_1} = \quad &[(2a' + c + d + 2e + s) - r \cdot \frac{a'+d+e}{m} - s \cdot \frac{a'+c+e}{m} \\
&- (\frac{a'(d+e)}{m} + \frac{a'(a'-1)}{m} + \frac{(c+e)(d+e)}{m} + \frac{(c+e)a'}{m})\frac{m-r}{m-1}]m!
\end{aligned}$$

and that of the collection C_0 is

$$
\begin{aligned}
T_{c,C_0} = \quad & [(2a' + c + d + 2e + s) - r \cdot \tfrac{a'+d+e}{m} - s \cdot \tfrac{a'+c+e}{m} \\
& - (\tfrac{(a'+e)d}{m} + \tfrac{(a'+e)(a'+e-1))}{m} + \tfrac{cd}{m} + \tfrac{c(a'+e)}{m}) \tfrac{m-r}{m-1}] m!.
\end{aligned}
$$

Hence, the difference between the average stacking Hamming weight of each share matrix of the shifted collections C_0 and C_1, denoted by D_A, is

$$
\begin{aligned}
D_A \quad & = (T_{1\cdots1,C_1} - T_{1\cdots1,C_0})p_{1\cdots1} + \cdots + (T_{0\cdots0,C_1} - T_{0\cdots0,C_0})p_{0\cdots0} \\
& = -\tfrac{e(p_{1\cdots1}+\cdots+p_{0\cdots0})}{m}\tfrac{m-r}{m-1} \\
& = -\tfrac{e}{m}\tfrac{m-r}{m-1}
\end{aligned}
$$

and the average contrast is

$$
\bar{\alpha} = \frac{D_A}{m} = -\frac{e}{m^2}\frac{m-r}{m-1}.
$$

Because the shifted scheme is not a DVCS anymore and $\bar{\alpha} \neq 0$, let $\bar{h} = T_{1\cdots1,C_1}p_{1\cdots1} + \cdots + T_{0\cdots0,C_1}p_{0\cdots0}$ and $\bar{l} = T_{1\cdots1,C_0}p_{1\cdots1} + \cdots + T_{0\cdots0,C_0}p_{0\cdots0}$, then it is known that the shifted scheme is a PVCS. This completes the proof of Theorem 1. □

Note that, after a shift, the value of the average contrast has a negative value $\bar{\alpha} < 0$, which means that the recovered secret image is the complementary image of the original one, and the absolute value of $\bar{\alpha}$ reflects how clear the image can be viewed visually.

The above Theorem 1 shows that, to align the transparencies when decrypting the DVCS, one does not need to align the transparencies precisely. So, when the participants of a DVCS want to align the transparencies, for example, the transparencies in Example 1, they can first align the transparencies precisely in the vertical direction, and then move the second transparencies to the right then to the left in the horizontal direction. Then they will get the recovered secret image for three times. Furthermore, this phenomenon also helps to determine the size of the subpixels printed on the transparencies.

For other visual cryptography schemes, such as the extended visual cryptography schemes in [2], the visual cryptography schemes for the general access structures [1] and the color visual cryptography schemes [15, 12], they all have the phenomenon stated above. The proof for these visual cryptography schemes can be modified from that of Theorem 1.

11.4 Misalignment with Less Than One Subpixel

In this section, we do further investigation on misalignment with less than one subpixel. We show that the secret image can still be recovered visually by a slight misalignment. We first give an example, a $(2,2)$-VCS, to show this phenomenon. Denote (d_x, d_y) as the horizontal and vertical misalignment deviations from the original position of the subpixel.

Example 3 *The images (a), (b), and (c) in Figure 11.2 show the recovered secret image of a (2,2)-VCS for three misalignment deviations, $(d_x, d_y) = (0,0)$, $(0.5,0)$, and $(1,2)$ (unit: subpixel).*

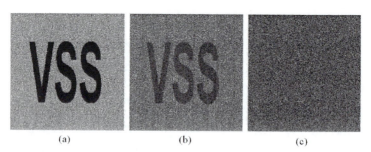

(a)   (b)   (c)

FIGURE 11.2
Recovered secret images of a (2,2)-VCS for three misalignment deviations (a) $(d_x, d_y) = (0,0)$, (b) $(d_x, d_y) = (0.5,0)$, and (c) $(d_x, d_y) = (1,2)$.

From Figure 11.2, we can observe that, the clearness of Figure 11.2 (b) is worse than the Figure 11.2 (a), and the secret on Figure 11.2 (c) is completely invisible.

Figure 11.2 shows the case that only two shares are superimposed. The misalignment problem will become more complex for stacking k shares for a (k, n)-VCS. Note that, the recovered image will disappear if any one out of k shares is not at the correct position. In fact the misalignment problem critically increases as k grows.

The next two subsections are organized as follows: first, we will investigate the conditions that the secret image can still be observed when the shares are slightly misaligned, whereby saying slight misalignment we mean deviation (d_x, d_y) satisfying $0 \le d_x, d_y \le 1$; and then we will investigate the misalignment tolerance of large and small pixels. To simplify the discussion, we only consider the case of $(2, 2)$-VCS.

11.4.1 Shares with Slightly Misalignment Can Still Recover the Secret Image

We consider a $(2, 2)$-VCS with deviation (d_x, d_y), where a pixel 1W1B (resp. 1B1W) represents a white pixel that contains a white (resp. black) and a black (resp. white) subpixel; and a pixel 2B0W represents a black pixel that contains two black subpixels. All possible four cases, $(a1)$, $(a2)$, $(b1)$, and $(b2)$, of stacked shares with deviation (d_x, d_y) are shown in Figure 11.3. The stacked results in Figure 11.3 are marked with red boxes. As shown in Figure 11.3, the stacked result is divided into eight parts. Denote A_1, A_2, A_3, A_4, A_1', A_2', A_3', and A_4' as the areas of the eight parts respectively, and they satisfy the following equations.

$$
\begin{aligned}
A_1 &= A_1' = d_x d_y \\
A_2 &= A_2' = (s - d_x) d_y \\
A_3 &= A_3' = d_x (s - d_y) \\
A_4 &= A_4' = (s - d_x)(s - d_y) \\
A_1 &+ A_2 + A_3 + A_4 = s^2
\end{aligned}
\tag{11.1}
$$

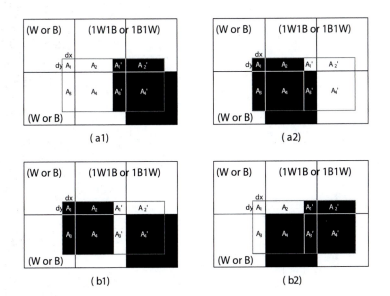

FIGURE 11.3
Stacked results of white pixels (a1: 1W1B+1W1B, a2: 1B1W+1B1W) and black pixels (b1: 1W1B+1B1W, b2: 1B1W+1W1B) for the same deviation (d_x, d_y).

Denote $s \times s$ as the size of a subpixel, and denote $A_{C,c}$ as the area in the stacked results with color $C \in \{W, B\}$ for case $c \in (a1), (a2), (b1)$, and $(b2)$. Note that, some areas are not deterministic in Figure 11.3, for example the area A_1 in Figure 11.3 $(a1)$ belongs to the adjacent pixel on the top left. It can be either black or white with equal probability 50%. We introduce a symbol $P(A)$ to denote it, i.e., $P(A)$ is the area of A to be either black or white with equal probability 50%. Furthermore, we denote $P(A_1, A_2)$ as the area that is either A_1 or A_2 with equal probability 50% since there exist such cases that one of two areas is black and the other is white, for example the areas A_1' or A_2' in Figure 11.3 $(a2)$. The symbols $P(A)$ and $P(A_1, A_2)$ satisfy the following equations:

$$\min\{P(A)\} = 0$$
$$\max\{P(A)\} = A$$
$$\bar{P}(A) = A/2$$
$$\min\{P(A_1, A_2)\} = \min\{A_1, A_2\} \tag{11.2}$$
$$\max\{P(A_1, A_2)\} = \max\{A_1, A_2\}$$
$$\bar{P}(A_1, A_2) = A_1/2 + A_2/2$$

According to Figure 11.3, it is easy to verify the following equations:

$$A_{W,a1} = P(A_1) + P(A_2) + P(A_3) + A_4$$
$$A_{B,a1} = P(A_1) + P(A_2) + P(A_3) + A_1' + A_2' + A_3' + A_4'$$
$$A_{W,a2} = P(A_1', A_2') + A_4'$$
$$A_{B,a2} = A_1 + A_2 + A_3 + A_4 + P(A_1', A_2') + A_3'$$
$$A_{W,b1} = P(A_1' + A_2') + A_3' \tag{11.3}$$
$$A_{B,b1} = A_1 + A_2 + A_3 + A_4 + P(A_1' + A_2') + A_4'$$
$$A_{W,b2} = P(A_1) + P(A_2) + P(A_3)$$
$$A_{B,b2} = P(A_1) + P(A_2) + P(A_3) + A_4 + A_1' + A_2' + A_3' + A_4'$$

The following Theorem 2 shows that the stacked shares with slight misalignment can still recover the secret image. Note that, by saying the secret image is recovered by its original color, we mean that a black (resp. white) pixel in the secret image is represented by a black (resp. white) pixel in the recovered secret image; and by saying the secret image is recovered by its complementary color, we mean that a black (resp. white) pixel in the secret image is represented by a white (resp. black) pixel in the recovered secret image. In order to consist with Definition 1 and Definition 2, we generalize the definition of contrast α and average contrast $\bar{\alpha}$ as follows:

$$\alpha = \frac{h - l}{2s^2} \ and \ \bar{\alpha} = \frac{\bar{h} - \bar{l}}{2s^2} \tag{11.4}$$

where $2s^2$ is the area of a pixel (including two subpixels) and the definitions of h, l, $\bar{h}$, and $\bar{l}$ are the same as that in Definition 1 and Definition 2.

Theorem 2 *For a misaligned $(2,2)$-VCS, denote (d_x, d_y) as the deviation of the stacked shares, then the secret image can still be recovered if (d_x, d_y) falls in the regions of R_1, R_2, and R_3 in Figure 11.4, where the properties of the misaligned scheme are shown in Table 11.3.*

Proof: Denote $l_{black\,area}^W$ (resp. $h_{black\,area}^B$) as the maximum (resp. minimum) black area of a white (resp. black) pixel, denote $\bar{l}_{black\,area}^W$ (resp. $\bar{h}_{black\,area}^B$) as the average black area of a white (resp. black) pixel, denote $h_{white\,area}^B$ (resp. $l_{white\,area}^W$) as the minimum (resp. maximum) white area of a black (resp. white) pixel, denote $\bar{h}_{white\,area}^B$ (resp. $\bar{l}_{white\,area}^W$) as the average white area of a black (resp. white) pixel. According to Figure 11.3, we have:

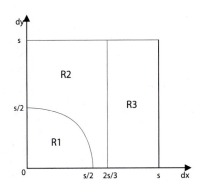

FIGURE 11.4
The regions of deviation (d_x, d_y) that can recover the secret image.

TABLE 11.3
The properties of the misaligned schemes.

regions	R_1	R_2	R_3
(d_x, d_y) *satisfy*	$(s - d_x)(s - d_y)$ $> s^2/2$ $0 \le d_x, d_y < s$	$0 \le d_x < 2s/3$ $0 \le d_y < s$	$2s/3 < d_x \le s$ $0 \le d_y < s$
misaligned scheme	$DVCS$	$PVCS$	$DVCS$
contrast	$\alpha_{R_1} =$ $\frac{s^2 - 2(d_x + d_y)s + 2d_x d_y}{2s^2}$	$\bar{\alpha}_{R_2} =$ $\frac{(2s - 3d_x)(s - d_y)}{4s^2}$	$\bar{\alpha}_{R_3} =$ $-\frac{(3d_x - 2s)(s - d_y)}{4s^2}$
recovered secret image	*original color*	*original color*	*complementary color*

$l^W_{black\,area}$ is the maximum value of the black area for a white pixel, which contains two cases $(a1)$ and $(a2)$, i.e.,

$$
\begin{aligned}
l^W_{black\,area} &= \max\{A_{B,a1}, A_{B,a2}\}\\
&= \max\{P(A_1) + P(A_2) + P(A_3) + A'_1 + A'_2 + A'_3 + A'_4,\\
&\quad\; A_1 + A_2 + A_3 + A_4 + P(A'_1, A'_2) + A'_3\}\\
&= A_1 + A_2 + A_3 + A_4 +\\
&\quad\; \max\{P(A_1) + P(A_2) + P(A_3), P(A'_1, A'_2) + A'_3\}\\
&= A_1 + A_2 + A_3 + A_4 + A_1 + A_2 + A_3\\
&= s^2 + A_1 + A_2 + A_3\\
&= 2s^2 - A_4
\end{aligned}
$$

(11.5)

$h^B_{black\,area}$ is the minimum value of the black area for a black pixel, which contains two cases $(b1)$ and $(b2)$, i.e.,

$$h^B_{black\,area} = \min\{A_{B,b1}, A_{B,b2}\} = s^2 + A_4 \tag{11.6}$$

$\bar{l}^W_{black\,area}$ is the average black area for a white pixel, which contains two cases $(a1)$ and $(a2)$, i.e.,

$$
\begin{aligned}
\bar{l}^W_{black\,area} &= \tfrac{\bar{A}_{B,a1} + \bar{A}_{B,a2}}{2}\\
&= \tfrac{1}{2}[(P(A_1) + P(A_2) + P(A_3) + A'_1 + A'_2 + A'_3 + A'_4)\\
&\quad\; + (A_1 + A_2 + A_3 + A_4 + P(A'_1, A'_2) + A'_3)]\\
&= \tfrac{1}{2}[(\bar{P}(A_1) + \bar{P}(A_2) + \bar{P}(A_3) + A'_1 + A'_2 + A'_3 + A'_4)\\
&\quad\; + (A_1 + A_2 + A_3 + A_4 + \bar{P}(A'_1, A'_2) + A'_3)]\\
&= s^2 + \tfrac{A_1}{2} + \tfrac{A_2}{2} + \tfrac{3A_3}{4}
\end{aligned}
$$

(11.7)

$\bar{h}^B_{black\,area}$ is the average black area for a black pixel, which contains two cases $(b1)$ and $(b2)$, i.e.,

$$\bar{h}^B_{black\,area} = \frac{\bar{A}_{B,b1} + \bar{A}_{B,b2}}{2} = s^2 + \frac{A_1}{2} + \frac{A_2}{2} + \frac{A_3}{4} + A_4 \tag{11.8}$$

$h^B_{white\,area}$ is the minimum value of the white area for a black pixel, which contains two cases $(b1)$ and $(b2)$, i.e.,

$$h^B_{white\,area} = \min\{A_{W,b1}, A_{W,b2}\} = 0 \tag{11.9}$$

$l^W_{white\,area}$ is the maximum value of the black area for a white pixel, which contains two cases $(a1)$ and $(a2)$, i.e.,

$$l^W_{white\,area} = \max\{A_{W,a1}, A_{W,a2}\} = s^2 \tag{11.10}$$

$\bar{h}^B_{white\,area}$ is the average white area for a black pixel, which contains two cases $(b1)$ and $(b2)$, i.e.,

$$\bar{h}^B_{white\,area} = \frac{\bar{A}_{W,b1} + \bar{A}_{W,b2}}{2} = \frac{A_1}{2} + \frac{A_2}{2} + \frac{3A_3}{4} \tag{11.11}$$

$\bar{l}^W_{white\,area}$ is the average white area for a white pixel, which contains two cases $(a1)$ and $(a2)$, i.e.,

$$\bar{l}^W_{white\,area} = \frac{\bar{A}_{W,a1} + \bar{A}_{W,a2}}{2} = \frac{A_1}{2} + \frac{A_2}{2} + \frac{A_3}{4} + A_4 \qquad (11.12)$$

According to Definition 1 (definition of a deterministic VCS), in order to deterministically recover the secret image by its original color, the values of $l^W_{black\,area}$ and $h^B_{black\,area}$ should satisfy $l^W_{black\,area} < h^B_{black\,area}$. Together with Equations (11.1), (11.5), and (11.6), we get (d_x, d_y) to satisfy $(s - d_x)(s - d_y) > s^2/2$, i.e., (d_x, d_y) falls in the region R_1. And the contrast α_{R_1} is

$$\alpha_{R_1} = \frac{h^B_{black\,area} - l^W_{black\,area}}{2s^2} = \frac{s^2 - 2(d_x + d_y)s + 2d_x d_y}{2s^2}$$

Similarly, in order to recover deterministically by its complementary color, the values of $h^B_{white\,area}$ and $l^W_{white\,area}$ should satisfy $h^B_{white\,area} > l^W_{white\,area}$. However, according to Equations (11.9) and (11.10), we get that $h^B_{white\,area} > l^W_{white\,area}$ does not hold. Hence, the secret image cannot be recovered deterministically by its complementary color.

According to Definition 2 (definition of a probabilistic VCS), in order to probabilistically recover the secret image by its original color, the values of $\bar{l}^W_{black\,area}$ and $\bar{h}^B_{black\,area}$ should satisfy $\bar{l}^W_{black\,area} < \bar{h}^B_{black\,area}$. Together with Equations (11.1), (11.7), and (11.8), we get (d_x, d_y) to satisfy $0 \le d_x < 2s/3$, $0 \le d_y < s$. By excluding the region R_1 we get to know that (d_x, d_y) falls in the region R_2. And the contrast $\bar{\alpha}_{R_2}$ is

$$\bar{\alpha}_{R_2} = \frac{\bar{h}^B_{black\,area} - \bar{l}^W_{black\,area}}{2s^2} = \frac{(2s - 3d_x)(s - d_y)}{4s^2}$$

Similarly, in order to probabilistically recover the secret image by its complementary color, the values of $\bar{h}^B_{white\,area}$ and $\bar{l}^W_{white\,area}$ should satisfy $\bar{h}^B_{white\,area} > \bar{l}^W_{white\,area}$. Together with Equations (11.1), (11.11), and (11.12), we get (d_x, d_y) to satisfy $2s/3 < d_x \le s$, $0 \le d_y < s$, i.e. (d_x, d_y) fall in the region R_3. And the contrast $\bar{\alpha}_{R_3}$ is

$$\bar{\alpha}_{R_3} = \frac{\bar{h}^B_{black\,area} - \bar{l}^W_{black\,area}}{2s^2} = -\frac{(3d_x - 2s)(s - d_y)}{4s^2}$$

$\square$

The above Theorem 2 is consistent with Theorem 1. According to Theorem 1 and Theorem 2, for the deviation $(d_x, d_y) = (s, 0)$, the secret image can be probabilistically recovered by its complementary color with average contrast $\bar{\alpha} = 1/4$.

The above Theorem 2 considers the deviations (d_x, d_y) less than one subpixel. In fact, when the value of d_x is between s and $2s$ for a misaligned $(2, 2)$-VCS, the secret image can also be recovered. The proof is left to the interested readers.

11.4.2   Large Subpixels Have Better Misalignment Tolerance

In this subsection, we investigate the misalignment tolerance of the large and small subpixels. And show that shares with large subpixels have better misalignment tolerance when recovering the secret image by its original color than that with small subpixels. Denote the size of the large and small subpixels as $s_2 \times s_2$ and $s_1 \times s_1$ respectively, where $s_2 > s_1$. According to Figure 11.5, the recovered secret image using the small subpixels (Figure 11.5(a)) has a higher resolution than that using the large subpixels (Figure 11.5(c)). It is observed that Figure 11.5(a) indeed has the refined resolution in detail of Lena image. We can clearly view the hair in Figure 11.5(a) but the area of hair is all black in Figure 11.5(c). The image quality using the medium-sized subpixel (Figure 11.5(b)) is between Figure 11.5(a) and Figure 11.5(c). Although using large subpixels in share has poorer resolution, it will be more robust to the misalignment error. Next, the misalignment tolerance of different-sized subpixels is formally analyzed.

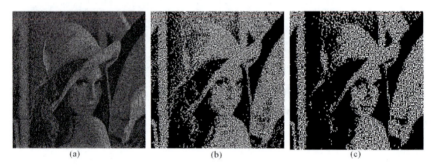

(a)   (b)   (c)

FIGURE 11.5
Recovered Lena images for (2,2)-VCS using the different sized subpixels: (a) the small-sized subpixel, (b) the medium-sized subpixel and (c) the large-sized subpixel.

Consider a misalignment deviation (d_x, d_y) in the $(2, 2)$-VCS. All the possible cases of stacked shares of a white secret pixel (1B1W) and a black secret pixel (2B0W) with a deviation (d_x, d_y) are shown in Figure 11.3. For the small subpixel ($s_1 \times s_1$), we define the whiteness $R_{W,a1}$ (resp. $R_{W,a2}$) as the ratio of black area in a stacked two-subpixel area for the case $a1$ (resp. $a2$) in Figure 11.3, and the darkness $R_{B,b1}$ (resp. $R_{B,b2}$) as the ratio of the black area in a stacked two-subpixel area for the case $b1$ (resp. $b2$) in Figure 11.3. For the large subpixel ($s_2 \times s_2$), the corresponding denotations are $R'_{W,a1}$, $R'_{W,a2}$, $R'_{B,b1}$ and $R'_{B,b2}$.

To simplify the discussion, we only consider the one dimension deviation, i.e., $(d_x, 0)$ or $(0, d_y)$. Then we have the following one-dimensional deviation lemma.

Lemma 3 (One-dimensional deviation) *The darkness and the whiteness ratios always satisfy* $R'_{W,a1} \geq R_{W,a1}$, $R'_{W,a2} \geq R_{W,a2}$ $R'_{B,b1} \geq R_{B,b1}$ *and* $R'_{B,b2} \geq R_{B,b2}$ *for the deviations* $(d_x, 0)$ *and* $(0, d_y)$ *where* $0 < d_x, d_y < s_1$.

Proof: We only consider the case with deviation $(d_x, 0)$ since the proof is the same for the case with deviation $(0, d_y)$. According to equation (11.1), when $d_y = 0$, we have

$$
\begin{aligned}
A_1 &= A'_1 = 0 \\
A_2 &= A'_2 = 0 \\
A_3 &= A'_3 = d_x s \\
A_4 &= A'_4 = (s - d_x)s \\
A_3 + A_4 &= s^2
\end{aligned}
\tag{11.13}
$$

We consider the relation of $R_{C,c}$ and $R'_{C,c}$ for $C \in \{W, B\}$ and $c \in (a1)$, $(a2)$, $(b1)$ and $(b2)$. Recall that $s_1 < s_2$, and $R_{C,c}$ and $R'_{C,c}$ can be calculated by the following formula:

$$
R_{C,c} = \frac{A_{C,c}}{2s_1^2} \text{ and } R'_{C,c} = \frac{A_{C,c}}{2s_2^2}
\tag{11.14}
$$

We have Table 11.4 to show the relation of $R_{C,c}$ and $R'_{C,c}$ for different values of C and c.

TABLE 11.4
The relation of the ratios $R_{C,c}$ and $R'_{C,c}$.

Stacked pixel	Area ($A_{C,c}$)	Ratio ($R_{C,c}$)	Ratio ($R'_{C,c}$)	Relation
White,a1	$A_{W,a1} = A_4$	$R_{W,a1} = \frac{s_1 - d_x}{2s_1}$	$R'_{W,a1} = \frac{s_2 - d_x}{2s_2}$	$R'_{W,a1} > R_{W,a1}$
	$A_{W,a1} = A_3 + A_4$	$R_{W,a1} = 1/2$	$R'_{W,a1} = 1/2$	$R'_{W,a1} = R_{W,a1}$
White,a2	$A_{W,a2} = A'_4$	$R_{W,a2} = \frac{s_1 - d_x}{2s_1}$	$R'_{W,a2} = \frac{s_2 - d_x}{2s_2}$	$R'_{W,a2} > R_{W,a2}$
Black,b1	$A_{B,b1} = A_3 + A_4 + A'_4$	$R_{B,b1} = 1 - \frac{d_x}{2s_1}$	$R'_{W,a1} = 1 - \frac{d_x}{2s_2}$	$R'_{B,b1} > R_{B,b1}$
Black,b2	$A_{B,b2} = A_3 + A'_3 + A'_4$	$R_{B,b2} = 1 - \frac{d_x}{2s_1}$	$R'_{W,a2} = 1 - \frac{d_x}{2s_2}$	$R'_{B,b2} > R_{B,b2}$
	$A_{B,b2} = A_3 + A_4 + A'_3 + A'_4$	$R_{B,b2} = 1$	$R'_{W,a2} = 1$	$R'_{B,b2} = R_{B,b2}$

The proofs of the relations in Table 11.4 are quite similar. We only take the proof of the relation between $R'_{W,a1}$ and $R_{W,a1}$, for example.

According to equation (11.3), for the stacked pixel of Figure 11.3 (a1), there are two cases of $A_{W,a1}$, i.e., I: $A_{W,a1} = A_4$ and II: $A_{W,a1} = A_3 + A_4$.

For the Case I, we have $R_{W,a1} = \frac{A_{W,a1}}{2s_1^2} = \frac{A_4}{2s_1^2} = \frac{s_1 - d_x}{2s_1}$ and $R'_{W,a1} = \frac{s_2 - d_x}{2s_2}$, by subtraction we have $R'_{W,a1} - R_{W,a1} = \frac{s_2 - d_x}{2s_2} - \frac{s_1 - d_x}{2s_1} = \frac{d_x(s_2 - s_1)}{2s_1 s_2} > 0$, i.e. $R'_{W,a1} > R_{W,a1}$.

For the Case II, we have $R'_{W,a1} - R_{W,a1} = \frac{s_2^2}{2s_2^2} - \frac{s_1^2}{2s_1^2} = 0$, i.e. $R'_{W,a1} = R_{W,a1}$. According to Table 11.4, the lemma follows. □

According to Lemma 3, for one-dimensional deviation, it is evident that if $R'_W \geq R_W$ and $R'_B \geq R_B$, the whiteness of the white secret pixel is whiter and the darkness of the black secret pixel is darker when using the large subpixels. We can conclude that the large subpixels have better misalignment tolerance than that of small subpixels. Actually, for two-dimensional deviation, the conclusion also holds. However, the proof is rather complex; we omit the proof here. Readers also can reach this conclusion via simulations, for example the Figure 11.6, and more simulations for two-dimensional deviation can be found in [16].

(a)   (b)

FIGURE 11.6
Recovered secret image for a $(2,2)$-VCS using two different-sized subpixels and $(d_x, d_y) = (0.5, 0.5)$, $(s_1/s_2) = 2$: (a) the small subpixel and (b) the large subpixel; two secret image (a printed text "VSS" and a halftoned Lena image) are tested.

From the preceding description and the results in Figure 11.5 and Figure 11.6, unfortunately, there exists another dilemma of using the large or small subpixels. Together with previous comparisons in Table 11.1, we summarize the advantages and disadvantages of large and small subpixels in Table 11.5.

In order to bring these conflicting goals in a kind of balance, we properly distribute two-sized subpixels in shares to develop their specialities and simultaneously avoid corresponding disadvantages. Our method is based on the trade-off between the usage of large and small subpixels. Both two-sized subpixels create a trade-off, which a large size subpixel leads to the high misalignment tolerance but the low resolution, while the small subpixel has the opposite properties. Finally, we successfully reduce the difficulty of aligning shares. Designing our misalignment tolerant VCS therefore delivers the fol-

TABLE 11.5

The advantages and disadvantages of large and small subpixels.

size of the subpixels	advantages	disadvantages
larger	easier to align and better misalignment tolerance	larger transparency size and lower resolution
smaller	smaller transparency size and higher resolution	harder to align and worse misalignment tolerance

lowing problems: (1) What is the appropriate percentage of the large subpixel in a share? (2) What is the appropriate size ratio of the large subpixel to the small subpixel? (3) How do we arrange the large and small subpixels in one share?

11.5 A Misalignment Tolerant VCS

11.5.1 The Algorithm

Concerning the first two problems, both items (the percentage and the size ratio) can be used to trade the quality of the recovered secret image for the misalignment tolerance. Different percentages and size ratios will be experimented and analyzed in Section 11.5.2. Consider the third problem, how to arrange the large and small subpixels in one share. In this chapter, we arrange them in a regular mask or a random mask. Subsequently, we describe the encrypting/decrypting algorithm of a (k, n)-threshold misalignment tolerant VCS, when given a percentage (the item in Problem (1)), a size ratio (the item in Problem 2) and an arrangement mask (the item in problem (3)).

Notation Used:

I A gray secret image of size $x \times y$.

γ^2 The size ratio of a large to a small subpixel, i.e., $\gamma^2 = (s_2/s_1)^2$.

I_S A small-scaled secret image of size $x \times y$. This image is obtained by halftoning I into a binary image. Note that, this image is used for embedding small subpixels.

I_B A large-scaled secret image of size $(x/\gamma) \times (y/\gamma)$. This image is obtained by reducing the secret image I to size $(x/\gamma) \times (y/\gamma)$ and then halftoning to a binary image. Note that, this image is used for embedding large subpixels.

$p_B(p_S)$ p_B and p_S are percentages of large and small subpixels, respectively.

M_{reg} A regular mask of size $(x/\gamma) \times (y/\gamma)$. This mask is only designed for $p_B = p_S = 50\%$. Figure 11.7(a) shows the regular mask with the alternate blocks, $\boxed{\text{B}}$ and $\boxed{\text{S}}$, where a block $\boxed{\text{B}}$ contains large subpixels and a block $\boxed{\text{S}}$ contains small subpixels.

M_{ran} A random mask of size $(x/\gamma) \times (y/\gamma)$. This mask is designed for any p_B and p_S. At this time, $\boxed{\text{B}}$ and $\boxed{\text{S}}$ are randomly chosen according p_B and p_S, as shown in Figure 11.7(b).

$D_B(\cdot)$ Let $A = [a_{i,j}]$, where $a_{i,j}$ is the element of the i-th row and j-th column in A, be a matrix in C_1 or C_0. The function $D_B(\cdot)$ divides a secret pixel s into m large subpixels $(bp)^i_j$ ($(bp)^i_j = 0$ denotes white and $(bp)^i_j = 1$ denotes black) in the i-th share, $1 \leq i \leq n$ and $1 \leq j \leq m$, defined as follows:

$$D_B(s) = \begin{cases} (bp)^i_j &= a_{i,j} \text{ in } A \in C_0 \text{ for } s \text{ is the white secret pixel.} \\ (bp)^i_j &= a_{i,j} \text{ in } A \in C_1 \text{ for } s \text{ is the black secret pixel.} \end{cases}$$

$D_S(\cdot)$ Similar to $D_B(\cdot)$, it divides a secret pixel into m small subpixels $(sp)^i_j$ in the i-th share, $1 \leq i \leq n$ and $1 \leq j \leq m$.

$A_{m_x,m_y}(\cdot)$ Arrangement function that randomly arranges m large (or small) subpixels into an $(m_x \times m_y)$-sized rectangle where $m = m_x \times m_y$.

$O^{(i)}$ n output shares, $i \in [1, n]$, of size $(x \times m_x) \times (y \times m_y)$.

I' The recovered secret image of size $(x \times m_x) \times (y \times m_y)$.

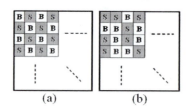

(a)   (b)

FIGURE 11.7
Regular and random masks for arranging the large and small subpixels: (a) regular mask and (b) random mask.

     Even though the shares use a regular mask, an attacker could not gain any information from the mask and the secrecy is not compromised.

Encrypting Algorithm:

Input: I, γ^2, p_B, p_S, C_0 and C_1 of a (k, n)-threshold VCS.

Output: $O^{(i)}$, $i \in [1, n]$.

Step 1:   Obtain the large-scaled and small-scaled secret images I_B and I_S from a secret image I.

Step 2: Use p_B and p_S to generate the M_{ran} mask or choose the M_{reg} mask for the case $p_B = p_S = 50\%$; let the chosen mask $M = M_{ran}$ or M_{reg}.

Step 3: For each block in the mask M, do the following:

Step 3-1: If the block is the large block $\boxed{B}$, for the secret pixel "s" in the corresponding block of the large-scaled image I_B, do the following:

Assign $(bp)^i_j = D_B(s)$ for $i \in [1, n]$ and $j \in [1, m]$;

Use $A_{m_x, m_y}((bp)^i_j)$ to create n $(m_x \times m_y)$-sized rectangles, where $m = m_x \times m_y$, and deliver them in the corresponding block to $O^{(1)}, O^{(2)}, \ldots, O^{(n)}$, respectively.

Step 3-2: If the block is the large block $\boxed{\text{S}}$, for the secret pixel "s" in the corresponding block of the large-scaled image I_S, do the following:

Assign $(bp)^i_j = D_S(s)$ for $i \in [1, n]$ and $j \in [1, m]$;

Use $A_{m_x, m_y}((sp)^i_j)$ to create n $(m_x \times m_y)$-sized rectangles, where $m = m_x \times m_y$, and deliver them in the corresponding block to $O^{(1)}, O^{(2)}, \ldots, O^{(n)}$, respectively.

Decrypting Algorithm:

Input: Any k shares from $O^{(i)}$, $i \in [1, n]$.

Output: I'.

Step 1: Print out k shares on transparencies. Stack and align them by hand with the approximate accuracy.

/* Note that, when stacking shares the tradition VCS needs the precise alignment; the proposed scheme has the misalignment tolerance and so that one just aligns them roughly. */

Step 2: Decrypt the secret directly by human eyes.

Step 3: Align shares gradually to get a refined secret image of I'.

/* One can first align transparencies precisely in the vertical direction, and then move gradually by hand without losing the secret in the horizontal direction, finally the recovered secret image with no deviation can be obtained.*/

For easily understanding the encrypting algorithm, we herein give an example to show how to encode a secret image.

Example 4 *Construct two shares by the $(2, 2)$ misalignment tolerant VCS, where $p_B = p_S = 50\%$, $\gamma^2 = 4$, and M_{reg} are used. We use a 16×16-pixel secret image I, one black horizontal on a white background (Figure 11.8).*

Since I is black and white now, so $I_S = I$, and I_B has the size of 8×8 pixels shown in Figure 11.8(a) and (b). By using M_{reg} (Figure 11.7(a)), we encrypt *the first pixel in I_B, which is a white secret pixel. Suppose the white pixel is divided into two subpixels (■□) in the first share $O^{(1)}$, and the corresponding two subpixels in the second share $O^{(2)}$ are also (■□). Subsequently, encrypt four small secret pixel in I_S (see Figure 11.8(c)), these four white secret pixels* $\begin{pmatrix} \square\square \\ \square\square \end{pmatrix}$ *are respectively encrypted into* $\begin{pmatrix} ■\square\square■ \\ \square■\square■ \end{pmatrix}$ *and* $\begin{pmatrix} ■\square\square■ \\ \square■\square■ \end{pmatrix}$ *in* $O^{(1)}$ *and* $O^{(2)}$. *According to the M_{reg} mask, encrypt all secret pixels in I_B and I_S. Finally, we obtain two shares $O^{(1)}$ and $O^{(2)}$ of 32×16 pixels.*

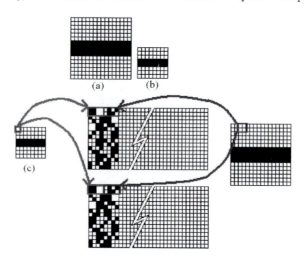

FIGURE 11.8
Encrypt a 16×16-pixel secret image by using a $(2,2)$ misalignment VCS, where $p_B = p_S = 50\%$, $\gamma^2 = 4$, and M_{reg} are used: (a) the small-scaled secret image I_S, (b) the large-scaled secret image I_B, and (c) two shares $O^{(1)}$ and $O^{(2)}$.

11.5.2 Simulations

In this section, we give some simulations to show the performance of the misalignment tolerant VCS for different deviations, size ratios, percentages, and masks. Example 5 uses fifty-fifty large and small subpixels with regular and random masks, respectively, such that we can try out the performance of the mask. In Example 6, we use different size ratios to study how the size ratio affects the visual quality and the misalignment tolerance.

Example 5 *Construct the $(2,2)$ misalignment tolerant VCS, where $p_B = p_S = 50\%$ and $\gamma^2 = 16$. Regular mask M_{reg} and random mask M_{ran} are tested, respectively. The printed-text "VSS" is used as a secret image.*

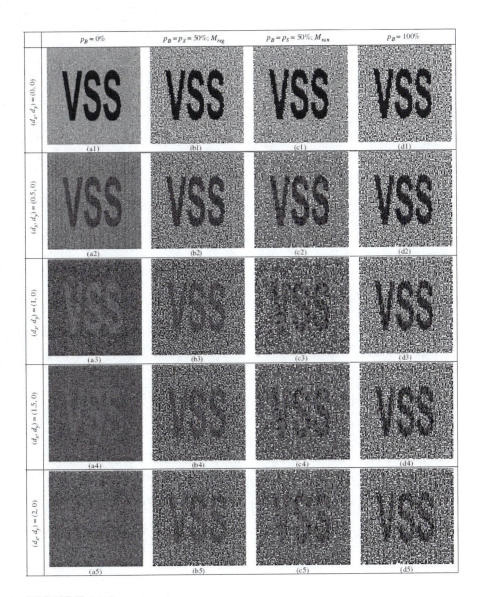

FIGURE 11.9

Recovered secret images for a $(2, 2)$ misalignment VCS using two-sized subpixels of $\gamma^2 = 16$: (a) $p_B = 0\%$, (b) $p_B = p_S = 50\%$, M_{reg}, (c) $p_B = p_S = 50\%$, M_{ran} and (d) $p_B = 100\%$, horizontal deviations: 0, 0.5, 1, and 2 are tested (unit: small subpixel).

Use the deviations $(0,0)$, $(0.5,0)$, $(1,0)$, $(1.5,0)$ and $(2,0)$ to test the misalignment tolerance of the following four schemes: (a) $p_B = 0\%$, (b) $p_B = p_S = 50\%$, M_{reg}, (c) $p_B = p_S = 50\%$, M_{ran} and (d) $p_B = 100\%$. Notice that (a) and (d) are just the traditional VCS with all small subpixels and large subpixels, respectively. The first scheme has the best resolution when aligning correctly, as shown in Figure 11.9 (a1), but the secret on the recovered secret image diminishes quickly when the deviation increases. The secret becomes invisible when the deviation is large enough (see Figure 11.9 (a5)). Comparing Figure 11.9 (a) and (d), the small subpixel gives the high resolution (see Figure 11.9 (a1) and (d1)), whereas the large subpixel enhances the misalignment tolerance. For example, the deviation is $(d_x, d_y) = (2,0)$, the secret in Figure 11.9 (a5) becomes invisible but we still visually view the secret "VSS" in Figure 11.9 (d5). As shown in Figure 11.9 (a3) and (a4), the secret image is recovered by its complementary color. This interesting phenomenon is compatible with the result in Section 11.3 and Section 11.4. From these stacked results, our two-sized subpixel approach actually provides the misalignment tolerance. Also, it is observed that Figure 11.9 (b) (M_{reg}) has the better visual quality than Figure 11.9 (c) (M_{ran}). This result is anticipated because the random arrangement of subpixels will introduce additional noise to the recovered secret image.

Example 6 *Consider the first scheme in Example 5 (a $(2,2)$ misalignment tolerant VCS that the large and small subpixels are half used, respectively, and regularly arranged in a share). Three size ratios, $\gamma^2 = 4$, 16, and 64, are tested. The printed-text "VSS" is used as a secret image.*

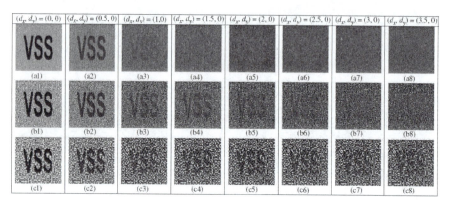

FIGURE 11.10
Recovered secret images for a $(2,2)$ misalignment tolerant VCS using $p_B = p_S = 50\%$, M_{reg} and three size ratios: (a) $\gamma^2 = 4$, (b) $\gamma^2 = 16$ and (c) $\gamma^2 = 64$, horizontal deviations: 0, 0.5, 1, 1.5, 2, 2.5, 3, and 3.5 are tested (unit: small subpixel).

Figure 11.10 (a–c) reveal the recovered secret images for $\gamma^2 = 4$, 16, and 64, respectively. The clearness of the recovered secret image diminishes when the misalignment increases. The misalignment tolerant VCS using $\gamma^2 = 64$ has the best misalignment tolerant capability, whereas it has the worst visual quality for no deviation. On the contrary, the misalignment tolerant VCS using $\gamma^2 = 4$ has a different characteristic. We may trade the misalignment tolerance for the image quality by the size ratio.

11.6 Conclusions and Discussions

This chapter is intended to show some recent results about the alignment problem of the VCS. We considered two kinds of misalignment, (1) misalignment with integer number of subpixels and (2) misalignment with less than one subpixel. In both cases, the secret image can be visually recovered. This phenomenon indicates that, the VCS, by itself, has some misalignment tolerance.

Then we compared the misalignment tolerance of large and small subpixels, and showed that the large subpixel had better misalignment tolerance. Based on this result, a misalignment tolerant VCS was given that traded the large and small subpixels in one share. Simulations were provided to show the performance of the misalignment tolerant VCS.

Because of the page limit, we cannot introduce more misalignment tolerant VCS. Interested readers can find more information about the alignment problem of the VCS in the following papers [8, 16, 13, 6, 9].

11.7 Acknowledgments

I gratefully acknowledge Prof. Chuankun Wu, as he introduced me to visual cryptography. We shared many pleasant memories on studying visual cryptography. This work was supported by NSFC grant No. 60903210.

Bibliography

[1] G. Ateniese, C. Blundo, A. De Santis, and D.R. Stinson. Visual cryptography for general access structures. In *Information and Computation*, volume 129, pages 86–106, 1996.

[2] G. Ateniese, C. Blundo, A. De Santis, and D.R. Stinson. Extended capabilities for visual cryptography. In *ACM Theoretical Computer Science*, volume 250 Issue 1-2, pages 143–161, 2001.

[3] E. Biham and A. Itzkovitz. Visual cryptography with polarization. In the *Dagstuhl Seminar on Cryptography*, September 1997, and in the RUMP Session of CRYPTO'98, 1997.

[4] S. Cimato, R. De Prisco, and A. De Santis. Probabilistic visual cryptography schemes. In *The Computer Journal*, volume 49, Number 1, pages 97–107, 2006.

[5] S. Droste. New results on visual cryptography. In *CRYPTO '96, Springer-Verlag LNCS*, volume 1109, pages 401–415, 1996.

[6] K. Kobara and H. Imai. Limiting the visible space visual secret sharing schemes and their application to human identification. In *ASIACRYPT '96, Springer-Verlag LNCS*, volume 1163, pages 185–195, 1996.

[7] H. Kuwakado and H. Tanaka. Size-reduced visual secret sharing scheme. In *IEICE Transactions on Fundamentals*, volume E87-A, No. 5, pages 1193–1197, 2004.

[8] F. Liu, C.K. Wu, and X.J. Lin. The alignment problem of visual cryptography schemes. In *Designs, Codes and Cryptography*, volume 50, pages 215–227, 2009.

[9] M. Nakajima and Y. Yamaguchi. Enhancing registration tolerance of extended visual cryptography for natural images. In *Journal of Electronic Imaging*, volume 13, pages 654–662, 2004.

[10] M. Naor and A. Shamir. Visual cryptography. In *EUROCRYPT '94, Springer-Verlag Berlin*, volume LNCS 950, pages 1–12, 1995.

[11] R. Ito, H. Kuwakado, and H. Tanaka. Image size invariant visual cryptography. In *IEICE Transactions on Fundamentals of Electronics, Communications and Computer Science*, volume E82-A, No. 10, pages 2172–2177, 1999.

[12] S. J. Shyu. Efficient visual secret sharing scheme for color images. In *Pattern Recognition*, volume 39, pages 866–880, 2006.

[13] W.Q. Yan, D. Jin, and M.S. Kankanhalli. Visual cryptography for print and scan applications. In *Proceedings of the 2004 International Symposium on Circuits and Systems*, volume 5, pages 572–575, 2004.

[14] C.N. Yang. New visual secret sharing schemes using probabilistic method. In *Pattern Recognition Letters*, volume 25, pages 481–494, 2004.

[15] C.N. Yang and C.S. Laih. New colored visual secret sharing schemes. In *Designs, Codes and Cryptography*, volume 20, pages 325–335, 2000.

[16] C.N. Yang, A.G. Peng, and T.S. Chen. Mtvss:(m)isalignment (t)olerant (v)isual (s)ecret (s)haring on resolving alignment difficulty. In *Signal Processing*, volume 89, pages 1602–1624, 2009.

12

Applications of Visual Cryptography

Bernd Borchert

Universität Tübingen, Germany

Klaus Reinhardt

Universität Tübingen, Germany

CONTENTS

12.1 Introduction

Naor and Pinkas in their seminal paper [13] suggested to use visual cryptography in a transparency-on-screen version. Their main purpose was authentication, in the sense that an online server is able to authenticate itself to a user sitting in front of the screen. Implicitly, this already suggests the following application of visual cryptography to the problem of manipulation of online transactions, like online money transfers, by trojans:

Main Method. In order to secure online transactions, like online money transfers, the user gets a numbered set of transparencies, each with a visual cryptography pattern printed on it, from the transaction server. Now the user is able to command online transactions in a secure way, see Figure 12.1 as follows. He fills out an online form containing the data for the intended transaction; in the case of a money transfer this would be the account number and bank number of the destination bank account and the amount of the money. This transaction data is submitted via Internet to the server. The server does not execute the transaction immediately because in that case it

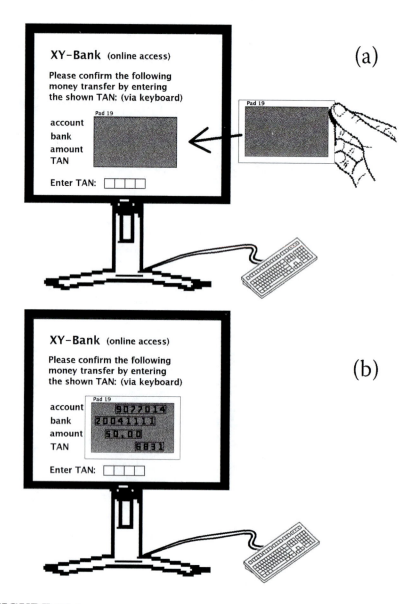

FIGURE 12.1
(a) The bank sends the information to be confirmed in an encrypted image to the user's computer and (b) the user is able read this information using the transparency he got from the bank.

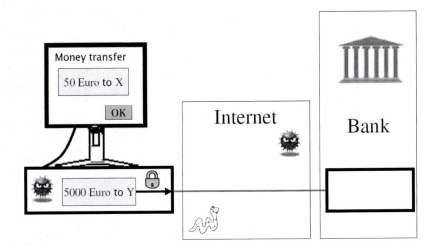

FIGURE 12.2
A man-in-the-middle manipulation attack by a trojan on an online money transfer.

would be an easy task for a man-in-the-middle to manipulate the transaction: the man-in-the-middle would just send his manipulated transaction to the server. In order to prevent such a manipulation, the server sends a visual message containing the transaction data to the user's screen — but of course this image is not sent openly but instead it is encoded via visual cryptography: if the user puts the transparency with a certain number on top of the encoded image on the screen he can see the message contained within the image, i.e., the transaction data. Note that the image on the screen is random to a man-in-the-Middle, as this a guaranteed by visual cryptography. The number of the transparency requested to be used is shown by the server on the user's screen together with the secret image. In order to finally confirm the transaction the user types a transaction number (TAN), which is additionally shown on the secret image message from the server, into a form on the screen and submits this TAN to the server. When the server receives the right TAN it executes the transaction, otherwise not.

Why does this method protect the transaction from being manipulated by a man-in-the-middle (which may be, for example, a trojan sitting on the user's PC)? Because a man-in-the-middle does not know the transparencies the user got from the server, the man-in-the-middle is not able to manipulate the image message sent from the server to the user. In other words, the user will see the transaction that is planned to be executed by the server and will only confirm such a transaction—a clandestine manipulation of the transaction by the trojan is impossible.

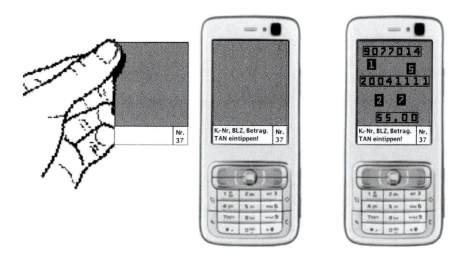

FIGURE 12.3
The main method is also applicable to mobile banking.

In the original purpose of secret sharing in [14], the order of the slides was not relevant. Later work [15] showed that a better contrast can be achieved with colors if the first slide can have non-transparent colors. Furthermore, practical applications will work in the way that the (first) slide is sent first from Alice to Bob over a secure channel (i.e., by surface mail) and used later as a key to decrypt an image received over an insecure channel.

We regard visual cryptography as a special case of the Cardano grille, which works on pixels instead of letters. In both cases we can describe the slide (grille) as a 2-dimensional array over $\{0, 1\}$, where 0 stands for "transparent" and 1 for "black." We describe the encrypted image as a 2-dimensional array over $\Sigma = \{0, 1\}$ or $\Sigma = \{0, 1, red, green, ...\}$ or $\Sigma = \{0, a, b, c, d, ...\}$, where 0 stands for "white" and 1 for "black." Colors are used for pixel-oriented applications and in case the areas are big enough, any alphabet of symbols can be used, which the receiver of the image can read through a transparent area.

A compromise between pixel- and symbol-orientation is the segment-based method described in [2], which works as demonstrated in Figure 12.5. It is applicable whenever the message consists of symbols that can be represented by a segment code, for example the 10 digits by the well-known 7-segment code. The encryption method is basically the same: Instead of pixels, longish and larger segments are encoded via two possible parallel positions.

Outline: Based on the above idea in [13], we describe techniques in Section 12.2 where the user can confirm a transaction as shown in Figure 12.1. Section 12.3 describes similar techniques that allow the user who received a slide from an account provider to securely enter a PIN or confirm transactions. The multiple use of a slide would be an economical and ecological asset and improve the convenience of the user who could, for example, leave the slide adjusted to the screen, but leads to security problems addressed in [13] and in Section 12.4. A further generalization concerning the slide leads to refractional (optical) cryptography, which is described in Section 12.5. Technical problems are discussed in Section 12.6. Section 12.7 describes Chaum's application of visual cryptography in elections. It verifies that a ballot was counted without giving the voter the possibility to show others what she voted for.

12.2 Trojan-Secure Confirmation of Transactions

Naor and Pinkas state the application to online transactions implicitly in their conference paper [13]. Explicitly it is stated in Appendix A of their full paper, which can be found on their homepages. Klein, in 2005, describes this Naor/Pinkas "transparency onto screen" idea as a main application of visual cryptography in [11]. Hogl independently re-invents visual cryptography and the Naor/Pinkas idea in the patent application [9]. Greveler refines some of the aspects of the Naor/Pinkas idea in [8]. Borchert and Reinhardt [3] discuss variants of the Naor/Pinkas idea.

We assume the computer can be infected with a trojan, which is able to eavesdrop and manipulate all input- and output information. Even after a secure login, a trojan (Malice) can manipulate a transaction, which is confirmed with the TAN or iTAN method in the following way as in Figure 12.2: Bob wants to instruct his banker Alice to transfer 50 dollars to X, but Malice chances this to "transfer 5000 dollars to Y." When Alice requests a confirmation by sending the message "To transfer 5000 dollars to Y enter the TAN No. 37," Malice changes it to "To transfer 50 dollars to X enter the TAN No. 37," and Bob will cluelessly enter the TAN No. 37.

To prevent this kind of attack, the authors proposed in [3] methods as in Figure 12.1 and Figure 12.4, with the idea that Eve is not able to produce a forged encrypted image of the original transaction. Here again, the image of message is shifted by a random offset in the x and y-direction to prevent Eve from concluding back to the slide.

However, the method in Figure 12.1 has the disadvantage that the user still needs TANs and that Eve might place the image of the original transaction in an unencrypted way on the screen, which will have the same appearance with the slide as if it would be the encrypted image of the original transaction. Thus, Bob might get fooled, if he did not check that there should be a "gray" pattern without any information before he places the slide.

This can be improved using the method in Figure 12.4; it makes sure that the user is able to see the black balls, which is only possible if (at least

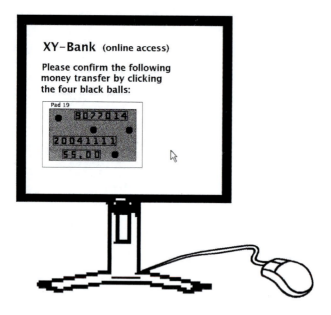

FIGURE 12.4
For confirmation, the user has to click the black balls placed between parts of
the transaction data.

most of) the encrypted image from Alice was sent unchanged to Bob. Since
Malice does not know the position of the parts of the transactions on the
picture, any attempt to alter a part of the transaction would most likely lead
to an incorrect image, which can easily be detected by Bob. A similar version
using the segment-based method in [2] is shown in Figure 12.6 and another
similar version for mobile phones is shown in Figure 12.3. Two versions using
Cardano Cryptography, where the user has to verify the transaction consiting
of an account number by following the blue path, are shown in Figure 12.7 and
implemented in [1]. In the 1-factor confirmation case, the user has to confirm
entering the numbers along the red path; in the 2-factor confirmation case,
the user has to confirm by typing his PIN on the keyboard below according
to the permutation of the digits shown within the red-edged holes.

12.3 Trojan-Secure Authentication Using a PIN

The purpose of this section is to apply visual cryptography in a way such that
the user can enter a password to the server in a way such that the trojan is
not able to get the password.

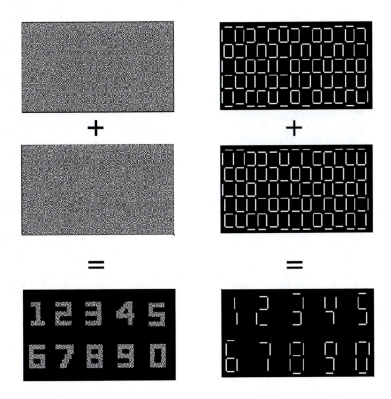

FIGURE 12.5
Pixel-based (left) versus segment-based (right) visual cryptography.

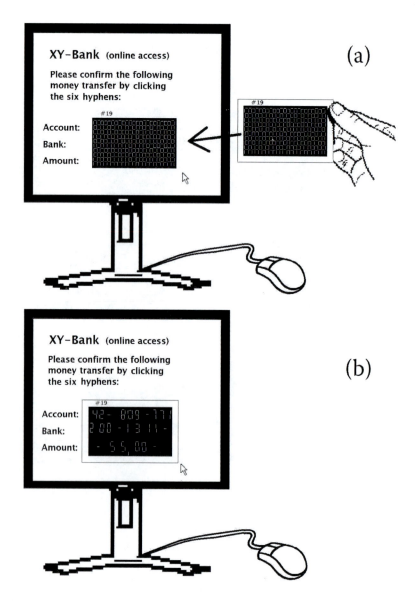

FIGURE 12.6
The main method using segment-based visual cryptography in (a) and (b).

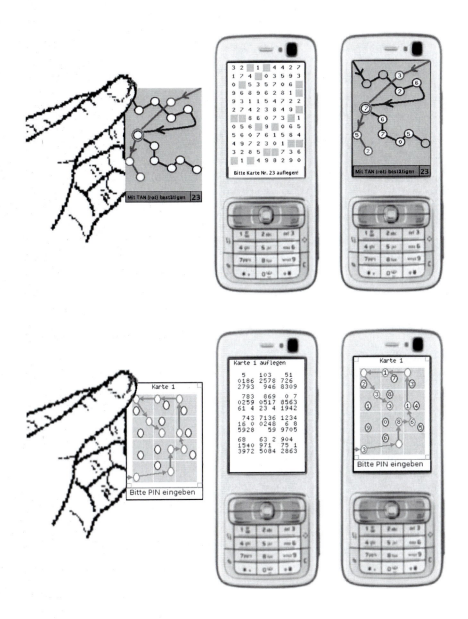

FIGURE 12.7
Cardano cryptography: above a 1-factor confirmation (user types 3752), below
a 2-factor confirmation (for example, in case his PIN is 1234, the user types
in 4136).

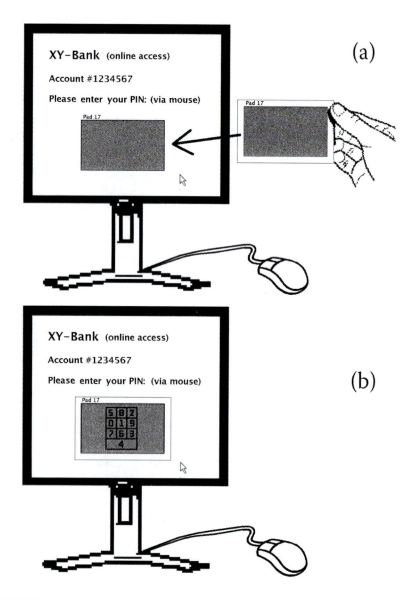

FIGURE 12.8
(a) To log in, the server sends an encrypted image of a permutated keyboard, which the user can only read after placing the slide over it. (b) The user enters the PIN by clicking at the positions according to their order in the PIN.

As shown in Figure 12.8, the trojan on the computer is not able to see the permutation chosen by Alice on the keys, which was randomly chosen by the server. Thus, the mouse-clicks of the user cannot be interpreted by the trojan. This method can be generalized to any alphabet and allows Bob to send short messages to Alice in a secure way.

Note here that this method becomes insecure if the message contains multiple occurrences of the same symbol, a PIN should thus be chosen without repetitions. In case Bob wants to send messages of length l that may contain repetitions, this could still be accomplished in a secure way by extending the alphabet to $\Sigma \cup \{r_1, r_2, ..., r_{l-1}\}$, where r_i indicates the repetition of the symbol at position i. In this way, for example, the message "*messages*" could be submitted as "$mesr_3agr_2r_4$" containing no repetitions in the extended alphabet.

In order to achieve 2-factor security for transactions, we combine the PIN method with the confirmation method of Section 12.2 as described in Figure 12.9.

Furthermore, to prevent the attack using the original transaction in an unencrypted way on the screen as in Section 12.2, we use the refined method of Figure 12.9.

12.4   Security versus Multiple Use

To achieve information theoretic security for a single use, we can divide the array into clusters of c pixels (resp. areas), where c is the size of the alphabet of the encrypted image. Only one pixel in each cluster has a 0 on the slide. To encrypt a pixel $p \in \Sigma$, place p at the position of the cluster with the 0 on the slide and fill the rest of the cluster with a random permutation of $\Sigma \setminus \{p\}$. Since each pixel-value in Σ occurs in each cluster of the encrypted image, each image is possible from the viewpoint of an a evesdropper.

In the model of a known plaintext attack, we assume that the a evesdropper Eve may receive the secret image later, then she can find out which position in each cluster has the o on the slide and thus the slide cannot be used securely a second time.

Known plaintext is relevant for authentication as considered in [13] as well as for confirmation as considered in Section 12.2; in both cases Bob has to be convinced that the message was sent by Alice. The problem of multiple usability is solved in [13] by dividing the slide in distinct areas, where each has to be big enough to contain the complete message; here we use an approach with a different distribution. To achieve information theoretic security use a slide n times, we propose the following two possibilities:

1. For one pixel use a cluster of nc pixels divided into n subclusters of c pixels

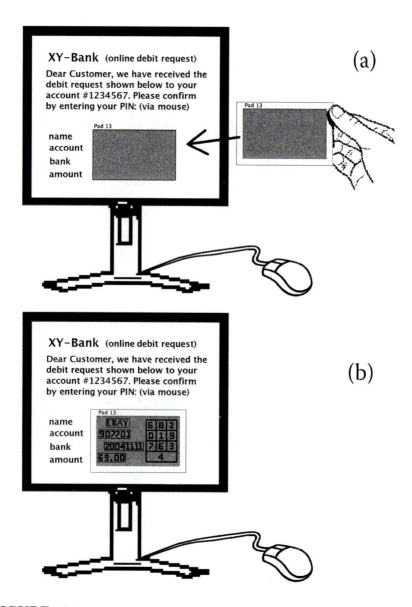

FIGURE 12.9
For confirmation, the user has to click his PIN using a permutation of digits on the right side in (a) and (b).

FIGURE 12.10
For confirmation, the user has to click his PIN using inverted numbers.

in which each subcluster has a 0 on the slide. To encrypt the i-th image, we use only the i-th subcluster as above and fill the rest with 1. This leads to a contrast of $\frac{1}{nc}$.

2. For one pixel use a cluster of c^n pixels, which we address as an n-dimensional array (but arrange 2-dimensionally). Only one pixel in each cluster has a 0 on the slide. To encrypt a pixel $p \in \Sigma$ in the i-th image, place p at each position of the cluster, which has the same i-th coordinate as the 0 on the slide then choose a random permutation of $\Sigma \setminus \{p\}$ and fill the rest of the cluster in a way such that pixels with the same i-th coordinate get the same color. This leads to a contrast of $\frac{1}{c^n}$.

In both cases each combination of images is consistent with any combination of encrypted images. In the first case, the contrast can be improved only to $\frac{1}{n}$ using refraction, whereas it can be improved to 1 using refraction in the second case.

In the case of an unknown plaintext attack, Eve is still able to obtain the secret images if the slide was used too often and if she can anticipate patterns in the picture. Let us consider the simple case $\Sigma = \{0, 1\}$, $c = 2$ and the slide was used twice. Then Eve can XOR both encrypted images resulting in an image that is 0 at pixels where both original images coincide and 1 at pixels where both original images differ, thus she can see the difference of the original pictures as shown in [12] page 35 and Figure 12.11. Now assume the original

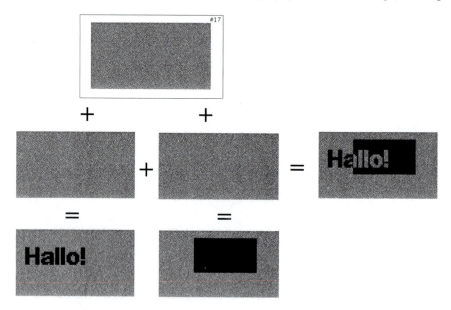

FIGURE 12.11
Superimposing two encrypted images for the same key-slide shows the difference of the original picture.

images depict messages using a Font F consisting of $f = |F|$ small symbols (symbol-pictures) having q pixels.

Assume furthermore the symbols are placed on fixed positions, then Eve can identify the pairs of symbols on corresponding positions as long as $f^2/2 << 2^q$. Then Eve can use, for example, the redundancy of natural languages to decipher the text. One measure to complicate this attack is to shift the picture by a random number of pixels to the left or the right. Then Eve will have to try out the position of the first symbol and consider $x \cdot y \cdot f^2/2$ combinations, where x and y are the differences of the shifts but smaller than the width and the height of the symbols. Further complications can be caused by filling the space around the text by partially random patterns, which Bob can easily distinguish from the symbols, but Eve will have to start analyzing parts in the middle of the image. Here she can only assume that about a quarter of the surface of the letters overlap, which means this method of attacking can be expected to be successfull if $x \cdot y \cdot f^2/2 << 2^{q/4}$.

Furthermore, the partially random patterns can compensate the statistical imbalance of the correlation of neighboring points. For example, Eve could look at pairs of pixels, where one is in some small distance above the other. If both are on a position having a 0 on the slide, then, given many encrypted images of texts, Eve could detect that they have the same color with a higher probability than other pairs. But the partially random patterns are made in a way such that this probability decreases in the overall image.

Let us now turn to the slides on the previous page: If the slide was used more than n times, then Eve can try the following attack: She chooses an area of $q = x \cdot y$ pixels somewhere in the image, then she tries each combination of positions of the 0 in each of the corresponding q clusters on the slide ($(2^n)^q$ possibilities) and checks if it is consistent with each encrypted image in the sense that there are 4 symbols in F overlapping the area. This takes $4 \cdot x \cdot y \cdot f$ steps. And thus $2^{n \cdot q} \cdot 4 \cdot x \cdot y \cdot f$ steps in total.

The number of possible subimages of a possible original image, where four symbols overlap, is approximately $(x \cdot y \cdot f)^4$. This means each observed encrypted image can help Eve to exclude a sufficient number of possible choices if $2^q \gg (x \cdot y \cdot f)^4$. We therefore estimate the number of steps for Eve as $\gg (x \cdot y \cdot f)^{4 \cdot n} \cdot 4 \cdot x \cdot y \cdot f = 4 \cdot (x \cdot y \cdot f)^{4 \cdot n + 1}$. Now assume we use $x = y = 10$ and a huge font F with many possibilities to depict a symbol that leads to $f = 1000$ and roughly $10^{20 \cdot n}$ steps for Eve. Considering many obvious and also less obvious improvements of the algorithm for Eve, we believe that the attack is still too expensive for Eve for $n = 3$.

12.5 Using Refraction

In [4] we generalize the slide from a 2-dimensional array over $\Sigma = \{0, 1\}$ to a 2-dimensional array over $\Sigma \subset \{1\} \cup \mathbb{R} \times \mathbb{R}$ with the idea that each pixel on the slide can either be black or contains a prism (x, y) that refracts the light from a region on the encrypted image or, from the perspective of the user as shown in Figure 12.12, refracts the view to a region that is shifted by (x, y) from the pixel, which is directly behind the pixel on the slide. For example $(0, 0)$ would correspond to the 0 in the case of usual Visual Cryptography just showing the pixel directly behind. One possible application would be to use clusters of $2 \cdot 2 = 4$ pixels for each pixel of the original and randomly choose one of the 4

$(1, 0)$	$(0, 0)$
$(1, -1)$	$(0, -1)$

pixels to be visible. For example [table above] would direct the view to the upper right pixel on the encrypted image. This corresponds to construction 2 in Section 12.4. Using the slide two times is information theoretically secure and the contrast is 1.

If we use lenses or fragments of lenses instead of prisms, the view can be focused on a point inside a pixel. This has the advantage that the positioning of the slide allows an error of up to half of a pixel. An example is shown in

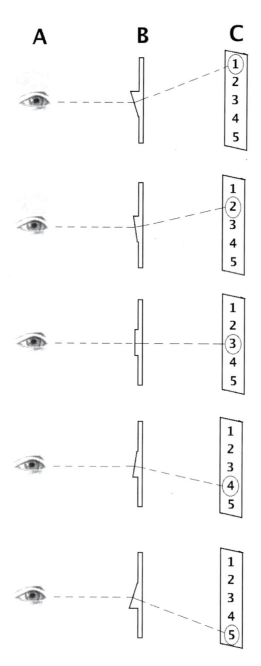

FIGURE 12.12
This example shows how the view from the observer (A) through prisms (B) is directed to areas 1,2,... or 5 on the encrypted image (C); the deviation depends on the slope of the prism.

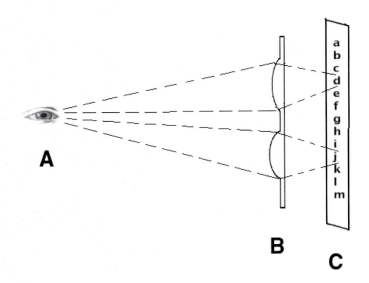

FIGURE 12.13
Some parts of the encrypted image (C) is magnified for the observer (A), while other parts are hidden. For example, d and j are in the focus while $b, c, e, f, h, i, k,$ and l are hidden.

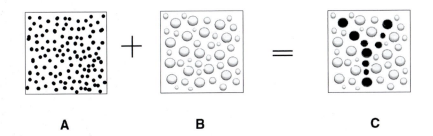

FIGURE 12.14
Lenses are placed randomly on the slide (b). This can be done by spraying a transparent liquid that becomes hard on the side. The area in the focus of the lenses in the encrypted image (a) is colored in the color of the original image at this region. The rest of (a) is filled such that colors in (a) are equally distributed so that the original image can not be obtained from (a) alone but only together with the slide (b).

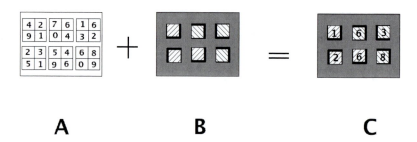

A   **B**   **C**

FIGURE 12.15
Each area of the slide (b) has fragments of lenses, which direct the view (c) in a magnifying manor to one of the symbols on the encrypted text (a).

Figure 12.13. While producing fragments of lenses and prisms on the slide might require expensive special machines, it will be much cheaper to produce complete lenses. Since the lenses do not need to have a perfect shape, it would be sufficient to place drops of a transparent liquid that becomes hard on the side. This can be done by either using a modified ink-printer or by spraying the liquid using physical randomness (and Alice can scan it before sending it to Bob). The use of such a slide is shown in Figure 12.14. Figure 12.15 shows an optical generalization of Cardano cryptography. The advantage to Cardano cryptography is that the letters are magnified and can be read in a more natural ordering. The advantage to pixel-based visual cryptography is that much less precision is needed to position the slide. The disadvantage is that, in order to achieve a sufficient level of security, more than four choices (like in Figure 12.15) would be required. A repeated use would be too insecure as the slide contains only little information. Furthermore, a bigger distance of slide and screen is required.

12.6 Technical Problems Concerning Adjustment and Parallaxes

A disadvantage of pixel-based visual cryptography to Cardano cryptography is that the slide has to be placed at an exact position. We propose to position the slide at the (for example, left lower) corner of the screen making use of its frame. Then use the mouse to position and stretch the encrypted image on the screen accordingly. The parallaxes is the shift of the pixel on the screen that can be seen through one point of the slide, which is caused by looking at it from

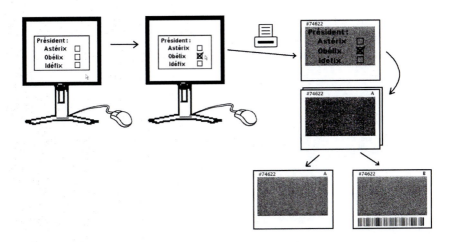

FIGURE 12.16
The voter enters the vote, verifies the image, and separates the slides.

a certain angle that differs from the right angle because of central projection and because the viewer has two eyes at different positions. This is estimated in [10] to be 0.25 mm in the case of a usual TFT screen, less for displays of new mobile phones, but might be more if we require a higher distance of slide and screen so that the use of refraction can take effect (depending on the size of the lenses). Slight misplacements of the slide cause effects like a bad contrast or even inverting the picture as described in [10] where the author also proposes methods to use interference effects at the frame of the slide as an aid for adjustment.

12.7 Voting with a Receipt Based on Visual Cryptography

The purpose is to give a voter a receipt, which allows her or anyone else to verify that her vote was counted in the final result. The difficulty comes from the requirement that the voter should not be able to prove to anyone else, what her vote was since this would make abuses such as vote selling possible.

The main procedural method in [6] lets the voter enter her ballot on a touch screen and then the voting machine produces two slides laminated together that show the ballot image with visual cryptography to the voter only as shown in Figure 12.16. Then the voter separates the two slides and goes to the poll worker. Here one of the slides is destroyed and the other one is scanned

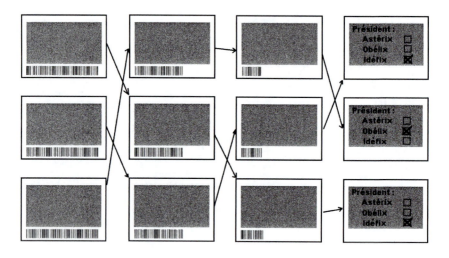

FIGURE 12.17
Each trustee strips one layer of the doll (represented by the barcode) and uses it to modify the image. The order is randomly permutated.

and uploaded to the official election website. Furthermore, the voter keeps her slide as her receipt.

For confidentiality, the voter has to trust that the voting machine has no memory. But how can the results be computed only from the scanned slides? To make this possible, the slide cannot be completely random (as usual in visual cryptography) but is produced by a pseudorandom generator from the cryptographic version of a nested doll that contains the necessary information to reproduce the ballot image inside. This means that confidentiality is not information theoretically secure but only computationally secure under the usual cryptographic assumptions.

The cryptographic version of a nested doll is produced by the voting machine from the serial number of the ballot by successively encrypting with the public keys of a sequence of trustees and printed on the slide as a barcode. Only all trustees together would be able to compute the ballot image using their secret keys. The result of the election is computed by a sequence of mix-operations as described in [5]: The first trustee gets a batch of scanned slides as input. For each vote, he removes the first layer of the doll and modifies the encrypted image using the removed layer of the doll. Then he uploads a batch containing a random permutation of the results to the official election website. All other trustees do the same with the batch from their predecessor (see Figure 12.17). In the batch produced by the last trustee, the dolls were used up and the image became the original ballot image, which can be seen (and thus counted) by anyone.

To verify that the trustees worked properly, a public random choice is used to audit the trustees, which have to reveal half of the connections to the following batch by publishing the removed layer of the doll, which again allows anyone to verify these connections. But the choice was made in a way such that there will be no complete path visible leading from a scanned slide to its ballot image. In [7], a refinement of the choice is described, which ensures that nothing can be learned about the ballots of even groups of voters.

But how can the voter be sure that she was not cheated by the voting machine printing a false second (and later destroyed) slide letting her see her ballot image but producing a different one after passing all the mixes? The idea is that after the voting machine printed the visual cryptography part on the laminated slides, the voter has still the choice to take the upper or lower slide as receipt and tell the choice to the voting machine before it continues by printing the barcode of the doll to it; in this way the voting machine would for each voter only have a 50% chance of cheating without being caught.

12.8 Conclusion

Visual cryptography is a fascinating technique and very intuitive to the user. However, it is surprising that within the last 15 years since its invention by Naor and Shamir only a few suggestions have been made to apply it to practical problems. In this paper we presented Naor and Pinkas technique to use visual cryptography in order to protect online transactions against manipulation and Chaum's idea to apply it to verify the correctness of the outcome of an election. Because of difficulties like adjustment, multiple use, and costs of special equipment, these suggestions did not lead to applications that are used for serious purposes. But in the future, further developments of the ideas presented in this paper, as well as new ideas, could spread practical applications of visual cryptography.

Bibliography

[1] Sebastian Beschke. *Implementierung der Cardano-TAN (cTAN)*, 2009. Studienarbeit.

[2] B. Borchert. Segment-based visual cryptography. Technical Report WSI–2007–04, Universität Tübingen (Germany), Wilhelm-Schickard-Institut für Informatik, 2007.

[3] B. Borchert and K. Reinhardt. *Abhör- und manipulationssichere Verschlüsselung für Online Accounts mittels Visueller Krytographie an der*

Bildschirmoberfläche, 2007. Patent application DE-10-2007-018802.3 (approved 2008).

[4] B. Borchert and K. Reinhardt. *Lichtbrechungs-Kryptographie*, 2010. Patent application DE-10-2010-031 960.0.

[5] David Chaum. Untraceable electronic mail, return addresses, and digital pseudonyms. *Commun. ACM*, 24(2):84–88, 1981.

[6] David Chaum. Secret-ballot receipts: True voter-verifiable elections. *IEEE Security & Privacy*, 2(1):38–47, 2004.

[7] Marcin Gomulkiewicz, Marek Klonowski, and Miroslaw Kutylowski. Rapid mixing and security of chaum's visual electronic voting. In Einar Snekkenes and Dieter Gollmann, editors, *ESORICS*, volume 2808 of *Lecture Notes in Computer Science*, pages 132–145. Springer, 2003.

[8] Ulrich Greveler. VTANs Eine Anwendung visueller Kryptographie in der Online-sicherheit. In *2. Workshop* "Kryptologie in Theorie und Praxis," Bremen, *Lecture Notes in Informatics (LNT)*, pages 210–214, 2007.

[9] Christian Hogl. *Verfahren und System zum bertragen von Daten*, 2005. Patent application DE-10-2010-031 960.0.

[10] Frank Hunszinger. *Implementierung und Untersuchung eines Verfahrens zur visuellen Kryptographie*, 2010. Diplomathesis.

[11] Andreas Klein. Eine Einführung in die visuelle Kryptographie. In *DMV Mitteilungen 1/2005*, 2005, 54-57.

[12] Andreas Klein. *Visuelle Kryptographie*. Springer, Berlin Heidelberg New York, 2007.

[13] Moni Naor and Benny Pinkas. Visual authentication and identification. In *Lecture Notes in Computer Science*, pages 322–336. Springer-Verlag, 1997.

[14] Moni Naor and Adi Shamir. Visual cryptography. In *EUROCRYPT*, pages 1–12, 1994.

[15] Moni Naor and Adi Shamir. Visual cryptography ii: Improving the contrast via the cover base. In *Proceedings of the International Workshop on Security Protocols*, pages 197–202, London, UK, 1997. Springer-Verlag.

13

Steganography in Halftone Images

Oscar C. Au

Hong Kong University of Science and Technology, China

Yuanfang Guo

Hong Kong University of Science and Technology, China

John S. Ho

Hong Kong University of Science and Technology, China

CONTENTS

13.1 Introduction

While gray-scale images can be readily displayed in computer monitors and other light-emitting displays, they also need to be displayed routinely in other reflective media such as newspaper, magazines, books, and other printed documents. However, in reflective media, the application of ink on the reflective media implies that only 1-bit images (with two tones: black and white) can be displayed. A problem arising from this is that straightforward 1-bit quantization on an image would lose most of the important image details. With such constraints, there is a class of image processing technique called image halftoning that converts an 8-bit image into an 1-bit image, which resembles the 8-bit image when viewed from a distance. Such 1-bit images are called halftone images [1]. Halftone image technologies are widely used in printed matters.

There are two main kinds of halftoning methods: ordered dithering [1] and error diffusion [2]. Ordered dithering uses straightforward 1-bit quantization with fixed pseudo-random threshold patterns to give halftone images with reasonable visual quality. Error diffusion also performs simply 1-bit quantization but allows the 1-bit quantization error to be fed back to the system and thus can achieve higher visual quality than ordered dithering.

Sometimes it is desirable to hide watermarking data in halftone images. Some halftone image watermarks are designed to be fragile and are useful for authentication and tamper detecting of the halftone images. Some halftone image watermarks are designed to be robust and are useful for copyright protection. In some applications, the data are to be embedded into a single halftone image and some special method can be used to read the hidden data. In other applications, visual patterns are hidden in two or more halftone images such that, when they are overlaid, the hidden visual patterns can be revealed. This kindly visual pattern hiding is also called visual cryptography. This chapter is about visual cryptography in error diffused halftone images.

The chapter is organized as follows. Section 13.2 introduces the basic error diffusion technique. Section 13.3 introduces a visual cryptography method for error diffused images called Data Hiding by Stochastic Error Diffusion (DHSED) [3]. Section 13.4 introduces an improved method called Data Hiding by Conjugate Error Diffusion (DHCED) [4]. Section 13.5 gives theoretical and empirical analysis of DHSED and DHCED. At last, Section 13.6 will give a summary of this chapter.

13.2 A Review of Error Diffusion

In this section, we will briefly introduce a halftoning method called Error Diffusion. The method that we will describe is by no means the only way to achieve halftoning, but is a popular approach that gives good visual quality while maintaining reasonable complexity. The error diffusion process converts a multitone image to a halftone one by distributing the error introduced at the current pixel to a neighborhood of yet unprocessed pixels. The neighborhood, as well as the weights in the distribution, is described by a set of positions and weights known as an error kernel. This diffusion of error across a region allows the local intensity of the halftone image to be preserved approximately.

Consider a multitone image with pixels defined over the range of 0 (black) to 255 (white). Let $h(k,l)$ be an error kernel defined over a neighborhood $\mathcal{N}$. For example, the common Steinberg kernel [2] is

$$\frac{1}{16} \begin{bmatrix} & * & 7 \\ 3 & 5 & 1 \end{bmatrix}$$

defined over the neighborhood $\mathcal{N} = \{(0,1),(1,-1),(1,0),(1,1)\}$. Here we have used $*$ to indicate the location of the current pixel. Let (i,j) be the current pixel location. Let $x(i,j)$ be the current multitone pixel to be processed. A modified pixel value $u(i,j)$ will be derived from $x(i,j)$ and 1-bit quantization is applied to $u(i,j)$ to give the output halftoned pixel value $y(i,j)$. Let $e(i,j)$ be the error between $u(i,j)$ and $y(i,j)$. In error diffusion, the error $e(i,j)$ is distributed to future pixels in its neighborhood. For the case of error diffusion using the Steinberg kernel, a portion $\frac{7}{16}$ of $e(i,j)$ will be distributed to $(i,j+1)$ corresponding to $(0,1)$ in $\mathcal{N}$. Similarly, $\frac{3}{16}$ of $e(i,j)$ goes to $(i+1,j-1)$, $\frac{5}{16}$ of it goes to $(i+1,j)$, and $\frac{1}{16}$ of it goes to $(i+1,j+1)$.

For location (i,j), an offset term $a(i,j)$ is defined as

$$a(i,j) = \sum_{k,l \in \mathcal{N}} e(i-k, j-l) \times h(k,l) \tag{13.1}$$

which is the total error propagated from past pixels to the current pixel. For the error at location $(i-k, j-l)$, a portion $h(k,l)$ of it is passed to the current pixel according to the error kernel. The modified pixel value $u(i,j)$ is then defined as

$$u(i,j) = x(i,j) + a(i,j) \tag{13.2}$$

The output halftone value $y(i,j)$ is defined as

$$y(i,j) = \begin{cases} 0, & u(i,j) < 128 \\ 255, & u(i,j) \geq 128 \end{cases} \tag{13.3}$$

and the error $e(i,j)$ is

$$e(i,j) = u(i,j) - y(i,j) \tag{13.4}$$

These steps are summarized in Figure 13.1, in which the block $Q(\cdot)$ is the 1-bit quantization and the block $h_{k,l}$ is the error kernel applied to past errors in the neighborhood to generate the current offset $a(i,j)$.

Consider the special case of X being of constant intensity A in a neighborhood around (i,j). As error diffusion can preserve the average image intensity, the probability distribution of a halftone pixel $y(i,j)$ is

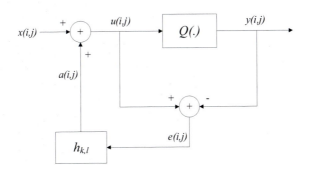

FIGURE 13.1
Error diffusion process.

$$P[y(i,j) = 255] = A/255 \tag{13.5}$$

$$P[y(i,j) = 0] = (255 - A)/255 \tag{13.6}$$

such that

$$\begin{aligned}
E[y(i,j)] &= 0 \bullet P[y(i,j) = 0] + 255 \bullet P[y(i,j) = 255] \\
&= 0 \bullet (255 - A)/255 + 255 \bullet A/255 \\
&= A \tag{13.7}
\end{aligned}$$

Typically, the percentages of white and black pixels in X are $A/255 \bullet 100\%$ and $(255 - A)/255 \bullet 100\%$ respectively, distributed evenly in X. In a way, one can argue that

$$P[y(i,j) = 255 | x(i,j)] = x(i,j)/255 \tag{13.8}$$

such that

$$E[y(i,j) | x(i,j)] = x(i,j) \tag{13.9}$$

Another common error diffusion kernel is the Jarvis kernel [5]

$$\frac{1}{48} \begin{bmatrix} & & * & 7 & 5 \\ 3 & 5 & 7 & 5 & 3 \\ 1 & 3 & 5 & 3 & 1 \end{bmatrix}$$

FIGURE 13.2
Original multitone "Lena" (X).

FIGURE 13.3
Halftone image generated by error diffusion with the Steinberg kernel.

FIGURE 13.4
Halftone image generated by error diffusion with the Jarvis kernel (Y_1).

which has a larger support than the Steinberg kernel. A typical image, Lena, is shown in Figure 13.2. The corresponding halftone images generated by Steinberg and Jarvis kernels are shown in Figures 13.3 and 13.4 respectively. It can be observed that different error diffusion kernels give rise to different textures in the halftone images. In general, Jarvis gives images with higher contrast while the Steinberg kernal gives smoother texture. Both are capable of generating halftone images that mimic the original images when viewed afar, though tiny details of the original image such as Lena's hair tend to be masked by the halftone image texture generated by the error kernel.

The size of all the images in this chapter are 512×512. Due to limited space, all remaining halftone figures are generated by the Jarvis kernel only, though the methods described in the chapter are applicable to any error diffusion kernels.

13.3   Data Hiding by Stochastic Error Diffusion (DHSED)

Data Hiding for halftone images is quite different from that for multitone images due to the fact that halftone pixels can take on only two values: 0 and 255. They contain high frequency noise but resemble the original multitone images when viewed afar. As such, normal data hiding techniques such as least significant bit (LSB) embedding technique [6] would not work on them because the resulting stego-images will be effectively the watermark image and would not resemble the original multitone images even when viewed afar. Thus, it is necessary to develop special data hiding techniques for halftone images. Although several data hiding technologies for halftone images have been proposed before, Data Hiding by Stochastic Error Diffusion (DHSED) is the first visual cryptography method based on error diffusion.

DHSED is a method that embeds a binary secret pattern into two halftone images derived from the same underlying multitone image. The binary pattern should be revealed when the two halftone images are superimposed. The idea of DHSED is to stochastically create a texture phase shift between the two halftone images at locations where the watermark is "active" or black (binary pattern value being zero). The resulting mismatch allows the watermark to become visible while maintaining the original halftone background. Let X be the original multitone image and W the binary secret pattern of the same size. Let Y_1 be a halftone image obtained by applying regular error diffusion to X. Let Y_2 be the halftone image obtained through DHSED. The problem, then, is to obtain Y_2 such that W is revealed when Y_1 and Y_2 are overlaid. We denote the individual pixels at location (i, j) of both halftone images as $y_1(i, j)$ and $y_2(i, j)$, respectively.

The second image Y_2 is generated by applying regular error diffusion to certain areas in X, but with different error conditions. These areas are obtained by referencing both Y_1 and W. Let W_b be the collection of the locations of all the black pixels in W, and W_w the collection of the white pixel locations. In constructing Y_2, we will force the pixel value at all locations belonging to W_w to be identical to Y_1. In other words, values at these locations are merely copied from Y_1 to Y_2. That is,

$$y_2(i,j) = y_1(i,j) \quad \forall (i,j) \in W_w \qquad (13.10)$$

For the remaining pixels in W_b, Y_2 needs to look natural and thus DHSED applies error diffusion with the same error kernel. However, DHSED seeks to make Y_2 different from Y_1 statistically so that when they are overlaid, pixels in W_b would tend to be darker. To achieve this, DHSED first morphologically dilate, W_b with a structuring element D consisting of a $(2L+1) \times (2L+1)$ matrix. We denote the dilated W_b as C.

$$C = W_b \oplus D \qquad (13.11)$$

which can be interpreted as a L-pixel expansion of W_b in all directions.

For the pixels outside C, DHSED copies Y_2 from Y_1 using (13.10) but forces the error for Y_2 to be zero, i.e. $e_2(i,j) = 0$ for $(i,j) \notin C$. Note that the error for Y_1 are nonzero in general for the same locations.

Let $E = C - W_b = C \cap W_w$ be the "border" of the secret pattern, obtained by removing W_b from the expanded region C. For the pixels in E, (13.1) and (13.2) are still applied while (13.3) and (13.4) are not. (13.10) will be used to replace (13.3) since $E \subset W_w$ and Y_2 pixels in W_w are copied from Y_1. We will use (13.12) to replace (13.4).

$$e_2(i,j) = \max\{\min\{u_2(i,j) - y_2(i,j), 127\}, -127\} \qquad (13.12)$$

which is basically (13.4) with a limiter. As Y_2 pixels in E are copied from Y_1 with artificial zero error outside C, there are chances that $u_2(i,j) - y_2(i,j)$ is outside ± 127. The limiter would then help to make the $e_2(i,j)$ reasonable.

For the pixels in W_b, DHSED uses regular error diffusion to generate Y_2 so that region W_b in Y_2 still has the same characteristic texture as regular error diffusion. But the "phase" of the texture will be different compared to the corresponding region in Y_1 since the error outside the region C is different in Y_1 and Y_2.

The overlaying operation is equivalent to applying the logical AND operation between images Y_1 and Y_2. Since the pixels in region W_w of Y_1 and Y_2 have been forced to be identical, the overlaid pixel values are simply the regular error diffused pixel values. However, in region W_b, although the texture in Y_1 and Y_2 maintains the same characteristic, there is an artificially introduced phase shift such that collocated Y_1 and Y_2 pixels tend to be statistically independent. As a result, the overlaying operation tends to give darker local intensity thus revealing the secret pattern W. A more detailed analysis of this will be given in Section 13.4.

Using Lena as the test image and Figure 13.5 as the secret binary pattern

USTUST
USTUST
USTUST

FIGURE 13.5
Secret pattern "UST" to be embedded in the halftone image (W).

FIGURE 13.6
DHSED-generated Y_2 ($L = 5$) of Lena with respect to X in Figure 13.2, W in Figure 13.5, and Y_1 in Figure 13.4.

FIGURE 13.7
Image Y obtained by overlaying Y_1 in Figure 13.4 and Y_2 in Figure 13.6.

W, Figure 13.4 is Y_1 and Figure 13.6 is Y_2 generated using DHSED with respect to Y_1 and W. A threshold $L = 5$ is used. Note that Figure 13.6 looks like Figure 13.4, which verifies that DHSED can give halftone images with good visual quality. Figure 13.7 shows the image obtained by overlaying Figure 13.6 and Figure 13.4. The secret pattern W is clearly visible in Figure 13.7 verifying that DHSED is an effective visual cryptography method.

13.4 Data Hiding by Conjugate Error Diffusion (DHCED)

DHSED as outlined in the previous section is both computationally and conceptually simple, but suffers from three major problems. First, Y_1 and Y_2 must be obtained from the same X. In other words, DHSED cannot embed a binary secret pattern in two halftone images obtained from two different multitone images. Second, when the images are overlaid, the contrast of the revealed secret pattern is relatively low. Third, occasional boundary artifacts may happen in Y_2 especially towards the right and bottom sides of the secret pattern where error is not diffused properly across the W_b boundary. Such boundary artifacts can lower the visual quality of Y_2 considerably.

In this section, we will introduce another method called Data Hiding by Conjugate Error Diffusion (DHCED) that addresses these three problems.

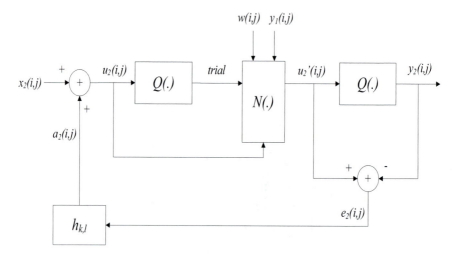

FIGURE 13.8
The DHCED (Data Hiding by Conjugate Error Diffusion) process.

The block diagram of DHCED is shown in Figure 13.8. Unlike DHSED, the problem now contains two original multitone images X_1 and X_2, where X_1 and X_2 may or may not be identical. Two halftone images Y_1 and Y_2 are to be generated from X_1 and X_2, respectively such that when Y_1 and Y_2 are overlaid, the binary secret pattern W is revealed. Similar to DHSED, Y_1 is generated using regular error diffusion on X_1. Then the DHCED process in Figure 13.8 is used to generate Y_2.

To explain the detailed operation of DHCED, we assign logical "1" to pixel value 255 and logical "0" to pixel value 0. The *conjugate* of a pixel is then equivalent to logical NOT, or toggling of the value. Recall that W_w and W_b are the collections of white and black pixel locations in W, respectively. As Y_1 is already generated by regular error diffusion, the secret pattern W at location (i, j) can be naively inserted in Y_2 as follows

$$y_2(i, j) = \begin{cases} y_1(i, j), & w(i, j) = 1 \\ y_1^c(i, j), & w(i, j) = 0 \end{cases} \tag{13.13}$$

where $(\cdot)^c$ denotes logical NOT. This is equivalent to logical XNOR between Y_1 and W. The resulting Y_2 will look like X_1 in W_w and the negative of X_1

in W_b because

$$E[y_2(i,j)|w(i,j) = 1, x_1(i,j)]$$
$$= E[y_1(i,j)|x_1(i,j)]$$
$$= 255 \bullet P[y_1(i,j) = 255|x_1(i,j)] + 0 \bullet P[y_1(i,j) = 0|x_1(i,j)]$$
$$= 255 \bullet P[y_1(i,j) = 255|x_1(i,j)]$$
$$= 255 \bullet [x_1(i,j)/255]$$
$$= x_1(i,j) \tag{13.14}$$

$$E[y_2(i,j)|w(i,j) = 0, x_1(i,j)]$$
$$= E[y_1^c(i,j)|x_1(i,j)]$$
$$= 0 \bullet P[y_1(i,j) = 255|x_1(i,j)] + 255 \bullet P[y_1(i,j) = 0|x_1(i,j)]$$
$$= 255 \bullet [1 - P[y_1(i,j) = 255|x(i,j)]]$$
$$= 255 - x_1(i,j) \tag{13.15}$$

But Y_2 should resemble X_2 instead of X_1 and thus this naive method does not work. Even if X_1 and X_2 are the same image, this naive method does not work in W_b as Y_2 should resemble X_1 instead of the negative of X_1.

Although this naive method does not work, DHCED follows a similar logic except that it *favors*, instead of *forces*, Y_2 to take on the values in (13.13). In other words, it treats the value in (13.13) as the *favored* value of $y_2(i,j)$.

Consider Figure 13.8 and any location $(i,j) \in W_b$. Basically, DHCED applies error diffusion on $x_2(i,j)$ including (13.1) to calculate $a_2(i,j)$, (13.2) to calculate $u_2(i,j)$ and, in the first $Q(\cdot)$ block, (13.3) to calculate the trial halftone value $y_2(i,j)$. In the $N(\cdot)$ block, the trial value is compared with the favored value that is obtained by XNOR of $w(i,j)$ and $y_1(i,j)$. If they are equal, no change needs to be done to $x_2(i,j)$ and $u_2(i,j)$ such that $u_2'(i,j) = u_2(i,j)$. If they are not equal, DHCED considers the possibility of forcing trial $y_2(i,j)$ to toggle to achieve the favored value. Recognizing that forced toggling is equivalent to applying a distortion to $x_2(i,j)$ followed by regular error diffusion, DHCED will execute the forced toggling only if the required distortion is not excessive.

Here are the details. Note that forcing the trial halftone value to toggle is equivalent to adding an offset value $\Delta u(i,j)$ to $u_2(i,j)$. If the trial value should be toggled from 0 to 255, then $u_2(i,j) < 128$ initially and we need an offset $\Delta u(i,j)$ such that the resulting value, which we called $u_2'(i,j)$, is $u_2'(i,j) \equiv u_2(i,j) + \Delta u(i,j) \geq 128$. Thus, there is a lower bound for $\Delta u(i,j)$: $\Delta u(i,j) \geq 128 - u_2(i,j) > 0$.

Likewise, if we want to toggle from 255 to 0, then $u_2(i,j) \geq 128$ initially and we need an offset $\Delta u(i,j)$ such that $u_2'(i,j) \equiv u_2(i,j) + \Delta u(i,j) \leq 127$. There is thus an upper bound for $\Delta u(i,j)$: $\Delta u(i,j) \leq 127 - u_2(i,j) < 0$.

The smallest $\Delta u(i,j)$ (in terms of magnitude) needed to achieve toggling is

$$\Delta u(i,j) = \begin{cases} 128 - u_2(i,j), & u_2(i,j) < 128 \\ 127 - u_2(i,j), & u_2(i,j) \geq 128 \end{cases} \tag{13.16}$$

We further note that adding a distortion $\Delta u(i,j)$ to $u_2(i,j)$ may be interpreted as a distortion to the original multitone image pixel $x_2(i,j)$. Defining $\Delta x(i,j) = \Delta u(i,j)$, the input to the normal quantizer $u_2'(i,j)$ may be written as

$$\begin{aligned} u_2'(i,j) &= u_2(i,j) + \Delta u(i,j) \\ &= x_2(i,j) + \sum_{k,l \in \mathcal{N}} e_2(i-k, j-l) \times h(k,l) + \Delta u(i,j) \\ &= x_2'(i,j) + \sum_{k,l \in \mathcal{N}} e_2(i-k, j-l) \times h(k,l) \end{aligned} \tag{13.17}$$

$$x_2'(i,j) = x_2(i,j) + \Delta u(i,j) = x_2(i,j) + \Delta x(i,j) \tag{13.18}$$

The interpretation of (13.18) is that the output halftone image using DHCED in fact represents X_2', not X_2, in the sense of that it can be obtained from X_2' directly using standard error diffusion. Thus, $|\Delta x|$ can be treated as a measure of the distortion introduced by the DHCED process. To control this distortion, we define a threshold T that determines whether or not the pixel should be toggled, allowing a trade-off between distortion and visual quality of the watermarked halftone image. If $|\Delta x|$ is less than T, the pixel toggling will be performed, and vice versa. If T decreases, the visual quality of the watermarked image Y_2 will improve at the price of lower contrast of the secret pattern when the two halftone images are superimposed.

Consider any location $(i,j) \in W_w$. If X_1 and X_2 are the same image, DHCED would copy $y_1(i,j)$ to $y_2(i,j)$ so that Y_2 values are effectively obtained by applying error diffusion to X_1. The error $e_2(i,j)$ will be computed as in normal error diffusion. When Y_1 and Y_2 are overlaid, the regular error diffused value $y_1(i,j)$ will be revealed. The overlaying operation will reveal a local intensity similar to the local intensity of X_1, which is typical for regular error diffusion.

If X_1 and X_2 are different, DHCED would not force $y_2(i,j)$ to be identical to $y_1(i,j)$ at $(i,j) \in W_w$. Instead, it merely treats $y_1(i,j)$ as the favored value of $y_2(i,j)$. And DHCED performs the same operation as in W_b.

The proposed DHCED for the case of identical X_1 and X_2 is simulated. Using Lena as the test image and Figure 13.5 as the secret binary pattern W, Figure 13.4 is Y_1 and Figure 13.9 is Y_2 generated using DHCED with respect to Y_1 and W. A threshold of $T = 10$ is used. Note that Figure 13.9 looks like Figure 13.4 which verifies that DHCED can give halftone images with good visual quality. Figure 13.10 shows the image obtained by overlaying Figure 13.9 and Figure 13.4. The secret pattern W is clearly visible in Figure 13.7 verifying that DHSED is an effective visual cryptography method.

FIGURE 13.9
DHCED-generated Y_2 $(T = 10)$ of Lena with respect to X in Figure 13.2, W in Figure 13.5, and Y_1 in Figure 13.4.

Comparing DHSED and DHCED, it can be observed that DHSED in Figure 13.6 has strong boundary artifacts along the embedded watermark especially on the right and bottom edges of W. With DHCED, the boundary artifacts are reduced considerably. In addition, the watermark in Figure 13.10 is significantly more visible than Figure 13.7 in terms of contrast, when Y_1 and Y_2 are overlaid.

DHCED for the case of different X_1 and X_2 is also simulated. Here Lena is used as X_1 and Pepper is used as X_2. The Y_1 with respect to X_1 is simply the one in Figure 13.4. The DHCED generated Y_2 from X_2 with respect to Y_1 and W is shown in Figure 13.12. A threshold of $T = 10$ is used. The overlaid image of Y_1 and Y_2 is shown in Figure 13.13. As expected, traces of both Lena and Pepper can be observed in the overlaid image. More importantly, the watermark W is revealed also, though the contrast of W in Figure 13.13 is not as good as in Figure 13.10.

13.5 Performance Analysis

In this section, we give an indepth analysis of both DHSED and DHCED. We will show theoretically why DHCED has better performance than DHSED.

FIGURE 13.10
Image Y obtained by overlaying Y_1 in Figure 13.4 and Y_2 in Figure 13.9.

FIGURE 13.11
Original multitone "Pepper" (X_2).

FIGURE 13.12
DHCED-generated Y_2 $(T = 10)$ of Pepper with respect to X_2 in Figure 13.11, W in Figure 13.5, and Y_1 in Figure 13.4.

FIGURE 13.13
Image Y obtained by overlaying Y_1 in Figure 13.4 and Y_2 in Figure 13.12.

Consider the case when X_1 and X_2 are the same image X. Consider a rectangular region R of constant intensity A in X. Suppose that the left half of R is in the white region W_w of W and the right half in the black region W_b of W. We assume that error diffusion is effective in both the left and right halves of R such that the average image intensity is preserved. Thus, the probability distribution of a halftone pixel $y_1(i,j)$ in the region R of Y_1 is

$$P[y_1(i,j) = 255] = A/255 \qquad (13.19)$$
$$P[y_1(i,j) = 0] = (255 - A)/255 \qquad (13.20)$$

such that the expected value is

$$E[y_1(i,j)] = 0 \bullet P[y_1(i,j) = 0] + 255 \bullet P[y_1(i,j) = 255] = A \qquad (13.21)$$

for $(i,j) \in R$. Typically, the percentage of white and black pixels in Y_1 in R are $A/255 \bullet 100\%$ and $(255 - A)/255 \bullet 100\%$ respectively, distributed evenly in R.

The generation of Y_2 using DHCED is effectively the application of error diffusion to the equivalent noisy multitone image. And the error diffusion should be effective, as usual. The distortion $\Delta x(i,j)$ introduced by DHCED is equally likely to be positive and negative and its magnitude is bounded by T. The average intensity in R should still be approximately A. Thus, the probability distribution of a halftone pixel $y_2(i,j)$ in the region R of Y_2 is approximately

$$P[y_2(i,j) = 255] \approx A/255 \qquad (13.22)$$
$$P[y_2(i,j) = 0] \approx (255 - A)/255 \qquad (13.23)$$

such that

$$E[y_2(i,j)] = 0 \bullet P[y_2(i,j) = 0] + 255 \bullet P[y_2(i,j) = 255] \approx A \qquad (13.24)$$

for $(i,j) \in R_A$.

Let $y(i,j) = y_1(i,j) \bigcap y_2(i,j)$ be the output pixel obtained by overlaying Y_1 and Y_2. The $y(i,j)$ would be white only if both $y_1(i,j)$ and $y_2(i,j)$ are white. For the proposed DHCED, the $y_1(i,j)$ and $y_2(i,j)$ are designed to be dependent.

In the left half of R, which is in W_w, DHCED would simply copy $y_1(i,j)$ as $y_2(i,j)$ such that $P[y_2(i,j) = y_1(i,j)] = 1$. Then

$$P[y(i,j) = 255] = P[y_1(i,j) \bigcap y_2(i,j) = 255] = P[y_1(i,j) = 255] = A/255 \qquad (13.25)$$

such that

$$E[y(i,j)] = 0 \bullet P[y(i,j) = 0] + 255 \bullet P[y(i,j) = 255] = A \qquad (13.26)$$

In other words, $y(i,j) = y_1(i,j)) = y_2(i,j)$ for the region W_w and the overlay operation does not change the halftone pixel values at all.

In the right half of R which is in W_b, the $y_1(i,j)$ and $y_2(i,j)$ are favored to be conjugate to each other $(y_2(i,j) \approx \overline{y_1(i,j)})$ such that $y_1(i,j)$ and $y_2(i,j)$ tend not to be black at the same time. To investigate the behavior of $y(i,j)$, we consider two different cases: (1) $255 \geq A > 127$ and (2) $127 \geq A \geq 0$.

For the case of $255 \geq A > 127$, there are more white pixels than black pixels in R of Y_1. The percentage of black pixels in the right half of R of Y_1 is about $(255 - A)/255 \bullet 100\%$. For example, if $A = 190$, about 25% of the pixels in the right half of R of Y_1 should be black. As $y_1(i,j)$ and $y_2(i,j)$ tend not to be black at the same time, the black pixels in the right half of R of Y_2 tend to be at different locations from those in Y_1. Consequently, the percentage of black pixels in the right half of R of Y tends to be doubled to $2 \bullet (255 - A)/255 \bullet 100\%$. In our example of $A = 190$, the approximately 25% black pixels in the right half of R of Y_1 and those of Y_2 tend to be at different locations such that there are about 50% black pixels in the right half of R of Y. Thus,

$$P[y(i,j) = 0] \approx 2P[y_1(i,j) = 0] = 2(255 - A)/255 \tag{13.27}$$
$$P[y(i,j) = 255] \approx (2A - 255)/255 \tag{13.28}$$

such that

$$E[y(i,j)] = 0 \bullet P[y(i,j) = 0] + 255 \bullet P[y(i,j) = 255]$$
$$\approx 2A - 255 = A - (255 - A) \leq A \tag{13.29}$$

In other words, the expected value should increase approximately linearly with A, from 1 (for $A = 128$) to 255 (for $A = 255$). The difference between the average intensity of the left and right halves of R for $255 \geq A > 127$ is

$$\Delta intensity = E[y(i,j)|(i,j) \in W_w] - E[y(i,j)|(i,j) \in W_b]$$
$$\approx A - (2A - 255) = 255 - A \tag{13.30}$$

For the case of $127 \geq A \geq 0$, there are fewer white pixels than black pixels in the right half of R of Y_1. Again the $y_1(i,j)$ and $y_2(i,j)$ tend not to be black at the same time in DHCED. This implies that the black pixels in the right half of R of Y_2 tend to fill up all the white pixel locations in the right half of R of Y_1, leading to all $y(i,j)$ being black. Thus, $P(y(i,j) = 0) \approx 1$ and

$$E[y(i,j)] = 0 \bullet P[y(i,j) = 0] + 255 \bullet P[y(i,j) = 255] \approx 0 \tag{13.31}$$

for $127 \geq A \geq 0$. And the difference between the average intensity of the left and right halves of R is, for $127 \geq A \geq 0$,

$$\Delta intensity \approx A - 0 = A \tag{13.32}$$

Consequently, the contrast between the left and right halves of R of Y can be expressed as, for $255 \geq A > 127$,

$$contrast = \frac{E[y(i,j)|(i,j) \in W_w] - E[y(i,j)|(i,j) \in W_b]}{E[y(i,j)|(i,j) \in W_w]}$$

$$\approx \frac{A - (2A - 255)}{A} = \frac{255 - A}{A} = \frac{255}{A} - 1 \qquad (13.33)$$

and, for $127 \geq A \geq 0$,

$$contrast = \frac{E[y(i,j)|(i,j) \in W_w] - E[y(i,j)|(i,j) \in W_b]}{E[y(i,j)|(i,j) \in W_w]}$$

$$\approx \frac{A - 0}{A} = 1 \qquad (13.34)$$

To summarize, for any (i,j) in the right half of R that is in W_b, the expected value of $y(i,j)$ is

$$E[y(i,j)] \approx \begin{cases} 0, & 127 \geq A \geq 0 \\ 2A - 255, & 255 \geq A > 127 \end{cases} \qquad (13.35)$$

The difference in average intensity of the left half and right half of R is

$$\Delta intensity \approx \begin{cases} A, & 127 \geq A \geq 0 \\ 255 - A, & 255 \geq A > 127 \end{cases} \qquad (13.36)$$

and the contrast between the left half and right half of R is

$$contrast \approx \begin{cases} 1, & 127 \geq A \geq 0 \\ \frac{255}{A} - 1, & 255 \geq A > 127 \end{cases} \qquad (13.37)$$

To verify these, we simulate DHCED using an artificial image called "Ramp" shown in Figure 13.14 as $X = X_1 = X_2$, in which the image intensity decreases gradually and linearly from 255 at the top row to 0 at the bottom row. The hidden pattern to be embedded is called "Column" and is black at the center and white on the left and right, as shown in Figure 13.15. The DHCED generated Y_1 and Y_2 are shown in Figures 13.16 and 13.17, respectively. The row-wise average intensity of Y_1 and Y_2 are plotted against the average intensity of the corresponding row in X in Figure 13.22. As expected, the row-wise average intensity of Y_1 and Y_2 are very similar to the corresponding intensity in X, which verifies (13.21) and (13.24).

Figure 13.18 is the overlaid image Y. We note that in the W_w region, Y is identical to Y_1 and Y_2, with intensity decreasing from top to bottom. In the W_b region, the intensity of Y is as high as X at the top. But as the intensity of X decreases from the top to bottom, we observe that the intensity of Y decreases at a fast pace to about the middle of the image (where intensity is about 127), and remains low in the lower half of the image.

To show the exact behavior of Y in the W_b region, we compute the average intensity in the W_b for each row of Y and plot it against the average intensity of

FIGURE 13.14
Original multitone image "Ramp" (X).

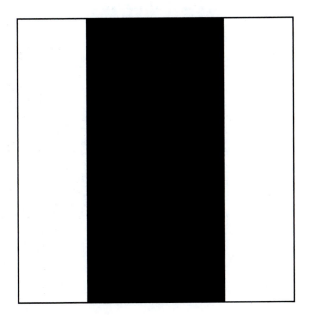

FIGURE 13.15
Secret pattern "Column" to be embedded in the halftone image (W).

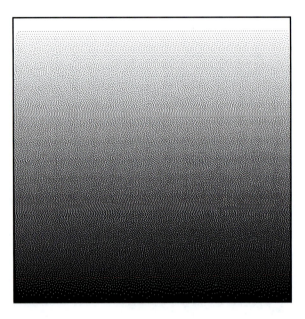

FIGURE 13.16
Halftone images generated by error diffusion with the Jarvis kernel (Y_1).

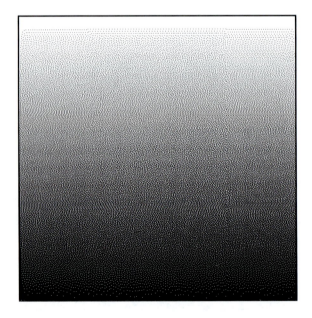

FIGURE 13.17
DHCED-generated Y_2 ($T = 10$) of Ramp with respect to X in Figure 13.14, W in Figure 13.15, and Y_1 in Figure 13.16.

FIGURE 13.18
Image Y obtained by overlaying Y_1 in Figure 13.16 and Y_2 in Figure 13.17.

the corresponding row in X in Figure 13.19. Three such curves are obtained with 3 values of T, namely $T = 5, 10, 15$. Also shown in the figure is the theoretical behavior as predicted by (13.35). It can be observed from the figure that as T increases, the average intensity curves appear to converge to the theoretical curve, as expected.

We also compute the difference between the average intensity in the W_b and the W_w regions for each row of Y, and plot this $\Delta intensity$ against the average intensity of the corresponding row in X in Figure 13.20. Similarly, the *contrast* is computed and plotted in Figure 13.21. Also shown in the two figures are the theoretical behavior predicted by (13.36) and (13.37). It can be observed that as T increases, the $\Delta intensity$ curves and the *contrast* curves appear to converge to the corresponding theoretical curves, as expected.

Next, we analyze DHSED and make a comparison with DHCED. Since DHSED forces Y_1 and Y_2 to be identical in W_w, both $y_1(i,j)$ and $y_2(i,j)$ are identical in the left half of R and the probability distribution is

$$P[y(i,j) = 255] = P[y_1(i,j) \cap y_2(i,j) = 255] = P[y_1(i,j) = 255] = A/255 \tag{13.38}$$

such that $E[y(i,j)] = A$. For any (i,j) in W_b, the corresponding pixel values in Y_1 and in Y_2 are error diffused with relative random phase. Thus, the local intensity for Y_1 and Y_2 is

$$E[y_2(i,j)] = E[y_1(i,j)] = A \tag{13.39}$$

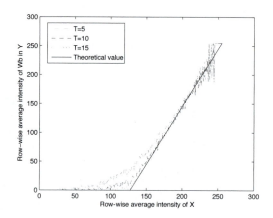

FIGURE 13.19
Row-wise average intensity of W_b in Y in Figure 13.18 vs row-wise average intensity of X in Figure 13.14 (Ramp).

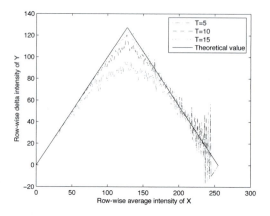

FIGURE 13.20
Row-wise Δintensity of Y in Figure 13.18 vs row-wise average intensity of X in Figure 13.14 (Ramp).

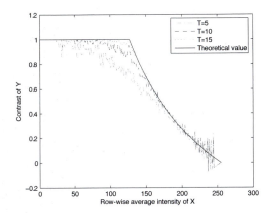

FIGURE 13.21
Contrast of Y in Figure 13.18 vs row-wise average intensity of X in Figure
13.14 (Ramp).

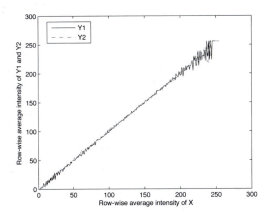

FIGURE 13.22
Row-wise average intensity of Y_1 in Figure 13.4 and Y_2 in Figure 13.9 vs
row-wise average intensity of X in Figure 13.14 (Ramp).

and $y_1(i,j)$ and $y_2(i,j)$ are approximately independent with

$$P[y(i,j) = 255] = P[y_1(i,j) \cap y_2(i,j) = 255]$$
$$= P[y_1(i,j) = 255] \bullet P[y_2(i,j) = 255] = (A/255)^2 \quad (13.40)$$
$$P[y(i,j) = 0] = 1 - (A/255)^2 \quad (13.41)$$

such that

$$E[y(i,j)] = A^2/255 \quad (13.42)$$

Thus, the intensity of DHSED is expected to be greater than or equal to that of DHCED, with equality if $A = 255$. The difference in average intensity is

$$\Delta intensity = E[y(i,j)|(i,j) \in W_w] - E[y(i,j)|(i,j) \in W_b]$$
$$= A - A^2/255 \quad (13.43)$$

The contrast is

$$contrast = \frac{E[y(i,j)|(i,j) \in W_w] - E[y(i,j)|(i,j) \in W_b]}{E[y(i,j)|(i,j) \in W_w]}$$
$$= \frac{A - A^2/255}{A} = \frac{255 - A}{255} \quad (13.44)$$

We also simulate DHSED on Ramp to verify the equations above. The hidden pattern is still the Column in Figure 13.15. The DHSED generated Y_1 and Y_2 are shown in Figures 13.16 and 13.23, respectively. In Figure 13.28, the row-wise average intensity of Y_1 and Y_2 are plotted against the average intensity of the corresponding row in X. As expected, they are very similar to X, which verifies (13.39).

Figure 13.24 is the overlaid image Y. Similar to DHCED, the Y of DHSED is identical to Y_1 and Y_2 in the W_w region, with intensity decreasing from top to bottom. In the W_b region, the intensity of Y is as high as X at the top row. But as the intensity of X decreases towards the bottom of the image, the intensity of Y decreases faster in the center than on the two sides. In other words, the intensity of Y decreases faster in W_b than in W_w, making W_b visible in Y.

Similar to DHCED, to show the exact behavior of DHSED in W_b in Y, we compute the average intensity in the W_b for each row of Y and plot it against the average intensity of the corresponding row in X in Figure 13.25. Three curves are obtained for 3 values of L, namely $L = 1, 5, 10$. Also shown is the theoretical curve according to (13.42). It can be observed that the 3 empirical curves match the theoretical curve very well. This also suggests that the choice of L does not have much effect on the intensity of W_b in Y.

We also compute the difference between the average intensity in W_b and W_w regions for each row of Y, and plot this $\Delta intensity$ against the average intensity of the corresponding row in X in Figure 13.26. Similarly, the *contrast* is computed and plotted in Figure 13.27. Also shown in the two figures are

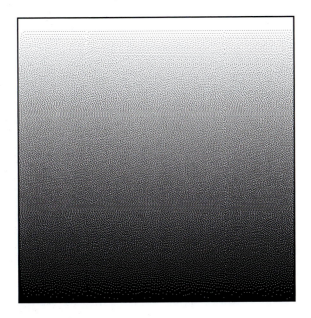

FIGURE 13.23
DHSED-generated Y_2 ($L = 5$) of Ramp with respect to X in Figure 13.14, W in Figure 13.15 and Y_1 in Figure 13.16.

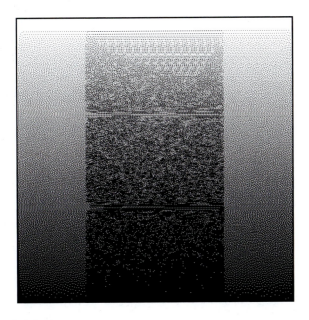

FIGURE 13.24
Image Y obtained by overlaying Y_1 in Figure 13.16 and Y_2 in Figure 13.23.

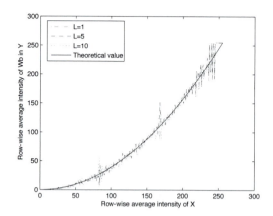

FIGURE 13.25
Row-wise average intensity of W_b in Y in Figure 13.24 vs row-wise average intensity of X in Figure 13.14 (Ramp).

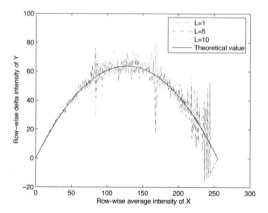

FIGURE 13.26
Row-wise Δintensity of Y in Figure 13.24 vs row-wise average intensity of X in Figure 13.14 (Ramp).

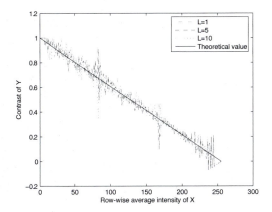

FIGURE 13.27
Contrast of Y in Figure 13.24 vs row-wise average intensity of X in Figure 13.14 (Ramp).

the theoretical behavior predicted by (13.43) and (13.44). It can be observed that the empirical $\Delta intensity$ curves and *contrast* curves are similar to the corresponding theoretical curves, as expected. Again, the choice of L has little effect.

Comparing DHCED and DHSED, they have the same Y_1. For their Y_2, their pixel values are identical in W_w but different in W_b. On overlaying the corresponding Y_1 and Y_2, the Y of both DHCED and DHSED are identical in W_w, but DHCED has lower $E[y(i,j)]$ in W_b than DHSED such that the black patterns of W would look darker in DHCED than in DHSED as predicted by Figure 13.29. And, the contrast of the revealed W in DHCED is higher than that in DHSED as predicted by Figure 13.30.

13.6 Summary

In this chapter, we introduce two ways to achieve steganography in halftone images, namely DHSED and DHCED. Both methods can embed a binary secret pattern into two halftone images that come from the same multitone image. When the two halftone images are overlaid, the secret pattern is revealed. DHCED can further embed a binary secret pattern into two halftone images from two different multitone images. DHSED operates by introducing different stochastic phases in the two images. DHCED operates by favoring certain conjugate values for each pixel and taking on the values only if the implied distortion is small enough. Both theoretical analysis and simulation

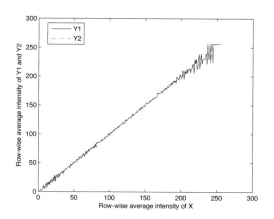

FIGURE 13.28
Row-wise average intensity Y_1 in Figure 13.4 and Y_2 in Figure 13.23 vs row-wise average intensity of X in Figure 13.14 (Ramp).

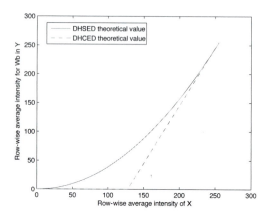

FIGURE 13.29
Theoretical average local intensity of W_b in Y for DHSED and DHCED vs row-wise average intensity of X in Figure 13.14 (Ramp).

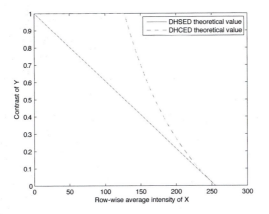

FIGURE 13.30
Theoretical contrast of Y for DHSED and DHCED vs row-wise average intensity of X in Figure 13.14 (Ramp).

results suggest that DHCED can give better contrast of the revealed secret pattern than DHSED.

Bibliography

[1] B.E. Bayers. An optimum method for two level rendition of continuous tone pictures. In *Proceedings of the IEEE International Communication Conference*, pages 2611–2615, 1973.

[2] R.W. Floyd and L. Steinberg. An adaptive algorithm for spatial grayscale. In *Proceedings of the Society of Information Display*, volume 17, pages 75–77, 1976.

[3] M.S. Fu and O.C. Au. Data hiding in halftone images by stochastic error diffusion. In *Proceedings of the IEEE International Conference on Acoustics, Speech, and Signal Processing*, volume 3, pages 1965–1968, May 2001.

[4] M.S. Fu and O.C. Au. Steganography in halftone images: Conjugate error diffusion. *Signal Processing*, 83(10):2171–2178, 2003.

[5] J.F. Jarvis, C.N. Judice, and W.H. Ninke. A survey of techniques for the display of continuous tone pictures on bilevel displays. *Computer Graphics and Image Processing*, 5(1):13–40, 1976.

[6] P. Moulin and R. Koetter. Data-hiding codes. In *Proceedings of the IEEE*, volume 93, pages 2083–2126, December 2005.

14

Image Secret Sharing

WeiQi Yan

Queen's University Belfast, UK

Jonathan Weir

Queen's University Belfast, UK

Mohan S. Kankanhalli

National University of Singapore, Singapore

CONTENTS

14.1 Introduction

Cryptography plays a very vital enabling role in our modern computing infrastructure. Almost all real-world applications require keys (such as passwords) for the purposes of confidentiality, authentication, and nonrepudiation. The strength of such cryptographic applications is based on the secrecy of a key. Therefore, the loss of a key can lead to disastrous consequences. Thus, many cryptographers have tackled the following problem:

Suppose a secret s (a key) is divided into $n > 1$ parts (called secret shares) and it satisfies these properties:

1. The secret key s can be easily restored from k ($k \leq n$) shares.

2.   The secret key s cannot be restored from $k - 1$ (or less) shares.

3.   The size of each share is not more than the size of the secret key s.

Such a scheme is referred to as a (k, n) threshold cryptography scheme or a secret sharing system[2][17]. It provides a backup mechanism to the secret key and it provides protection against the loss of a key. Secret sharing is also regarded as a mechanism to transfer secret information by public channels in cryptography [4]. Blakley based his secret sharing scheme on hyperplanes [17] and Shamir provided a solution based on the Lagrange interpolation [2]. Asmuth and Bloom scheme is based on the Chinese Remainder Theorem [5]. The details of these methods are available in [1] [12]. These traditional secret sharing schemes primarily concentrate on bit strings and do not take the specific content of these bits into account. However, with the increasing emphasis on security and digital rights management of multimedia data, the connection between multimedia and cryptography is becoming stronger. In this context, we present our novel ideas on color image sharing in which we utilize the concept of secret sharing from cryptography and employ it to protect a secret color image. As we shall see, the ideas cannot be directly applied so we need to take into account that the data under consideration describes color images and is not any generic bit stream. In our scheme, a secret color image is divided into n shares. Each share is an innocuous image totally unrelated to the secret image. We utilize k (or more) shares in order to perfectly reconstruct the secret image. However, having access to $k - 1$ (or less) shares will not reveal the secret color image.

We envisage several useful applications for a color image sharing scheme. Suppose we have a secret color image that we desire to protect. If we employ traditional cryptographic techniques, then we need to encrypt the image and store the image on a secure server. We then need to pay attention on the security of the key used for encryption. This server would then become a single focus of attack from a potential adversary. However, with an image sharing scheme, we can divide the information in the image into several shares and keep them on separate servers. This would allow for a lot more redundancy in the protection since breaking one server will not reveal the secret image. Another application would be that of data hiding. Suppose we would like to transmit a secret image over a noisy and insecure channel. We could divide the image information into several shares that are basically innocuous images. These images could be transmitted and at the other end, the secret image could be reconstructed from the threshold number of shares. Another useful application would be that of a military command and control system based on the Clark-Wilson security model [11]. Suppose we want the battlefield plans to be made only if k out of n commanders agree. In which case, we could divide the battle terrain map into n shares and distribute it to the commanders. Only if k of them get together can they restore the terrain map and agree to a battle plan. In general, our scheme would be advantageous in any situation where a group of mutually suspicious individuals or processes need to cooperate and every threshold subset of the group needs to be given the veto power.

The problem of color image sharing is formally defined as follows. In order to transfer a color image $\mathcal{I}$ through a public channel securely, the information of the color image is divided into n pieces and embedded into images I_i, $(i = 1, 2, \cdots, n)$, and we call the images I_i, $(i = 1, 2, \cdots, n)$ as shares. With the knowledge of any $k(k < n)$ shares I_i, $(i = 1, 2, \cdots, k)$, the restoration of the original color I image is easy; with the knowledge of any $k - 1(k < n)$ shares, the restoration of the original image $\mathcal{I}$ is impossible (i.e., any image is equally likely to be reconstructed).

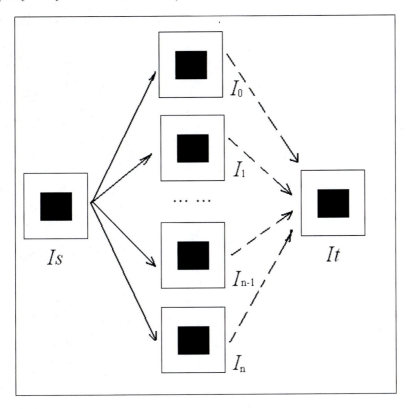

FIGURE 14.1
Principle of image sharing.

In Figure 14.1, the left-most image is the original color image that we desire to keep as a secret and it is divided into several blocks (subimages). We process each of these blocks into n shares in order to create the subimages shares. If required, we can compose these shares I_i, $(i, = 1, 2, \cdots, n)$ together and compute the blocks in the right-most image. Block-based processing is done for breaking correlation. Note that our scheme will require a minimum of k shares in order for I_r to be exactly equal to I_s. If less than k shares are used, then I_s cannot be restored. The threshold is indispensable in color image sharing.

14.2   State of the Art

Applying visual cryptography techniques to color images is a very important area of research because it allows the use of natural color images to secure some types of information. Due to the nature of a color image, this again helps to reduce the risk of alerting someone to the fact that information is hidden within it. It should also allow high quality sharing of these color images. Color images are also highly popular and have a wider range of uses when compared to other image types. Many of the techniques presented within this section use halftone technologies on the color images in order to make them work with visual cryptography. That is why color visual cryptography is presented within this section.

In 1996, Naor and Shamir published a second article on visual cryptography "Visual Cryptography II: Improving the Contrast via the Cover Base" [23]. The new model contains several important changes from their previous work; they use two opaque colors and a completely transparent one.

The first difference is the order in which the transparencies are stacked. There must be an order to correctly recover the secret. Therefore, each of the shares needs to be predetermined and recorded so recovery is possible. The second change is that each participant has c sheets, rather than a single transparency. Each sheet contains red, yellow, and transparent pixels. The reconstruction is done by merging the sheets of participant I and participant II, i.e., put the i-th sheet of II on top of the i-th sheet of I and the $(i + 1)$-th of I on top of the i-th of II.

The two construction methods are monochromatic construction and bichromatic construction. In the monochromatic construction, each pixel in the original image is mapped into c subpixels and each participant holds c sheets. In each of participant I sheets, one of the subpixels is red and the remaining $c - 1$ subpixels are transparent. In each of participant II sheets, one of the subpixels is yellow, the other $c - 1$ subpixels are transparent. The way the sheets of participant I and II are merged is by starting from the sheet number 1 of participant I, then putting sheet number 2 of participant II on top of it, then sheet number 2 of participant I on top of that, and so on.

The order in which subpixels of participant I are colored red constitutes a permutation π on $\{1, \cdots, c\}$ and the order which the subpixels of participant II are colored yellow constitutes a permutation σ. π and σ are generated as follows: π is chosen uniformly at random from the set of all permutations on c's elements. If the original pixel is yellow, then $\pi = \sigma$, therefore each red subpixel of the i-th sheet of participant I will be covered by a yellow subpixel of the same position of the i-th sheet of participant II. If the original pixel is red, then $\sigma(i) = \pi(i+1)$ for $1 \leq i \leq c-1$ and $\sigma(c) = \pi(1)$, therefore each yellow subpixel of the i-th sheet of participant II will be covered by a red subpixel of the same position of the $(i+1)$-th sheet of participant I except the c-th sheet. In practice, the first sheet of participant I is not necessarily stored since it is always covered by other sheets.

A very primitive example of color image sharing appeared in [24]. In this example, each pixel of the color secret image is expanded to a block of 2×2 subpixels. Each one of these blocks is filled with red, green, blue, and white (transparent) colors, respectively. Taking symmetries into account, 24 different possibilities for the combination of two pixels can be obtained. It is claimed that if the subpixels are small enough, the human visual system will average out the different possible combinations to 24 different colors. To encrypt a pixel of the colored image, round the color value of that pixel to the nearest representable color. Select a random order for the subpixels on the first share and select the ordering on the second share such that the combination produces the required color.

The advantage of this scheme is that it can represent 24 colors with a resolution reduction of 4, instead of $24^2 = 576$. The disadvantage is that the 24 colors are fixed once the basic set of subpixel colors is fixed.

Another primitive scheme was also presented [29] and extended more recently [34]. Verheul and Van Tilborg's scheme provides a c-color (k, n)-threshold scheme. This scheme uses the black pixel to superimpose on the result of two color pixels, superimposition, if they give a resultant color that is not in the original color palette. This can be achieved by making sure the superimposed color pixels result in a noncolor palette color, one of which is changed to a black pixel or by ensuring that one of the color pixels is changed to black before the superimposing operation [10]. Yang and Laih improve on the pixel expansion aspect of the Verheul and Van Tilborg scheme and their (n, n)-threshold scheme is optimal since they match the following lower bound placed on pixel expansion, formulated in [10]:

$$m \geq \begin{cases} c \cdot 2^{n-1} - 1, & \text{if } n \text{ is even} \\ c \cdot 2^{n-1} - c + 1, & \text{if } n \text{ is odd} \end{cases} \tag{14.1}$$

Hou et al. [18] proposed a novel approach to share color images based on halftoning. With this halftone technology, different gray levels can be simulated simply by altering the density of the printed dots. Within bright parts of the image the density is sparse, while in the darker parts of the image, it is dense. This is very helpful in the visual cryptography sense because it is able to transform a gray-scale image into a black and white image. This allows for traditional visual cryptography techniques to be applied. Similarly, the color decomposition method is used for color images, which also allows the proposed scheme to retain all the advantages of traditional visual cryptography, such as no computer participation required for the decryption/recovery of the secret.

Hou himself also provided one of the first color decomposition techniques to generate visual cryptograms for color images [19]. Using this color decomposition, every color within the image can be decomposed into one of three primary colors: cyan, magenta, or yellow. This proposal is similar to traditional visual cryptography with respect to the pixel expansion that occurs. One pixel is expanded into a 2×2 block where two color pixels are stored along with two transparent (white) pixels.

However, [21] examined the security of Hou's [19] scheme, and while the scheme is secure for a few specific two-color secret images, the security cannot be guaranteed for many other cases.

Improving this pixel expansion and also working out the optimal contrast of color visual cryptography schemes have been investigated [10]. In the paper, they prove that contrast-optimal schemes are available for color visual cryptography (VC) and then further go on to prove the optimality with regard to pixel expansion.

A lossless recovery scheme outlined by [20] considers halftoning techniques for the recovery of color images within visual cryptography. The scheme generates high quality halftone shares that provide lossless recovery of the secrets and reduces the overall noise in the shares without any computational complexity. Their proposed method starts by splitting the color channels into its constituent parts, cyan (C), magenta (M), and yellow (Y). Each channel has grayscale halftoning applied to it. Error diffusion techniques discussed in [35] are then applied to each halftone channel. A circularly symmetric filter is used along with a Gaussian filter. This provides an adequate structure for the dot placement when constructing the shares.

Efficiency within color visual cryptography [25] is also considered which improves on the work done by [34, 3]. The proposed scheme follows Yang and Laih's color model. The model considers the human visual system's effect on color combinations out of a set of color subpixels. This means that the set of stacked color subpixels would look like a specific color in original secret image. As with many other visual cryptography schemes, pixel expansion is an issue. However Shyu's scheme has a pixel expansion of $\lceil log_2 c \rceil$, which is superior to many other color visual cryptography schemes especially when c, the number of colors in the secret image becomes large. An area for improvement however would be in the examination of the difference between the reconstructed color pixels and the original secret pixels. Having high quality color VC shares would further improve on the current schemes examined within this survey, this includes adding a lot of potential for visual authentication and identification.

Chang et al. [6] present a scheme based on smaller shadow images, which allows color image reconstruction when any authorized k shadow images are stacked together using their proposed revealing process. This improves on the following work [32], which presents a scheme that reduces the shadow size by half. Chang et al.'s technique improves on the size of the share in that, as more shares are generated for sharing purposes, the overall size of those shares decreases.

In contrast to color decomposition, Yang and Chen [33] propose an additive color mixing scheme based on probabilities. This allows for a fixed pixel expansion and improves on previous color secret sharing schemes. One problem with this scheme is that the overall contrast is reduced when the secrets are revealed.

In most color visual cryptography schemes, when the shares are superimposed and the secret is recovered, the color image gets darker. This is due to the fact that when two pixels of the same color are superimposed, the resultant pixel gets darker. Cimato et al. [9] examine this color darkening by proposing a scheme that has to guarantee that the reconstructed secret pixel has the exact same color as the original. Optimal contrast is also achieved as part of their scheme. This scheme differs from other color schemes in that it considers only 3 colors when superimposing, black, white, or one pixel of a given color. This allows for perfect reconstruction of a color pixel, because no darkening occurs, either by adding a black pixel or by superimposing two colors that are identical, that ultimately results in a final darker color.

A technique that enables visual cryptography to be used on color and grayscale images is developed in progressive color visual cryptography [13]. Many current state-of-the-art visual cryptography techniques lead to the degradation in the quality of the decoded images, which makes it unsuitable for digital media (image, video) sharing and protection. In [13], a series of visual cryptography schemes have been proposed that not only support gray-scale and color images, but also allow high quality images including that of perfect (original) quality to be reconstructed.

The annoying presence of the loss of contrast makes traditional visual cryptography schemes practical only when quality is not an issue which is relatively rare. Therefore, the basic scheme is extended to allow visual cryptography to be directly applied on grayscale and color images. Image halftoning is employed in order to transform the original image from the grayscale or color space into the monochrome space which has proved to be quite effective. To further improve the quality, artifacts introduced in the process of halftoning have been reduced by inverse halftoning.

With the use of halftoning and a novel microblock encoding scheme, the technique has a unique flexibility that enables a single encryption of a color image but enables three types of decryptions on the same ciphertext. The three different types of decryptions enable the recovery of the image of varying qualities. The physical transparency stacking type of decryption enables the recovery of the traditional VC quality image. An enhanced stacking technique enables the decryption into a halftone quality image. A progressive mechanism is established to share color images at multiple resolutions. Shares are extracted from each resolution layer to construct a hierarchical structure; the images of different resolutions can then be restored by stacking the different shared images together.

The advantage is that this scheme allows for a single encryption, multiple decryptions paradigm. In the scheme, secret images are encrypted/shared once, and later, based on the shares, they can be decrypted/reconstructed in a plurality of ways. Images of different qualities can be extracted, depending on the need for quality as well as the computational resources available. For instance, images with loss of contrast are reconstructed by merely stacking the shares; a simple yet effective bit-wise operation can be applied to restore the halftone image; or images of perfect quality can be restored with the aid of the auxiliary look-up table. Visual cryptography has been extended to allow for multiple resolutions in terms of image quality. Different versions of the original image of different qualities can be reconstructed by selectively merging the shares. Not only this, a spatial multiresolution scheme has been developed in which images of increasing spatial resolutions can be obtained as more and more shares are employed.

This idea of progressive visual cryptography has recently been extended [14] by generating friendly shares that carry meaningful information and that also allows decryption without any computation at all. Purely stacking the shares reveals the secret. Unlike [13] and [7] which require computation to fully reconstruct the secret, the scheme proposed in [15] has two types of secrets, stacking the transparencies reveals the first, but computation is again required to recover the second-level secret. Fang's scheme is also better than the polynomial sharing method proposed in [26] by Thien and Lin. The method proposed in [26] is only suitable for digital systems and the computational complexity for encryption and decryption is also a lot higher.

Currently, one of the most robust ways to hide a secret within an image is by typically employing visual cryptography. The perfect scheme is extremely practical and can reveal secrets without computer participation. Recent state-of-the-art watermarking [8] can hide a watermark in documents that require no specific key in order to retrieve it. We take the idea of unseen visible watermarks and apply a secure mask to them and incorporate it for use within the VC domain, thus improving the overall security that is currently one of its weaknesses.

Weir et al. [31] also provide a mechanism for secret sharing using color images as a base. The technique relies on visual cryptography as a mechanism for sharing the secret. Many smaller secrets can be embedded within a color image and a final share can be created in order to reveal all of the secrets. The color image is visually similar to the original image before embedding due to the high Peak-Signal-to-Noise Ratio (PSNR) achieved after embedding.

Another recent novel application for color image sharing and using color images for secret sharing was presented by Weir and Yan [30]. Using Google Maps, along with its Street View implementation, personally identifiable information can be obscured using visual cryptography techniques and can be accurately recovered by authorized individuals who may need the information. Specifically for law enforcement agencies who many find this type of information helpful for a particular case. This type of practical application is very important for the progression of visual cryptography, which presents a unique way of using these techniques in a real-world situation.

14.3 Approaches for Image Sharing

14.3.1 Shamir's Secret Sharing Scheme

Shamir's secret sharing scheme is based on the Lagrange interpolation [2]. Given a set of points $(x_i, y_i)(i = 0, 1, 2, ..., k)$, the Lagrange interpolation polynomial $L^k(x)$ can be constructed using:

$$f^k(x) = \sum_{i=0}^{k} y_i \prod_{j=0; i \neq j}^{k} \frac{x - x_i}{x_j - x_i} \tag{14.2}$$

Given a secret, it can be easily shared using this interpolation scheme. If $GF(q)$ denotes a Galois field $(q > n)$, the following polynomial is constructed by choosing proper coefficients $a_0, a_1, \cdots, a_k$ from $GF(q)$, which satisfy:

$$f^k(x) = s^* + \sum_{i=0}^{k} a_i x^i \tag{14.3}$$

where s^* is the secret key. The coefficients are randomly chosen over the integers $[0, q)$ and the details are provided in [2]. Suppose $s_i = f(a_i), (i = 0, 1, \cdots, k)$, each S_i is known as a *share* and they all can be delivered to different persons.

Now we would like to reconstruct the original secret. Suppose k people have provided their shares s_i, $(i = 0, 1, \cdots, k)$. The following Lagrange interpolation polynomial is utilized to reconstruct the original secret:

$$P^k(x) = \sum_{i=0}^{k} s_i \prod_{j=0; i \neq j}^{k} \frac{a - a_i}{a_j - a_i} \tag{14.4}$$

where addition, subtraction, multiplication, and division are defined over $GF(q)$.

$$P^k(a_i) = s_i; i = 0, 1, 2, \cdots, k; s^* = P^k(0); \tag{14.5}$$

Thus, we can obtain the original secret s^*[1][2].

14.3.2 Color Image Sharing Based on the Lagrange Interpolation

When an image is treated as a secret, we can share the secret based on the Lagrange interpolation. We consider the image to be a matrix; and Lagrange interpolation is generalized for matrices.

We assume a grayscale image corresponds to a matrix A. Given matrix $A_i = (a_{w,h}^i)_{W \times H}$, where w and h are integer $(w = 1, 2, \cdots, W; h =$

$1, 2, \cdots, H$; $i = 0, 1, \cdots, k$), k is the number of shares. We define the matrix operations of Lagrange polynomials with degree k in the Galois field:

$$L^k(x) = \sum_{i=0}^{k} A_i \prod_{j=0;i\neq j}^{k} \frac{x - x_i}{x_j - x_i} \tag{14.6}$$

where (x_i, A_i) is the feature point, $A_i(\ (i = 0, 1, \cdots, k))$ are matrices of size $W \times H$, the elements being nonnegative, x_i is a real number and $x_i \leq x_j (i \leq j)$.

Our novel idea for color image sharing using Lagrange interpolation is the following. We utilize the secret image and a few $(< n)$ other chosen innocuous images to build the Lagrange interpolation polynomial. We then construct new images based on this interpolation to obtain a total of n images. We now can use all the innocuous images and the new reconstructed images as the shares of the secret image. And the secret image is not distributed but it can be always reconstructed using the requisite number of shares. So our novelty is in the construction of new shares based on the secret image and some chosen innocuous images. We now provide the details of the scheme. In Figure 14.2, assume that the secret image that we desire to share is at the position $x = 0$.

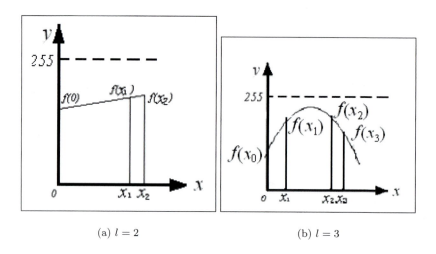

(a) $l = 2$ (b) $l = 3$

FIGURE 14.2
Image sharing based on the Lagrange interpolation in (a) and (b).

In Figure 14.2(a), we position an innocuous image at x_2. We can now compute many new shares with the parameters in the interval $(x_2 - d, x_2 + d)$, $d > 0$, (we have used $d_{max} = \frac{x_2 - x_0}{4}$). Without loss of generality, we select the parameter at the position x_1 as our newly created share (based on 0 and x_2). We now consider the images at x_2 and x_1 as the shares of the secret image at 0. For the reconstruction of the secret, we can collect the two shares, we calculate the coefficients of Lagrange interpolation polynomial first and then compute the image at position 0. In Figure 14.2(b), we put the shares in x_1, x_2 and obtain the secret color image at 0; and in Figure 14.2(b), we put the shares at x_1, x_2, x_3 and restore the secret image at position 0. For a color image, this operation can be performed for each color channel. Figure 14.3 shows our experimental results for the Lagrange interpolation scheme of a color image.

(a) Original image and shares 1 and 2

(b) Original image and shares 1, 2, and 3

FIGURE 14.3
(See color insert.) Experimental results of image sharing based on the Lagrange interpolation in (a) and (b).

In Figure 14.3(a), the original color image is the secret image to be shared, at least two shares are needed to reconstruct the original image, i.e., it is a $(2, n)$ scheme, thus $k = 2$. Thus, share 2 has been generated using the original color image and share 1 (which is an innocuous image). Now, share 1 and share 2 can be distributed independently and the original color image can be reconstructed anytime if we obtain both the shares. In Figure 14.3(b), the original color image is the secret color image and at least three shares are required to restore the secret color image. This is because $k = 3$, therefore it is a $(3, n)$ secret image sharing scheme. Notice that the secret image is somewhat visible in the generated shares. We will fix this problem later with the block-based approach.

However, the Lagrange polynomial based image sharing has a potential practical problem. It cannot yield too many shares. This is because the Lagrange polynomial curve with a high degree has severe oscillations once the degree of the polynomial is greater than nine [16]. The consequence of this oscillations phenomenon is that it is impossible to constrain the pixel values to lie between 0 and 255. As a result, the resultant shares are not proper images anymore as shown in Figure 14.4.

(a) Original image   (b) Shares 1   (c) Shares 2

FIGURE 14.4
(See color insert.) The image sharing by using a high degree polynomial interpolation in (a)-(c).

In Figure 14.4, the resultant image Figure 14.4(a) is the original secret image, Figure 14.4(b) is the innocuous image and Figure 14.4(c) is the generated share while the rest of the shares are all-white images. We use a polynomial of degree 8 for generating Figure 14.4(c). Figure 14.4(c) has obvious color overflow problems because some pixel values are more than 255 and some are less than zero, hence the image quality is severely degraded. Note that this is not a problem in Shamir's original secret sharing scheme because they consider the secret to be shared as a binary integer, and thus a share can take on any value. In our case, this binary integer has some constraints because they denote image pixel values. In order to overcome this serious limitation, it is obvious that some form of a piece-wise polynomial interpolation is required in order to bound the degree of the polynomial and thus constrain the oscillations. We have developed a new color image sharing scheme based on moving lines, which is a rational implicit curve.

14.3.3   Color Image Sharing Based on Moving Lines

It is well known that the equation of a line in the homogenous form in projective geometry [22] is:

$$aX + bY + cW = 0 \tag{14.7}$$

where a, b, and c are not all zero, (X, Y, W) are the homogenous coordinates of points whose Cartesian coordinates are:

$$(x, y) = (X/W, Y/W) \tag{14.8}$$

It is obvious that X, Y, W cannot all be zero. Let P denote the triple (X, Y, W) and L denote the triple (a, b, c). We refer to the line L that is:

$$\{(X, Y, W)|L \cdot P = (a, b, c) \cdot (X, Y, W) = aX + bY + cW = 0\} \tag{14.9}$$

Thus, we can see that a point P lies on a line $L = (a, b, c)$ only and if only $P \cdot L = 0$, where $P \cdot L$ is the dot product.

Now we consider the line L containing two points $P_1 = (X_1, Y_1, W_1)$ and $P_2 = (X_2, Y_2, W_2)$, and also the point P at which two lines $L_1 = (a_1, b_1, c_1)$ and $L_2 = (a_2, b_2, c_2)$ intersect. Because of the duality principle (of points and lines), we have these cross products:

$$L = P_1 \times P_2, P = L_1 \times L_2 \tag{14.10}$$

A homogeneous point whose coordinates are functions of a variable t (i.e., it is parameterized by a variable t) is denoted as:

$$P[t] = (X[t], Y[t], W[t]) \tag{14.11}$$

which actually is the rational curve:

$$x = \frac{X[t]}{W[t]}; y = \frac{Y[t]}{W[t]}; \tag{14.12}$$

If the functions are of the following form:

$$X[t] = X_i \phi_i[t]; Y[t] = Y_i \phi_i[t]; W[t] = W_i \phi_i[t] \tag{14.13}$$

With $\{\phi_i[t]\}$ being a given set of blending function, then equation (14.11) defines a curve:

$$P[t] = \sum P_i \phi_i[t] \tag{14.14}$$

where the homogeneous points are $P_i = (X_i, Y_i, W_i)$. Likewise,

$$L[t] = (a[t], b[t], c[t]) \tag{14.15}$$

denotes the family of lines $a[t]x + b[t]y + c[t] = 0$.

In order to obtain the intersection of two pencils, we notice the following four lines $L_{0,0}$, $L_{0,1}$, $L_{1,0}$, and $L_{1,1}$ from which two pencils are defined:

$$L_0[t] = L_{0,0}(1 - t) + L_{0,1}t, L_1[t] = L_{1,0}(1 - t) + L_{1,1}t \tag{14.16}$$

The points, at which they intersect for parameter values $t =$

$0, \Delta t, 2\Delta t, \cdots, 1$ are on a curve. The parameter $t = 0, \Delta t, 2\Delta t, \cdots, 1$ is adaptively given and it may take any real number value in the interval $[0,1]$ for a piece of curve and it also can be extended to the infinite interval $[-\infty, +\infty]$ for the whole curve. The curve turns out to be a conic section, which can be expressed as a rational Bernstein–Bezier curve $P[t]$ as follows:

$$P[t] = L_0[t] \times L_1[t] \tag{14.17}$$

It is clear that $P[t]$ is a quadratic rational Bezier curve [27][28] whose control points are:

$$P_0 = L_{0,0} \times L_{1,0}, P_1 = \frac{L_{0,0} \times L_{1,1} + L_{0,1} \times L_{1,0}}{2}, P_2 = L_{0,1} \times L_{1,1} \tag{14.18}$$

The graph of a quadratic curve generated by moving lines is shown in Figure 14.5(a).

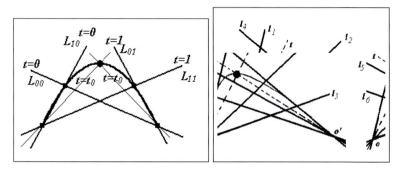

(a) The graph of a quadratic curve   (b) The graph of a quartic curve

FIGURE 14.5
Intersection of two pencils of lines in (a) and (b).

In general, the curve comprising of $L_{0,0}$, $L_{0,1}$, $L_{0,2}$, and $L_{1,0}$, $L_{1,1}$ is:

$$L_0[t] = L_{0,0}(1-t) + 2L_{0,1}(1-t)t + L_{0,2}t^2, L_1[t] = L_{1,0}(1-t) + L_{1,1}t \tag{14.19}$$

then $P[t] = L_0[t] \times L_1[t]$, $t \in [0, 1]$ is a cubic rational Bezier curve [16].
Without loss of generality, the moving lines consisting of $L_{0,0}$, $L_{0,1}$, $\cdots$, $L_{0,p}$, and $L_{1,0}$, $L_{1,1}$, $\cdots$, $L_{1,q}$ are:

$$L_0[t] = \sum_{i=0}^{p} L_{i,0} B_i^p(t), L_1[t] = \sum_{i=0}^{q} L_{i,1} B_i^q(t) \tag{14.20}$$

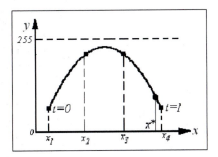

FIGURE 14.6
Image sharing scheme based on moving lines.

then $P[t] = L_0[t] \times L_1[t]$, $t \in [0,1]$ represents a rational Bezier curve of degree $p+q$ as shown in Figure 14.5(b) [16].

As in the case of Lagrange interpolation, we now construct the scheme of image sharing based on moving lines as shown in Figure 14.6. In Figure 14.6, the x-coordinate depicts the given position of the original secret color image, the y-coordinate indicates the values of the gray-scale pixel value, t is the parameter for moving lines generation, and x^* is the position of the new computed share. For image sharing, we compute the shares using the above implicit curve. We now present the detailed steps for our moving lines based on the color image sharing scheme.

Steps of sharing a color image:

1. Suppose we are given one secret image I_0 to be shared, $k+1$ images $I_0, I_1, \cdots, I_k$ as the innocuous images and a group of corresponding parameters $x_0, x_1, \cdots, x_k$ $(x_0 \le x_1 \le \cdots \le x_k)$ as input;

2. Calculate an arbitrary new share corresponding to a parameter within the interval of the given parameters (x_1, x_k);

3. The images (greater than k) $I_0, I_1, \cdots, I_{i-1}, I_p, I_i, I_{i+1}, \cdots, I_k$; $0 \le i-1 \le p \le i \le k$ and their corresponding parameters $x_0, x_1, \cdots, x_{i-1}, x_p, x_i, x_{i+1}, \cdots, x_k$ $(x_0 \le x_1 \le \cdots \le x_{i-1} \le x_p \le x_i \le x_{i+1} \le \cdots \le x_k)$ are the output computed image shares.

Steps of reconstructing the color image:

1. Select k image shares $I_0, I_1, \cdots, I_k$ and their corresponding position parameters $x_0, x_1, \cdots, x_k$ $(x_0 \le x_1 \le \cdots \le x_k)$ and the position parameter of the color image x_0 as the input;

2. Calculate the implicit curve corresponding to pixels in each share according to the moving lines scheme $P[t] = L_0[t] \times L_1[t]$, $t \in [0,1]$;

3. Reconstruct the secret color image by the scheme of moving lines at the position of x_0 as output.

Note that each of the steps has to be applied separately for each color channel. The use of a rational curve has several advantages, such as it does not have the oscillations phenomenon of polynomials with a high degree; a rational curve is easy to be controlled and is able to express conic curves. Also a polynomial curve with a high degree cannot guarantee that the interpolation values can be constrained to a given range [16]. Thus, when a rational curve is employed to share a color image, the scheme based on moving lines yields more shares than that of the one based on the Lagrange interpolating polynomial. With more shares, the secret color image can be shared in a more secure manner with greater flexibility. However, this greater flexibility comes at an increased cost of computation compared to that of the Lagrange interpolations scheme.

14.3.4   Improved Algorithm

If we carefully examine Figure 14.3, we can see that the profile of the secret image is visible in the constructed image shares. The reason is that the correlation of the secret image is not broken during the image sharing process. So far, we selected only one X-position parameter for the secret image, thus only one share (i.e., the newly created one) is closely related to the secret image and the other shares are totally independent. Thus, the innocuous images do not contribute towards image hiding. We now modify the earlier approach slightly to make all the shares involved in data hiding by utilizing a block-based approach with multiple parameters:

Steps of block-based image sharing:

1. Divide the secret image into blocks.

2. Designate different parameters (i.e., different innocuous image position) for each block.

3. Share the secret image using the earlier approach.

4. Write down the parameters and positions embedded in the corresponding shares of each block for secret restoration.

The restoration handling is the inverse procedure of the encoding procedure.

In order to clearly explain the scheme, we illustrate the improved approach in Figure 14.7. In Figure 14.7, the secret image is divided into four blocks, each block is embedded into the given innocuous image share at a different spatial position. The secret image can be reconstructed by applying the scheme blockwise again. This method effectively breaks the correlation of the original secret image with the secret image being divided into many blocks and these blocks being hidden in different shares at different locations. Thus, the secret image is no longer visible in the shares.

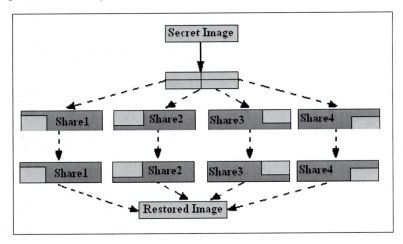

FIGURE 14.7
Improved algorithm of image sharing.

14.4 Experiment and Evaluation

Both of the algorithms proposed in this chapter can theoretically be employed for image sharing, but actually they are different in terms of practical implementation. Color image sharing based on Lagrange interpolation allows for only a limited number of shares due to the oscillations phenomenon. Color image sharing based on moving lines can theoretically generate an unlimited number of shares. In practice, generating more shares will require more computation. Thus, as a trade-off, we usually share a color image into at most five shares.

FIGURE 14.8
(See color insert.) The experimental results of image sharing by moving lines.

We now present some experimental results on images of size 128×128. The shares and original secret image based on quadric curves of moving lines are

FIGURE 14.9
(See color insert.) The experimental results of image sharing by moving lines.

FIGURE 14.10
(See color insert.) The experimental results of image sharing by moving lines.

shown in Figure 14.8, which is a $(2, n)$ sharing example. In Figure 14.9, we illustrate a $(3, n)$ scheme based on cubic curves of moving lines.

Figure 14.9 is the result of $(3, n)$ scheme based upon moving lines to share a given original image, with share 1 and share 2 being innocuous. Share 3 is the new computed share. By using shares 1, 2, 3, and their parameters, the original color image can be restored.

Figure 14.10 is the result of a $(4, n)$ scheme based on a moving line to share a color image by given original secret image, share 1, share 2, and share 3. Share 4 is the new computed share. When shares 1, 2, 3, 4, and their parameters are used for the moving lines based scheme, the original color image can be perfectly restored.

For a sensitive application, it is better to compute the shares by performing block-by-block (e.g., 8×8 pixel blocks) processing. The advantage of such an approach is that distributing the reconstructed blocks into the various innocuous images can break the correlation of the secret image in the image share. Thus, instead of taking $n - 1$ innocuous images and creating the n-th share (which can possibly reveal some correlations), we can take n innocuous images and distribute the secret image blocks within them. In fact, the secret color images need not be of the same size as that of the shares. Actually, if the shares have a larger size, the hiding of the color image will be more secure.

Figure 14.11 shows an example of breaking the correlation between neighboring blocks of the secret image. In this $(4, n)$ image sharing scheme, we have divided the original image into four equal-sized rectangular blocks and have shared these blocks using a quadric curve generated by the moving lines tech-

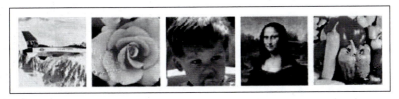

FIGURE 14.11
(See color insert.) Breaking the correlation of neighboring blocks in an image.

nique. What is different here from the earlier examples (where we created a whole new image as a share) is that we embed the information of each rectangular block into one of the image shares. The embedded information of share 1 is in the lower-left corner, that of share 2 is in the upper-left position while that of share 3 is embedded is in the upper-right region, and that of share 4 is in the lower-right corner. Thus, the information of the original secret image is shared among all the image shares. While this example is a simple illustration of how the correlation can be broken, it is clear that more sophisticated secret image subdivision and sharing schemes can be devised using the same principle.

14.5 Conclusion

In this chapter, we introduce the novel concept of color image sharing based on Lagrange interpolation and moving lines. The security of secret sharing is built upon the computation of polynomials. Given enough shares, the coefficients of the interpolating polynomial can be exactly determined. However, if an attacker does not possess the threshold number of shares and the image shares, he will not be able to reconstruct the polynomial. Thus, he will not able to restore the secret image properly.

Secret sharing based on Lagrange interpolation is often utilized to share binary strings, but it is difficult to use for color image sharing since it yields only a limited number of shares. We therefore have developed a new color image sharing scheme based on moving lines that does not have that limitation. An implicit curve generated by moving lines is a rational curve; we believe it can therefore be directly applied for sharing compressed-domain images.

Our future work will focus on further investigation of color image sharing in the compressed domain. The possible research directions in the future are:

1. Color image sharing based on compressed domain processing. Ideas related to DCT, DFT, and DWT can be employed in color image sharing.

2. Color image sharing based on visual cryptography. If visual cryptography could be implemented in color images, we could share color images with visual cryptography. Since visual cryptography is perfectly secure, color image sharing based on visual cryptography will be extremely robust.

Bibliography

[1] A. Salomaa. *Public-Key cryptography*. Springer, Berlin Heidelberg, 1990.

[2] A. Shamir. How to share a secret. *Communications of the ACM*, 22(11):612–613, 1979.

[3] C. Blundo, A. De Bonis, and A. De Santis. Improved schemes for visual cryptography. *Designs, Codes and Cryptography*, 24(3):255–278, 2001.

[4] C. Liu. *Introduction to combinatorial mathematics*. McGraw-Hill, New York, 1968.

[5] C.A. Asmuth and J. Bloom. A modular approach to key safeguarding. *IEEE Transactions on Information Theory*, 29:208–210, 1983.

[6] C.-C. Chang, C.-C. Lin, C.-H. Lin, and Y.-H. Chen. A novel secret image sharing scheme in color images using small shadow images. *Information Sciences*, 178(11):2433–2447, 2008.

[7] S.-K. Chen and J.-C. Lin. Fault-tolerant and progressive transmission of images. *Pattern Recognition*, 38(12):2466–2471, 2005.

[8] S.-C. Chuang, C.-H. Huang, and J.-L. Wu. Unseen visible watermarking. In *ICIP (3)*, pages 261–264, 2007.

[9] S. Cimato, R. De Prisco, and A. De Santis. Colored visual cryptography without color darkening. *Theoretical Computer Science*, 374(1-3):261–276, 2007.

[10] S. Cimato, R. De Prisco, and A. De Santis. Optimal colored threshold visual cryptography schemes. *Designs, Codes and Cryptography*, 35(3):311–335, 2005.

[11] D. Gollmann. *Computer Security*. John Wiley & Sons, Chichester, 1999.

[12] D.E. Denning. *Cryptography and data security*. Addison-Wesley, London, 1982.

[13] J. Duo, W.-Q. Yan, and M. S. Kankanhalli. Progressive color visual cryptography. *SPIE Journal of Electronic Imaging*, 14(3), 2005.

[14] W.-P. Fang. Friendly progressive visual secret sharing. *Pattern Recognition*, 41(4):1410–1414, 2008.

[15] W.-P. Fang and J.-C. Lin. Visual cryptography with extra ability of hiding confidential data. *Journal of Electronic Imaging*, 15(2):023020, 2006.

[16] G. Farin. *Curves and Surfaces for Computer aided Geometric Design: A Practical Guide (3rd Edition)*. Academic Press, Boston, USA, 1992.

[17] G.R. Blakley. Safeguarding cryptographic keys. In *Proc. of AFIPS 1979 National Computer Conference*, volume 48, pages 313–317, Aelington, 1979.

[18] Y.C. Hou, C.Y. Chang, and S.F. Tu. Visual cryptography for color images based on halftone technology. In Proceedings of the 5th World Multiconference on Systemics, Cybernetics, and Informatics (SCI 2001), Orlando, FL, Vol. XIII, pp. 441–445.

[19] Y.-C. Hou. Visual cryptography for color images. *Pattern Recognition*, 36:1619–1629, 2003.

[20] N. K. Prakash and S. Govindaraju. Visual secret sharing schemes for color images using halftoning. In *Proceedings of Computational Intelligence and Multimedia Applications*, 3:174–178, December 2007.

[21] B.W. Leung, F.Y. Ng, and D.S. Wong. On the security of a visual cryptography scheme for color images. *Pattern Recognition*, 2008, 42(5): 929–940 (2009).

[22] M.A. Penna and R.R. Patterson. *Projective Geometry and Its Applications to Computer Graphics*. Prentice Hall, New Jersey, 1986.

[23] M. Naor and A. Shamir. Visual cryptography II: Improving the contrast via the cover base. In *Proceedings of the International Workshop on Security Protocols*, pages 197–202, London, UK, 1997. Springer-Verlag.

[24] V. Rijmen and B. Preneel. Efficient color visual encryption for shared colors of benetton. *EUCRYPTO '96*, 1996.

[25] S. J. Shyu. Efficient visual secret sharing scheme for color images. *Pattern Recognition*, 39(5):866–880, 2006.

[26] C.-C. Thien and J.-C. Lin. An image-sharing method with user-friendly shadow images. *IEEE Transactions on Circuits and Systems for Video Technology*, 13(12):1161–1169, 2003.

[27] T.W. Sederberg and F.L. Chen Implicitization using moving curves and surfaces. In *Proceedings of SIGGRAPH'95*, pages 301–308, Los Angeles, 1995. ACM Press.

[28] T.W. Sederberg, T. Saito, D.X. Qi, and K.S. Klimaszewski. Curve implicitization using moving lines. *Computer Aided Geometric Design*, (11):687–706, 1994.

[29] E.R. Verheul and H.C.A. Van Tilborg. Constructions and properties of k out of n visual secret sharing schemes. *Des. Codes Cryptography*, 11(2):179–196, 1997.

[30] J. Weir and W Yan. Resolution variant visual cryptography for street view of Google maps. In *IEEE International Symposium on Circuits and Systems, 2010. ISCAS 2010*, May 2010.

[31] J. Weir, W.Q. Yan, and D. Crookes. Secure mask for color image hiding. In *Third International Conference on Communications and Networking in China, 2008. ChinaCom 2008*, pages 1304–1307, Aug. 2008.

[32] C.-N. Yang and T.-S. Chen. New size-reduced visual secret sharing schemes with half reduction of shadow size. In *ICCSA (1)*, pages 19–28, 2005.

[33] C.-N. Yang and T.-S. Chen. Colored visual cryptography scheme based on additive color mixing. *Pattern Recognition*, 41(10):3114–3129, 2008.

[34] C.-N. Yang and C.-S. Laih. New colored visual secret sharing schemes. *Designs, Codes and Cryptography*, 20(3):325–336, 2000.

[35] Z. Zhou, G. R. Arce, and G. Di Crescenzo. Halftone visual cryptography. *IEEE Transactions on Image Processing*, 15(8):2441–2453, 2006.

15

Polynomial-Based Image Sharing

Shiuh-Jeng Wang

Central Police University, Taiwan

Chen-Hsing Yang

National Pingtung University of Education, Taiwan

Yu-Ting Chen

Central Police University, Taiwan

CONTENTS

15.1   Introduction

The concept of secret sharing comes from the method of secret key management and was first seen in the following works [1, 7]. The secret's owner wants to share the secret with other participants, but no participants can obtain the secret alone. When some of the participants work together, the secret is then revealed. From a technical viewpoint, this secret sharing scheme may also be referred to as the (t, n)-threshold, where t denotes the threshold value that will reveal the secret and n is the total number of holding shadows. In this method, when a secret is given, it must be divided into n shadows and it is then reconstructed by t or more shadows, possessed by the shadow holders; no information can be conjectured by fewer than t shadows. In the wake of this cryptographic application proposal, many related methods were further proposed to enrich secret sharing diversity in both theoretical and practical arenas [4, 6, 8, 12].

Visual security was proposed by Naor and Shamir [6] in 1994, where an image could be reconstructed by superimposing two shares. The shares issued in this scheme are made up of random binary patterns. The target secret, recognized by the human visual system, is finally stacked by the (t, n)-threshold scheme. One of the main advantages, in this scheme, is that it does not require complicated computations, which traditional numeric cipher-text in secret decryption does, while the sizes of the enlarged shares, and target secrets are left the same. Thien and Lin [8] proposed an applied secret image scheme for image embedding to improve upon Naor and Shamir's method. The concealment of a secret image avoids cipher-based attacks on the basis of stego-image imperceptibility. In 2006, Horng et al. uncovered another way to cheat the (t, n)-threshold under visual cryptography (VC) applications [3]. In their proposal, a scenario was successfully put forward showing that shareholders were able to collaborate with each other by sending fake shares to other shareholders. Accordingly, the final target secret image was different from the genuine one, corrupting the secret sharing system.

Steganography is a kind of data hiding technique that provides another method of security protection for digital image data. The purpose of steganography is to embed secret data in preselected meaningful images, called cover images, with people being visually unaware of the secret's existence. Numerous schemes have been developed to achieve successful data hiding [2, 5]. To prevent the fake stego-image from the dishonest participants, Lin and Tsai [4] authored an authentication-based steganography study, where their detection technique tested whether the offered shares were genuine or not. In their scheme, parity checks were applied to prevent the modified shares from malicious attacks. Later on, in 2007, Yang et al. [12] enhanced the mechanism of Lin and Tsai by proposing a revised scheme in which more detailed fake shares were created so that they would be detected by those malicious shareholders in the target secret image (t, n)-threshold system, thus acting as a deterrent.

In 2008, Wang et al. [10] took advantage of the CRC (Cyclic Redundant Code) in conjunction with the hash function to make authentication-based steganography more robust, in order to prevent the attacks of fake share offerings. In addition to making it more robust, both the capacity and the image quality remain comparable to the scheme suggested by the scheme in [12]. In this chapter, we introduce the polynomial-based image sharing scheme by reviewing some preliminaries and related works and depicting Wang et al.'s approach. Moreover, some improvements of Wang et al.'s approach are proposed in this chapter.

The presentation of this chapter is organized as follows. In Section 15.2, the concept of the polynomial-based sharing scheme is introduced. In Section 15.3, we introduce some preliminaries and related works about the polynomial-based image sharing scheme. Wang et al.'s scheme and its improvements are presented in Section 15.4. Observations and experiments of benchmark images are presented and discussed in Section 15.5. Finally, conclusions are offered in Section 15.6.

15.2 Polynomial-Based Sharing Scheme

15.2.1 Shamir's Secret Sharing Scheme

A (t, n)-threshold scheme is a method of sharing a secret among n participants such that any subset consisting of t participants can reconstruct the secret, but no subset of smaller size can reconstruct the secret. The first method was provided in 1979 by Shamir [7] and is known as Shamir's secret sharing scheme. In the initial state, a dealer chooses a large prime number p, and makes p public. The following calculation is performed on Z_p: A secret, $s \in Z_p$ exists, so the dealer randomly generates a $(t-1)$th degree polynomial, in Z_q:

$$q(x) \equiv x + q_1 x + q_2 x^2 + \cdots + q_{t-1} x^{t-1} (mod p)$$

where $q(0) = s$ is the secret. The dealer distributes the shadow, $y_i \equiv q(x_i)(mod\ p)$ to node P_i, where x_i is the ID number for P_i.

Suppose that t participants get together to reconstruct the polynomial. Also, assume that their pairs are $(x_1, y_1), (x_2, y_2), \cdots, (x_t, y_t)$. The polynomial $q(x)$ can be reconstructed by solving the following equation:

$$\begin{bmatrix} 1 & x_1 & \cdots & x_1^{t-1} \\ 1 & x_2 & \cdots & x_2^{t-1} \\ \vdots & \vdots & \ddots & \vdots \\ 1 & x_t & \cdots & x_t^{t-1} \end{bmatrix} \begin{bmatrix} s \\ q_1 \\ \vdots \\ q_{t-1} \end{bmatrix} \equiv \begin{bmatrix} y_1 \\ y_2 \\ \vdots \\ y_t \end{bmatrix} (mod\ p)$$

The equation has a unique solution if the determinant of the matrix is nonzero mod p. It can be shown that the determinant of the matrix is nonzero mod p if all x_i, $i = 1, 2, \cdots, t$, are different. After the polynomial is reconstructed, the secret s is obtained by the output of $q(0)$.

15.2.2 Lagrange Interpolation Scheme

An efficient method to obtain the sharing polynomial f is to use the Lagrange interpolation. Assume that the pairs (x_1, y_1), $(x_2, y_2), \cdots, (x_t, y_t)$ are used to reconstruct the polynomial $q(x)$. For $k = 1, 2, \cdots, t$, let

$$l_k(x) = \prod_{i=1, i \neq k} \frac{x - x_i}{x_k - x_i} \, mod \, p$$

We know that

$$l_k(x_j) = \begin{cases} 1, when \ k = j \\ 0, when \ k \neq j \end{cases}$$

Then, the Lagrange interpolation polynomial is formed as follows:

$$q'_x = \sum_{k=1}^{t} y_k l_k \, mod \, p$$

Because $y_i \equiv q'(x_i) (mod\,p)$ for $i = 1, 2, \cdots, t$ and $q'(x)$ has degree $t - 1$, $q'(x)$ is equal to the polynomial $q(x)$. Finally, the secret s can be obtained by evaluating $q(x)$ at $x = 0$. That is,

$$s \equiv \sum_{k=1}^{t} y_k \prod_{j=1, j \neq b}^{t} \frac{-x_i}{x_k - x_i} \, mod \, p$$

In fact, all coefficients in $q(x)$ can be seen as secrets. We show an example for the Lagrange interpolation polynomial.

Example 1.

Take a $(2, 4)$-threshold scheme. In the share generation phase, assume that the pixels of the target secret image S indexed in order of left-to-right and top-to-down is S_i, $I = 0, 1, 2, \cdots, n$. The first two pixels, for instance, $S_0 = 100$ and $S_1 = 200$ are set in this case. First of all, a polynomial of $q(x) = 200x + 100 \mod 251$ is given according to the S'_is. Compute the shares y'_js associated with each participants as $y_1 = q(x_1{=}1) = 49$, $y_2 = q(x_2{=}2) = 249$, $y_3 = q(x_3{=}3) = 198$, and $y_4 = q(x_4{=}4) = 147$, where x'_is are the identity numbers of the four participants. Next, consider the recovery of the pixels of the target secret image in this case. In the $(2, 4)$-threshold scheme, any two participants who own the shares y'_js can gather to recover the secret, for example, the participants nos. x_2 and x_4. There are two pairs of $(x_2, y_2) = (2, 249)$ and $(x_4, y_4) = (4, 147)$ offered from the participants. In such a way that a polynomial of $q'(x)$ is constructed as follows:

$$g(x) = (249 \times \frac{(x-4)}{(x-1)} + 147 \times \frac{(x-2)}{(x-4)}) mod251$$
$$= (249 \times (-2)^{-1} \times (x-4) + 147 \times (2)^{-1} \times (x-2)) mod251$$
$$= (200x + 100) mod251$$

Following up the result, the first two pixels of the target secret image, $S_0 = 100$ and $S_1 = 200$ are obtained from the coefficients of $q'(x) = 200x + 100$ $mod251$.

15.3 Preliminaries and Related Works

15.3.1 Thien-Lin Scheme

The (t, n)-threshold scheme in visual cryptography was proposed by Naor and Shamir [6], where the target secret (image) could only be recognized by the human eye if there were exactly t or more than t share-images stacked. However, the share-images generated could only reveal one target secret. When there are more than 2 target secrets, the numbers of share-images required, increases. As a result, the occupied share-images, held by participants, also increase. Inspired by these observations, Thien and Lin (Thien-Lin scheme) [8] proposed a specific method to accommodate a higher number of share-images, with the normal storage requirements, using LaGrange polynomial construction. In this method, there are a total of n share-images, with each of them being only $\frac{1}{t}$ times the target secret image. This is also done in the (t, n)-threshold scheme.

15.3.2 Lin-Tsai Scheme

In [4], Lin and Tsai (Lin-Tsai scheme) proposed a scheme of sharing image, where the requirements of capacity and detecting ability are further considered when compared to the study of [8]. Similarly, a polynomial of $q(x) = (a_0 + a_1 x + a_2 x^2 + \cdots + a_{i-1} x^{i-1})$ of degree t-1 is applied upon the concept of the (t, n)-threshold scheme. The $q(x)$ is then devised for exact reconstruction when t pairs of $(x_i, q(x_i))$ are offered among n share holders, where x_i is one of the pixels of a pixel-value and $q(x_i)$ is the share held in the share holder. Accordingly, the pixels of the target secret image can be obtained from the coefficients of the new one of $q'(x)$ via the (t, n)-threshold polynomial construction. The procedures in [4] are briefly given as follows:

$x_i = (x_{i1}, x_{i2}, \cdots, x_{i8})$	$w_i = (w_{i1}, w_{i2}, \cdots, w_{i8})$
$v_i = (v_{i1}, v_{i2}, \cdots, v_{i8})$	$u_i = (u_{i1}, u_{i2}, \cdots, u_{i8})$

FIGURE 15.1
The format of B_{ij}, where x_i, w_i, v_i, and u_i are represented as binary pattern.

Notations:

$q(x)$: The $(t-1)$-degree polynomial of $q(x) = (a_0 + a_1 x + a_2 x^2 + \ldots + a_{t-1} x^{t-1}) \bmod p$, where $p = 251$.

S: The target secret image with the size of $m \times m$. The pixel $S_i \in [0, 250]$ in S, $1 \leq i \leq m^2$. It is caught by up-to-down and left-to-right in order.

I_j: The jth cover image with the size of $2m \times 2m$ pixels. There are m^2 blocks, B'_{ij}s, where each one is the size of 2×2 containing four pixels as x_i, w_i, v_i, and u_i in the order of left-to-right and up-to-down and $1 \leq i \leq m^2$ and $1 \leq j \leq n$. The pixel format of B_{ij} is shown in Figure 15.1.

$\hat{I}_j$: The jth stego-image with the size of $2m \times 2m$. There are m^2 blocks, $\hat{B}_{ij}'$s. $\hat{B}_{ij}$ is stego-block when the secret is embedded the B_{ij}. The corresponding pixels in $\hat{B}_{ij}$ to B_{ij} is $\hat{x}_i$, $\hat{w}_i$, $\hat{v}_i$, and $\hat{u}_i$, respectively.

b_i: Generate a random bit-string with the form of $(b_1, b_2, \cdots, b_{m^2})$, where $1 \leq i \leq m^2$.

The procedure to generate the stego-image, named as Proc-Lin-Tsai, is briefly given in the following:

Procedure Proc-Lin-Tsai
Input: S, I_j, $j = 1, 2, \cdots, n$.
Output: $\hat{I}$, $j = 1, 2, \cdots, n$.

Step 1. Generate the random bit string $(b_1, b_2, \cdots, b_{m^2})$.

Step 2. Set $a_0 = S_i$, $i \in [1, m^2]$. For each S_i chooses a $(t-1)$-degree polynomial $q(x) = (a_0 + a_1 x + a_2 x^2 + \cdots + a_{i-1} x^{i-1}) \bmod p$, where randomly choose $t-1$ value $a_1, a_2, \cdots, a_{i-1}$ under the modulo p. Calculate $y_i = q_i(x_i)$, where x_i is obtained and defined from x_i in the I_j and represent y_i as the bit string with the form of $y_i = (y_{i1}, y_{i2}, \cdots, y_{i8})$.

Step 3. Generate the stego-block, $\hat{B}_{ij}$, comprised of the four bit strings as
$\hat{x}_i = (x_{i1}, x_{i2}, \cdots, x_{i8})$,
$\hat{w}_i = (w_{i1}, w_{i2}, \cdots, w_{i5}, b'_i, y_{i1}, y_{i2})$,

$\hat{x}_l = (x_{l1}, x_{l2}, \cdots, x_{l8})$	$\hat{w}_l = (w_{l1}, w_{l2}, \cdots, w_{l5}, b'_l, y_{l1}, y_{l2})$
$\hat{v}_l = (v_{l1}, v_{l2}, \cdots, v_{l5}, y_{l3}, y_{l4}, y_{l5})$	$\hat{u}_l = (u_{l1}, u_{l2}, \cdots, u_{l5}, y_{l6}, y_{l7}, y_{l8})$

FIGURE 15.2

The format of stego-block $\hat{B}_{ij}$ with the size of 2×2 gray-level pixels.

$\hat{x}_l = (x_{l1}, x_{l2}, \cdots, x_{l6}, y_{l1}, y_{l2})$	$\hat{w}_l = (w_{l1}, \cdots, w_{l5}, h_l, y_{l3}, y_{l4})$
$\hat{v}_l = (v_{l1}, \cdots, v_{l4}, x_{l7}, x_{l8}, y_{l5}, y_{l6})$	$\hat{u}_l = (u_{l1}, \cdots, u_{l6}, y_{l7}, y_{l8})$

FIGURE 15.3

Stego-block $\hat{B}_{ij}$ used in Yang et al. which is revised from Figure 15.2, where hash is used in the pixel of $\hat{w}_i = (w_{i1}, \cdots, w_{i5}, h_i, y_{i3}, y_{i4})$.

$\hat{v}_i = (v_{i1}, v_{i2}, \cdots, v_{i5}, y_{i6}, y_{i7}, y_{i8})$,
and $\hat{u}_i = (u_{i1}, u_{i2}, \cdots, u_{i5}, y_{i6}, y_{i7}, y_{i8})$,
where b'_i is the parity bit, which is chosen to make $\hat{w}_i$ including the corresponding bit-string b_i. The format of stego-block $\hat{B}_{ij}$ is shown in Figure 15.2.

Step 4. Collect $\hat{B}_{ij}$ generated in the Step 3 to create stego-image $\hat{I}_j$.

15.3.3 Yang et al.'s Scheme

As observed from the Lin-Tsai scheme, the fake shares can be easily made by the manner of fixing the parity bit b'_i. For example, if $\hat{a}_i = (w_{i1}, w_{i2}, \cdots, w_{i5}, b'_i, y_{i1}, y_{i2}) = (0, 0, 1, 1, 1, 0, 1, 1)$, the parity bit b'_i is all the same when the original $\hat{w}_i$ is changed to a new $\hat{w}_i$ as $\hat{w}_i = (1, 0, 1, 1, 1, 0, 0, 1)$. In such a way, the malicious attackers can easily modify $\hat{w}_i's$ in the stego-image to pass the authentication detection. As a result, the revelation of the target secret image is not a right one. Next, Yang et al. [12] proposed a revised version to set a new check function, e.g., hash function, instead of parity bit in [4]. The stego-block $\hat{B}_{ij}$ in [4] is then revised as the format shown in Figure 15.3. As analyzed in Yang et al., the probability of passing the detection with the fake share offers is $(\frac{1}{2})^{m^2}$.

15.3.4 Inverted Pattern LSB Scheme

In order to conceal the existence of shadow data, steganographic methods are used to keep the shadows imperceptible. The steganographic method is a kind of data hiding technique that embeds secret data in preselected meaningful images, called cover images, with people being visually unaware of the secret's existence. One common technique to achieve the data hiding is the LSB-based approach, in which secret data are embedded by directly replacing the least significant bits (LSBs) with equal bits of the secret for each pixel.

Moreover, some papers applied various strategies to improving the quality of stego-images generated by LSB-based approaches. In 2008, Yang [11] proposed an Inverted-Pattern-LSB information hiding algorithm (IPLA for short) to further highlight the quality of the stego-image. His method combines the idea of processing secret messages before embedded and processing the stego-image after embedding. Before embedding, the secret is transformed into a suitable format so as to benefit the next embedding step. Moreover, the bits that are used to record the transformation are the critical key in the course of the secret revelation. The IPLA algorithm is shown below, which was applied in Wang et al.'s scheme.

Inverted-Pattern-LSB Algorithm (H, S, k, n)
Input: A cover image H and a secret string S for k-bit LSB substitution, z is the number of sections.
Output: A stego-image and an inverted pattern P.

Step 1. Partition both H and S into z sections evenly. Let $H = H_0, H_1, \cdots, H_{z-1}$, $S = S_0, S1, \cdots, S_{z-1}$ and $R = R_0, R_1, \cdots, R_{z-1}$, where R is the replaced string of H. Also, let the inverted pattern $P = p_0, p_1, \cdots, p_{z-1}$, where p_i is a bit, for $i = 0, 1, \cdots, z - 1$.

Step 2. For $i = 0$ to $z - 1$
If $MSE_{s_i R_i} \leq MSE_{s_i \bar{R}_i}$
Then $p_i = 0$
Else $p_i = 1$

Step 3. For $i = 0$ to $z - 1$
If $p_i = 0$
Then Embed S_i into H_i
Else embed $\bar{S}_i$ into H_i

Step 4. Return the inverted pattern P.

15.3.5   Scalable Secret Image Sharing

The conventional secret image sharing (SIS) scheme, such as the Thien-Lin scheme [8], only has the threshold property that recovers either the entire image or nothing. For the (t, n)-SIS scheme, the target secret image could only be reconstructed by only t or more than t qualified participants. This limits its possible applications. Recently, Wang and Shyu recommended adding the scalable decoding capability (the scalability) into the threshold scheme [9]. The so-called scalability is that the amount of secret information is proportional to the number of shadows used in reconstruction. Wang and Shyu constructed a polynomial based $(2, n)$ scalable SIS (SSIS) scheme, which not only had the threshold property but also the scalability. The clarity of an image in [9] is measured in terms of three modes: the multisecret mode (spatially partitioning the target image into disjoint subimages), the priority mode (dividing the target image according to the bitplanes), and the progressive mode (combining the multisecret mode and the priority mode). To extend Wang and Shyu's $(2, n)$-SSIS scheme, Yang and Huang proposed (t, n)-SSIS schemes where a qualified set of participants consists of any t participants [13]. Two approaches were proposed for a general construction for any t, $2 \leq t \leq n$. For $t = 2$, Approach 1 has the lesser shadow size than Wang and Shyu's $(2, n)$-SSIS scheme, and Approach 2 is reduced to Wang and Shyu's $(2, n)$-SSIS scheme. The following is the shadow constructing algorithm of their Approach 1.

Shadow Constructing Algorithm

Input: a secret image O.

Output: n shadows S_i, $i = 1, 2, \cdots, n$.

Step 1. Divide image O into $\binom{n}{t}$ subimages O_x, $x = 1. 2, \cdots, \binom{n}{t}$ by one of the three modes.

Step 2. For every image O_x, use a polynomial-based (t, t)-SIS scheme to create t subshadows $(O_x^1, O_x^2, \cdots, O_x^t)$.

Step 3. Set $S_1 = S_2 = \cdots = S_n = \Phi$.

Step 4. Choose a binary matrix $B_{n,t} = [b_{i,j}]$.

Step 5. For $j = 1$ to $\binom{n}{t}$
Set $y = 1$;
For $i = 1$ to n
If $b_{i,j} = 1$ then $\hat{O}_j^i = O_y^y$ and $y = y + 1$.
else $\hat{O}_j^i = \Phi$.

Step 6. $S_1 = \bigcup_{j=1}^{\binom{n}{t}} O_j^i$, $i = 1, 2, \cdots, n$

The binary matrix $B_{n,t} = [b_{i,j}]$ is a $n \times \binom{n}{t}$ matrix, where $b_{i,j} \in [0, 1]$, $1 \leq i \leq n$, and $1 \leq i \leq \binom{n}{t}$. Every column vector has a Hamming weight t. For example, the matrix $B_{n,t} = B_{4,3}$ of a $(3, 4)$-SSIS scheme is

$$B_{4,3} = \begin{bmatrix} b_{1,1} & b_{1,2} & b_{1,3} & b_{1,4} \\ b_{2,1} & b_{2,2} & b_{2,3} & b_{2,4} \\ b_{3,1} & b_{3,2} & b_{3,3} & b_{3,4} \\ b_{4,1} & b_{4,2} & b_{4,3} & b_{4,4} \end{bmatrix} = \begin{bmatrix} 1 & 1 & 1 & 0 \\ 1 & 1 & 0 & 1 \\ 1 & 0 & 1 & 1 \\ 0 & 1 & 1 & 1 \end{bmatrix}$$

The corresponding matrix $[\hat{O}^i_j]$ is created as follows.

$$[\hat{O}^i_j] = \begin{bmatrix} \hat{O}^1_1 & \hat{O}^1_2 & \hat{O}^1_3 & \hat{O}^1_4 \\ \hat{O}^2_1 & \hat{O}^2_2 & \hat{O}^2_3 & \hat{O}^2_4 \\ \hat{O}^3_1 & \hat{O}^3_2 & \hat{O}^3_3 & \hat{O}^3_4 \\ \hat{O}^4_1 & \hat{O}^4_2 & \hat{O}^4_3 & \hat{O}^4_4 \end{bmatrix} = \begin{bmatrix} O^1_1 & O^1_2 & O^1_3 & \phi \\ O^2_1 & O^2_2 & \phi & O^4_1 \\ O^3_1 & \phi & O^3_3 & O^2_4 \\ \phi & O^3_2 & O^3_3 & O^3_4 \end{bmatrix}$$

Then, four shadows S_1, S_2, S_3, and S_4 are generated as follows.

$$S_1 = \hat{O}^1_1 \cup \hat{O}^1_2 \cup \hat{O}^1_3 \cup \hat{O}^1_4 = O^1_1 \cup O^1_2 \cup O^1_3 \cup \phi = O^1_1 \cup O^1_2 \cup O^1_3;$$

$$S_2 = \hat{O}^2_1 \cup \hat{O}^2_2 \cup \hat{O}^2_3 \cup \hat{O}^2_4 = O^2_1 \cup O^2_2 \cup \phi \cup O^4_1 = O^2_1 \cup O^2_2 \cup O^1_4;$$

$$S_3 = \hat{O}^3_1 \cup \hat{O}^3_2 \cup \hat{O}^3_3 \cup \hat{O}^3_4 = O^3_1 \cup \phi \cup O^2_3 \cup O^2_4 = O^3_1 \cup O^2_3 \cup O^2_4;$$

$$S_4 = \hat{O}^4_1 \cup \hat{O}^4_2 \cup \hat{O}^4_3 \cup \hat{O}^4_4 = \phi \cup O^3_2 \cup O^3_3 \cup O^3_4 = O^3_2 \cup O^3_3 \cup O^3_4;$$

15.4   Wang et al.'s Scheme

In this section, we describe in detail Wang et al.'s scheme [10] to introduce the standard method of polynomial-based image sharing. Some notations in Wang et al.'s scheme have been trimmed for easy description. Also, the Galois Field $GF(2^8)$ operations are applied to their scheme such that the all gray-level values 0~255 can be completely represented. Therefore, the secret image is completely lossless. The detailed approaches in Wang et al.'s scheme are introduced in the following subsections.

15.4.1   Secret Image Sharing

Based on the study in the literature [8], a secret image sharing technique was applied in our scheme. In their proposal, the size of shadows could be dramatically reduced, with n shares being embedded into n different host images to avoid the attention of hackers. However, if malicious participants deliver fake stego-images, the receivers may recover a fake secret image, thus misleading those participating in the secret sharing scheme, due to poor detection. Thus, yet another secret sharing function was invented, that we also applied in our proposal, using detection authentication to enhance security. Below, a revised algorithm is presented to achieve both high capacity and extended applications, where participants can recover two or more secret images from a stego-image. The notations for the Revised Algorithm of High-Capacity and Applications (**RAHA** for short) are given prior to all the steps shown.

Notations:

S: The target secret image with size $m \times m$.

t: More than or equal to t shadows possessed by the participants that can recover the target secret image.

n: The number of participants.

S_i: Each pixel of the target secret image $S, i = 1, 2, \cdots, m^2$.

$B_{i,j}$: The gth block (size of 1×8) in the jth cover image, where $j = 1, 2, \cdots, n$ and $g = 1, 2, \cdots, \left\lceil \frac{m^2}{t} \right\rceil$. Each block is made of 8 pixels.

$B_{g,j}^{(d)}$: The dth pixel of block $B_{i,j}$. The bit string of each pixel $B_{g,j}^{(d)}$ is shown as the form $(x_{d1}, x_{d2}, \cdots, x_{d7}, x_{d8})$ where $1 \le d \le 8$.

$\hat{B}_{g,j}$: The gth block in the jth stego-image.

f_j: The feature value of the jth stego-image (or the jth cover image) when the block $B_{g,j}^{(d)}$ ($or B_{g,j}$) is processed, where the stego-image size is of $2m \times 2m$. $f_j = (x_{11}, x_{12}, x_{13}, x_{14}, x_{21}, x_{22}, x_{23}, x_{24})$ is created from the first two pixels in block $B_{g,j}^{(d)}(or B_{g,j})$, for $j = 1, 2, \cdots, n$. If f_j is equal to one of feature values $f_1, f_2, \cdots, f_{j-1}$, keep looking at the next possible pair of pixels in block $B_{g,j}^{(d)}(or B_{g,j})$ following a fixed order. (Note that there are $C(8, 2) \times 2 = 56$ possible choices in a block.)

$PV\_y_j^{(g)}$: The gth pixel in jth shadows, $g = 1, 2, \cdots, \left\lceil \frac{m^2}{t} \right\rceil$ and $j = 1, 2, \cdots, n$. The bit-string format of $PV\_y_j^{(g)}$ is $(y_{g1}, y_{g2}, \cdots, y_{g8})$.

p: 256 for the Galois Field $GF(2^8)$.

Revised Algorithm of High-Capacity and Applications (RAHA)
Part I: Aim at the high capacity approaching:
For $g = 1, 2, \cdots, \left\lceil \frac{m^2}{t} \right\rceil$, do a loop as follows:

Step 1. Generate the polynomial $q_g(x)$ of $t - 1$ degrees as follows:
$q(x) = (a_0 + a_1 x + a_2 x^2 + ... + a_{i-1} x^{i-1})$
where $a_0 = S_i, a_1 = S_i + 1, \cdots, a_{t-1} = S_{i+t-1}, i = (g-1)t$, and all operations are over the Galois Field $GF(2^8)$.

Step 2. Get block $B_{g,j}$ from the jth cover image. Set f_j as the feature value of block $B_{g,j}, j = 1, 2, \cdots, n$.

Step 3. Compute $PV\_y_j^{(g)} = q_g(f_i), j = 1, 2, ..., n$.

Step 4. Assign each $PV\_y_j^{(g)}$ to the jth shadow, $j = 1, 2, \cdots, n$.

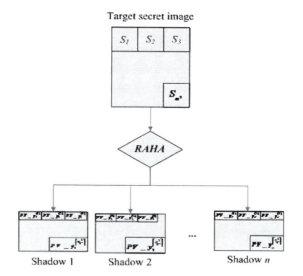

FIGURE 15.4
Demonstration of steps in RAHA.

The steps stated in RAHA are also demonstrated in Figure 15.4.

Part II. Aim at the recovery of the original secret image.

Step 1. Collect at least t shadows among n shadows. Renumber t of them as $1, 2, \cdots, t$ in order.
For $g = 1, 2, \cdots, \left\lceil \frac{m^2}{t} \right\rceil$, do a loop of the following steps:

Step 2. Chose the t feature value $f_j, j = 1, 2, \cdots, t$ forms the collected stego-images.

Step 3. Let $x_j = f_j$, for $j = 1, 2, \cdots, t$. Reconstruct $q_g(x)$by using the pair of $(x_j, PV\_y_j^{(g)})$ as follows:

$$q_g(x) = \sum_{b=1}^t PV\_y_b^{(g)} \prod_{j=1, j\neq b}^t \frac{x-x_j}{x_b-x_j} modp$$

Step 4. Restore t pixels of the secret image from the t coefficients in the polynomial function $q_g(x) = (a_0 + a_1x + \cdots + a_{t-1}x^{t-1})$.

15.4.2 Set-Up Authentication

In Thien and Lin's proposal [8], the size of shadows could be dramatically reduced, with n shares being embedded into n different host images to avoid the attention of hackers. However, if malicious participants deliver fake stego-images, the receivers may recover a fake secret image, thus misleading those participating in the secret sharing scheme, due to poor detection. Thus, yet another secret sharing function was invented in Wang et al.'s scheme to enhance security by introducing an authentication procedure. The CRC (Cyclic Redundancy Code Check) and hash function are therefore considered to prevent the fake shares from the malicious participants who offer in the course of revealing the secret image, where the CRC is implemented by CRC-8. Based on the CRC adoption, once the participants receive the stego-image, the CRC will be launched to check all pixels in the stego-image prior to the reconstruction of secret image. The CRC-based detecting authentication procedure for each block in the stego-image is presented as follows:

Step 1. Let $G(x)$ be the polynomial generator. Set/Reset the authentication code of CRC-8 as $C_g = (c_{g1}, c_{g2}, \cdots, c_{g8}) = $ "00000000," $g = 1, 2, \cdots, \left\lceil \frac{m^2}{t} \right\rceil$. MAC (Message Authentication Code) with hash implementation is used as $H_k(\hat{B}_{g,j} \| g \| ID_{stego-image} \| y_{g8})$, where K is a secret key. Set $B(x) = (\hat{B}_{g,j} \| Hk(\hat{B}_{g,j} \| g \| ID_{Stego-image} \| y_{g8}))(x)$, where $\hat{B}_{g,j}$ is the stego-block, and $ID_{stego-image}$ is the identity of a stego-image, *i.e.*, the block message $\hat{B}_{g,j}$ is then followed by the MAC of $H_k(\hat{B}_{g,j} \| g \| ID_{stego-image} \| y_{g8})$.

Step 2. Obtain remainder polynomial $R(x) = B(x) mod G(x)$.

Step 3. Assign the coefficient of $R(x)$ to C_g.

Step 4. Store C_g to the block $\hat{B}_{g,j}$ of the stego-image.

Step 5. Compare the $R(x)$ and C_g. If the result is the same, the block $\hat{B}_{g,j}$ of the stego-image is a legal one. Otherwise reject it whenever the compared result is different.

Example 2.
Consider a gray-level block with a size of 8 pixels. Assume that the $\hat{B}_{g,j} = $ "A1A2A3A4" (it is caught by up-to-down and left-to-right in order), where $C_g = $ "00000000", and the output of HMAC is $H_k(\hat{B}_{g,j} \| g \| ID_{stego-image} \| y_{g8})$ = "D08932564F74499D4EC45CA9EBD54B64". Set $Bx = (\hat{B}_{g,j} \| H_k(\hat{B}_{g,j} \| g \| ID_{stego-image} \| y_{g8}))(x) = $ (A1A2A3A4D08932564F74499D4 EC45CA9EBD54B64)(x). Let the generator polynomial $G(x) = 100110001(x) = x^8 + x^5 + x^4 + 1$. The CRC code is then generated as $C_g = R(x) = B(x) mod G(x) = (c_{g1}, c_{g2}, \cdots, c_{g8}) = $ "00110000". Then store C_g into $\hat{B}_{g,j}$ of the stego-image. In the request of detecting the stego-image, pick up the embedded C_r and Let $T = Cr = $ "00110000" and reset $C_g = $ "00000000." Run the CRC-based detecting authentication procedure again to obtain $R(x) = (00110000)(x)$. Compare if T is equal to the coefficient of $R(x)$ as "00110000." Clearly, in this case, they are consistent with each other. In a word, the block $\hat{B}_{g,j}$ of the stego-image remains the safe one from the malicious attacks.

15.4.3   Main Algorithms

In this subsection, Wang et al.'s main algorithm, Secret Embedding and Detecting Algorithm (SEDA for short) and Target Secret Image Recovery Algorithm (TSIRA) are presented. More notations are given together with the above-mentioned notations as follows.

Notations:

I_j: The jth cover image with the size $2m \times 2m$ pixels.

$\hat{I}_j$ The jth stego-image with the size of $2m \times 2m$.

$B_{g,j}^{(d)}$ The dth pixel value of the gth block $B_{g,j}$ in the jth cover image.

$\hat{B}_{b,j}^{(d)}$ The dth pixel value of the gth block $\hat{B}_{g,j}$ in the jth stego-image.

K: A secret key K used in the hash function.

Secret Embedding and Detecting Algorithm: The SEDA aims at the secret embedding and detecting authentication.

Input: $S, I_j, j = 1, 2, \cdots, n, k$.

Output: $\hat{I}_j, j = 1, 2, \cdots, n$.

Step 1. Get the distinct feature-value $f_j = (x_{11}, x_{12}, x_{13}, x_{14}, x_{21}, x_{22}, x_{23}, x_{24})$ from $B_{g,j}$ of the cover image I_j, where $j = 1, 2, \cdots, n$. If f_i is equal to one of the feature-values $f_1, f_2, \cdots, f_{j-1}$, then keep looking for the next pair of pixels in block $B_{g,j}$.

Step 2. Call RAHA to compute $PV\_y_j^{(g)} = q_g(f_i)$ and let each $PV\_y_j^{(g)}$ be the bit-string format as $(y_{g1}, y_{g2}, \cdots, y_{g8})$.

Step 3. Apply IPLA to generate a pattern-string. Inspecting each block $\hat{B}_{g,j}$, set the bit string of $(y_{r1}, y_{r2}, \cdots, y_{r8})$ to be the input of IPLA, for $g = 1, 2, \cdots, \left\lceil \frac{m^2}{t} \right\rceil$. The output of IPLA, a k-bit pattern-string of $(p_1, p_2, \cdots, p_{\frac{m^2}{t}})$ is then generated after $\left\lceil \frac{m^2}{t} \right\rceil$ blocks $\hat{B}'_{g,j}s$ are given.

Step 4. According to CRC-based detecting authentication procedure, generate $\left\lceil \frac{m^2}{t} \right\rceil$ authentication codes of $(c_{g1}, c_{g2}), \cdots, c'_{g8}s$, for $g = 1, 2, \cdots, \left\lceil \frac{m^2}{t} \right\rceil$.

Step 5. Arrange all pixels in block $\hat{B}_{g,j}$, for $g = 1, 2, \cdots, \left\lceil \frac{m^2}{t} \right\rceil$, as the following rule:
$$\hat{B}_{g,j}^1 = (x_{11}, \cdots, x_{16}, y_{g1}, c_{g1}),$$
$$\hat{B}_{g,j}^2 = (x_{21}, \cdots, x_{26}, y_{g2}, c_{g2}),$$
$$\hat{B}_{g,j}^3 = (x_{31}, \cdots, x_{36}, y_{g3}, c_{g3}),$$
$$\hat{B}_{g,j}^4 = (x_{41}, \cdots, x_{46}, y_{g4}, c_{g4}),$$

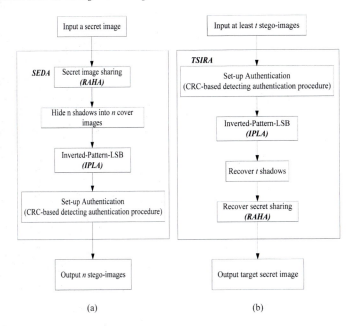

(a)   (b)

FIGURE 15.5
A secret sharing system with target secret embedding of high capacity and
detection authentication: (a) Flowchart of target secret embedding procedures;
(b) Recovery of target secret forms the stego-images in our secret sharing
systems.

$$\hat{B}_{g,j}^5 = (x_{51}, \cdots, x_{56}, y_{g5}, c_{g5}),$$
$$\hat{B}_{g,j}^6 = (x_{61}, \cdots, x_{66}, y_{g6}, c_{g6}),$$
$$\hat{B}_{g,j}^7 = (x_{71}, \cdots, x_{76}, y_{g7}, c_{g7}),$$
$$\hat{B}_{g,j}^8 = (x_{81}, \cdots, x_{86}, y_{g8}, c_{g8}),$$

Step 6. Construct $\hat{I}_j$ by gaining $\left\lceil \frac{m^2}{t} \right\rceil$ blocks $\hat{B}_{g,j}$ in order, $g = 1, 2, \cdots, \left\lceil \frac{m^2}{t} \right\rceil$.

Target Secret Image Recovery Algorithm: The TSIRA aims at the
recovery of the original secret image.

Input: A set of at least t stego-images and a secret key K.

Output: A target secret image S if all the stego-images are authenticated to
be genuine.

Step 1. Renumber t of the inputted stego-images as $\hat{I}_j$, $j = 1, 2, \cdots, t$, in
order.

Step 2. Use a secret key K and call the CRC-based detecting authentication procedure in the stego-image $\hat{I}_j$. If the comparison for the authentication code C_g is then correct, proceed to the next step, otherwise, a fake stego-image is detected.

Step 3. For each $PV\_y_j^{(g)}$ and block $\hat{B}_{g,j}$, apply IPLA to compute the pattern-string $(p_1, p_2, \cdots, P_{\frac{m^2}{t}})$ in the IPLA. Follow the sign of $p_1, p_2, \cdots, P_{\frac{m^2}{t}}$ to recover the bit-string of $(y_{g1}, y_{g2}, \cdots, y_{g8})$ of in the $\hat{B}_{g,j}$, for $g = 1, 2, \cdots, \left\lceil \frac{m^2}{t} \right\rceil$.

Step 4. Pick up f_j and calculate $PV\_y_j^{(g)}$ in $\hat{B}_{g,j}$, $j = 1, 2, \cdots, t$. Follow RAHA to recover the polynomial with the t pairs of $(f_j, PV\_y_j^{(g)})$'s, $j = 1, 2, \cdots, t$.

Step 5. Call RAHA to recover the polynomial with the t pair of $(f_j, PV\_y_j^{(g)})$'s, $j = 1, 2, \cdots, t$. Obtain secrets $S_i, i = (g-1)t + 1, (g-1)t + 2, \cdots, (g-1)t + (t-1)$.

Step 6. Combine all S_i in order to reveal the target secret image S.

Wang et al.'s scheme with the proposed algorithms to fulfill the sharing system with the target secret high capacity and detection authentication are sketched as shown in Figure 15.5.

15.5 Experimental Results

Capacity and detecting authentication procedure are presented and discussed in this section via the observations of experiments for benchmark images. Two samples of $(2, 4)$-threshold and $(4, 6)$-threshold schemes are taken into consideration, in order to highlight the contributions made by Wang et al.'s proposed method.

15.5.1 Fidelity Analysis

The target secret image is shown as Figure 15.6 (a). The benchmarks of the cover images are Bridge, Butterfly, Toys, and Airplane, respectively, shown in Figures 15.6 $(b) - (e)$. The size of the target secret image is $\frac{1}{4}$ of each cover image mentioned above. The corresponding four stego-images are shown in Figures 15.6 $(b') - (e')$ after SEDA, where the cover images still look the same as the stego-images to the human eye. The measurements of peak signal noise ratio (PSNR) to discriminate between cover images in Figures 15.6 $(b)-(e)$ and stego-images in Figures 15.6 $(b') - (e')$ were evaluated as 43.42, 43.43, 43.41, and 43.44, respectively. The target secret image, shown in Figure 15.6 (a) could be revealed when the $(2,4)$-threshold was applied, via the four stego-images shown in Figures 15.6 (b')–(e'). The experiments to embed Figure 15.6 (a) and the resulting PSNR are summarized in Table 1. In addition, the revelation of the target secret image was the same as Figure 15.6(a) when the $(2,4)$-threshold was applied in the stego-images, as shown in Figure 15.6(b')–(e') and offered in Wang et al.'s scheme. It can be seen that significantly improved fidelity was achieved in Wang et al.'s scheme with the higher PSNRs, when compared to other studies [4, 12].

Now considering another case, the $(4,6)$-threshold scheme is used to embed two target secret images, as shown in Figure 15.7 (a) and (b). The six cover images are shown in Figure 15.7 $(c) - (h)$. The size of each target secret image is $\frac{1}{4}$ of each cover image. After SEDA, the corresponding six stego-images can be seen in Figure 15.7 (c')–(h'). The PSNR measurements for the discriminations between the six cover images in Figure 15.7 (c)–(h) and stego-images in Figure 15.7 (c')–(h') were evaluated as 43.39, 43.43, 43.42, 43.39, 42.95 and 43.05, respectively, as shown in Table 2. Accordingly, the quality of the stego-image was visually acceptable, when compared to the corresponding cover images upon the high PSNRs. On the other hand, the target secret images, shown in Figures 15.7 (a) and (b) could also be exactly revealed when the $(4,6)$-threshold was applied, via the six stego-images shown in Figures 15.7 (c')–(h').

As seen below, colored benchmark images are further considered to the embedding system via SEDA. The target secret image is shown in Figure 15.8 (a). These show the experimental results, using full color cover images. The PSNR measurements were evaluated as 43.39, 43.12, 43.41, and 43.40, respectively, for the discriminations between the cover and stego-images, shown in Figures 15.8 (b)–(e) and (b')–(e'). The comparisons, between the PSNR measurements in Wang et al.'s and some other studies are presented in Table 15.3. The $(2,4)$-threshold scheme was applied, in this case, among the stego-images shown in Figures 15.8 (b')–(e'). The target secret image can be seen to be totally revealed, and identical to the original, shown in Figure 15.8(a).

TABLE 15.1

PSNR comparisons for past schemes and our scheme.

	Bridge	Butterfly	Toys	Airplane
Lin-Tsai-scheme (PSNR)	39.11	39.13	39.23	39.16
Yang et al. scheme (PSNR)	41.61	41.46	41.44	41.69
Wang et al. scheme (PSNR)	43.42	43.43	43.41	43.44

15.5.2 Evaluating Authentication

In this section, the detection ability of the Lin-Tsai scheme, Yang et al.'s scheme and Wang et al.'s are discussed. These are then compared, as shown in Figures 15.9–15.11.

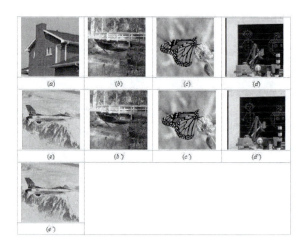

FIGURE 15.6
(a) the target secret images; (b)–(e) the four cover-images; (b′)–(e′) the four stego-images.

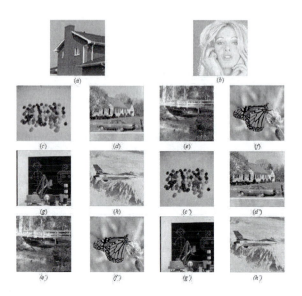

FIGURE 15.7
(a) and (b) the target secret images; (c)–(h) the four cover-images; (c′)–(h′) the four stego-images.

TABLE 15.2
PSNR measurements when embedding the two target secret images in Figure 15.5 (*a*) and (*b*) comparisons for past schemes and our scheme.

	Ball	House	Bridge	Butterfly	Toys	Airplane
PSNR	43.39	43.43	43.2	43.39	42.95	43.05

TABLE 15.3
PSNR comparisons with the Lin-Tsai scheme.

	Milk	Tiffany	Baboon	Airplane
Lin-Tsai-scheme	39.07	39.08	39.08	39.07
Wang et al. scheme	43.39	43.12	43.41	43.40

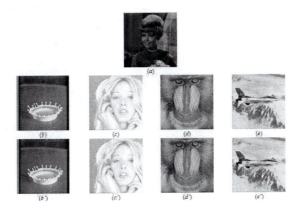

FIGURE 15.8
(*a*) the target images; (*b*)–(*e*) the four cover images; (*b′*)–(*e′*) the four stego-images.

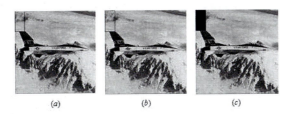

FIGURE 15.9
Minor bit adjustments in the stego-image of ”*airplane.*” (a) Lin-Tsai scheme, (b) Yang et al. scheme, and (c) our scheme.

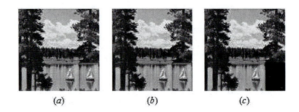

(a)   (b)   (c)

FIGURE 15.10
Partial area adjustments in the stego-images of "*sailboat.*" (a) Lin-Tsai scheme, (b) Yang et al.'s scheme, and (c) our scheme.

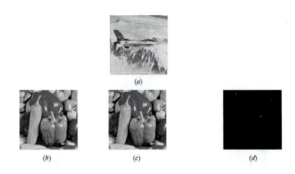

FIGURE 15.11
Replacement of "*airplane*" stego-image with "*pepper.*" (a) original stego-image of "Airplane", (b) replaced by "pepper" Lin-Tsai scheme, (c) replaced by "pepper" in Yang et al.'s scheme, and (c) detected in our scheme.

Case I. Minor bit adjustments in the stego-images, with results compared in Figure 15.9.

Case II. Partial area adjustment of a fake stego-image that can succeed the fake scenarios in [4, 12], but this case is blocked in Wang et al.'s scheme, as shown in Figure 15.10.

Case III. Replacement of a fake stego-image that was visually imperceptible was detected in Wang et al.'s , shown in Figure 15.11. As observed in Figures 15.9–15.11, the three cases not detected by the schemes in [4, 12] were easily made out when Wang et al.'s was applied. The probability of passing a fake stego-image, in the scheme of Yang et al. [12], was $(\frac{1}{2})^{m^2}$, where one bit parity-check and hash were applied in a block (4-pixel) and $\frac{(4m)^2}{4} = m^2$ non-overlapping blocks were set in the stego-image. On the other hand, in Wang et al.'s, a block had an 8-pixel format, such that the probability of this block succeeding was $(\frac{1}{2})^8$ while the probability of a fake stego-image succeeding in

Wang et al.'s method-2 was $((\frac{1}{2})^8)^{\frac{m^2}{2}} = (\frac{1}{2})^{4m^2}$ only, where there were $\frac{4m^2}{8} = \frac{m^2}{2}$ blocks in total in the stego-image. The summary comparisons are shown in Table 15.4.

TABLE 15.4
PSNR comparisons with the Lin-Tsai scheme.

	High capacity	Techniques used in authentications	Probability of cheating (Numerical analysis)
Thien-Lin scheme [8]	YES	N/A	N/A
Lin-Tsai scheme [4]	N/A	Parity bit check	N/A
Yang et al. scheme [12]	N/A	Hash Function	$(1/2)^{m^2}$
Wang et al. scheme [10]	YES	CRC and Hash fuction	$(1/2)^{4m^2}$

15.6 Conclusions

In this chapter, we introduce some polynomial-based image sharing schemes as well as our proposed scheme. These image sharing schemes are based on the (t, n)-threshold scheme to share the secret image to n participants. Sometimes, some authentication approaches and steganographic approaches were applied. Algorithms, such as IPLA, RAHA, SEDA, and TSIRA, were incorporated into our scheme; the RAHA and IPLA were applied [8, 11], aiming to guarantee higher capacity, and SEDA and TSIRA were used to enable the robust authentication requirement, to reveal the original target secret image. Compared to some other studies [4, 8, 12], Wang et al.'s scheme behaved better in the terms of capacity with secret embedding in a stego-image and in detection/authentication to block offers of fake shares.

Bibliography

[1] G. R. Blakley. Safeguarding cryptographic keys. In *Proceedings of the National Computer Conference*, volume 48, pages 313–317. AFIPS Press, 1979.

[2] C.-C., Chang, C.-L., Lin, and H. Chou. Perfect hashing schemes for mining traversal patterns. *Fundamenta Informaticae*, 70:185–202, November 2005.

[3] G. Horng, T.H. Chen, and D.S. Tsai. Cheating in visual cryptography. In *Designs, Codes and Cryptography*, volume 38, pages 219–236. Kluwer Academic Publishers, 2006.

[4] C.C. Lin and W.H. Tsai. Secret image sharing with steganography authentication. In *The Journal of Systems Software*, volume 24, pages 405–414, 2004.

[5] S.P. Mohanty and B.K. Bhargava. Invisible watermarking based on creation and robust insertion-extraction of image adaptive watermarks. In *Transactions on Multimedia Computing, Communications, and Applications*, volume 5 (12), pages 612–613, 1979.

[6] M. Naor and A. Shamir. Visual cryptography. In *EUROCRYPT '94, Springer-Verlag Berlin*, volume LNCS 950, pages 1–12, 1995.

[7] A. Shamir. How to share a secret. *Communication of the ACM*, 22:612–613, November 1979.

[8] C. C. Thien and J. C. Lin. Secret image sharing. In *Computers & Graphics*, volume 26, pages 765–770, 2002.

[9] R.-Z. Wang and S.-J. Shyu. Scalable secret image sharing. *Image Commun.*, 22:363–373, April 2007.

[10] S.-J. Wang, I.-S. Lin, Y.-L. Hsieh, and C.-Y. Weng. Secret sharing systems with authentication-based steganography. In *Proceedings of the 2008 International Conference on Intelligent Information Hiding and Multimedia Signal Processing*, IIH-MSP '08, pages 1146–1149, Washington, DC, USA, 2008. IEEE Computer Society.

[11] C.-H. Yang. Inverted pattern approach to improve image quality of information hiding by LSB substitution. *Pattern Recognition*, 41:2674–2683, August 2008.

[12] C.-N. Yang, T.-S. Chen, K. H. Yu, and C.-C. Wang. Improvements of image sharing with steganography and authentication. *Journal of Systems and Software*, 80:1070–1076, July 2007.

[13] C.-N. Yang and S.-M. Huang. Constructions and properties of k out of n scalable secret image sharing. *Optics Communications*, 283(9):1750–1762, 2010.

16

Image Sharing with Steganography and Authentication

Zhao-Xia Yin

Anhui University, China

Chia-Chen Lin

Providence University, Taiwan

Chin-Chen Chang

Feng Chia University, Taiwan

CONTENTS

16.1 Introduction

More and more data are being transmitted via the public access network, the Internet, due to its favorable characteristics such as low cost, speed, and reliability. Although the Internet is a public access network, certain data require secrecy, such as commercial or military images, to protect them from illegitimate users during transmission. Steganography is a technique for transmitting secret data without being noticed. This technique hides secret messages into cover media to avoid malicious attacks. The cover media can be digital images, digital videos, source codes, or even HyperText Markup Language (HTML) codes. The digital image used to embed the secret data is called a cover image, which becomes a stego-image once the secret data are embedded. With steganography, malicious attackers do not know that a stego-image carries secret data. Therefore, they will not try to extract the data or otherwise trespass on it. However, there is a weakness common to all steganographic techniques. If one of the stego-media is lost or corrupted, the secret data cannot be revealed exactly and completely. Therefore, several secret sharing techniques have been proposed to overcome this weakness. With regard to the concept of secret sharing, the well-known (k, n)-threshold schemes pioneered by Shamir [20] and Blakley [2], respectively, have four characteristics in common:

(1) $k \leq n$.

(2) The secret is shared among n participants.

(3) Only k or more participants can reconstruct the original secret.

(4) When the number of participants is $(k - 1)$ or less, they cannot recover the original secret.

Based on the (k, n)-threshold schemes, in 1995 Naor and Shamir [17] developed the first secret image sharing technique to safeguard and share image-based secrets. Their approach, called visual secret sharing (VSS), makes it possible to publicly transmit secret images efficiently and safely. The detailed principles of such a technique follow:

(1) The secret image is divided into n noise-like images, called shadows or shares, with each noise-like shadow containing no original image information.

(2) n noise-like shares can be transmitted to n participants via the Internet instead of the secret image itself, so that each participant keeps only one single share. Thereby, VSS can prevent attackers from gaining access to the secret image directly.

(3) When any k or more shadows are brought together co-operatively, the secret image can be restored without any computations based on the properties of the human visual system.

Following this pioneering research, many VSS schemes have been proposed over the past decade. Shamir's polynomial-based secret sharing approach was adopted into secret image sharing by Thien and Lin in 2002 [23]. Then, Wang and Su [25] made some improvements to decrease the size of the shared images based on Thien and Lin's scheme [23]. Studies in secret image sharing have focused on different active topics: decreasing share size [4,5,25], dealing with attacks by malicious users, and detecting fake shadows [14,18,29]. The research domain has been also extended from gray images to color images [1,3,4,6,7,21]. And, the generated shares become meaningful instead of noise-like from the start [24]. Aside from these issues, Yang [27], Cimato et al. [8], and Wang et al. [26] separately proposed a new VSS scheme, called a ProbVSS scheme, to deal with problems such as pixel expansion and computational complexity that typically arise in image sharing. The ProbVSS scheme is based on probabilistic concepts and is designed solely for binary images. By taking advantage of Boolean operations, this scheme overcomes the shortcomings inherent in high computational complexity and large pixel expansion at the same time.

In 2004, Lin and Tsai proposed another kind of secret sharing scheme incorporating both steganography and authentication [15]. In Lin-Tsai's scheme, the secret image is divided into shadows and data hiding is used to embed the shares and corresponding authentication codes into the cover image. During the reconstruction and verifying phase, the hidden authentication codes are then extracted to verify the integrity of the stego-image. However, the size of the cover image becomes fourfold that of the secret image. The Lin-Tsai scheme combines steganography and authentication features to avoid transmitting an erroneous stego-image or intentionally providing a false image by using a parity check bit. However, by the principle of the parity check itself, the authentication ability is weak.

Yang et al. proposed a new scheme to overcome some of the weaknesses in the Lin-Tsai scheme in 2007 [28]. Their objective was to prevent cheating by participants. They also improved the authentication ability and the quality of the reconstructed grayscale secret image. Although the Yang et al. scheme can restore a distortion-free secret image, it allows for a significant probability that a fake stego-image can be authenticated successfully, and the cover image size is still four times that of the original secret image. Building on their predecessors' experience, in 2008 Chang et al. proposed a secret sharing scheme that incorporates authentication based on the Chinese Remainder Theorem (CRT) [9]. They successfully improved authentication capability by combining CRT. However, in Chang et al.'s scheme, pixel expansion remains a problem and the size of the cover image should be at least twice that of the secret image. In addition, generating authentication bits with CRT increased computational complexity.

This chapter proposes an enhanced scheme with the aim of maintaining the effective authentication but eliminating the pixel expansion problem while retaining an acceptable computational cost. Our scheme uses two techniques in combination to reduce cover image size. The first is an error diffusion technique [13,22] that helps to transform a grayscale image into a binary image. The second, called ELUT, is an edge lookup inverse halftoning technique that relies on edge detection and a lookup table to generate a reconstructed grayscale image from a halftone image [11]. Inverse halftoning is a kind of commonly used technique to reconstruct grayscale images from halftone images. In 2005, a new edge-based lookup table scheme for inverse halftoning was proposed by Chung and Wu [11], by which, the quality of the reconstructed grayscale images was improved. In this chapter we call Chung and Wu's scheme ELUT for short.

Experimental results confirm that no pixel expansion occurs in the proposed scheme and the visual quality of the stego-image carrying the shadows is much better. Moreover, the proposed scheme successfully reduces computational complexity while maintaining an acceptable authentication ability, which is much better than the performance of either Lin-Tsai's scheme [15] or Yang et al.'s scheme [28].

The rest of this chapter is organized as follows. In Section 16.2, we briefly introduce the Lin-Tsai scheme [15], the Yang et al. scheme [28], and the Chang et al. scheme [9]. In Section 16.3, we briefly review the techniques adopted in the proposed scheme. Full details of our proposed scheme are explained in Section 16.4. Section 16.5 gives some experimental results. Finally, our conclusions are presented in Section 16.6

16.2   Related Work

In this section three existing schemes, the Lin-Tsai scheme [15], Yang et al.'s scheme [28] and Chang et al.'s scheme [9] are introduced, in turn.

16.2.1   Lin and Tsai's Scheme

Lin-Tsai's scheme is a (k,n)-threshold polynomial-based scheme that combines steganography and authentication to enable the sharing of secret images [15]. Their scheme uses parity check to achieve authentication, and each pixel in a secret image is used to produce n shared pixels for n participants by using the polynomial function in Equation 16.1.

$$F(x) = (c_0 + c_1 x + ... + c_{k-1} x^{k-1}) \bmod p. \tag{16.1}$$

Later, both the parity check bits and the shared pixels are embedded into the corresponding four-pixel blocks in the cover image. For simplicity, assume that I_{ij} is the ij-th pixel in the secret image and $B_{ij}^1, B_{ij}^2, ..., B_{ij}^n$ are the n cover blocks in the cover image, where B_{ij}^k is the block with position (i, j) in the k-th cover for $1 \leq k \leq n$, which contains four pixels, Y_{ij}^k, V_{ij}^k, W_{ij}^k, and Z_{ij}^k, as shown in Figure 16.1.

Y_{ij}^k $y_7 y_6 y_5 y_4 y_3 y_2 y_1 y_0$	V_{ij}^k $v_7 v_6 v_5 v_4 v_3 v_2 v_1 v_0$
W_{ij}^k $w_7 w_6 w_5 w_4 w_3 w_2 w_1 w_0$	Z_{ij}^k $z_7 z_6 z_5 z_4 z_3 z_2 z_1 z_0$

FIGURE 16.1
The ij-th block B_{ij}^k of the k-th cover.

Here, the binary representations of the pixels Y_{ij}^k, V_{ij}^k, W_{ij}^k and Z_{ij}^k, are $(y_1 y_2 ...y_8)$, $(v_1 v_2 ...v_8)$, $(w_1 w_2...w_8)$ and $(z_1 z_2...z_8)$, respectively. The steps in this scheme are as follows.

Step 1: Set the modulus p used in the polynomial in 16.1 to 251. If the pixel $I_{ij} \geq 250$, it is directly truncated to 250. Apparently, this operation can distort the secret image.

Step 2: Take the pixel I_{ij} as the coefficient c_0 and choose $k - 1$ integers as other coefficients $c_1, c_2, ..., c_{k-1}$ randomly to construct a polynomial in Equation 16.1. Then takes the pixel value Y_{ij}^k for $1 \leq k \leq n$ as the input x of the polynomial to compute the shared pixel $F(Y_{ij}^k) = S_{ij}^k = (s_1 s_2...s_8)$, where $(s_1 s_2...s_8)$ means eight bits of the shared pixel S_{ij}^k.

Step 3: Embed the shared pixel S_{ij}^k into the block B_{ij}^k by replacing the eight bits $v_7 v_8$, $w_6 w_7 w_8$ and $z_6 z_7 z_8$ with $s_1 s_2...s_8$ as shown in Figure 16.2.

Step 4:   Generate a check bit b according to parity check policy to achieve
the authentication capability shown in Figure 16.2.

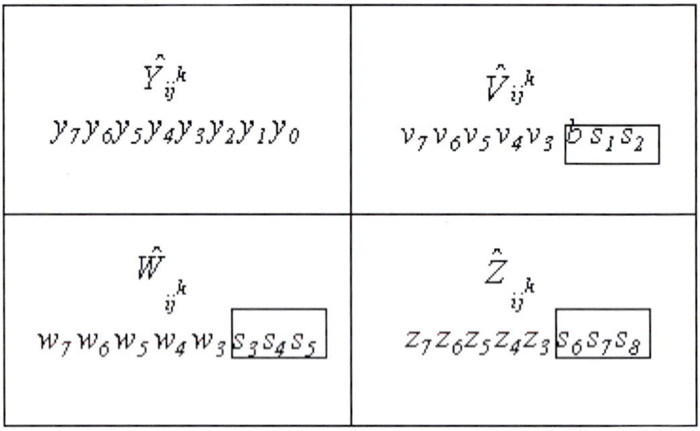

FIGURE 16.2
The block of the k-th stego-image in Lin-Tsai's scheme.

Unfortunately, in Lin-Tsai's scheme, the parity bit of the upper-right pixel
shown in Figure 16.2 is chosen to make this pixel an even or odd parity as
a binary parity sequence. Dishonest participants can derive the parity information from their own stego-images, and thus can easily and maliciously
counterfeit a stego-image. For instance, assume that the upper-right pixel $\bar{V}_{ij}^k$
= (11011 111). A dishonest participant can modify it to (11011 010), which
still meets the odd parity, but the 8-bit input of $(k\text{-}1)$-degree polynomial becomes changed. Thus, it successfully passes authentication but the $(k\text{-}1)$-degree
polynomial cannot be obtained. In addition, the dishonest participant can also
modify three other pixels, $\bar{Y}_{ij}^k$, $\bar{W}_{ij}^k$, and $\bar{Z}_{ij}^k$, to change the input and output
of $(k\text{-}1)$-degree polynomial without influencing the upper-right pixel $\bar{V}_{ij}^k$, thus
passing authentication. Therefore, their scheme has a weak authentication
process, which may allow a fake stego-image to pass the authentication check
quite easily.

16.2.2   Yang et al.'s Scheme

To overcome the weaknesses in Lin-Tsai's scheme, Yang et al. [28] proposed
an improved version in 2007. In Yang et al.'s scheme, the modulus value p is

set to the Galois Field GF(2^8), that is, $p = g(x) = x^8 + x^4 + x^3 + x^1 + x^0$. This approach allows the acquisition of a secret image without distortion.

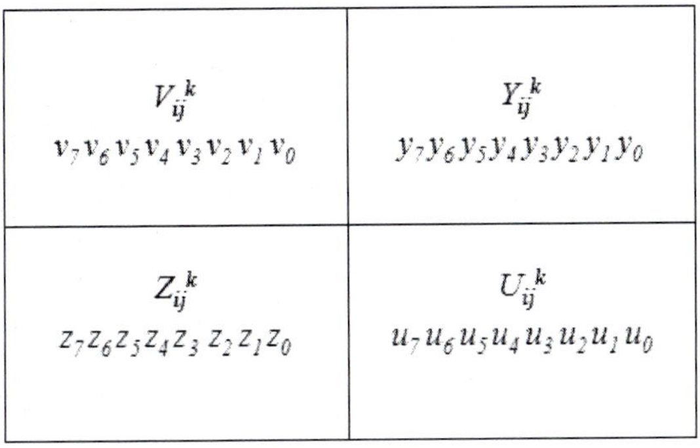

FIGURE 16.3
The cover blocks used in Yang et al.'s scheme.

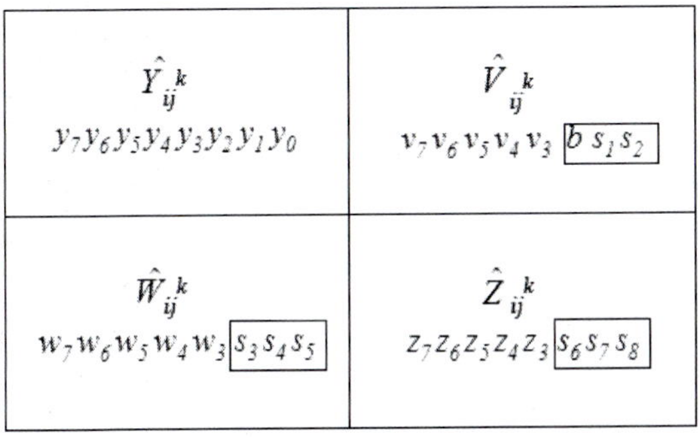

FIGURE 16.4
The block of a stego-image in Yang et al.'s scheme.

In Yang et al's scheme, let the pixel I_{ij} of the secret image be used to produce the n shared pixels S_{ij}^1, S_{ij}^2,...,S_{ij}^n, to be embedded into the cover blocks $C_{ij}^1, C_{ij}^2,...,C_{ij}^n$, where each cover block is represented as $C_{ij}^k = \{V_{ij}^k, Y_{ij}^k, Z_{ij}^k, U_{ij}^k\}$ for $1 \le k \le n$ as shown in Figure 16.3. Pixel I_{ij} of the secret image is first embedded into the coefficient c_0 to construct a polynomial as shown in Equation 16.1. Next, the shared pixel S_{ij}^k is computed by taking the pixel value V_{ij}^k of the corresponding cover image block C_{ij}^k as the input x of the polynomial shown in Equation 16.1. Let the binary representation of the shared pixel S_{ij}^k be ($s_1\ s_2...s_8$). Hence, the shared pixel S_{ij}^k is embedded into its corresponding cover image block C_{ij}^k by replacing the eight bits v_1 v_0 y_1 y_0 z_1 z_0 u_1 u_0 with s_1 s_2 ...s_8, and at the same time, the bits z_3 z_2 of pixel Z_{ij}^k are replaced by $v_1 v_0$ of pixel V_{ij}^k in order to keep the complete pixel value V_{ij}^k for recovering the corresponding secret pixel as shown in Figure 16.4.

However, Yang et al.'s scheme cannot avoid the fact that four least significant bits (LSBs) of the pixel value Z_{ij}^k in each block must be modified. Such modification may make the stego-images different from the cover images visually and the risk of being observed by attackers is increased.

To further improve Lin-Tsai's authentication ability, Yang et al. used the hash-based message authentication code (HMAC) in Equation 16.2.

$$H_k((\bar{C}_{ij}^k - b)\|id), \qquad (16.2)$$

where id is the block ID; C_{ij}^k is 31 bits except the check bit b; $H(\ )$ and k denote a one-way hash function and the secret key, respectively. Next, as shown in Equation 16.3, the 160-bit HMAC output and executes the XOR operation to obtain the authentication bit b.

$$b = H_k((\bar{C}_{ij}^k - b)\|id) \qquad (16.3)$$

However, the average probability of detecting any malicious modification is just 50% because of the characteristic of the parity check itself. In other words, the fake stego-images still have a very high probability of being authenticated successfully.

16.2.3   Chang et al.'s Scheme

In 2008, Chang et al. proposed an enhanced secret sharing scheme based on Lin-Tsai's scheme and Yang et al.'s scheme. They used the concept of Thien and Lin's secret image sharing [23] to ensure that no distortion is introduced into the secret image and applied the concept of the Chinese Remainder Theorem (CRT) to improve authentication ability. As a result, their scheme can achieve high authentication ability. The flowchart of Chang et al.'s scheme [9] is shown in Figure 16.5.

Four check bits p_1, p_2, p_3, p_4 can be calculated and embedded into the stego-block (shown in Figure 16.5). The details are shown in the steps below.

Step 1:   The residues set of the block $\bar{C}_{ij}^k$ in Figure 16.6 is determined.

$$R_{ij,1}^k = (x_7 x_6 x_5 x_4 x_3 s_1 s_2);$$
$$R_{ij,2}^k = (v_7 v_6 v_5 v_4 s_3 s_4);$$
$$R_{ij,3}^k = (w_7 w_6 w_5 w_4 w_3 s_5 s_6);$$
$$R_{ij,4}^k = (z_7 z_6 z_5 z_4 z_3 s_7 s_8);$$
$$R_{ij,5}^k = i;$$
$$R_{ij,6}^k = j.$$

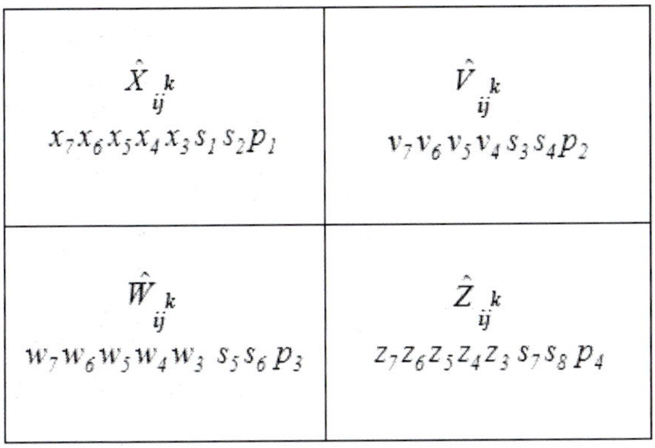

$\hat{X}_{ij}^{\,k}$ $x_7 x_6 x_5 x_4 x_3 s_1 s_2 p_1$	$\hat{V}_{ij}^{\,k}$ $v_7 v_6 v_5 v_4 s_3 s_4 p_2$
$\hat{W}_{ij}^{\,k}$ $w_7 w_6 w_5 w_4 w_3\; s_5 s_6 p_3$	$\hat{Z}_{ij}^{\,k}$ $z_7 z_6 z_5 z_4 z_3\; s_7 s_8 p_4$

FIGURE 16.5
The block of a stego-image in Chang et al.'s scheme.

Step 2: Decide six moduli that are pairwise relatively prime $M_{ij,1}^k$, $M_{ij,2}^k$, $M_{ij,3}^k$,..., and $M_{ij,6}^k$, where each modulus $M_{ij,a}^k$, for $1 \le a \le 6$, is the prime number greater than $R_{ij,a}^k$.

Step 3: Calculate the Y_{ij}^k integer by Equation 16.4. Suppose that the binary representation of the Y_{ij}^k is $(y_1 y_2...y_e)$. Note that the number of binary bits of Y_{ij}^k has to be a multiple of 4.

$$Y = (\sum_{i=1}^{r} R_i \times \frac{M}{M_i} \times I_i) \qquad (16.4)$$

Step 4: Four interim authentication bits a_1, a_2, a_3, and a_4 are computed by Equation 16.5.

$$a_1 a_2 a_3 a_4 = (y_1 y_2 y_3 y_4) \oplus (y_5 y_6 y_7 y_8) \oplus ... \oplus (y_{e-3} y_{e-2} y_{e-1} y_e). \quad (16.5)$$

Step 5: Denote b_1, b_2, b_3, and b_4 as the current four watermark bits; four check bits p_1, p_2, p_3, and p_4 can be computed by Equation 16.6.

$$p_1 p_2 p_3 p_4 = (a_1 a_2 a_3 a_4) \oplus (b_1 b_2 b_3 b_4). \qquad (16.6)$$

Step 6: Replace the four LSBs $x_0 v_0 w_0 z_0$ with the computed check bits p_1 p_2 p_3 p_4 as presented in Figure 16.6.

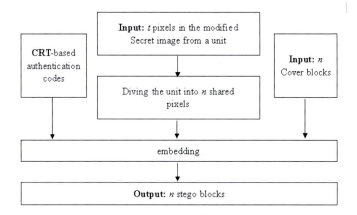

FIGURE 16.6
The flowchart of Chang et al.'s scheme.

16.3   Adopted Techniques in the Proposed Scheme

To give sufficient background knowledge, this section briefly introduces the related techniques adopted in our proposed scheme, including error diffusion, interpolation, canny edge detector, and edge lookup inverse halftoning as well as the Shamir scheme for secret sharing, respectively.

16.3.1   Error Diffusion Technique

The error diffusion technique is typically used to convert a multiple-level color image into a two-level color image. There are many kinds of idiographic error diffusion strategies. The common concept behind diffusion is the diffusion of errors to neighboring pixels; in this way, no image luminance is lost. The diffused image generated based on an error diffusion strategy is called an error filter. Each error filter has a set of kernel weights. Assume that $GI(x,y)$ is the value of a pixel in position (x,y) in a grayscale secret image. Figure 16.7 is a flowchart of the error diffusion technique.

In the proposed scheme, the Floyd and Steinberg error diffusion strategy is adopted and the kernel weights are $Wr = \frac{7}{16}, Wb = \frac{5}{16}, Wbl = \frac{3}{16}, Wbr = \frac{1}{16}$, as shown in Figure 16.8.

After the quantization procedure, a pixel $GI(x,y)$ at position (x,y) in grayscale image GI becomes $HI(x,y)$ and has a value of either 0 or 255. During the quantization procedure, a threshold TH is used to determine $HI(x,y)$ according to Equation 16.7 and the quantization error is determined by Equa-

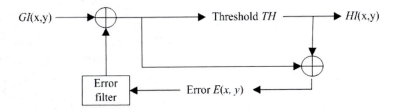

FIGURE 16.7
Error diffusion architecture.

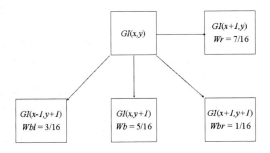

FIGURE 16.8
The kernel weights of Floyd and Steinberg's error filter.

tion (5.8).

$$HI(x,y) = \begin{cases} 255, & if\, GI(x,y) \geq TH \\ 0, & otherwise \end{cases} \tag{16.7}$$

$$E(x,y) = GI(x,y) - HI(x,y) \tag{16.8}$$

Next, error $E(x,y)$ is diffused over four neighboring pixels, $GI(x, y+1)$, $GI(x+1, y-1)$, $GI(x+1, y)$, and $GI(x+1, y+1)$, according to Equation 16.9 and the kernel weights (denoted as W) of the error filter are shown in Figure 16.8.

$$GI(x + i, y + j) = GI(x,y) + E(x,y) \times W, \tag{16.9}$$

where $i, j \in \{0, 1\}$, $W \in \{\frac{1}{16}, \frac{3}{16}, \frac{5}{16}, \frac{7}{16}\}$.

Based on Figure 16.8 and Equation 16.9, we can see that when pixels are on the border of the grayscale image, special cases occur. The four kinds of pixels listed below should be considered in distinguishing these cases.

Case 1: The pixels located at $(x, 1)$, where $x = 1, 2,..., H\text{-}1$.
Case 2: The pixels located at (H, y), where $y = 1, 2,... , W\text{-}1$.
Case 3: The pixels located at (x, W), where $x = 1, 2,..., H\text{-}1$.
Case 4: The pixel located at (H, W).

The positions of all four kinds of pixels are represented in Figure 16.9, with the white squares showing the positions of the pixels located at the border of an image. To deal with these special cases, we adopt a special skill, called "excursion" in the procedure, as shown in Figure 16.10.

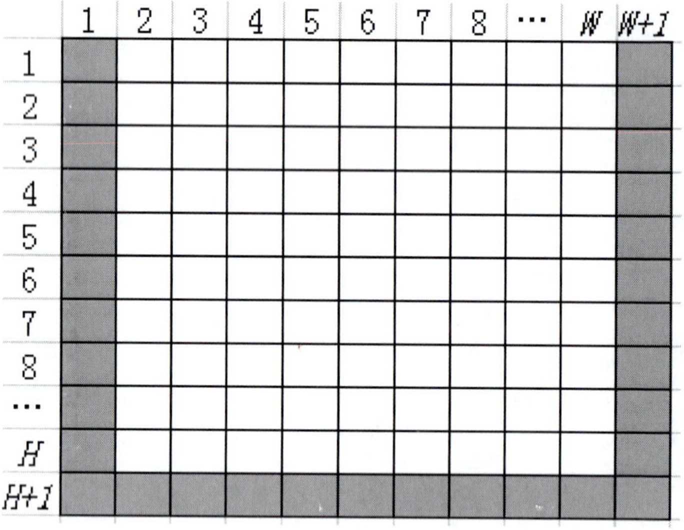

FIGURE 16.9
The positions of pixels located at the border in an image.

When applied to these pixels, Floyd and Steinberg's error filter is slightly modified to ignore the nonexistent pixels because the number of these pixels is small compared with the total number of pixels in the whole image. To give a clearer explanation, the following paragraphs give one instance for each case.

At first, set $E(1, 1) = 0$. For case 1, the pixels at position $(x, 1)$, where $x = 1, 2,..., H\text{-}1$ are considered. We take the pixel at position $(1, 1)$ for example. The value of $HI(1, 1)$ is determined by Equation 16.7. However, there is no pixel at position $(2, 0)$ corresponding to $GI(x{+}1, y{-}1)$. Thus, the neighboring pixels that accepted error diffusion are shown in Figure 16.11 for Case 1.

For Case 2, we take the pixel at position $(H, 1)$ for example. The pixel at position $(H, 1)$ is called $GI(H, 1)$ and has an error value of $E(H, 1)$. However,

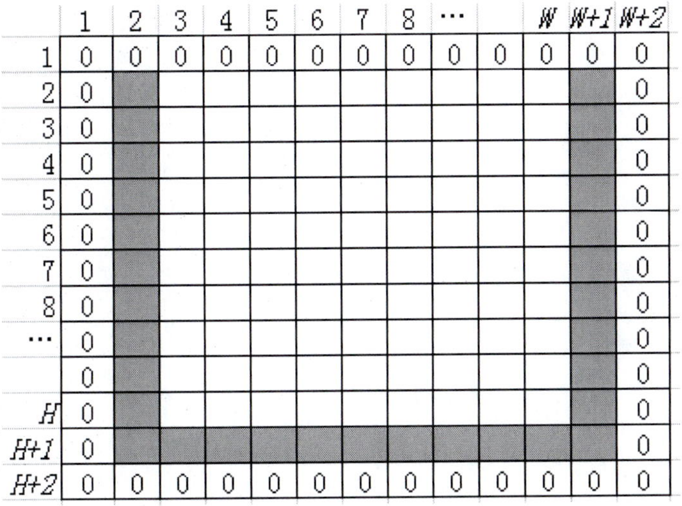

	1	2	3	4	5	6	7	8	···	*W*	*W+1*	*W+2*
1	0	0	0	0	0	0	0	0	0	0	0	0
2	0											0
3	0											0
4	0											0
5	0											0
6	0											0
7	0											0
8	0											0
···	0											0
	0											0
H	0											0
H+1	0											0
H+2	0	0	0	0	0	0	0	0	0	0	0	0

FIGURE 16.10
The "excursion" skill.

there are no pixels at positions $(H+1, 0)$, $(H+1, 1)$, and $(H+1, 2)$ corresponding to $GI(x+1, y-1)$, $GI(x+1, y)$, and $GI(x+1, y+1)$. Thus, the neighboring pixels that accepted error diffusion are shown in Figure 16.12 for Case 2.

For Case 3, we take the pixel at position $(1, W)$ for example. There are no pixels at positions $(1, W+1)$ and $(2, W+1)$ corresponding to $GI(x, y+1)$ and $GI(x+1, y+1)$. Thus, the neighboring pixels that accepted error diffusion are shown in Figure 16.13 for Case 3.

And last, for Case 4, the pixel $GI(H, W)$ at position (H, W), there are no pixels at positions $(H, W+1)$, $(H+1, W-1)$, $(H+1, W)$, and $(H+1, W+1)$ corresponding to $GI(x, y+1)$, $GI(x+1, y-1)$, $GI(x+1, y)$, and $GI(x+1, y+1)$. Thus, there are no neighboring pixels that accepted error diffusion for this case.

We next describe how to use the Floyd and Steinberg error diffusion strategy to transform a gray-scale image into a binary image. The detailed steps are set forth as follows:

Input: A gray-scale secret image GI sized $H \times W$.

Output: A binary image HI sized $H \times W$.

Step 1: Set a threshold $TH = 127$

Step 2: for $i=1$ H, $j=1: W$

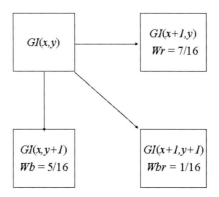

FIGURE 16.11
The neighboring pixels that accepted error diffusion for Case 1.

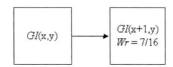

FIGURE 16.12
The neighboring pixels that accepted error diffusion for Case 2.

         Loop: if $(i = 1)$ or $i = H + 2)|(j = 1)$ or $(j= W+2) \rightarrow$ EGI$(i,j) = 0$;
         else $EGI(i, j) = GI(i - 1, j - 1)$;

Step 3:   for $i = 2{:}H + 1$, $j = 2{:}W + 1$
               Loop: if $EGI(i, j) \geq TH \rightarrow$ EHI$(i,j) = 255$;
               else EHI$(i,j) = 0$;
               E(i,j) = EGI(i,j)-EHI(i,j);
               EGI$(i,j + 1)$ = EGI$(i,j + 1)$ +E$(i, j) \times Wr$;
               EGI$(i + 1,j)$ = EGI$(i,j + 1)$ +E$(i, j) \times Wb$;
               EGI$(i + 1,j + 1)$ = EGI$(i + 1,j + 1)$ +E$(i, j) \times Wb$;
               EGI$(i + 1,j$ - $1)$ = EGI$(i + 1,j$ - $1) + E(i, j) \times Wbl$;

Step 4:   for $i = 1{:}H$, $j = 1{:}W$
               HI(i,j) = EHI$(i + 1, j + 1)$.

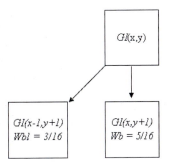

FIGURE 16.13
The neighboring pixels that accepted error diffusion for Case 3.

16.3.2 Interpolation Technique

In computer computations, a discrete number system rather than a real number system is applied due to the limitations of machine representation. Therefore, a continuous signal must be sampled for computer storage and calculation. However, sampling is always destructive to the original information and new samples often must be reselected when the sampling scale is changed. Let us denote R as a real signal performed in the real number system as shown in Figure 16.14(a); denote D as the digital signal created by R, which is performed in the discrete number system shown in Figure 16.14 (b). The digital signal D' shown in Figure 16.14(c) is generated by sampling digital signal D.

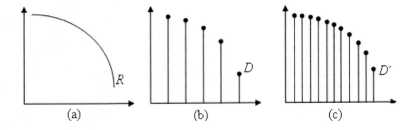

FIGURE 16.14
Example of sampling an image: (a) Real signal R; (b) Sampled signal D of R; (c) Sampled signal D' of D.

Figure 16.14 shows that the re-sampled signal D' can be easily retrieved when the real signal R is always available. However, the original signal is usually discarded after receiving the digital signal D. As a result, the resampled signal D' must be created from the former signal D. Interpolation is a classical solution for calculating the new samples from another digital signal. Take bilinear interpolation for example. A new sample in signal D' is expected to be an average value of the two neighboring original samples in D. Assuming that $(x, \mathrm{f}(x))$ and $(y, \mathrm{f}(y))$ are two neighboring samples in D, we can calculate a new sample $(z, \mathrm{f}\,(z))$ in D' according to Equation 16.10.

$$f(z) = \frac{y - z}{y - x} \times f(x) + \frac{z - x}{y - x} \times f(y). \tag{16.10}$$

16.3.3 Canny Edge Detector

Edge detection is a kind of commonly used technique in image processing. The areas in an image with strong intensity contrasting from one pixel to the next are called edges. The purpose of edge detection is filtering out useless information in an image and significantly reducing the amount of data, while preserving the important structural properties.

Since the pioneering work by Roberts in 1965 [19], a mass of schemes have been developed for detecting edges. The Canny edge detection algorithm is known to many as the optimal edge detector proposed by J. Canny in 1986 [12], which finds edges by looking for local maxima of the gradient of the input image. The gradient is calculated using the derivative of a Gaussian filter. The Canny edge detector contains several adjustable parameters, listed as follows, which can affect the computation time and effectiveness of the algorithm itself. The first parameter is the size of the Gaussian filter, the smoothing filter used in the first stage, which directly affects the results of the Canny detection algorithm. Generally, smaller filters cause less blurring, and allow detection of small, sharp lines. On the contrary, larger filters cause more blurring, smearing out the value of a given pixel over a larger area of the image. Two thresholds are the other parameters. Actually, a too high threshold may make missing important information. On the other hand, a threshold set too low will falsely identify irrelevant information as important. Therewith, the detection would be more likely to be fooled by noise. The Canny edge detection method uses two thresholds, to detect strong and weak edges, and includes the weak edges in the output only if they are connected to strong edges.

The Canny edge detector works in a multistage process. First of all, the image is smoothed by Gaussian convolution. Afterwards, to highlight regions of the image with high first spatial derivatives, a simple 2-D first derivative operator is applied to the smoothed image and edges give rise to ridges in the gradient magnitude image. Then, the algorithm tracks along the top of these ridges and sets to zero all pixels that are not actually on the ridge top so as to give a thin line in the output, a process known as nonmaximal suppression. The tracking process exhibits hysteresis controlled by two thresholds, denoted by T_1 and T_2, where $T_1 > T_2$. Tracking can only begin at a point on a ridge higher than T_1 and then continues in both directions out from that point until the height of the ridge falls below T_2. The Canny edge detector is more likely to detect true weak edges and less likely than the others to be fooled by noise.

16.3.4 Edge Lookup Inverse Halftoning Technique

The inverse halftoning technique is used to reconstruct a gray-scale image from an input halftone image. Based on the lookup table (LUT) [10,16] technique, in 2005 Chung and Wu [11] proposed a new edge-based lookup table (LUT) scheme that improves the quality of the reconstructed grayscale image. In the following paragraphs we call Chung and Wu's scheme ELUT for short. The ELUT scheme first applies the LUT-based inverse halftoning scheme as a preprocessing step to transform the input halftone image to a base grayscale image, and then the edges are extracted and classified from the base grayscale image. According to these classified edges, a novel edge-based LUT is built up to reconstruct the grayscale image, i.e., the ELUT scheme. Figure 16.15 is a flowchart of Chung and Wu's ELUT scheme.

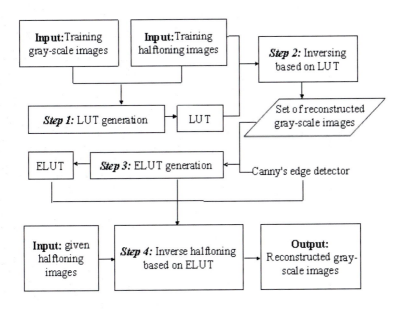

FIGURE 16.15
The flowchart of Chung and Wu's ELUT scheme.

Input: A set of n training image pairs (GI_i, HI_i), where GI_i and HI_i are denoted as the i-th grayscale image and its corresponding halftone image, respectively.

Step 1:   Generate the lookup table (LUT) by using the five substeps that
          follow. An array LUT[ ] is used to map the given halftone image to
          the corresponding grayscale image.

  (a) For initialization, $i = 1$, $\text{LUT}[k] = 0$, where $0 \leq k \leq 2^{16} - 1$.

  (b) Divide images double HI_i and GI_i into overlapping 4×4
       blocks and denote them as BH_{ij} and BG_{ij}. In other words,
       BH_{ij} and BG_{ij} are the j-th halftone block of halftone im-
       age HI_i and the j-th grayscale block of grayscale image
       GI_i, respectively.

  (c) Calculate index k for each halftone block BH_{ij} and update
       the value of the intermediate $\text{LUT}[k]$ by using Equation
       16.10. Here, $BG_{ij}(3,3)$ is a representative pixel for each
       grayscale block.

$$\begin{cases} k = \sum_{u=1}^{4} \sum_{v=1}^{4} 2^{(v-1)+4 \times (u-1)} \times BH_{ij}(u,v) \\ LUT[K] = LUT[K] + BG_{ij}(3,3) \end{cases},$$

$$(16.11)$$

  (d) $i = i + 1$. If $i \leq n$, go to Step b. Otherwise, compute and
       archive the final LUT[ ]. N[k] is another array used to store
       the number of halftone blocks that obtain the same index
       value k.

  (e) For $0 \leq k \leq 2^{16} - 1$, $LUT[k] = \frac{LUT[k]}{N[k]}$.

Step 2:   Replace the element $BH_{ij}(3,3)$ in every block BH_{ij} by LUT[k],
          where k is the index value of block BH_{ij} calculated in Step c. When
          the replacement procedure is finished, retrieve the grayscale image
          with pixel values corresponding to LUT[k].

Step 3:   After applying Step 2 to a set of n training images in succession,
          a set of reconstructed grayscale images called GI_i', $i = 1, 2,..., n$,
          can be retrieved. Adopt Canny's edge detector to each reconstructed
          grayscale image GI_i' to generate an edge map EM_i, for $i = 1, 2,..., n$.
          Each edge map EM_i consists of a set of 4×4 blocks. Therefore, the
          j-th block of the edge map EM_i is denoted as BE_{ij}. By combining
          the lookup table LUT generation procedure described in Step 1 with

a set of edge maps EM_i, an edge-based LUT, called an ELUT, is generated. Note that in this step, we call the edge map EM. The order of the edge pattern of the ij-th block is denoted as BE_{ij}. The index value for ij-th block is I_{ij} for $0 \leq k \leq 2^{16} - 1$. Finally, the mean grayscale value can be derived from ELUT$[I_{ij}, BE_{ij}]$, where $0 \leq EM_{ij} \leq 38$ based on the number of edge patterns reported by Chung and Wu [11]. In Chung and Wu's scheme, the value of $GI'_{ij}(3, 3)$ is determinate as Equation 16.12.

$$GI'_{ij}(3,3) = ELUT[I_{ij}, BE_{ij}] \qquad (16.12)$$

Step 4: Input a halftone image HI, which will be converted to the grayscale image GI'. The detailed procedure is shown in Figure 16.16.

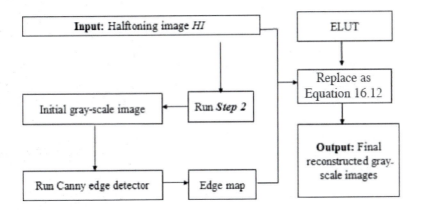

FIGURE 16.16
Procedure for generating final grayscale image in Step 4.

16.3.5 Shamir Scheme for Secret Sharing

The proposed scheme for secret image sharing is based on the (k,n)-threshold secret sharing scheme proposed by Shamir (1979) [20]. In this section we describe how to use the Shamir scheme to generate secret sharing. In the following steps we use the $(k-1)$-degree polynomial shown in Equation 16.16 to generate n shares for a group of n secret sharing participants from a secret integer value s for the threshold k.

$$F(x_i) = s + a_1 \times x_i^1 + a_2 \times x_i^2 + ... + a_{k-1} \times x_i^{k-1} \qquad (16.13)$$

where $i = 1, 2, ..., n$.

Step 1: Select the number k, $k \leq n$.

Step 2: Choose the $(k\text{-}1)$ integers $a_1, a_2, ..., a_{k-1}$ randomly.

Step 3: For the i-th secret sharing participant, choose a value of x_i freely, i
= 1, 2,..., n. Note that all x_i must be distinct from one another.

Step 4: For each chosen x_i, compute the corresponding $F(x_i)$ by using
Equation 16.13.

Step 5: Deliver each pair of $(x_i, F(x_i))$ as a secret share to each participant.

In this secret sharing process, the values of a_i, $i = 1, 2, ..., k - 1$, need not be
kept after all secret shares are generated. As well as the secret value s, each
a_i can be recovered from the n secret shares in the secret recovery steps listed
below.

Step 1: Collect at least k secret shares from the n ones to form a system of
equations as shown in Equation 16.13. Note that the xi and $F(x_i)$
in Equation 16.13 with $i =$1, 2, ..., k can all be extracted from the
k secret shares.

Step 2: Use a polynomial interpolation technique, e.g., Lagrange scheme, to
solve s and a_i, $i = 1, 2, ..., k - 1$. Reconstruct the $(k - 1)$-degree
polynomial F(x) described by Equation 16.14.

$$
\begin{aligned}
F(x) = {} & F(x_1)\frac{(x - x_2)(x - x_3)...(x - x_k)}{(x_1 - x_2)(x_1 - x_3)...(x_1 - x_k)} \\
& + F(x_2)\frac{(x - x_1)(x - x_3)...(x - x_k)}{(x_2 - x_1)(x_2 - x_3)...(x_2 - x_k)} \\
& + ... \\
& + F(x_k)\frac{(x - x_1)(x - x_2)...(x - x_{k-1})}{(x_k - x_1)(x_k - x_2)...(x_k - x_{k-1})}
\end{aligned}
\qquad (16.14)
$$

Step 3: Compute the solution for the secret value s by Equation 16.15.

$$s = F(0)$$

$$=F(x_1)\frac{(x-x_2)(x-x_3)...(x-x_k)}{(x_1-x_2)(x_1-x_3)...(x_1-x_k)}$$

$$+F(x_2)\frac{(0-x_1)(0-x_3)...(0-x_k)}{(x_2-x_1)(x_2-x_3)...(x_2-x_k)}$$

$$+...$$

$$+F(x_k)\frac{(0-x_1)(0-x_2)...(0-x_{k-1})}{(x_k-x_1)(x_k-x_2)...(x_k-x_{k-1})} \tag{16.15}$$

$$=(-1)^{k-1}[F(x_1)\frac{x_2 x_3...x_k}{(x_1-x_2)(x_1-x_3)...(x_1-x_k)}$$

$$+F(x_2)\frac{x_1 x_3...x_k}{(x_2-x_1)(x_2-x_3)...(x_2-x_k)}$$

$$+...$$

$$+F(x_k)\frac{x_1 x_2...x_{k-1}}{(x_k-x_1)(x_k-x_2)...(x_k-x_{k-1})}]$$

16.4 Proposed Scheme

Following on the details described in the previous discussions, this section presents a detailed and complete description of the proposed scheme. In our scheme, a halftone image *HI* is created from the grayscale secret image *GI*, sized $H \times W$ with 8 bits per pixel, by using an error diffusion technique called EDT. The transmitted stego-image is called *SI* and the reconstructed grayscale secret image is called *GI'*. The proposed scheme includes two procedures: the first is the sharing and embedding phase, and the second is the reconstruction and verifying phase. General flowcharts of the phases of our scheme appear in Figure 16.17 and Figure 16.18.

16.4.1 Sharing and Embedding Phase

A detailed algorithm for the sharing and embedding phase is described in this section.

Input: The secret grayscale image *GI* and cover image *CI*.

Output: Stego-image SI.

Step 1: Apply the error diffusion technique (EDT) to the grayscale image *GI* to retrieve a halftone image *HI*. Obviously, the width and

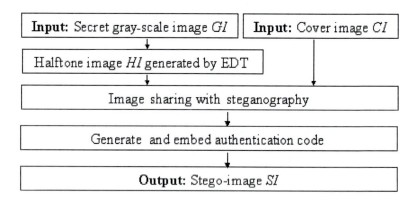

FIGURE 16.17
The flowchart of the work by the sender.

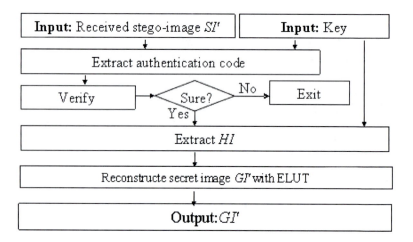

FIGURE 16.18
The flowchart of the work by the recipient.

height of HI are W and H. Moreover, each pixel in halftone image HI contains only 1 bit.

Step 2: Divide halftone image HI into Z nonoverlapping 6-bit blocks, $Z = \lfloor \frac{H \times W}{6} \rfloor$. HI_z ($z = 1,2,...,Z$) is the z-th block of HI and the value of HI_z is denoted as P_z illustrated in Figure 16.19.

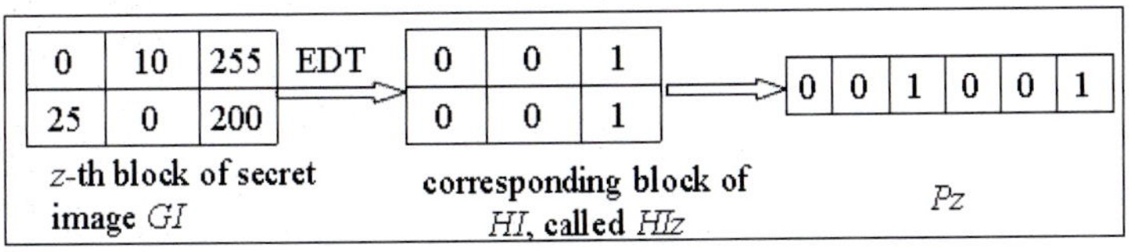

0	10	255	EDT
25	0	200	

0	0	1
0	0	1

0	0	1	0	0	1

z-th block of secret image GI corresponding block of HI, called HI_z P_z

FIGURE 16.19
The z-th block of GI and the corresponding HI_z, P_z.

Step 3: This is the basic process for secret image sharing with P_z as the secret, $z = 1, 2,..., Z$, $Z = \lfloor \frac{H \times W}{6} \rfloor$. This procedure contains five substeps, which are performed as follows.

(a) Take the value X_i formed from 2-MSBs (Most Significant Bits) of the four pixels of each cover CB_i as the value x specified in Equation 16.16, $i = 1, 2, ..., n$, as shown in Figure 16.20, $X_i = (c_7 c_6 d_7 d_6 g_7 g_6 h_7 h_6)_D$.

$$F(x_i) = (s + a_1 \times x_i^1 + a_2 \times x_i^2 + ... + a_{k-1} \times x_i^{k-1}) \bmod 2^6 \tag{16.16}$$

(b) Take the secret P_z as the value s specified in Equation 16.16.

(c) With a random number generator, generate a set of $k - 1$ integers $a_1, a_2, ..., a_{k-1}$ in Equation 16.16, where $k \le n$.

(d) For each X_i, compute the corresponding value of $F(X_i)$ by Equation 16.16 to form a secret share $(X_i, F(X_i))$ for each participant in the secret sharing group, $F(X_i) = f_5\ f_4\ f_3\ f_2\ f_1\ f_0$.

(e) Hide the six data bits of $F(X_i)$ in the Least Significant Bits of the four pixels C_i, D_i, G_i, and H_i of the corresponding cover block CB_i as shown in Figure 16.21.

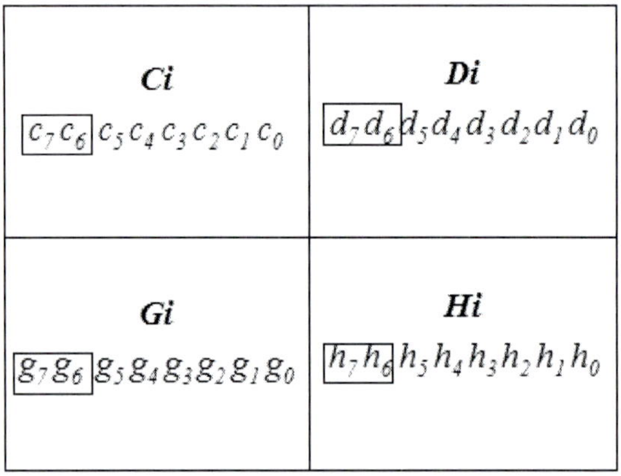

FIGURE 16.20
The four pixels of each cover block CBi.

Step 4: To prevent illicit attempts, a simple authentication ability is added here. Compute check bits pi by using Equation 16.17 and embed pi into the remaining LSBs of C_i, D_i, G_i, H_i from the same cover block CB_i, where $i = 1, 2, ..., t$. Note that t is the total number of check bits, which is discretionary and could be decided by the sender according to authentication strength. For simplicity, we take $t = 2$ as an example to explain the procedure as shown in Figure 16.22. Figure 16.23 illustrates a flowchart of

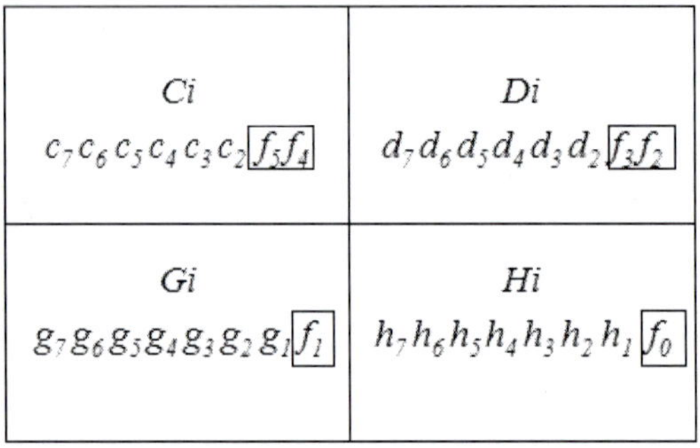

FIGURE 16.21
Hiding the six data bits of $F(Xi)$.

this phase.

$$\begin{cases} T = c_7c_6c_5c_4c_3c_2f_5f_4 \parallel d_7d_6d_5d_4d_3d_2f_3f_2 \parallel g_7g_6g_5g_4g_3g_2f_1 \parallel h_7h_6h_5h_4h_3h_2f_0 \\ P = T \bmod 2^t \\ P = \{p_i \mid i \in \{1, 2, ...t\}, p_i \in \{0, 1\}\} \end{cases}$$

$$(16.17)$$

C_i $c_7\,c_6c_5\,c_4\,c_3\,c_2f_5f_4$	D_i $d_7\,d_6\,d_5\,d_4\,d_3\,d_2f_3f_2$
G_i $g_7\,g_6g_5\,g_4\,g_3\,g_2\,\boxed{p_2}\,f_1$	H_i $h_7\,h_6\,h_5\,h_4\,h_3\,h_2\,\boxed{p_1}\,f_0$

FIGURE 16.22
Hiding the check bits p_1, p_2.

16.4.2 Reconstruction and Verifying Phase

This section describes the proposed secret recovery scheme and authentication scheme. Recall that after performing the secret sharing process for a group of n participants, each participant obtains one stego-image SI'_j. The proposed process for stego-image authentication and secret image recovery is summarized in the following paragraphs.

Input: The random number generator used to generate integers $a_1, a_2, ..., a_{k-1}$ in Equation 16.16, the secret key t used for generating the check bits (e.g., $t = 2$), and a set of at least k stego-images SI'_j, say m ones, with $k \le m \le n$, $j = 1, 2, ..., m$.

Output: A report of failure of secret reconstruction, or the recovered secret image GI' if all the stego-images are authenticated as genuine.

Step 1: Use the random number generator to generate $a_1, a_2, ..., a_{k-1}$ in

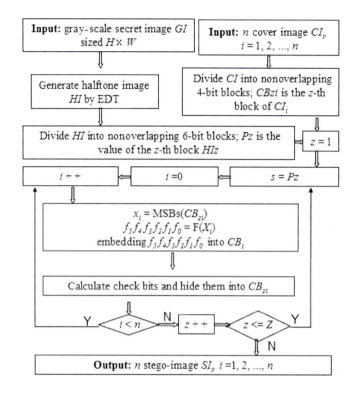

FIGURE 16.23
The flowchart of Step 4.

Equation 16.16.

Step 2: Divide each received stego-image SI'_j, $j = 1, 2,..., m$, into 2×2 blocks CB'_1, CB'_2, ..., CB'_z and denote the four pixels in each block CB'_z as C'_z, D'_z, G'_z, and H'_z as shown in Figure 16.24.

Step 3: For each stego-image SI'_j from participant j, $j \in \{1, 2, ..., m\}$, perform the following substeps for stego-image authentication.

   (a) $t = 2$, extract the 2-bit check bits from each block CB'_z, as shown in Figure 16.25. $P_2 = g'_1$, $P_1 = h'_1$.

   (b) Calculate the data for verification according to Equation 16.18.

$C'j$ $c_7'c_6'c_5'c_4'c_3'c_2'c_1'c_0'$	$D'j$ $d_7'd_6'd_5'd_4'd_3'd_2'd_1'd_0'$
$G'j$ $g_7'g_6'g_5'g_4'g_3'g_2'g_1'g_0'$	$H'j$ $h_7'h_6'h_5'h_4'h_3'h_2'h_1'h_0'$

FIGURE 16.24
The four pixels of each stego block $CB'j$.

$$\begin{cases} T = C'z \parallel D' \parallel G'z - g_1' \parallel Hz' - h_1' \\ P = T' \bmod 2^t \\ P' = \{p_i' \mid i \in \{1, 2, ...t\}, p_i' \in \{0, 1\}\} \end{cases} \quad , \quad (16.18)$$

(c) If for each block, $P_2' = p_2$, $P_1' = p_1$, then regard the stego-image as having passed authentication and continue; otherwise, decide that the stego-image has been tampered with.

$C'j$ $c_7'c_6'c_5'c_4'c_3'c_2'c_1'c_0'$	$D'j$ $d_7'd_6'd_5'd_4'd_3'd_2'd_1'd_0'$
$G'j$ $g_7'g_6'g_5'g_4'g_3'g_2'\boxed{g_1'}g_0'$	$H'j$ $h_7'h_6'h_5'h_4'h_3'h_2'\boxed{h_1'}h_0'$

FIGURE 16.25
The 2-bit check bits carried in $CB'z$ when $t = 2$.

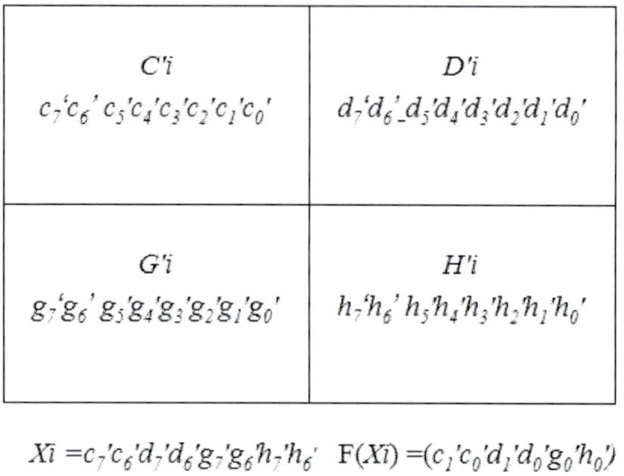

$C'i$	$D'i$
$c_7{}'c_6{}'\,c_5{}'c_4{}'c_3{}'c_2{}'c_1{}'c_0{}'$	$d_7{}'d_6{}'\_d_5{}'d_4{}'d_3{}'d_2{}'d_1{}'d_0{}'$
$G'i$	$H'i$
$g_7{}'g_6{}'\,g_5{}'g_4{}'g_3{}'g_2{}'g_1{}'g_0{}'$	$h_7{}'h_6{}'\,h_5{}'h_4{}'h_3{}'h_2{}'h_1{}'h_0{}'$

$$Xi = c_7{}'c_6{}'d_7{}'d_6{}'g_7{}'g_6{}'h_7{}'h_6{}' \quad F(Xi) = (c_1{}'c_0{}'d_1{}'d_0{}'g_0{}'h_0{}')$$

FIGURE 16.26
The bits of X_i and $F(X_i)$ carried in the stego block.

Step 4: If m' stego-images have passed authentication in Step 3, and $m' \geq k$, then continue. For each collected stego-image SI'_i, SI'_i = SI_i; otherwise, stop the program and report failure of the secret recovery step.

Step 5: For each $z =1, 2, ..., Z$, $Z = \lfloor \frac{H \times W}{6} \rfloor$, perform the following substeps to recover the secret data P_z.

(a) For each $SI'_i, i \in \{1, 2, ...m'\}$, extract the 2 most significant bits (MSBs) of the four pixels of each cover as the value X_i specified in Equation 16.16, $X = (c_7'c_6'd_7'd_6'g_7'g_6'h_7'h_6')_D$; extract the data bits of $F(X_i)$ from the LSBs of pixels C'_i, D'_i, G'_i, and H'_i, as shown in Figure 16.26, as a value of $F(X_i)$ appearing in Equation 16.16.

(b) By using the scheme described in Section 16.3.5, compute the corresponding value of y as the value P_z.

(c) For each $z = 1, 2, ..., Z$, $Z = \lfloor \frac{H \times W}{6} \rfloor$, perform last steps to obtain all of the P_z for the blocks of halftone image HI of the secret image GI.

Step 6: Apply the edge- and LUT-based inverse halftoning algorithm (ELUT, introduced in Section 16.3.4) to the halftone image HI

to retrieve the gray-scale image GI', i.e., the secret image.

16.5   Experimental Results

The feasibility of the proposed scheme was confirmed by the experimental results discussed in this section. We conducted three experiments, one to estimate the visual quality of the stego-images; the second to estimate the visual quality of the reconstructed secret images; and the last to evaluate the capability of identifying the tampered region. The test images shown in Figure 16.27 are four grayscale images "*Jet-F16*", "*Baboon*", "*Lena*", and "*Pepper*".

For our experiments, we chose the peak-signal-to-noise ratio (PSNR) as the criterion to measure the visual quality of the grayscale images, which is defined in Equation 16.19 and Equation 16.20.

$$PSNR = 10 log_{10} \frac{255^2}{MSN} \tag{16.19}$$

$$MSN = \frac{1}{H \times W} \sum_{h=1}^{H} \sum_{w=1}^{W} (P_{hw} - P'_{hw})^2 \tag{16.20}$$

H and W is the height and width of the images P and P' respectively. P_{hw} and P'_{hw} are the corresponding pixel values, and MSE is the mean-square-error between the two images.

In the first experiment, to compare the visual quality (i.e., PSNRs) between the cover images and the corresponding stego-images in Lin-Tsai's scheme, Yang et al.'s scheme, Chang et al.'s scheme, and ours, we took the image "*Jet-F16*" as the secret image and the images "*Lena*", "*Pepper*", and "*Baboon*" as the cover images, as shown in Figure 16.27. Figure 16.28 demonstrates the experimental results. The PSNRs of different schemes reveal that the visual quality of the stego-images generated by our scheme is the best of the schemes compared in our experiments.

To evaluate the visual quality of the recovered gray-scale images, we compared the PSNRs of the reconstructed grayscale images and the original secret image in Figure 16.29. From Figure 16.29, we can see the PSNR of Jet-F16, which is extracted from "*Lena*", "*Baboon*" and "*Pepper*", is larger than 30 dB. It means that not only does the revealing function of our scheme work well but also that the visual quality of the secret grayscale image is acceptable.

In addition, storage space and stego-image transmission time are important issues in secret sharing. Consequently, the pixel expansion problem should be eliminated or minimized with any useful secret image sharing scheme. In other

A. Jet-F16 (secret image)	B. Baboon (cover image)
C. Lena (cover image)	D. Pepper (cover image)

FIGURE 16.27
The test images.

words, the size of the cover images used in the sharing algorithm should be no larger than that of their secret images. However, the cover images produced during previous research, which are based on a polynomial approach [9,15,28], are much larger than their secret images. The cover-to-secret image size ratios of the existing schemes and the proposed scheme are listed in Table 16.1 for comparison. From Table 16.1, we can see the cover size/secret size of our scheme is the smallest and the most satisfactory.

To evaluate the capacity to identify a region of the secret image that has been tampered with, we compared the proposed scheme, Lin-Tsai's scheme, Yang et al.'s scheme, and Chang et al.'s scheme using some manipulated stego-images that satisfy the parity check policy (shown in Figure 16.30).

Both Figure 16.30 and Table 16.2 show the experimental results for the

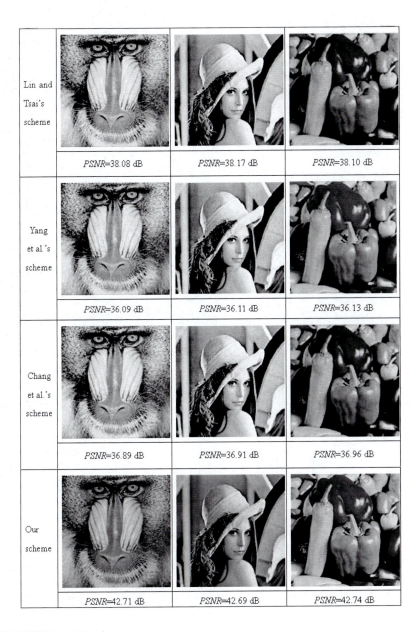

FIGURE 16.28
The experimental results for comparing the PSNR among the past work and
ours.

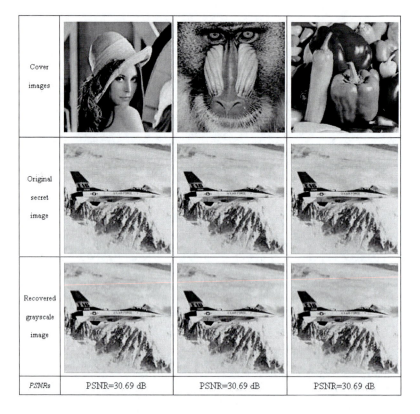

Cover images			
Original secret image			
Recovered grayscale image			
PSNRs	PSNR=30.69 dB	PSNR=30.69 dB	PSNR=30.69 dB

FIGURE 16.29
The visual quality of the reconstructed grayscale image.

various schemes. These results reveal that, among Lin-Tsai's scheme, Yang et al.'s scheme, and the proposed scheme, our scheme provides the best authentication capability. In Lin-Tsai's scheme, there is only one pixel for each block undergoing parity check. As a result, the authentication capability is not very good.

In Yang et al.'s scheme, four pixels of each block all undergo the procedure using a hash function. However, their final check bit is computed by using an XOR operation. So, their probability of successful authentication should remain at 50%. In our scheme, two check bits are adopted and the probability of successful authentication could be increased to 75%. On the other hand, compared with Chang et al.'s scheme, in which four check bits are adopted and authentication is implemented by using CRT (the Chinese Remainder Theorem), our proposed scheme has a lower detection ratio while maintaining lower computation complexity. However, because the set of shadows and the reconstructed secret image are generated through simple operations, no com-

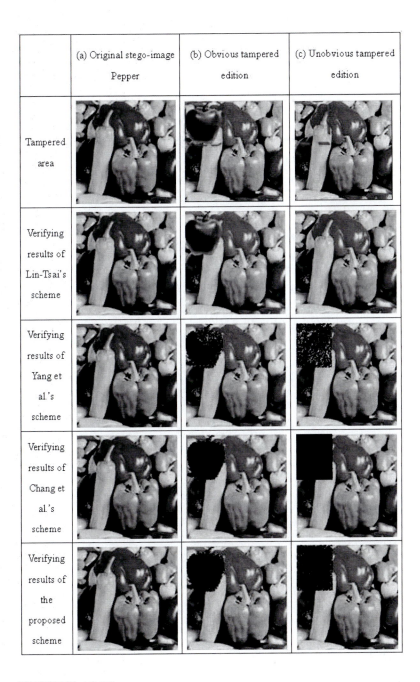

FIGURE 16.30
Three tampered stego-images.

TABLE 16.1

The cover-to-secret image size ratios of the existing schemes and the proposed scheme.

Schemes	Cover size / secret size
Lin-Tsai's scheme	4
Yang et al.'s scheme	4
Chang et al.'s scheme	2
Our scheme	2 / 3

TABLE 16.2

The detection ratios of different schemes.

Schemes	Detection ratios (DR)
Lin-Tsai's scheme	0%
Yang et al.'s scheme	52%
Chang et al.'s scheme	97%
Our scheme	79%

putational complexity or pixel expansion occurs with our scheme. Moreover, in practice, the nearly 80% detection ratio is sufficiently effective.

16.6 Conclusions

In this chapter, we propose a novel secret image sharing and authentication scheme in which the set of shadows and the reconstructed secret image are generated through simple operations, and no computational complexity or pixel expansion occur. The PSNR value of the reconstructed secret image is larger than 30 dB, and the visual quality of the reconstructed secret gray-scale image is acceptable.

Some research reported in the literature uses steganography as with Shamir's secret sharing scheme [9,15,28]. However, with these approaches the cover images must be many times larger than their secret images. For a secret image of $H \times W$ pixels, for earlier schemes (introduced in Section 16.2), the cover image should be $2H \times 2W$ pixels or $2H \times W$ pixels in size. Such a restriction requires more storage capacity in the cover image and consumes a larger bandwidth during transmission. Experimental results confirm that each shadow generated by our scheme is six times smaller than the secret image. Moreover, the ratio of cover image size and secret image size is reduced to 2/3. This is the primary advantage of our scheme over past work. To compare visual quality between cover images and their corresponding stego-images with the earlier schemes and our proposed scheme, we used several images for our experiments. The experimental results are shown in Figure 16.29. The PSNRs of the various schemes compared demonstrate that the visual quality of the stego-images generated by our scheme is the best. This is the second advantage of our scheme over earlier work. The third advantage of our scheme over past work is that, based on the check bits embedded, our scheme provides an effective solution for verifying the reliability of the set of collected shadows in the context of low computational complexity. Furthermore, experiments confirm that our proposed scheme is suitable for real-time applications.

Bibliography

[1] A. Adhikari, and S. Sikdar, A New (2, n)-Visual Threshold Scheme for Color Images, *INDOCRYPT 2003, Lecture Notes in Computer Science*, Vol. 2904, pp. 148–161, 2003.

[2] G. R. Blakley, Safe guarding Cryptographic Keys, *Proceedings of the National Computer Conference, American Federation of Information Processing Societies, Seoul, Korea*, pp. 313–317, 1979.

[3] C. Blundo, A. D. Santis, and M. Naor, Visual Cryptography for Grey Level Images, *Information Processing Letters*, Vol. 27, pp. 255–259, 2000.

[4] C.-C. Chang, C.-C. Lin, C.-H. Lin, and Y.-H. Chen, A Novel Secret Image Sharing Scheme in Color Images Using Small Shadow Images, *Information Sciences*, Vol. 178, No. 11, pp. 2433–2447, 2008.

[5] Y. F. Chen, Y. K. Chan, C. C. Huang, et. al, A Multiple-Level Visual Secret-Sharing Scheme Without Image Size Expansion, *Information Sciences*, Vol. 177, No. 21, Part II, pp. 4696–4710, 2007.

[6] C. C. Chang, C. C. Lin, T. H. N. Le, and B. H. Le, A Probabilistic Visual Secret Sharing Scheme for Grayscale Images with Voting Strategy, *Intelligent Information Hiding and Multimedia Signal Processing*, pp. 184–188, 2008.

[7] C. C. Chang, C. C. Lin, T. H. N. Le, and B. H. Le, A New Probabilistic Visual Secret Sharing Scheme for Color Images, *International Symposium on Electronic Commerce and Security*, pp. 1305–1308, 2008.

[8] S. Cimato, R. De Prisco, and A. De Santis, Probabilistic Visual Cryptography Schemes, *The Computer Journal*, Vol. 49, No. 1, pp. 97–107, 2006.

[9] C.-C. Chang, Y.-P. Hsieh, and C.-H. Lin, Sharing Secrets in Stego-Images with Authentication, *Pattern Recognition*, Vol. 41, No. 10, pp. 3130–3137, 2008.

[10] P. C. Chang, C. S. Yu, and T. H. Lee, Hybrid LMS-MMS Inverse Halftoning Technique, *IEEE Trans. Image Process.*, Vol. 10, No. 1, pp. 95–103, 2001.

[11] K. L. Chung and S. T. Wu, Inverse Halftoning Algorithm Using Edge-Based Lookup Table Approach, *IEEE Trans. Image Process.*, Vol. 14, No. 10, pp. 1583–1589, 2005.

[12] J. Canny, A Computational Approach to Edge Detection, *IEEE Trans. Pattern Analysis and Machine Intelligence.*, Vol. 8, No. 6, pp. 679–698, 1986.

[13] J. B. Feng, I. C. Lin, and Y. P. Chu, Halftone Image Resampling by Interpolation and Error-Diffusion, *Proceedings of the 2nd International Conference on Ubiquitous Information Management and Communication*, pp. 409–413, 2008.

[14] G. Horng, T. Chen, and D. Tasi, Cheating in Visual Cryptography, *Designs, Codes and Cryptography*, Vol. 38, pp. 219–236, 2006.

[15] C.-C. Lin and W.-H. Tsai, Secret Image Sharing with Steganography and Authentication, *Journal of Systems and Software*, Vol. 73, No. 3, pp. 405–414, 2004.

[16] M. Mese and P. P. Vaidyanathan, Look up Table (LUT) Scheme for Inverse Halftoning, *IEEE Trans. Image Process.*, Vol. 10, No. 10, pp. 1566–1578, 2001.

[17] M. Naor and A. Shamir, Visual Cryptography, *Advances in Cryptology - Eurocrypt' 94, Lecture Notes in Computer Science*, Vol. 950, pp. 1–12, 1995.

[18] R. D. Prisco and A. D. Santis, Cheating Immune (2, n)-Threshold Visual Secret Sharing, *Lecture Notes in Computer Science*, Vol. 4116, pp. 216–228, 2006.

[19] L. G. Roberts, Machine Perception of Three-Dimensional Solids, *Optical and Electro-Optical Information Processing*, pp. 159–197, 1965.

[20] A. Shamir, How to Share a Secret, *Communications of ACM*, Vol. 22, no. 11, pp. 612–613, 1979.

[21] S. J. Shyu, Efficient Visual Secret Sharing Scheme for Color Images, *Pattern Recognition Letters*, Vol. 39, No. 5, pp. 866–880, May 2006.

[22] A. Shiozaki, Digital Half-Toning by Error Diffusion with Perturbation, *Electronics Letters*, Vol. 32, No. 18, pp. 1655–1656, 1996.

[23] C.-C. Thien and J.-C. Lin, Secret Image Sharing, *Computers and Graphics*, Vol. 26, No. 5, pp. 765–770, 2002.

[24] C.-C. Thien and J.-C. Lin, An Image-Sharing Scheme with User-Friendly Shadow Images, *IEEE Transactions on Circuits and Systems for Video Technology*, Vol. 13, No. 12, pp. 1161–1169, 2003.

[25] R.-Z. Wang and C.-H. Su, Secret Image Sharing with smaller Shadow Images, *Pattern Recognition Letters*, Vol. 27, No. 6, pp. 551–555, 2006.

[26] A. Wang, L. Zhang, N. Ma, and X. Li, Two Secret Sharing Schemes Based on Boolean Operations, *Pattern Recognition*, Vol. 40, pp. 2776–2785, 2007.

[27] N. Yang, New Visual Secret Sharing Schemes Using Probabilistic Schemes, *Pattern Recognition Letters*, Vol. 25, No. 4, pp. 481–494, 2004.

[28] C.-N. Yang, T.-S. Chen, K.H. Yu, and C.-C. Wang, Improvements of Image Sharing with Steganography and Authentication, *Journal of Systems and Software*, Vol. 80, No. 7, pp. 1070–1076, 2007.

[29] R. Zhao, J.-J. Zhao, F. Dai, and F.-Q. Zhao, A new Image Secret Sharing Scheme to Identify Cheaters, *Computer Standards and Interfaces*, Vol. 31, No. 1, pp. 252–257, 2009.

17

Two-Decoding-Option Image Sharing Method

Ching-Nung Yang

National Dong Hwa University, Taiwan

Chuei-Bang Ciou

National Dong Hwa University, Taiwan

Tse-Shih Chen

National Dong Hwa University, Taiwan

CONTENTS

17.1   Introduction

An image secret sharing scheme (ISSS) divides a secret image into some shadow images (referred to as shadows) in a way that requires the shadows in a certain privileged coalitions for the secret reconstruction. However, the secret image cannot be revealed if they are not combined in the prescribed way. A typical ISSS is often a (k, n)-threshold scheme, where k is the threshold value to reveal the secret and n is the number of total shadows. One can reconstruct a secret image by k or more shadows, while he cannot conjecture any information from less than k shadows. There are two major categories in ISSS: one is the visual cryptography scheme (VCS) and the other is the polynomial-based ISSS (PISSS).

In the (k, n)-VCS, a secret image is encrypted into n shadows by expanding each secret pixel into m (the pixel expansion) subpixels. Notice that the difference between the pixel and the subpixel is that the "pixel" denotes the secret pixel located in the secret image, and the "subpixel" means the pixel located in shadows. Actually, the size of a subpixel is the same as that of the secret pixel. Therefore, shadows are, in general, expanded. Any k participants may photocopy their shadows on transparencies and stack them on an overhead projector to visually decode the secret through the human visual system (HVS) without hardware and computation. However, stacking $k - 1$ or fewer shadows will not gain any information. The first VCS encrypted a halftone (black-and-white) secret image into noise-like shadows [11]; subsequently, most VCSs were dedicated to reducing the pixel expansion [5, 7, 16, 17, 18, 2].

Indeed the visual quality of the VCS is poor, which comes from its intrinsic property using the OR-operation for decoding. Contrarily, the PISSS can recover the secret image without any distortion, while it needs the computation. By directly adopting Shamir's secret sharing scheme [12], a (k, n)-ISSS takes the secret pixel as the constant term in $(k-1)$-degree polynomials to share the secret. To gain small shadows, Thien and Lin used all coefficients in a $(k - 1)$-degree polynomial to generate shadows with size $1/k$ times to the secret image [13]. Afterwards, Wang and Su [15] further reduced the shadow size by using the Huffman code. Shadows in [13, 15] are noise-like, which shadows are often suspect to censors. It would be better to design a (k, n)-ISSS with the ability of steganography, i.e. shadows look like a cover image (a pre-selected meaningful image). Actually, we can construct a (k, n)-ISSS being provided with meaningful and meaningless shadows according to our need. Some (k, n)-ISSSs with meaningful shadows were proposed [14, 21, 8, 19, 1, 4, 6]. For example, two user-friendly (k, n)-ISSSs [14, 21] produced the shadow with a shrunken secret image on it. However, the portrait on shadows had already leaked the secret information, and thus this scheme is, strictly speaking, not a secret sharing scheme. Other (k, n)-ISSSs [8, 19, 1, 4] can present any cover images on shadows. Lin and Tsai's scheme [8] had the authentication capability to detect the faked shadows. The schemes in [19, 1, 4] improved Lin and Tsai's scheme to solve the dishonest participant problem of authentication, enhance the detection ratio of manipulated shadows, and improve the visual qualities of shadows.

A new type of ISSS with two decoding options was introduced recently, where the secret image is revealed both by stacking the transparencies and by computation. This scheme is referred to as two-in-one ISSS (TiOISSS). TiOISSS can decode secret images for preview by the HVS when a computer is temporarily unavailable. When the computer is available during the decoding scene, we then spend more computation to obtain a high-quality image for high-end applications. Two TiOISSSs [6, 9] have been proposed. Both schemes can stack shadows to decode a halftone secret image by the HVS in the first stage, and then can perfectly reconstruct the gray-level secret image in the second stage. A possible application scenario of TiOISSS is described below. In a distributed multimedia system, n shadows of PISSS can be delivered in a distributed system where each shadow is stored in any distributed storage node. The failure of $(n - k)$ shadows during transmission does not affect the reconstruction phase, as the secret image can be perfectly restored using k shadows. Suppose a fake shadow is received. The receiver spends considerable computation and finally finds the received shadow is wrong. Because the reconstruction phase of PISSS is very computationally intensive, we can apply the TiOISSS to save the computational time for verifying the validity of shadows. The receiver can first verify the shadows by visually previewing the secret without computation. After the successful verification, the receiver then recovers the original gray-level secret image by computation. In this chapter, we will briefly describe two TiOISSSs [6, 9], and also introduce our recent research result on TiOISSS published in [20].

17.2 Preliminaries

17.2.1 PISSS

A polynomial-based (k, n) secret sharing scheme was first proposed by Shamir [12], in which the secret data is encrypted into n shadows. Any k shadows can be used to reconstruct the secret, but any $k - 1$ or fewer shadows learn no information. By taking the secret data as g_0 (constant term) in the following $(k-1)$-degree polynomial $g(x)$ where p is a prime number, we could construct n shadows $(x_i, g(x_i))$ by choosing n different x_i, $i \in [1, n]$.

$$g(x) = (g_0 + g_1 x + ... + g_{k-1} x_{k-1}) \bmod p. \tag{17.1}$$

Any k shadows (without loss of generality, we use k shadows $(x_i, g(x_i)), i \in [1, k]$) can be used to reconstruct the $(k - 1)$-degree polynomial $g(x)$ by Lagrange interpolation as follows. Afterwards, the secret data g_0 can be determined from $g_0 = g(0)$.

$$\begin{aligned}
g(x) = g(x_1) &\frac{(x-x_2)(x-x_3)...(x-x_k)}{(x_1-x_2)(x_1-x_3)...(x_1-x_k)} \\
+ g(x_2) &\frac{(x-x_1)(x-x_3)...(x-x_k)}{(x_2-x_1)(x_2-x_3)...(x_2-x_k)} \\
+ \cdots + g(x_k) &\frac{(x-x_1)(x-x_2)...(x-x_{k-1})}{(x_k-x_1)(x_k-x_2)...(x_k-x_{k-1})} \bmod p.
\end{aligned} \tag{17.2}$$

Through Shamir's secret sharing scheme, we could take every secret pixel as g_0 in a $(k - 1)$-degree polynomial $g(x)$ to construct n random grayscale values in shadows to construct a (k, n)-PISSS. At this time, the prime number

$p = 251$ is chosen such that $g(x)$ is constrained between 0 and 250 and suitable to represent the conventional 8-bit grayscale or color images. Notice that the possible values of an 8-bit gray pixel are from 0 to 255, so the grayscale values (> 250) need to be modified to 250 and will cause distortion. Obviously, we can use the Galois Field $GF(2^8)$ rather than the ordinary arithmetic (mod 251) to achieve a lossless scheme. Thien and Lin further reduced shadows with size $1/k$ times that of the secret image [13] by embedding the secret data in all coefficients of $g(x)$. The formal encoding of Thien and Lin's scheme is briefly described below.

A secret image is first divided into τ non-overlapping k-pixel blocks, and every j-th ($0 \leq j \leq \tau - 1$) block includes the secret pixel values $p_{jk}, p_{jk+1}, ..., p_{jk+k-1}$. The $(k-1)$-degree polynomial $S_j(x)$ represents a shadow pixel associated with the j-th block on shadows.

$$S_j(x) = (p_{jk} + p_{jk+1}x + p_{jk+2}x^2 + ... + p_{jk+k-1}x^{k-1})\mathrm{in}GF(2^8), \quad (17.3)$$

where x is often an image identification and $0 \leq j \leq \tau - 1$. The value of $S_j(x)$ is generated using the original pixel values $p_{jk}, p_{jk+1}, ..., p_{jk+k-1}$ included in the j-th block. In this chapter, the Galois Field $GF(2^8)$ was chosen to achieve a lossless secret image. Because k pixels are processed each time, the size of the shadow image is $1/k$ of the secret image. By reversing the encoding, the polynomial in (17.3) can be reconstructed from k shadow pixels; hence, the blocks can be recovered and finally the secret image is reconstructed.

17.2.2 VCS

The first VCS was Naor–Shamir's (k, n)-VCS to encrypt a halftone secret image into noise-like shadows. The authors used the whiteness (the number of white subpixels in a m-subpixel block) to distinguish the black color from the white color, i.e., "$m - h$"B"h"W (respectively "$m - l$"B"l"W) represents a white (respectively black) color, where $h > l$. A black-and-white (k, n)-VCS can be designed using two base $n \times m$ matrices B_1 and B_0 with elements "1" and "0" denoting black and white subpixels. When sharing a black (respectively white) secret pixel, the dealer randomly chooses one row of the matrix in the set C_1 (respectively C_0), which includes all matrices obtained by permuting the columns in B_1 (respectively B_0) to a relative shadow. Let $\mathrm{OR}\,(B_i|r)$, $i = 0, 1$, denote the "OR"-ed vector of any r rows in B_i, and $H(\cdot)$ be the Hamming weight of a vector. The base matrices of the (k, n)-VCS should satisfy the following conditions:

(V-1). $H\,(\mathrm{OR}\,(B_1|r)) \geq (m - l)$ and $H\,(\mathrm{OR}\,(B_0|r)) \leq (m - h))$ for $r = k$, where $0 \leq l < h \leq m$.

(V-2). $H\,(\mathrm{OR}\,(B_1|r)) = H\,(\mathrm{OR}\,(B_0|r))$ for $r \leq (k - 1)$.

The first condition is often referred to as the contrast condition, and the secret image can be recognized due to their different contrasts of black and white colors. The second condition is the security condition that assures the (k, n)-VCS of perfect secrecy.

Example 1. Construct a (2, 2)-VCS of $h = 1$, $l = 0$, and $m = 2$ using $B_1 = \begin{bmatrix} 10 \\ 01 \end{bmatrix}$ and $B_0 = \begin{bmatrix} 10 \\ 10 \end{bmatrix}$. The secret is a printed-text image $\boxed{\text{VCS}}$.

A (2, 2)-VCS with $B_1 = \begin{bmatrix} 10 \\ 01 \end{bmatrix}$ and $B_0 = \begin{bmatrix} 10 \\ 10 \end{bmatrix}$ has $H\left(\text{OR}\left(B_1|2\right)\right) = 2$, $H\left(\text{OR}\left(B_0|2\right)\right) = 1$, and $H\left(\text{OR}\left(B_1|1\right)\right) = H\left(\text{OR}\left(B_0|1\right)\right) = 1$, which satisfy (V-1) and (V-2) conditions. In the reconstructed image, the black color is 2B0W and the white color is 1B1W (or 1W1B). However, each shadow contains 1B1W and 1W1B with the same frequencies so that one cannot see anything from his own shadow. Figure 17.1(a) is a black-and-white secret image $\boxed{\text{VCS}}$. Two noise-like shadows (Shadow 1, Shadow 2) and the reconstructed image by stacking two shadows (Shadow 1 + Shadow 2) are shown in Figures 17.1(b), 17.1(c) and 17.1(d). The printed-text secret $\boxed{\text{VCS}}$ can be revealed by HVS. □

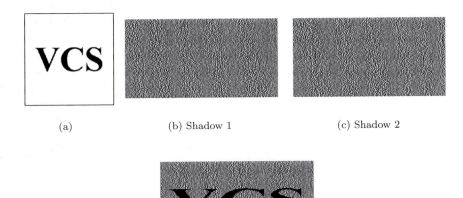

(a)   (b) Shadow 1   (c) Shadow 2

(d) Shadow 1+Shadow 2

FIGURE 17.1
A (2, 2)-VCS of $h = 1, l = 0$, and $m = 2$: (a) the secret image (b) and (c) two shadows (d) the reconstructed image.

17.3   Previous Works

Two TiOISSSs [6, 9] are briefly reviewed in the sequel. Jin and Lin's scheme [6] applied VCS on a halftone secret image, which is obtained from a prescribed halftoning transformation. Lin and Lin's scheme [9] combined the VCS and PISSS. Both TiOISSSs have stacking-to-see capability. By computation, both schemes could reconstruct the original gray-level secret image. The TiOISSSs in this chapter can be extended to share the chromatic secret image by applying the approaches on C (cyan), M (magenta), and Y (yellow) bit planes, respectively. So, we only discuss the gray-level secret image.

17.3.1   Jin et al.'s TiOISSS

In [6], the gray-level secret image is transformed into a binary image by a digital halftoning method, in which every secret pixel is represented by nine black-and-white pixels $(p_c, p_0, p_1, \ldots, p_7)$ in a 3×3-block as shown. Excluding the center pixel p_c, every 8-tuple $(p_0, p_1, \ldots, p_7)$ is uniquely decoded to the 256 possible grayscale values (0 to 255), and simultaneously the pattern of nine pixels can simulate shades of gray. This halftoning method is most similar to a patterning technique, which uses the black dots in a block to represent the intensity levels. The color of the center pixel p_c can be chosen according to the halftone version of the original grayscale image to enhance the resolution. Afterwards, the (k, n)-VCS is performed on this halftone secret image to construct n shadows. Suppose the gray-level secret image and the halftone images are I and I'. We first obtain I' from I by the halftoning technique (see Table 2 in [6], which provides the complete mapping between I and I', also gives the nine grayscale levels in I' to simulate the grayscale levels), and then arrange the pixels in a 3×3-block. Thus, we have $|I'| = 9 \times |I|$. Apply a (k, n)-VCS with the pixel expansion m on I', and then every shadow of Jin et al's TiOISSS has the image size $9 \times m \times |I|$. Jin et al's TiOISSS has two decoding options. The first is stacking any k out of n shadows to get a vague halftone image by VCS, and the second is to reconstruct the gray-level secret image by a look-up table from the arrangement of $(p_0, p_1, \ldots, p_7)$.

Because we have two secret images (I and I'), we herein define the pixel expansion of TiOISSS as the ratio of shadow size relative to the size of the original gray-level secret image. So, the pixel expansion of Jin et al's (k, n)-TiOISSS is

$$m_{JIN} = 9m. \qquad (17.4)$$

17.3.2   Lin and Lin's TiOISSS

Jin et al.'s TiOISS scheme has a terrible pixel expansion $9m$. For example, Jin et al.'s $(2, 2)$-TiOISS has the pixel expansion $9m = 18$ when using $(2, 2)$-VCS with $m=2$. Recently, Lin and Lin proposed a TiOISSS [9] based on the VCS and PISSS to reduce the pixel expansion. A halftone secret image is encrypted into n shadows by (k, n)-VCS with the pixel expansion m, which every row contains b "1" and w "0" and $m = b + w$. It is obvious that every m-subpixel has $\begin{pmatrix} m \\ w \end{pmatrix}$ combinations, and it can be used to represent about $\log_2 \begin{pmatrix} m \\ w \end{pmatrix}$ bits. Notice that in the conventional VCS, we can randomly permute all m columns in matrices. However, when using the $\begin{pmatrix} m \\ w \end{pmatrix}$ combinations to represent the information and simultaneously satisfy the security condition, we could only randomly choose the combination for one shadow and the remaining $(n-1)$ shadows are then determined according to the base matrices. Therefore, in a shadow, only $(|I'|/n)$ digits can be used to share $\log_2 \begin{pmatrix} m \\ w \end{pmatrix}$ bits. On the other hand, we need $(8 \times |I|/k)$ bits to share the gray-level secret image I by (k, n)-PISSS. To assure we have enough space to hide the secret, the gray-level secret image is compressed by Jpeg or other compression techniques to reduce the information as $(8 \times |I|/(k \times R))$ bits where R is the compression ratio. In [9], the authors consider the case in which the halftone secret image and the gray-level secret image have the same size, i.e., $|I| = |I'|$.

As a result, $(|I'|/n) \times \log_2 \begin{pmatrix} m \\ w \end{pmatrix} \geq (8 \times |I|/(k \times R))$, and thus the compression ration R should satisfy the following requirement to hide all information of a gray-level secret image.

$$R \geq \left(8n \bigg/ k \times \log_2 \begin{pmatrix} m \\ w \end{pmatrix} \right) \tag{17.5}$$

Since $|I| = |I'|$, finally, the pixel expansion of Lin and Lin's (k, n)-TiOISSS is

$$m_{LIN}^{(C)} = m. \tag{17.6}$$

For example, the pixel expansion of Lin and Lin's $(2, 2)$-TiOISSS when using $(2, 2)$-VCS with $m = 2$ is $m_{LIN}^{(C)} = 2$ less than $m_{JIN} = 18$. However, the gray-level secret image can be perfectly reconstructed in Jin and Lin's scheme, while Lin and Lin's scheme only recovers the compressed image. The greater compression ratio will degrade the visual quality of the secret image.

Example 2. Construct Lin and Lin's $(2, 4)$-TiOISSS and Jin and Lin's $(2, 4)$-TiOISSS using $B_1 = \begin{bmatrix} 1100 \\ 0110 \\ 0011 \\ 1001 \end{bmatrix}$ and $B_0 = \begin{bmatrix} 1100 \\ 1100 \\ 1100 \\ 1100 \end{bmatrix}$.

Since $\begin{pmatrix} m \\ w \end{pmatrix} = \begin{pmatrix} 4 \\ 2 \end{pmatrix} = 6$ (there are six combinations: (1100), (0011), (1010), (0101), (1001), (0110)), $\log_2 \begin{pmatrix} m \\ w \end{pmatrix} = \log_2 6 = 2.585$ bits. Suppose the secret image I is a 512 512 gray-level image. Lin and Lin's TiOISS (2, 4) scheme needs a compression ratio $R = \left(8n \Big/ k \times \log_2 \left(\dfrac{m}{w} \right) \right) = 32/(2 \times 2.585) = 6.19$. We first compress the gray-level secret image I to a compressed image I_C to obtain the less embedded bits. Notice that even though $|I_C|$ has a lesser file size (the embedded bits), it has the same physical size of the original image $|I|$. Finally, the shadow size of Lin and Lin's (2, 4)-TiOISSS is 1024 1024 (note: $m \times |I'|$), and the pixel expansion is $m_{LIN}^{(C)} = m = 4$. On the other hand, we need a $1536 \times 1536 (9 \times |I|)$ halftone image for Jin et al.'s (2, 4)-TiOISSS. The shadow size of Jin et al.'s (2, 4)-TiOISSS is $3072 \times 3072 (9m \times |I|)$, and the pixel expansion is $m_{JIN} = 9m = 36$. □

17.4   A New (k, n)-TiOISSS

The VCS and PISSS have their respective features. As is known, the VCS has the vague reconstructed image and PISSS has a perfect reconstruction. The VCS has the distinctive stacking-to-see capability, while PISSS spends the computation for reconstruction. It is reasonable to adopt the stacking-to-see property of the VCS into PISSS to achieve a two-in-one scheme where the secret image is revealed both by stacking the transparencies and by computation. Our new TiOISSS [20] is also a combination of the VCS and PISSS, which is somewhat similar to Lin and Lin's scheme, but the way is completely different to that in Lin and Lin's scheme.

Jin et al.'s TiOISSS is a lossless version (i.e., no distortion in the secret image) but has a large pixel expansion. Lin and Lin's TiOISSS is a compressible version. It compresses the secret image such that the shadow size has enough space to hide the information of a compressed image. Although Lin and Lin's scheme reduces the pixel expansion of Jin et al.'s scheme, the reconstructed image has distortion. Obviously, Lin and Lin's approach can be extended to the lossless version by expanding the halftone image with the size $|I'| = (n/k) \times |I| \times 8/\log_2 \left(\dfrac{m}{w} \right)$ to hide the original secret image. For the lossless version of Lin and Lin's TiOISSS, the pixel expansion $m_{LIN}^{(L)}$ is

$$m_{LIN}^{(L)} = (n/k) \times m \times 8/\log_2 \left(\frac{m}{w} \right). \tag{17.7}$$

For example, the pixel expansion of Lin and Lin's (2, 4)-TiOISSS in Example 2 (i.e. $k = n = 2$, $m = 4$, $w = 2$) is $m_{LIN}^{(L)} = 24.76$.

Jin et al.'s scheme cannot be used in a compressible version because it uses a look-up table to recover the grayscale value of the pixel. The proposed TiOISSS has two versions—the lossless version and the compressible version.

For simplicity, we first describe the proposed TiOISSS, which reconstructs a lossless secret image. The compressible version is just an easy extension of the lossless TiOISSS, and will be discussed in Section 17.4.2.

17.4.1 Design Concept

Our design concept adopts the gray subpixel into the VCS, and this grayscale values simultaneously represents the output of the $(k-1)$-degree polynomial in PISSS. In VCS, it is evident that when a subpixel is stacked by the white subpixel, its intensity is kept unchanged. While stacking two gray subpixels, we get a grayer color (a dark version of the color). Therefore, if we replace black subpixels with gray subpixels in the shadow, we still can use the whiteness in every m subpixel to distinguish the black color from the white color in the reconstructed image. Here, we adopt the widely accepted definition of color superimposition in [10] to define the color mixing function $C(\cdot)$ when stacking two subpixels with the grayscale values between 0 and 255.

The grey level of the resultant pixel by stacking the two pixels can be expressed (approximately) as follows, in which each mixed color is produced by a color mixing function $C(\cdot)$.

$$g_3 = C(g_1, g_2) = \text{Int}((g_1 \times g_2)/255),$$

where $\text{Int}(\cdot)$ function maps a real number to the nearest integer. The values of g_1, g_2, g_3 are any grayscale values between 0 and 255, and "0" (respectively "255") is a black (respectively, white) color.

It is easy to verify $g_3 < g_1$ and $g_3 < g_2$, and this implies that stacking two gray pixels of g_1 and g_2 results in a grayer pixel of g_3. For example, $g_1 = C(g_1, 255)$ shows the grayscale value unchanged when stacking with the white pixel and $255 = C(255, 255)$ shows that the stacked result is a white color when stacking two white pixels.

Example 3. Consider Example 1, and randomly use gray subpixels $g_i \in [0, 255]$ instead of black subpixels in B_1 and B_0 and do not change the white subpixel.

As a replacement in B_1 and B_0, we have $B_1' = \begin{bmatrix} g_i 0 \\ 0 g_j \end{bmatrix}$ and $B_0' = \begin{bmatrix} g_i 0 \\ g_j 0 \end{bmatrix}$, respectively. In the reconstructed image, it is observed that the stacked result in the black area (using B_1') is $1g_i 1g_j$ and the stacked result in the white area (using B_0') is $1g_k 1W$ or $1W1g_k$, where $g_k < g_i$ and $g_k < g_j$. Through the whiteness, we can still visually reveal the secret. Each shadow contains $1g_i 1W$ (or $1W1g_i$) which is a gray-and-white and noise-like image, so that one cannot see anything from any shadow. In Figure 17.2(a) and Figure 17.2(b) are two noise-like shadows (Shadow 1, Shadow 2) and Figure 17.2(c) shows the reconstructed image (Shadow 1 + Shadow 2). It is observed that the secret $\boxed{\text{VCS}}$ is also revealed but is a little blurred when compared with Figure 17.1(d). □

We call this VCS with the matrices B_1' and B_0' the gray-subpixel based VCS (GVCS). Matrices B_1' and B_0' are the same as B_1 and B_0 except that the

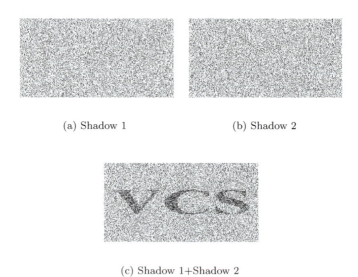

(a) Shadow 1   (b) Shadow 2

(c) Shadow 1+Shadow 2

FIGURE 17.2
A $(2, 2)$-GVCS of $h = 1, l = 0$, and $m = 2$: (a) and (b) two shadows (c) the reconstructed image.

gray subpixels replaced the black subpixels. Hence, GVCS holds the contrast and security conditions (V-1) and (V-2), and it is still a VCS. In this chapter, we use the notation (k, n, m, g)-GVCS, to denote a (k, n)-GVCS with B_1' and B_0', in which every row has g gray subpixels and $(m - g)$ white subpixels. The whiteness, $(m - g)$ white subpixels in every m-subpixel block, can be used to distinguish the black color from the white color. In GVCS, the grayness of the nonwhite subpixels in the reconstructed image is different from the pure blackness in the VCS, and it will distort the clarity (see Figure 17.1(d) and Figure 17.2(c)).

In order to design a TiOISSS based on GVCS and PISSS, we should carefully observe the distinguishing characteristics of both image secret sharing schemes. All characteristics of GVCS and PISSS are opposite; hence, combining GVCS and PISSS creates the following problem. It is obvious that we need two secret images—a halftone secret image for GVCS and a gray-level secret image for PISSS. In our TiOISSS, the gray-level secret image is shared by PISSS. Then, we divide the halftone secret image into shadows by GVCS, in which the grayscale values in B_1' and B_0' are chosen according to the outputs of the $(k - 1)$-degree polynomials in PISSS. Also, we need to determine the sizes of both secret images such that there are enough gray subpixels in GVCS to represent the outputs of PISSS.

17.4.2 The Lossless TiOISSS

We construct a (k,n)-TiOISSS based on a (k,n)-PISSS and a (k,n,m,g)-GVCS to share a gray-level secret image I. For the lossless version, we first determine $(8 \times |I|/k)$ bits by PISSS. So, $|I|/k$ gray supixels in each shadow of (k,n,m,g)-GVCS are required to hide $(8 \times |I|/k)$ bits. Since there are g gray subpixels in every m subpixels of (k,n,m,g)-GVCS, the halftone secret image I' for GVCS should be $|I'| = |I|/(k \times g)$. So, the shadow size is $m \times |I'|$. For the lossless version, the pixel expansion of our (k,n)-TiOISSS $m_{PRO}^{(L)}$ is

$$m_{PRO}^{(L)} = m/(k \times g). \tag{17.8}$$

The formal encoding and decoding algorithms are described as follows. Some notations are defined first.

Notation Used

$\mathbf{P}(\cdot)$ encryption of (k,n)-PISSS.

$\mathbf{P^{-1}}(\cdot)$ decryption of (k,n)-PISS.

I the gray-level secret image with the size $|I|$, which is used as the input of $\mathbf{P}(\cdot)$.

P_i the output shadows of $\mathbf{P}(I)$, $i \in [1,n]$, with the size $(|I|/k)$.

$\mathbf{G}(\cdot)$ encryption of (k,n,m,g)-GVCS with B_1' and B_0', and the values of gray subpixels are chosen according to the gray pixels in P_i.

$\mathbf{G^{-1}}(\cdot)$ decryption of (k,n,m,g)-GVCS (stack shadows and visually decode the secret by HVS).

$H(\cdot)$ halftoning function, transform and resize a gray-level image to a halftone image.

I' a halftone secret image with the size $|I'| = |I|/(k \times w)$ obtained from $I' = \mathbf{H}(I)$.

G_i the output shadows of $\mathbf{G}(I')$, $i \in [1,n]$, with the size $((m \times |I|)/(k \times g))$.

Encryption Algorithm of the Lossless Version of Our (k,n)-TiOISSS

Input: the gray-level secret image I; the parameters k, n, m, g; matrices B_1' and B_0'.

Output: n shadows $G_i, i \in [1,n]$.

1-1) Encrypt the secret image to obtain $P_i = \mathbf{P}(I)$, $i \in [1,n]$.

1-2) Obtain I' from $I' = \mathbf{H}(I)$.

1-3) Output n shadows $G_i = \mathbf{G}(I')$, $i \in [1,n]$.

Decryption Algorithm of the Lossless Version of Our (k,n)-TiOISSS

Input: any k out of n shadows $G_{i_1}, G_{i_2}, \ldots, G_{i_k}$.

Output: the halftone secret image I' (Phase 1); the gray-level secret image I (Phase 2).

/* Phase 1: does not need computation; visually decode the secret I' by stacking k shadows;

Phase 2: needs computation; decode the secret I by using Lagrange interpolation */

Phase 1 (preview phase):

2-1) $\mathbf{G^{-1}}(G_{i_1}, G_{i_2}, \ldots, G_{i_k})$;

/* stack k shadows to visually preview the halftone secret image */

Phase 2 (perfect-reconstruction phase):

2-2) Obtain $(P_{i_1}, P_{i_2}, \ldots, P_{i_k})$ from $(G_{i_1}, G_{i_2}, \ldots, G_{i_k})$ by discarding the white subpixels.

2-3) $I = \mathbf{P^{-1}}(P_{i_1}, P_{i_2}, \ldots, P_{i_k})$.

/* use Lagrange interpolation to reconstruct the gray-level secret image */

Our TiOISSS contains two decoding phases: the preview phase and the perfect-reconstruction phase. A halftone secret image can be visually previewed by simply stacking shadows in Phase 1. The preview phase may be used when a computer is temporarily not available, or in a scenario verifying whether the shadows are correct or not in a distributed multimedia system, as mentioned in the introduction. On the other hand, by extracting the gray subpixels from shadows we may perfectly reconstruct the gray-level secret image when the computer finally is available (or after successful verification in a distributed multimedia system). We call the second decoding phase the perfect-reconstruction phase because we can gain a lossless secret image.

Considering security, the proposed (k, n)-TiOISSS is a combination of two (k, n)-threshold schemes: the GVCS and the PISSS. So our scheme still retains the threshold property. An attacker could not stack less than k shadows to retrieve the black-and-white secret. Also, he cannot use the gray values of less than k shadows to reconstruct the $(k-1)$-degree polynomial. Thus, combining these two ISSSs together assures the secrecy of the threshold scheme.

Example 4 shows the proposed $(2, 2)$-TiOISSSs using different w and m to demonstrate different shadow sizes and the resolutions of the reconstructed images in the preview phase.

Example 4. Construct two $(2, 2)$-TiOISSSs using base matrices: (1) $B_1 = \begin{bmatrix} 10 \\ 01 \end{bmatrix}$ and $B_0 = \begin{bmatrix} 10 \\ 10 \end{bmatrix}$ (2) $B_1 = \begin{bmatrix} 110 \\ 101 \end{bmatrix}$ and $B_0 = \begin{bmatrix} 110 \\ 110 \end{bmatrix}$, respectively. The secret image is 512×512 Lena from the USC-SIPI image database.

A 512×512 gray-level Lena image I is shown in Figure 17.3(a). By $(2, 2)$-PISSS, we obtain two 512×256 gray-level noise-like shadows P_1 and P_2, as shown in Figure 17.3(b) and Figure 17.3(c).

Case (1) $B_1 = \begin{bmatrix} 10 \\ 01 \end{bmatrix}$, $B_0 = \begin{bmatrix} 10 \\ 10 \end{bmatrix}$:

Since $m = 2$ and $g = 1$, we then resize and halftone I to get a halftone image I' by $\mathbf{H}(\cdot)$, and the size is $|I'| = |I|/(k \times g) = |I|/2$. This 512×256 halftone Lena is shown in Figure 17.4(a). Output two shadows $G_i = \mathbf{G}(I')$, $i = 1, 2$, by using $(2, 2, 2, 1)$-GVCS, and the values of gray subpixels are chosen

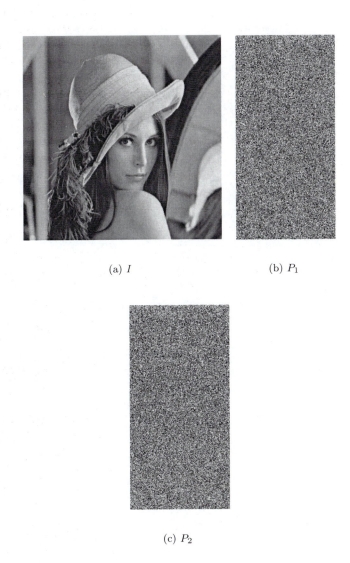

(a) I (b) P_1

(c) P_2

FIGURE 17.3
A $(2, 2)$-PISSS: (a) 512×512 gray-level Lena secret image (b) and (c) 512×256 gray-level noise-like shadows.

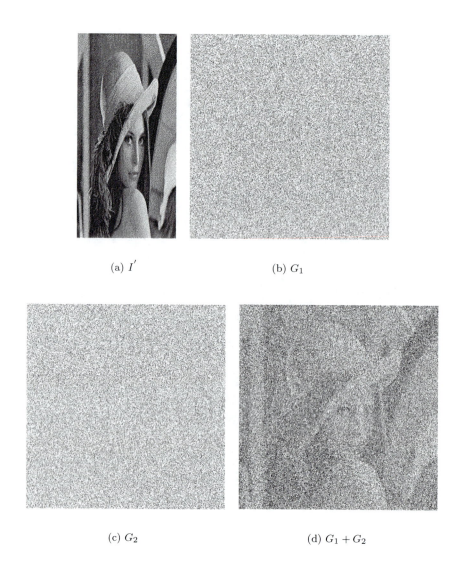

(a) I' (b) G_1

(c) G_2 (d) $G_1 + G_2$

FIGURE 17.4

The proposed (2, 2)-TiOISSS using base matrices with $h = 1, l = 0$, and $m = 2$: (a) 512×256 halftone image (b) and (c) two 512×512 gray-and-white shadows (d) the previewed image.

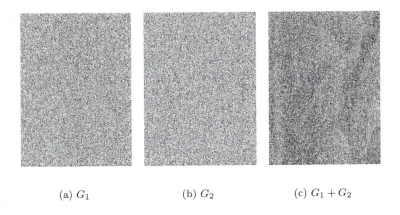

(a) G_1 (b) G_2 (c) $G_1 + G_2$

FIGURE 17.5
The proposed $(2, 2)$-TiOISSS using base matrices with $h = 1, l = 0$, and $m = 3$: (a) and (b) two 512×512 gray-and-white shadows (c) the previewed image.

according to the gray pixels in P_1 and P_2. In Figure 17.4(b) and Figure 17.4(c) are two gray-and-white shadows G_1 and G_2 of the size $m \times |I'| = |I|$ (512×512-pixels). Figure 17.4(d) is the previewed image by stacking G_1 and G_2 without computation. In the second phase of decoding, we can obtain P_1 and P_2 from the gray pixels of G_1 and G_2, and then reconstruct the gray-level Lena in Figure 17.4(a) by $I = \mathbf{P}^{-1}(P_1, P_2)$.

Case (2) $B_1 = \begin{bmatrix} 110 \\ 101 \end{bmatrix}$, $B_0 = \begin{bmatrix} 110 \\ 110 \end{bmatrix}$:

Since $m = 3$ and $g = 2$, we need a halftone image I' of the size is $|I'| = |I|/(k \times g) = |I|/4$. Output two shadows $G_i = \mathbf{G}(I')$, $i = 1, 2$, by using $(2, 2, 3, 2)$-GVCS, and the values of gray subpixels are chosen according to the gray pixels in P_1 and P_2. In Figure 17.5(a) and Figure 17.5(b) are two gray-and-white shadows G_1 and G_2 of the size $m \times |I'| = 0.75 \times |I|$ (512×384-pixels). Figure 17.5(c) is the stacked result by stacking G_1 and G_2. By the same approach in Case (1), we can also reconstruct the original gray-level Lena I. $\square$

All the above schemes can visually reveal the secret by simply stacking shadows in the preview phase. The original gray-level secret image can be perfectly reconstructed in the perfect-reconstruction phase. Of two $(2, 2)$-TiOISSSs in Example 4, Figure 17.4(d) possess the better resolution of the previewed result, but Figure 17.5(c) has a lesser shadow size of 512×384 pixels.

Because our TiOISSS uses PISSS in Phase 2 for decoding, we can use the compression approach in [9] to hide the compressed image into shadows.

Suppose $\mathbf{J}(\cdot)$ is a Jpeg-compression function, where $I_C{=}\mathbf{J}(I)$, and the compression ratio is $R =$ (the number of bits in I/the number of bits in I_C)≥ 1. At this time, the gray-level secret image I is compressed to I_C such that the information bits in I_C can be embedded into the halftone secret image I'. By replacing I with I_C in our lossless TiOISSS, we could get the compressible TiOISSS. The formal encoding/decoding algorithm of the compressible version of our (k, n)-TiOISSS is omitted for brevity.

In the lossless version of the proposed TiOISSS, since $|I'| = |I|/(k \times g)$, $k \geq 2$ and $g \geq 1$, then $|I'| < |I|$. This observation implies that our TiOISSS could embed all information bits of lossless I into I', where $|I'| < |I|$, for any values of (k, n, m, g). So, actually, our TiOISSS does not need compression on I. However, to be fairly compared with the compressible version of Lin and Lin's scheme, we use a compression ratio $R = R_{LIN} = (n/k) \times 8/\log_2\left(\dfrac{m}{w}\right)$ such that we have the same image quality of the reconstructed image (a Jpeg-compressed version) as Lin and Lin's scheme; at this time the size of our halftone secret image can be further reduced to $|I'| = |I|/(k \times g \times R)$. Then, the shadow size is $m \times |I'| = (m \times |I|/(k \times g \times R))$ and the pixel expansion of the compressible version $m_{PRO}^{(C)}$ is

$$m_{PRO}^{(C)} = m/(k \times g \times R). \qquad (17.9)$$

17.5   Experimental Results and Comparisons

Example 5 shows three $(2, 4)$-TiOISSSs, Jin et al.'s scheme [6], Lin and Lin's scheme [9] and our scheme [20], respectively; this example demonstrates the lossless version. The compressible versions of Lin and Lin's $(2, 4)$-TiOISSS and our $(2, 4)$-TiOISSS are given in Example 6, where a compression ratio $R = 6.5 > R_{LIN}{=}(4/2) \times 8/\log_2\left(\dfrac{4}{2}\right){=}6.19$ is used.

Example 5. Construct three lossless versions of $(2, 4)$-TiOISSSs: (1) Jin et al.'s scheme (2) Lin and Lin's scheme (3) our scheme (the lossless version) using $B_1 = \begin{bmatrix} 1100 \\ 0110 \\ 0011 \\ 1001 \end{bmatrix}$ and $B_0 = \begin{bmatrix} 1100 \\ 1100 \\ 1100 \\ 1100 \end{bmatrix}$.

A 512×512 gray-level Lena in Figure 17.3(a) is used as a secret image. By $(2, 4)$-PISSS, we first obtain four 512×256 gray-level shadows $P_1 - P_4$. To share all information bits in P_i, Jin et al's scheme and Lin and Lin's scheme need the 1274×1274 halftone secret image and the 1536×1536 halftone image, while our scheme needs the 256×256 halftone secret image (since $k = g = 2$, so $|I'| = |I|/(k \times g) = |I|/4$). The shadow sizes of Jin et al.'s scheme, Lin and Lin's scheme and our scheme are 3072×3072 pixels, 2548×2548 pixels and 512×512 pixels, respectively. Finally the pixel expansions are $m_{JIN} = (3072 \times 3072)/(512 \times 512) = 36$, $m_{LIN}^{(L)} = (2548 \times 2548)/(512 \times 512) = 24.76$ and $m_{PRO}^{(L)} = (512 \times 512)/(512 \times 512) = 1$. $\square$

Example 6. Consider Lin and Lin's scheme and our scheme in Example 5 but use the compressible versions with a compression ratio $R = 6.5$.

By $\mathbf{J}(\cdot)$, we compress the original Lena into the compressed image I_C with PSNR $= 38.88$ dB. Now, Lin and Lin's $(2, 4)$-TiOISSS has enough space in I' ($|I'| = |I|$) to embed the information of I_C. By using I_C as the original gray-level secret image in our TiOISSS, we only need the 201×201 halftone secret images for sharing information bits. For the compression version, the pixel expansions of $(2, 4)$-TiOISSS are $m_{LIN}^{(C)} = m = 4$ and $m_{PRO}^{(C)} = m/(k \times g \times R) = 1/R = 1/6.5 = 0.154$. $\square$

When comparing the pixel expansions among these three TiOISSSs [6, 9, 20], we consider the lossless version. This is a fair approach since all three schemes can reconstruct the same quality of the reconstructed image. From Equations (17.4), (17.7), and (17.8), it is obvious that our pixel expansion $m_{PRO}^{(L)} < m_{JIN}$ (since $k \geq 2$ and $g \geq 1$, so $m/(k \times g) < m/2 < 9m$). When comparing $m_{LIN}^{(L)}$ and $m_{PRO}^{(L)}$, one can verify that our TiOISSS has the lesser pixel expansion than Lin and Lin's TiOISSS for most values of k, n, and m. In the following theorem, we theoretically prove that $m_{PRO}^{(L)} < m_{LIN}^{(L)}$ when both TiOISSSs use Naor–Shamir optimal (n, n)-VCS, Naor–Shamir optimal $(3, n)$-VCS, and Naor–Shamir $(2, n)$-VCS [11].

Theorem: The pixel expansion $m_{PRO}^{(L)} = m/(k \times g)$ of the proposed (k, n)-TiOISSS is lesser than the pixel expansion $m_{LIN}^{(L)} = (n/k) \times m \times 8/\log_2\left(\dfrac{m}{w}\right)$ of Lin and Lin's TiOISSS when using Naor–Shamir optimal (n, n)-VCS, Naor–Shamir optimal $(3, n)$-VCS, and Naor–Shamir $(2, n)$-VCS.

Proof: When using Naor–Shamir optimal (n, n)-VCS (note: $k = n$) with $m = 2^{(n-1)}$, $w = g = m/2$, we have

$$
\begin{aligned}
m_{LIN}^{(L)} &= (n/k) \times m \times 8/\log_2\left(\frac{m}{w}\right) = m \times 8/\log_2\left(\frac{m}{w}\right) \\
&= 8m/\log_2\left(\frac{m!}{(m/2)! \times (m/2)!}\right) = 8m/\log_2\left(\frac{m \times (m-1) \times \cdots \times (m/2+1)}{(m/2) \times (m/2-1) \times \cdots \times 1}\right) \\
&> 8m/\log_2\left(m^{m/2}\right) = 8m/\left((m/2) \times \log_2(m)\right) = 16/\left(\log_2\left(2^{(n-1)}\right)\right) \\
&= 2^4/(n-1) > 2/n = m/(k \times g) = m_{PRO}^{(L)} \quad \text{(since } g = m/2 \text{ and } k = n\text{)}.
\end{aligned}
$$
$$(17.10)$$

When using Naor–Shamir $(3, n)$-VCS with $m = 2n - 2$, $w = g = m/2$, we have

$$m_{LIN}^{(L)} = (n/3) \times m \times 8/\log_2 \binom{m}{w} = ((8 \times m \times n)/3)/\log_2 \binom{m}{m/2}$$

$$= ((8 \times m \times n)/3)/\log_2 \left(\frac{m \times (m-1) \times \cdots \times (m/2+1)}{(m/2) \times (m/2-1) \times \cdots \times 1} \right)$$

$$> ((8 \times m \times n)/3)/\log_2 \left(m^{m/2} \right) = (16n/3)/\log_2 (m) = (16n/3)/$$
$$(\log_2 (2n-2))$$

$$> ((4n-4)/3)/(\log_2 (2n-2)) = (2/3) \times (2n-2)/(\log_2 (2n-2))$$

$$> 2/3 = m/(k \times g) = m_{PRO}^{(L)} \text{ (since } g = m/2 \text{ and } k = 3\text{)}.$$
$$(17.11)$$

When using Naor–Shamir $(2, n)$-VCS with $m = n, w = m - 1, g = 1$, we have

$$m_{LIN}^{(L)} = (n/2) \times m \times 8/\log_2 \binom{m}{w} = (4 \times m \times m)/\log_2 \binom{m}{m-1}$$

$$= (4m^2)/\log_2 (m) = (4m)(m/\log_2 (m)) > 4m$$

$$> m/2 = m/(k \times g) = m_{PRO}^{(L)} \text{ (since } g = 1 \text{ and } k = 2\text{)}.$$
$$(17.12)$$

From Equations (17.10), (17.11), and (17.12), we obtain $m_{LIN}^{(L)} > m_{PRO}^{(L)}$. The proof is completed. $\square$

Table 17.1 lists the pixel expansions of some (k, n)-TiOISSSs. Our scheme (respectively, Lin and Lin's scheme) could choose different g (respectively, w) in the same m to reduce the shadow size. For example, both $(2, 2)$-TiOISSSs can use $B_1 = \begin{bmatrix} 110 \\ 101 \end{bmatrix}, B_0 = \begin{bmatrix} 110 \\ 110 \end{bmatrix}$ instead of $B_1 = \begin{bmatrix} 10 \\ 01 \end{bmatrix}, B_0 = \begin{bmatrix} 10 \\ 10 \end{bmatrix}$ to reduce the pixel expansions from $m_{PRO}^{(L)} = 1$ and $m_{LIN}^{(L)} = 16$ to $m_{PRO}^{(L)} = 3/4$ and $m_{LIN}^{(L)} = 15.14$, respectively. Our $(2, 3)$-TiOISSS could reduce the pixel expansion from $3/2$ to $3/4$ by replacing $B_1 = \begin{bmatrix} 100 \\ 010 \\ 001 \end{bmatrix}, B_0 = \begin{bmatrix} 100 \\ 100 \\ 100 \end{bmatrix}$ with $B_1 = \begin{bmatrix} 110 \\ 011 \\ 101 \end{bmatrix}, B_0 = \begin{bmatrix} 110 \\ 110 \\ 110 \end{bmatrix}$, but the pixel expansion $m_{PRO}^{(L)} = 22.71$ of Lin and Lin's scheme cannot be further reduced since $\binom{m}{w} = \binom{3}{2} = \binom{3}{1}$.

Our pixel expansions $m_{PRO}^{(L)} = m/(k \times g)$. when using Naor–Shamir $(2, n)$-VCS, $(3, n)$-VCS and (n, n)-VCS, are $n/2$, $2/3$ and $2/n$, respectively. From Equations (17.4), (17.7), and (17.8), it is observed that $m_{LIN}^{(L)}$ is n-intensive and the value will increase when n increases whereas the other two schemes are n-invariant. On the other hand, Lin and Lin's scheme and our scheme obtain more effective performance for large k than Jin et al.'s scheme because their pixel expansions are inversely proportional to the value of k.

File sizes of shadows in Table 17.1 are represented in bits when using a 512×512 gray-level secret image. The values in parentheses imply that we send the raw data (use one bit "1" (or "0") denoting the black (or white) color) to transmit the black-and-white shadows for Jin et al.'s scheme and Lin and Lin's scheme. For example, when using the bitmap file format, the number of bits per pixel, which is the color depth of the image, typically can be 1, 4, 8, 16, 24, and 32. Even though both schemes use the color depth 1, our TiOISSS (use the color depth 8) still has the lesser file sizes of shadows for all cases in Table 17.1. For example, the file sizes of shadows for Jin et al.'s (3, 6)-TiOISSS and Lin and Lin's (3, 6)-TiOISSS are 23592960 bits ($\approx$23.6 Mbits) and 5258608 bits ($\approx$5.3 Mbits) when using one bit to represent the color, while our (3, 6)-TiOISSS just needs only 1398101 bits ($\approx$1.4 M Kbits) when using 8 bits to represent the color to transmit a gray-and-white shadow. Notice that when all schemes are using the same file format of the color depth 8, Jin et al.'s (3, 6)-TiOISSS and Lin and Lin's (3, 6)-TiOISSS require 188743680 bits ($\approx$188.7 Mbits) and 42068864 bits ($\approx$42.1 Mbits), respectively, which are significantly larger than 1.4Mbits.

By the contrast definition $((h - l)/(m + l))$ of VCS in [3], the contrasts of the previewed images are calculated and shown. Since the previewed image is used for verification in a distributed multimedia system; therefore, our primary task is to choose the suitable m and g to obtain the reduced shadow size to make the transmission and storage more efficient.

To elucidate on the good results of the proposed scheme among the existing TiOISSSs, some properties are evaluated: (1) the resolution of the reconstructed image, (2) the decoding complexity, (3) the shadow images (including the image pattern, the shadow size, the file size of shadow), and (4) the pixel expansion. A comparison among three TiOISSSs in [6, 9, 20] is listed in Table 17.2. All three schemes have dual modes of decoding—one for preview (or when a computer is temporarily unavailable) and the other for perfectly reconstructing the original gray-level secret image. Also, they all have a distinctive stacking-to-see property. Our scheme and Lin and Lin's scheme use Lagrange interpolation for perfect reconstruction with the computational complexity $O(k)$ and no additional memory space, while Jin et al.'s scheme uses a 8×8 ROM lookup table for reconstruction without computation. Our gray-and-white shadows have the benefit of reducing the physical size and the file size of the shadow when compared with other two TiOISSSs.

17.6 Conclusion

Our TiOISSS is a hybrid—half is the VCS and half is PISSS—with each specialties employed: the easy decoding of the VCS in Phase 1 and the perfect reconstruction of PISSS in Phase 2. Our combination of the VCS and PISSS is different from that in Lin and Lin's TiOISSS [9]. Lin and Lin use the combinations in the m-subpixel block to represent the output value of PISSS,

TABLE 17.1

Pixel expansions, file sizes of shadows, and contrasts of the previewed images for some (k,n)-TiOISSSs

(k,n)	B_1	B_0	(m,w,g)	Jin et al's TiOISSS#1		Lin and Lin's TiOISSS#2		the proposed TiOISSS		contrast of previewed image#4
				m#1	file size of shadow	m#2	file size of shadow	m#3	file size of shadow	
$(2,2)$#5	$\begin{bmatrix}10\\01\end{bmatrix}$	$\begin{bmatrix}10\\10\end{bmatrix}$	$(2,1,1)$	18	37748736 (4718592)	16	33554432 (4194304)	1	2097152	1/2
$(2,2)$	$\begin{bmatrix}110\\101\end{bmatrix}$	$\begin{bmatrix}110\\110\end{bmatrix}$	$(3,1,2)$	27	56623104 (7077888)	15.14	31750880 (3968860)	3/4	1572864	1/3
$(2,3)$#5	$\begin{bmatrix}100\\010\\001\end{bmatrix}$	$\begin{bmatrix}100\\100\\100\end{bmatrix}$	$(3,2,1)$	27	56623104 (7077888)	22.71	47626320 (5953290)	3/2	3145728	1/4
$(2,3)$	$\begin{bmatrix}110\\011\\101\end{bmatrix}$	$\begin{bmatrix}110\\110\\110\end{bmatrix}$	$(3,1,2)$	27	56623104 (7077888)	22.71	47626320 (5953290)	3/4	1572864	1/3
$(2,4)$#5	$\begin{bmatrix}1000\\0100\\0010\\0001\end{bmatrix}$	$\begin{bmatrix}1000\\1000\\1000\\1000\end{bmatrix}$	$(4,3,1)$	36	75497472 (9437184)	32	67108864 (8388608)	2	4194304	1/6
$(2,4)$	$\begin{bmatrix}1100\\0110\\0011\\1001\end{bmatrix}$	$\begin{bmatrix}1100\\1100\\1100\\1100\end{bmatrix}$	$(4,2,2)$	36	75497472 (9437184)	24.76	51925480 (6490685)	1	2097152	1/5
$(3,3)$#6	$\begin{bmatrix}1001\\0101\\0011\end{bmatrix}$	$\begin{bmatrix}1010\\1100\\0110\end{bmatrix}$	$(4,2,2)$	36	75497472 (9437184)	12.38	25962736 (3245342)	2/3	1398101	1/4
$(3,4)$#6	$\begin{bmatrix}111000\\110100\\110010\\110001\end{bmatrix}$	$\begin{bmatrix}000111\\001011\\001101\\001110\end{bmatrix}$	$(6,3,3)$	54	113246208 (14155776)	14.81	31058816 (3882352)	2/3	1398101	1/7
$(3,4)$	$\begin{bmatrix}11111000\\11100110\\11010101\\11001011\end{bmatrix}$	$\begin{bmatrix}01111100\\01110011\\01100111\\00011111\end{bmatrix}$	$(9,3,6)$	81	169869312 (21233664)	15.02	31499216 (3937402)	1/2	1048576	1/9

$(3,5)$#6	11110000 11101000 11100010 11100001	00001111 00010111 00011011 00011110	$(8,\ 4,\ 4)$	72	150994944 (18874368)	17.40	36490440 (4561305)	2/3	1398101	1/11
$(3,6)$#6	1111100000 1111010000 1111001000 1111000100 1111000010 1111000001	0000011111 0000101111 0000110111 0000111011 0000111101 0000111110	$(10,\ 5,\ 5)$	90	188743680 (23592960)	20.06	42068864 (5258608)	2/3	1398101	1/14
$(4,4)$#7	10001110 01001101 00101011 00010111	01001101 01000011 00111001 00010111	$(8,\ 4,\ 4)$	72	150994944 (18874368)	10.44	21894264 (2736783)	1/2	1048576	1/8

#1: $m_{JIN}=9m$ #2: $m_{LIN}^{(L)}=(n/k)\times m\times 8/\log_2\binom{m}{w}$ #3: $m_{PRO}^{(L)}=m/(k\times g)$ #4: contrast$=((h-l)/(m+l))$

#5: Naor–Shamir $(2, n)$-VCS #6: Naor–Shamir $(3, n)$-VCS #7: Naor–Shamir (n, n)-VCS

TABLE 17.2
A comparison among three TiOISSSs.

TiOISSS			Our scheme [20]	Jin et al.'s scheme [6]	Lin and Lin's scheme [9]
the resolution	Phase 1		U	U	U
	Phase 2		∞	∞	∞
the decoding complexity	Phase 1		H	H	H
	Phase 2	the space complexity	NO	L-T (8×8 ROM)	NO
		the computational complexity	L-I ($O(k)$)	NO	L-I ($O(k)$)
the shadow images	the pattern		G-W	B-W	B-W
	the shadow size		S	M	L
	the file size		S	M	L
the pixel expansion			S	M	L
the progressive decryption			YES	YES	YES

Notation: ∞: Perfect reconstruction of the original image (note: for lossless version); U: the unacceptable image quality of reconstructed image; H: stack shadows and use HVS to preview the secret; L-I: use Lagrange interpolation for reconstruction; L-T: use a lookup table for reconstruction; G-W: gray-and-white noise-like shadows; B-W: black-and-white noise-like shadows; S, M, L: the values of shadow size, file size, and pixel expansion are small, medium, and large scales, respectively.

while the value is embedded into GVCS to construct our TiOISSS. Finally, our scheme reduces the shadow size and the file size of the shadow.

Bibliography

[1] C.C. Chang, Y.P. Hsieh, and C.H. Lin. Sharing secrets in stego images with authentication. *Pattern Recognition*, 41:3130–3137, 2008.

[2] S. Cimato, R. De Prisco, and A. De Santis. Probabilistic visual cryptography schemes. *The Computer Journal*, 49:97–107, 2006.

[3] P.A. Eisen and D.R. Stinson. Threshold visual cryptography schemes with specified whiteness. *Designs, Codes and Cryptography*, 25:15–61, 2002.

[4] Z. Eslami, S.H. Razzaghi, and J.Z. Ahmadabadi. Secret image sharing based on cellular automata and steganography. *Pattern Recognition*, 43:397–404, 2010.

[5] R. Ito, H. Kuwakado, and H. Tanaka. Image size invariant visual cryptography. *IEICE Trans. on Fund. of Elect. Comm. and Comp. Sci.*, E82-A:2172–2177, 1999.

[6] D. Jin, W.Q. Yan, and M.S. Kankanhalli. Progressive color visual cryptography. *Journal of Electronic Imaging*, 14:033019–1–033019–13, 2005.

[7] H. Kuwakado and H. Tanaka. Size-reduced visual secret sharing scheme. *IEICE Trans. on Fund. of Elect. Comm. and Comp. Sci.*, E87-A:1193–1197, 2004.

[8] C.C. Lin and W.H. Tsai. Secret image sharing with steganography and authentication. *Journal of Systems & Software*, 73:405–414, 2004.

[9] S.J. Lin and J.C. Lin. VCPSS: A two-in-one two-decoding-options image sharing method combining visual cryptography (VC) and polynomial-style sharing (PSS) approaches. *Pattern Recognition*, 40:3652–3666, 2007.

[10] F. Liu, C.K. Wu, and X.J. Lin. Color visual cryptography schemes. *IET Information Security*, 2:151–165, 2009.

[11] M. Naor and A. Shamir. Visual cryptography. In *Advances in Cryptology-EUROCRYPT94 LNCS 950*, pages 1–12, 1994.

[12] A. Shamir. How to share a secret. *Communications of the Association for Computing Machinery*, 22:612–613, 1979.

[13] C.C. Thien and J.C. Lin. Secret image sharing. *Computers & Graphics*, 26:765–770, 2002.

[14] C.C. Thien and J.C. Lin. An image-sharing method with user-friendly shadow images. *IEEE Trans. on Circ. and Sys. for Video Tech.*, 13:1161–1169, 2003.

[15] R.Z. Wang and C.H. Su. Secret image sharing with smaller shadow images. *Pattern Recognition Letters*, 27:551–555, 2006.

[16] C.N. Yang. New visual secret sharing schemes using probabilistic method. *Pattern Recognition Letters*, 25:481–494, 2004.

[17] C.N. Yang and T.S. Chen. Size-adjustable visual secret sharing schemes. *IEICE Trans. on Fund. of Elect. Comm. and Comp. Sci.*, E88-A:2471–2474, 2005.

[18] C.N. Yang and T.S. Chen. New size-reduced visual secret sharing schemes with half reduction of shadow size. *IEICE Trans. on Fund. of Elect, Comm. and Comp. Sci.*, E89-A:620–625, 2006.

[19] C.N. Yang, T.S. Chen, K.H. Yu, and C.C. Wang. Improvements of image sharing with steganography and authentication. *Journal of Systems & Software*, 80:1070–1076, 2007.

[20] C.N. Yang and C.B. Ciou. Image secret sharing method with two-decoding-options: lossless recovery and previewing capability. *Image and Vision Computing*, 28:1600–1610, 2010.

[21] C.N. Yang, K.H. Yu, and R. Lukac. User-friendly image sharing using polynomials with different primes. *International Journal of Imaging Systems and Technology*, 17:40–47, 2007.

Index